ADVANCES IN
X-RAY ANALYSIS

Volume 20

ADVANCES IN X-RAY ANALYSIS

Volume 20

Edited by
Howard F. McMurdie
National Bureau of Standards
Washington, D.C.

and

Charles S. Barrett, John B. Newkirk, and Clayton O. Ruud
University of Denver
Denver, Colorado

Sponsored by
University of Denver
Department of Chemistry
and
Metallurgy and Materials Science Division
Denver Research Institute

PLENUM PRESS • NEW YORK AND LONDON

The Library of Congress cataloged the first volume of this title as follows:

Conference on Application of X-ray Analysis.
Proceedings. 6th-   1957-   [Denver]

　　v. illus. 24-28 cm. annual.
No proceedings published for the first 5 conferences.
Vols. for 1958-   called also: Advances in X-ray analysis, v. 2-
Proceedings for 1957 issued by the conference under an earlier name: Conference on Industrial Applications of X-ray Analysis. Other slight variations in name of conference.
Vol. for 1957 published by the University of Denver, Denver Research Institute, Metallurgy Division.
Vols. for 1958-   distributed by Plenum Press, New York.
Conferences sponsored by University of Denver, Denver Research Institute.
　　1. X-rays—Industrial applications—Congresses.   I. Denver. University. Denver Research Institute   II. Title: Advances in X-ray analysis.
TA406.5.C6   58-35928

Library of Congress Catalog Card Number 58-35928
ISBN 978-1-4613-9983-4   ISBN 978-1-4613-9981-0 (eBook)
DOI 10.1007/978-1-4613-9981-0

Proceedings of the Twenty-Fifth Annual Conference on Applications of
X-Ray Analysis held in Denver, August 4-6, 1976

Plenum Press, New York
A Division of Plenum Publishing Corporation
227 West 17th Street, New York, N.Y. 10011

FOREWORD

X-ray diffraction as a method of qualitative analysis for crystalline phases has been long accepted, and has had constant improvement in method and equipment. It has also been made more useful by the growth and improvement of the data collection available as reference standards. In recent years some attempts have been made to use the method to a greater extent by furnishing results on a quantitative basis. This has proved to be difficult because of the problem of comparing the relative intensities of the diffraction peaks from one phase to another.

This year the initial session of invited papers focuses primarily on this problem. The subject is approached both by the use of internal comparison standards and by calculation of intensities. In addition, the identification of crystalline phases by X-ray diffraction of single crystals is discussed in an invited paper. This method, with its advantages of the use of very small samples, is becoming increasingly feasible because of the development of simple equipment and the availability of a growing data bank.

Other X-ray diffraction developments discussed at the Conference include stress analysis, use of computers for searching the JCPDS powder diffraction file, texture analysis, and applications to specific fields.

Spectroscopy topics covered at the conference included a discussion of methods of concentration of materials for fluorescence analysis, soft X-ray spectra, and equipment for fluorescence analysis.

The Denver Conference on the Applications of X-Ray Analysis continues to be the source of information on new techniques and applications for the practical use of X-rays.

Howard F. McMurdie

PREFACE

This volume is the proceedings of the 1976 Denver Conference on the Applications of X-Ray Analysis, 25th in the series. It was co-sponsored by the Joint Committee on Powder Diffraction Standards (JCPDS) and the University of Denver. The conference co-chairmen in residence were C. S. Barrett, J. B. Newkirk and C. O. Ruud. The invited conference co-chairman, H. F. Murdie of the National Bureau of Standards, represented the JCPDS and organized the Plenary Session entitled "X-Ray Diffraction for Chemical Analysis."

In addition to the plenary and regular contributed sessions, three special sessions were organized. The chairmen and titles for these are listed below.

B. D. Cullity	X-Ray Stress Analysis
D. E. Leyden	Chemical Preconcentration for X-Ray Fluorescence Analysis
R. W. Armstrong	X-Ray Topography

Interest in X-ray stress analysis was so great that three sessions evolved from papers contributed.

A number of invited speakers also enhanced the content of the conference. Their names and the titles of their papers are listed below:

L. E. Alexander, "Forty Years of Quantitative Diffraction Analysis."

B. D. Cullity, "Some Problems in X-Ray Stress Measurements."

L. K. Frevel, "Quantitative Matching of Powder Diffraction Patterns."

C. R. Hubbard and D. K. Smith, "Experimental and Calculated Standards for Quantitative Analysis by Powder Diffraction."

D. E. Leyden, "Advances in the Preconcentration of Dissolved Ions in Water Samples."

K. Mather, "X-Ray Diffraction Examination of the Phases in Expansive Cements."

E. Meieran, "Current Status of X-Ray Topography."

A. D. Mighell, "The Single Crystal vs. the Powder Method for Identification of Crystalline Materials."

For the first time two tutorial workshops were offered the day preceding the conference. The attendance at both exceeded expectations; moreover, they were very well received by the attendees. The tutorial workshop titles and their chairmen are as follows:

Workshop on Automated Powder Diffraction

J. D. Hanawalt, "History and Development of Search Procedures"

D. K. Smith, "Specimen Preparation Techniques for Qualitative Powder Diffractometry"

C. R. Hubbard, "Provision of Calibration Standards and Reference Patterns"

R. Jenkins, "On the Obtaining of Adequate Instrumental Data for Qualitative Powder Diffractometry"

Workshop on X-Ray Fluorescence Sample Preparation and Preconcentration

R. W. Gould, "Sample Preparation by Grinding and Pressing"

F. Claisse, "Glass Disc Fusion Techniques for Sample Preparation"

B. P. Fabbi, "Samples by Fusion Briquetting"

J. R. Rhodes, "Particulate Samples on/in Filters"

D. E. Leyden, "Preconcentration of Ions from Solution Using Reagents Immobilized via Silylation"

W. J. Campbell, "Preconcentration on Ion Exchange Impregnated Filters"

C. M. Davis, "Preconcentration by Precipitation and Coprecipitation"

Finally it must be recognized that the session co-chairmen and the conference aids made the major contributions to the meeting running smoothly. Their names are as follows:

SESSION CO-CHAIRMEN

L. E. Alexander	L. K. Frevel	A. D. Mighell
R. W. Armstrong	J. D. Hanawalt	P. K. Predecki
R. Baro	C. R. Hubbard	J. R. Rhodes
D. W. Beard	D. E. Leyden	P. A. Russell
D. Chandra	K. Mather	D. K. Smith
B. D. Cullity	H. F. McMurdie	D. M. Smith
B. P. Fabbi	E. Meieran	

CONFERENCE AIDS

Herb Acree	Mildred Cain	Patricia Karr
Larry Apple	Robert Clark	Ronnie Masters
Jacqueline Battles	John Cronin	William Nonidez
David Butler	Bruce Jablonski	Betty Ungerman

On behalf of the conference committee in residence and the attendees, I thank the JCPDS and those persons whose names are listed above
for making possible a very successful conference. Also the JCPDS
graciously supplied travel funds for some of the speakers whose attendance would have otherwise been impossible.

C. O. Ruud

UNPUBLISHED PAPERS

The following papers were presented at the Conference but for
various reasons are not included in this volume.

"Measurement of Gold-Plate Thickness by X-Ray Fluorescence," R. M.
Rusnak, Bendix Research Laboratories, 20800 Civic Center Drive, South-
field, Michigan 48076

"Fast Simultaneous Thickness Measurements of Gold and Nickel Layers on
Copper Substrates by X-Ray Fluorescence," J. R. Maldonado and D. Maydan,
Bell Laboratories, 600 Mountain Ave., Murray Hill, New Jersey 17974

"On the Evaluation of Data in Nondispersive X-Ray Fluorescence Analysis
of Solid and Liquid Mineralogical Samples," J. Parus, W. Ratyński, T.
Zóltowski, J. Kownacki, J. Kierzek and J. Tys, Institute of Nuclear
Research, Dorodna 16, 03-195, Warsaw, Poland

"K-Line Analysis of Heavier Elements by Energy Dispersive X-Ray Fluorescence Spectrometry," C. R. Phillips and J. M. Mathiesen, Finnigan
Corporation, 845 West Maude Avenue, Sunnyvale, California 94086

"Texture Analysis by Orientation Distribution Functions," H.-J. Bunge,
Institut für Metallkunde und Metallphysik, Technische Universität
Clausthal, Clausthal-Zellerfeld, Germany

"Activated Carbon as a Trace Metal Enrichment Substrate for X-Ray Fluorescence Analysis," B. M. Vanderborght, J. Verbeeck and R. E. Van Grieken,
Department Scheikunde, Universitaire Instelling Antwerpen, Universiteit-
splein 1, 2610 Wilrijk, Belgium

"Preconcentration of Trace Metal Ions by Combined Complexation-Anion
Exchange, Part II. Cobalt, Zinc, Cadmium, Copper and Lead with 8-
Hydroxyquinoline-5-sulfonic Acid," J. E. Going and G. Wesenberg, Mid-
west Research Institute, 425 Volker Blvd., Kansas City, Missouri 64110,
and D. Berge, Department of Chemistry, University of Wisconsin, Oshkosh,
Wisconsin 54901

"X-Ray Analysis in the Wood Treating Industry," L. H. Leiner and E. W. Kifer, Koppers Company, Inc., 440 College Park Drive, Monroeville, Pennsylvania 15146

"Current Status of X-Ray Topography," E. Meieran, Intel Corporation, 3065 Bowers Ave., Santa Clara, California 95051

"Image Size, Intensity and Dislocation Visibility Within Borrmann (Anomalous Transmission) X-Ray Topographs." M. C. Narasimhan, Allied Chemical Corporation, P.O. Box 1021R, Morristown, New Jersey 07960; B. Roessler, Brown University, Providence Rhode Island 02912; and C. Cm. Wu and R. W. Armstrong, University of Maryland, College Park, Maryland 20742

"Industrial Applications of Powder X-Ray Diffraction Analysis to the TiH_x-$KClO_4$ Pyrotechnic System," D. B. Sullenger, R. R. Eckstein and R. S. Carlson, Mound Laboratory, Miamisburg, OH 45342

"The Method of Known Additions Revisited," R. A. Lohnes and T. Demirel, Civil Engineering Dept., Town Engineering Building, Iowa State University, Ames, Iowa 50011

"Recent German Activities in the Field of X-Ray Stress Analysis," E. Macherauch and U. Wolfstieg, Institut für Werkstoffkunde I, Universität Karlsruhe (TH), D-7500 Karlsruhe, Germany

"Recent Trends of Stress Measurement by X-Ray in Japan," Y. Sakamoto, Central Research Laboratories, Sumitomo Metal Industries, Ltd., 1-3, Nishinagasu-Hondori, Amagasaki, Japan

"Anomalous Absorption and Reflection Features in Ultrasoft X-Ray Spectrometry," D. M. Barrus, R. L. Blake, and A. J. Burek, Los Alamos Scientific Laboratory, Mail Stop 436, P.O. Box 1663, Los Alamos, New Mexico 87544

"High Speed X-Ray Pulse Generator," W. Hershyn, T. R. Emery, and W. S. Knodle III, Watkins-Johnson Company, 440 Mount Hermon Road, Scotts Valley, California 95066

"Industrial Application of the X-Ray Residual Stress Technique," C. J. Lambright, American Analytical Corporation, 10110 Royalton Road, North Royalton, Ohio 44133

"Applications of Residual Stress Analysis to the Solution of Engineering Problems," R. E. Herfert, Metallics Research Dept., Orgn. 3771, Zone 62, Northrop Corporation, 3901 W. Broadway, Hawthorne, California 90250

"Practical Applications of X-Ray Stress Analysis," D. Kirk, Department of Chemistry and Metallurgy, Lanchester Polytechnic, Priory Street, Coventry CV1 5FB, England

"Fundamental Parameter Method for Quantitative Elemental Analysis with Monochromatic X-Ray Sources," J. W. Otvos, G. E. A. Wyld and T. C. Yao, Shell Development Co., P.O. Box 1380, Houston, Texas 77001

"Applications of the Fundamental Parameter Method for the Analyses of Fused Geological Samples," C. Palme and E. Jagoutz, Max-Planck-Institut für Chemie, Abteilung Kosmochemie, Saarstrasse 23, 65 Mainz, West Germany

"A Detailed Mathematical Background Radiation Analysis for XRF Trace Element Measurements Employing a Multichannel Analyzer," E. Jagoutz and C. Palme, Max-Planck-Institut für Chemie, Abteilung Kosmochemie, Saarstrasse 23, 65 Mainz, West Germany

"A New Generation X-Ray Detector for Stress Analysis," C. O. Ruud, R. E. Sturm and C. S. Barrett, University of Denver Research Institute, Denver, Colorado 80208; and G. D. Farmer, MERDC, Fort Belvoir, Virginia 22060

"External Beam PIXE Analysis of Trace Elements in Neoplastic and Non-Neoplastic Human Tissues," R. Sarper, Z. Karcioglu, Emory University Medical School; R. W. Fink, School of Chemistry, Georgia Institute of Technology; and A. Katsanos and A. Hadjiantoniou, Nuclear Research Center, Athens, Greece

CONTENTS

X-RAY TOPOGRAPHY

X-RAY DIFFRACTION STRESS ANALYSIS

X-RAY FLUORESCENCE

Contents

X-RAY INSTRUMENTATION

Contents

FORTY YEARS OF QUANTITATIVE DIFFRACTION ANALYSIS

Leroy E. Alexander

Carnegie-Mellon University

Pittsburgh, Pennsylvania 15213

INTRODUCTION

In this paper an attempt is made to review in rather broad perspective the origins and history of quantitative methods in diffraction analysis, at the same time leaving an in-depth examination of the present state of the art to other better qualified contributors to this conference. Space limitations preclude mention of many significant contributions, for which I am very sorry. It will be possible to review only a number of pivotal historical events, while also taking note of certain other researches that seem representative of historical and present-day trends.

The birth of quantitatively meaningful analysis in the mid-1930s depended upon a realization of, and allowance for, the alteration of the diffracted intensities resulting from absorption of x-rays by the specimen. Furthermore, advances in the art achieved during the past forty years have been closely related to improvements in the treatment of the absorption factor.

The history of the quantitative discipline divides itself naturally into three eras with dominant features as follows:

Early era (1936-1950). Empirical basis, photographic technique, "unlimited" supply of sample.

Middle era (1950-1970). Development of theoretical groundwork, diffractometric technique.

Recent era (1970-1976). Development of microanalysis, automated data acquisition and processing, emphasis on environmental pollutants.

EARLY ERA (1936-1950)

The early era commenced with Clark and Reynolds' (1) development of a photographic-microphotometric method, which incorporated an admixed internal standard, for the measurement of quartz in mine dusts. The use of an internal standard was adapted from its prior application in ultraviolet spectroscopy (2). Using essentially the same method, Ballard, Oshry, and Schrenk (3) made extensive analyses of quartz in ores, rocks, and mine dusts. Debye-Scherrer patterns were microphotometered and the ratio of the heights of a quartz and standard line was determined directly from the photometer trace. The quartz concentration was then determined by reference to a calibration curve prepared from known mixtures of quartz and internal standard, which was CaF_2 or NiO. Criticisms that can be leveled at this widely used method include (a) limitations inherent in the photometric measurement of photographic densities, (b) neglect of extinction and microabsorption effects, and (c) the use of peak rather than integrated intensities. Nevertheless, experience has demonstrated that the method can yield accuracies ranging from 5 to 10% of the amount present for higher concentrations of the analyte.

An important modification of the foregoing photographic procedure was introduced by Cohen and coworkers (4,5) for the analysis of re-tained austenite in steel. They also microphotometered the patterns, but they employed integrated line intensities and utilized a line of the coexisting martensite phase, which served as an internal standard. Present-day analysis of austenite is still patterned after the basic method of Cohen (6,7).

MIDDLE ERA (1950-1970)

The invention of the parafocusing counter diffractometer by Friedman (8) revolutionized the quantitative evaluation of diffraction patterns. A reflection intensity could now be recorded digitally as a counting rate subject to a well-defined statistical accuracy. Further, the simpler diffraction geometry made possible a straightforward ex-pression of the theory underlying quantitative analysis, besides facilitating the correction of systematic errors. The displacement of photographic by counter techniques in this country was lent great impetus by the North American Philips Company, which between 1945 and 1955 pioneered and marketed widely the first commercial powder diffrac-tometers.

In the course of a program directed toward applying diffractometric techniques to the measurement of free silica in industrial dusts, Alexander and Klug (9) formulated the theoretical basis for quantitative analysis of mixed polycrystalline phases. They showed that for a flat specimen thick enough to give maximum diffracted intensity the integrated intensity of reflection i of component J is given by

$$I_{iJ} = \frac{K_{iJ}x_J}{\rho_J[x_J(\mu_J - \mu_M) + \mu_M]} \tag{1}$$

where x_J and ρ_J are respectively the weight fraction and density of the analyte, μ_J and μ_M are the mass absorption coefficients of component J and the matrix, and K_{iJ} is a constant whose value depends on the diffracting power of component J and the geometry of the apparatus. Equation (1) implicitly assumes that extinction, microabsorption, and preferred orientation may be neglected, acceptable approximations for very fine particulate matter such as the respirable dusts that comprised the bulk of the specimens analyzed in the Mellon Institute program (10,11).

Since in Equation (1) $x_J(\mu_J - \mu_M) + \mu_M$ is simply $\bar{\mu}$, the mass absorption coefficient of the entire specimen, we may restate the equation as

$$I_{iJ} = K_{iJ} x_J / \rho_J \bar{\mu} \tag{2}$$

with the same assumptions just given. If a constant weight fraction x_S of an internal standard is added to the sample, the weight fraction of component J is now given in terms of the intensities I_{iJ} and I_{kS} (line k of the internal standard) by the direct proportionality

$$x_J = k' \times (I_{iJ} / I_{kS}) \tag{3}$$

The validity of Equation (3) is demonstrated in Figure 1 by experimental data from reference mixtures of the analyte quartz (J) and internal standard calcium fluoride (S). Integrated intensities are preferred to peak intensities to compensate for possible variations in line breadth arising from lattice distortions or differences in crystallite size. In the course of their analytical investigations Alexander and associates (12) demonstrated theoretically and experimentally the importance of minimizing fluctuations in the experimental intensities caused by the presence of crystallites larger than about 5 microns in mean dimension.

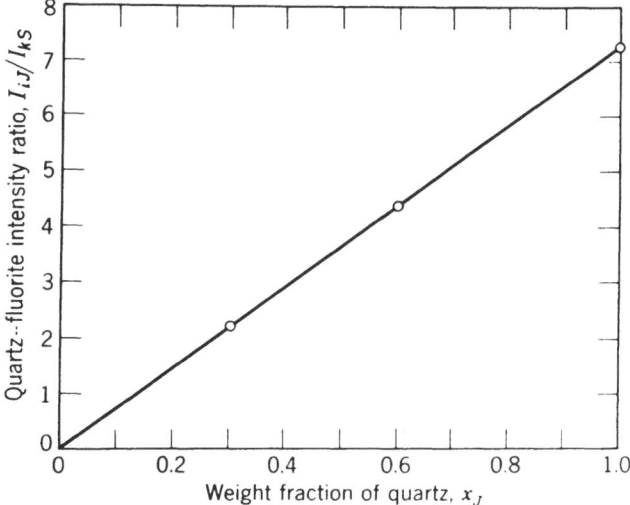

Figure 1.   Linearity of typical curve for quartz analysis using fluorite as internal standard. [L. E. Alexander and H. P. Klug, Anal. Chem. 20, 886 (1948)]

Equation (2) for pure component J may be written

$$(I_{iJ})_0 = K_{iJ}/\rho_J \mu_J \tag{4}$$

and division of Equation (2) by Equation (4) yields

$$x_J = \bar{\mu} I_{iJ}/\mu_J (I_{iJ})_0 \tag{5}$$

Employing Equation (5), Leroux, Lennox, and coworkers (13-15) devised an important method of analysis involving direct measurement of $\bar{\mu}$ rather than an internal standard to compensate for x-ray absorption by the specimen. Their method dispenses with the problems associated with admixing an internal standard and measuring its diffracted intensity. Nevertheless, it also suffers from special limitations: (a) sensitivity to small variations in the diffraction geometry, (b) unsuitability for specimens that strongly absorb the x-ray beam, and (c) susceptibility to errors in the direct measurements of $\bar{\mu}$, particularly of specimens giving appreciable small-angle scattering (16).

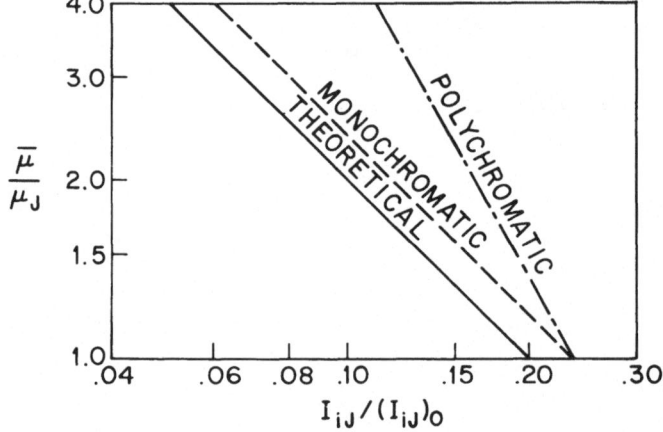

Figure 2.   Comparison of experimental and theoretical diffraction-absorption curves. [After D. H. Lennox, Anal. Chem. 29, 766 (1957)]

Figure 2 from a paper by Lennox (14) compares the theoretical logarithmic diffraction-absorption plot of Equation (5), $\bar{\mu}/\mu_J$ versus $I_{iJ}/(I_{iJ})_0$, with experimental plots for $x_J = 0.20$ using monochromatic and polychromatic incident beams. Compared with the monochromatic beam, the polychromatic produces a conspicuous deviation from the theoretical slope of -1, while both curves show a displacement parallel to the.

abscissa. In other experiments Leroux et al. (13,15) observed a
similar, but smaller, departure in slope from -1 for monochromatic
x-rays. These deviations from theory were attributed in part to lack
of true monochromaticity of the x-ray beam and in part to variations
in the geometric shape of the irradiated portion of the specimen (15).

The diffraction-absorption technique was modified to some advantage
by Williams (17), who mounted a thin specimen on a diffracting substrate, such as a fine-grained metal, to permit the specimen absorption
to be measured under geometrical conditions unchanged from those
affecting the analytical diffraction measurements.

In 1958 Copeland and Bragg (18) announced some further important
contributions to the science of quantitative diffraction analysis.
First, they pointed out that when only a very limited number of
analyses of a given type is to be performed, one may dispense with an
internal standard and add instead a known weight fraction y_J of the
analyte, which is already present in an unknown weight fraction x_J. If
the sample also contains some other constituent L having a strong line
suitable for reference intensity measurement, the new intensity ratio
of lines iJ and kL will be

$$\frac{I_{iJ}}{I_{kL}} = \frac{K_{iJ}\rho_L(x_J + y_J)}{K_{kL}\rho_J x_L}$$

$$= \text{Constant} \times (x_J + y_J) \qquad (6)$$

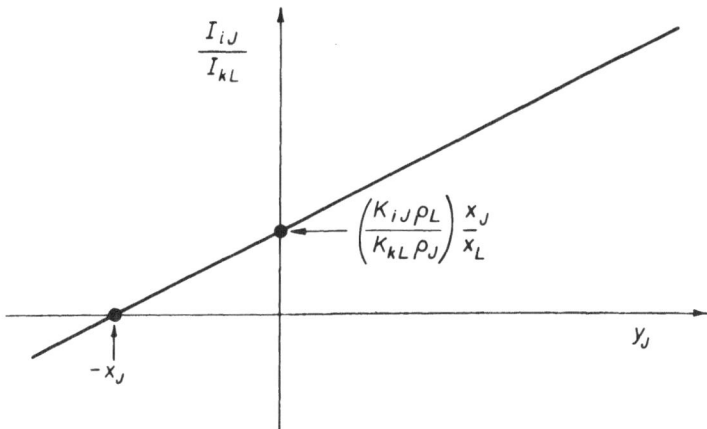

Figure 3. Analysis by dilution with a known weight fraction of the
           analyte. [H. P. Klug and L. E. Alexander, X-Ray Diffraction
           Procedures, 2nd Ed., p. 554, Wiley, New York (1974).]

Figure 3 shows that a plot of I_{iJ}/I_{kL} against y_J is a straight line that
does not intersect the origin and whose abscissa intercept is the negative of x_J, the concentration of component J in the original sample.

This well-known "dilution method" is most useful for samples containing relatively small concentrations of the analyte.

Copeland and Bragg (18) also treated the general case of an N-component sample producing a diffraction pattern with superposed "analytical" lines. In order to determine N components, m lines must be measured (m ≥ N), furnishing sufficient data to permit solution of m simultaneous equations,

$$I_i = \sum_J I_{iJ} = \sum_J C_{iJ} x_J \quad (i = 1, 2, 3, \cdots m) \tag{7}$$

where

$$C_{iJ} = K_{iJ} / \rho_J \bar{\mu} \tag{8}$$

as can be seen by comparison with Equation (2). Each constant C_{iJ} corresponds to a particular line i of component J.

It is seen that the m Equations (7) represent a scheme of direct analysis, such as that of Leroux et al. From experimental values of $\bar{\mu}$ for the sample and of known and experimental values, respectively, of ρ_J and $(I_{iJ})_0$ for pure compoment J ($x_J = 1$), K_{iJ} can be calculated from

$$I_{iJ} = K_{iJ} x_J / \rho_J \bar{\mu}$$

with x_J set equal to unity. This is seen to be a term from Equations (7) applicable to a line of pure component J and from which the constant C_{iJ} can be calculated using Equation (8). After all the C_{iJ}'s have been determined for the pure components, Equations (7) can be used to perform multicomponent analysis.

Table 1. Analysis of Test Mixtures for $Ca(OH)_2$

Diluent	Actual Percent	Percent Found
$\beta Ca_2 SiO_4$	17.9	19.6
$Ca_3 Si_2 O_7 \cdot 3H_2O$	39.4	37.4
Portland Cement	9.9	9.9
$Ca_3 SiO_5$	29.8	29.4
$CaSiO_3 \cdot H_2O$	39.8	39.9
MgO	64.5	63.0

[R. H. Bragg, in Handbook of X-Rays, Chap. 12, McGraw-Hill, New York (1967)]

This method can also be applied with an internal standard (added component $N+1$), in which case the intensities I_i of Equations (7) are replaced by intensity ratios I_i/I_{N+1} and the coefficients C_{iJ} by ratios $C_{iJ}/C_{m+1,N+1}$. In a typical analytical situation the number of analytes seldom exceeds three and not all the lines to be measured are composite, with the result that some of the C_{iJ}'s are zero and the analytical solutions are reasonably precise. However, it is evident that for larger numbers of superpositions and unknowns the potential accuracy is likely to be very poor.

From Copeland and Bragg's work on the analysis of Portland cement compositions for $Ca(OH)_2$ using $Mg(OH)_2$ as internal standard, Table 1 compares the percentages of $Ca(OH)_2$ found with the known amounts contained in test mixtures using six different diluents. This case constitutes a simplification of the general Equations (7) and (8) to two equations in two unknowns, the only composite line consisting of the 4.90 Å $Ca(OH)_2$ and 4.76 Å $Mg(OH)_2$ lines. It is evident that the degree of accuracy achieved is excellent.

RECENT ERA (1970-1976)

A useful extension of the internal-standard theory has been given by Chung (19,20) in the form

$$x_J = x_S \left(\frac{I_{kS}}{I_{iJ}} \right)_{50:50} \times \left(\frac{I_{iJ}}{I_{kS}} \right) \qquad (9)$$

where $(I_{kS}/I_{iJ})_{50:50}$ is the intensity ratio of selected lines k and i of a 50:50 binary mixture ($x_S = x_J = 0.5$) of components S and J. Equation (9) shows that for any multicomponent system a plot of I_{iJ}/I_{kS} versus x_J intersects the origin and is linear with the slope

$$x_S (I_{kS}/I_{iJ})_{50:50}$$

This permits the slope of the calibration curve to be precalculated, thus eliminating the preparation of a series of standard mixtures. Chung refers to the internal standard S as the "flushing agent" because in effect it flushes the sample free of x-ray absorption effects.

Chung's formulation simplifies for binary mixtures, and, furthermore, if Linde "A" synthetic alumina is employed as the reference substance S, rough quantitative analyses for many analytes may be performed by referring to Berry's (21) tables of peak $(I_{iJ}/I_{kS})_{50:50}$ values for several hundred compounds. Table 2 compares Chung's diffractometrically determined reference intensity ratios for six compounds with published values from the Powder Diffraction File (PDF). The agreement is good for the first three compounds but rather poor for the last three, thus emphasizing that for the most reliable results the reference intensity ratios must be measured under the same experimental conditions that prevail in the analytical measurements.

Table 2. Values of the Reference Intensity Ratio
$(I_{iJ}/I_{kS})_{50:50}$ for Selected Compounds

	PDF Card	Reference Intensity Ratio	
		By Counting	PDF Table
ZnO	5-664	4.35	4.5
KCl	4-587	3.87	3.9
LiF	4-857	1.32	1.3
$CaCO_3$	5-586	2.98	2.0
TiO_2 (rutile)	21-1276	2.62	3.4
TiO_2 (anatase)	21-1272	3.39	4.3

[F. H. Chung, J. Appl. Cryst. 7, 519-525 (1974)]

Clark and Preston (22) have recently proposed a "generalized" dilution method for eliminating the effects of x-ray absorption (both macro- and micro-), but their procedure suffers from some stringent limitations, in particular the great loss of intensity diffracted by the analyte as a consequence of the dilution. Sahores (23) has devised a method based on measurements of the Compton-scattering intensity to compensate for the absorption factor of the sample. Sahores' method permits analysis of very small specimens and has the great advantage of eliminating the effects of perferred orientation and microabsorption; its principal drawback appears to be the difficulty of accurately measuring the weak Compton intensities. In addition to their application in quantitative phase analysis, the methods of Chung (19,20) and Sahores (23) permit quantitative estimation of the amorphous fraction of materials, which is of special value in the study of polymers.

Undoubtedly the greatest current effort is being devoted to the analysis of environmental pollutants, a prominent example of which is the determination of free silica in respirable dusts. Two analytical steps are involved, first, quantitative sampling of the respirable dust content of air including its deposition on fine-pore (0.8-5.0 μ) organic- or silver-membrane filters and, second, diffractometric analysis of the thin, membrane-supported specimens. Because of their appreciable x-ray scattering, organic-membrane filters have now been largely replaced by silver filters. It is the usual practice to keep the sample thickness so small that the weight of quartz per unit area can be determined by direct comparison of the intensities diffracted by the "unknown" and the reference specimens. For somewhat thicker specimens the attenuation of the 2.36 Å reflection from the silver substrate can be used to correct for x-ray absorption by the method of Leroux et al. (13-15). Typical sample weights of 100 μg to a few mg permit detection limits as low as 1 μg for low-absorbing matrices (for example, coal dust). For further details of the analysis of air-borne dusts on a micro scale the reader is referred to work by Leroux and associates (24) and by Bumsted (25).

Quantitative analytical procedures for the measurement of asbestos minerals have been developed in response to the growing awareness of hazardous concentrations in the environment, both in water supplies of specific industry-related geographical locales and more generally, but at lower levels, in the atmosphere. Some years ago Crable (26) employed membrane filters and external standards to differentiate and quantitatively estimate chrysotile, amosite, and crocidolite in the range of 1 to 10 mg in various mineral matrices. More recently Rickards (27) used refined procedures to analyze chrysotile to detection limits of 10 to 100 μg, with standard deviations of 10 and 2%, respectively, for dilution-with-analyte and external-standard techniques.

Energy-dispersive diffractometry (EDD) might be expected to offer important advantages over conventional scanning diffractometry (CSD) in quantitative analysis, especially higher-speed data acquisition with reduction in analysis time. However, the practical application of EDD is sharply restricted because of (a) the much lower resolution of solid-state [e.g., Si(Li)] detectors compared with the gas-ionization detectors of CSD, typically by a factor of ten, and (b) the larger detector dead time of about 140 compared with 7-9 μsec for CSD counters. The first-named, and more severe, limitation effectively restricts analysis by EDD to mixed cubic phases. Ferrell (28) concluded that EDD is inapplicable even to the identification of mixed-mineral patterns. In a comparative study of the analysis of quartz on silver-membrane filters by CSD and EDD Mauer (29), using a silver-target x-ray tube, found that the high background scatter from the silver membrane resulted in a disadvantageous peak-to-background ratio with the EDD technique.

Despite the foregoing damaging criticisms, EDD can be used very effectively for the measurement of retained austenite because of the simplicity of the diffraction patterns involved. Voskamp (30) achieved a six-fold reduction in the analysis time for austenite in hardened, low-alloyed carbon steels without loss of accuracy (see Table 3).

Table 3. Measurement of Retained Austenite in Hardened, Low-Alloyed Carbon Steels by CSD and EDD

Sample	CSD MoK Radiation Time 2400 sec.		EDD Polychromatic Radiation Time 400 sec.	
	Found (%)	Abs. Error (%)	Found (%)	Abs. Error (%)
A1	21.3	0.9	22.4	0.8
A2	19.3	0.7	20.4	0.9
A3	18.4	0.8	17.8	0.7
A4	17.7	0.3	17.5	0.6
A5	14.3	0.5	14.3	0.6
A6	9.4	0.4	9.0	0.4
A7	8.7	0.5	8.1	0.3
A8	8.2	0.2	8.0	0.5
A9	7.3	0.2	8.1	0.4
A10	5.9	0.4	5.9	0.3

[A. P. Voskamp, in Advances in X-Ray Analysis, Vol. 17, pp. 124-138, Plenum Press (1974)]

The recent era has witnessed a proliferation of computer-control systems for the automatic collection of powder diffraction intensities (see representative references 31-35) and increasing availability of computer programs for on- and off-line processing of output intensities in quantitative analysis. Thus we may cite the analysis of retained austenite (7) and rock minerals (23), and Jenkins' software package for general quantitative analysis (36). It is self-evident that increasing availability of completely automated data collection and processing systems constitutes the key to any future wide-scale expansion in the application of quantitative diffraction disciplines.

In the present paper it has been unfortunately necessary to omit discussion of two crucially important aspects of quantitative analysis, namely, the procurement of a representative sample and the preparation therefrom of a satisfactory analytical specimen. An adequate treatment would constitute a very large manuscript and include such topics as sampling bulk materials and environmental contaminants, reducing samples to optimal particle size, effecting homogeneous mixing of sample and internal standard, mounting the specimen so as to preserve homogeneity and effectively combat preferred orientation, and estimating and dealing with extinction and microabsorption effects. Some selected references pertinent to these topics are appended to this paper (37-41).

In conclusion, I cannot resist hazarding a few prognostications about future trends. There can be little doubt of the continuing, and probably accelerating, expansion in automated data acquisition and processing, especially through the medium of larger-core minicomputers supplemented with quick access storage. We may expect the lower limit of detection to fall below the 1 μg level with the aid of higher-energy x-ray sources and high-flux monochromators. Synchrotron sources may well open up new, useful ranges of x-ray energies. Extension of EDD techniques are to be anticipated as the technology progresses to provide smaller detector dead times and sharper energy resolution. The recent development in Japan of a microdiffractometer of novel design* holds promise of further reduction in specimen size and detection limits. This instrument detects entire diffraction cones at either forward- or back-reflection angles and employs very small specimen-to-source and specimen-to-receiver distances.

----

*Rigaku Corporation, Kandasurugadai, Chiyoda-ku, Tokyo, Japan.

REFERENCES

1. G. L. Clark and D. H. Reynolds, "Quantitative Analysis of Mine Dusts," Ind. Eng. Chem., Anal. Ed. 8, 36-40 (1936).

2. G. Scheibe, Chemische Spektralanalyse, physikalische Methoden der analytischen Chemie, Vol. 1, p. 108, Akademische Verlagsgesellschaft, Leipzig (1933).

3. J. W. Ballard, H. I. Oshry, and H. H. Schrenk, "Quantitative Analysis by X-ray Diffraction. I. Determination of Quartz," U. S. Bur. Mines Repts. Invest., 3520 (June 1940).

4. F. S. Gardner, M. Cohen, and D. P. Antia, "Quantitative Determination of Retained Austenite by X-rays," Trans. AIME 154, 306-317 (1943).

5. B. L. Averbach and M. Cohen, "X-ray Determination of Retained Austenite by Integrated Intensities," Trans. AIME 176, 401-415 (1948).

6. R. E. Ogilvie, "Retained Austenite by X-rays," Norelco Reporter 6, No. 3, 60-61 (1959).

7. C. J. Kelly and M. A. Short, "A Paper Tape Controlled X-ray Diffractometer for the Measurement of Retained Austenite," in K. F. J. Heinrich, C. S. Barrett, J. B. Newkirk, and C. O. Ruud, Editors, Advances in X-Ray Analysis, Vol. 15, pp. 102-113, Plenum Press (1972).

8. H. Friedman, "Geiger Counter Spectrometer for Industrial Research," Electronics 18, April 1945, pp. 132-137.

9. L. E. Alexander and H. P. Klug, "Basic Aspects of X-ray Absorption in Quantitative Diffraction Analysis of Powder Mixtures," Anal. Chem. 20, 886-889 (1948).

10. H. P. Klug, L. E. Alexander, and E. Kummer, "X-ray Diffraction Analysis of Crystalline Dusts," J. Ind. Hyg. Toxicol. 30, 166-171 (1948).

11. H. P. Klug, L. E. Alexander, and E. Kummer, "Quantitative Analysis with the X-ray Spectrometer," Anal. Chem. 20, 607-609 (1948).

12. L. E. Alexander, H. P. Klug, and E. Kummer, "Statistical Factors Affecting the Intensity of X-rays Diffracted by Crystalline Powders," J. Appl. Phys. 19, 742-754 (1948).

13. J. Leroux, D. H. Lennox, and K. Kay, "Direct Quantitative X-ray Analysis by Diffraction-Absorption Technique," Anal. Chem. 25, 740-743 (1953).

14. D. H. Lennox, "Monochromatic Diffraction-Absorption Technique for Direct Quantitative X-ray Analysis," Anal. Chem. 29, 766-770 (1957).

15. J. Leroux and M. Mahmud, "Influence of Goniometric Arrangement and Absorption in Qualitative and Quantitative Analysis of Powders by X-ray Diffractometry," Appl. Spectrosc. 14, 131-134 (1960).

16. S. Ergun and V. H. Tiensuu, "Determination of X-ray Absorption Coefficients of Inhomogeneous Materials," J. Appl. Phys. 29, 946-949 (1958).

17. P. P. Williams, "Direct Quantitative Diffractometric Analysis," Anal. Chem. 31, 1842-1844 (1959).

18. L. E. Copeland and R. H. Bragg, "Quantitative X-ray Diffraction Analysis," Anal. Chem. 30, 196-208 (1958).

19. F. H. Chung, "A New X-ray Diffraction Method for Quantitative
     Multicomponent Analysis," in C. L. Grant, C. S. Barrett, J. B.
     Newkirk, and C. O. Ruud, Editors, Advances in X-Ray Analysis,
     Vol. 17, pp. 106-115, Plenum Press (1974).

20. F. H. Chung, "Quantitative Interpretation of X-ray Diffraction
     Patterns of Mixtures. I. Matrix-Flushing Method for Quantitative
     Multicomponent Analysis," J. Appl. Cryst. 7, 519-525 (1974).

21. L. G. Berry (ed.), Inorganic Index to the Powder Diffraction File,
     JCPDS, Philadelphia, 1970, pp. 1189-1196.

22. N. H. Clark and R. J. Preston, "Dilution Methods in Quantitative
     X-ray Diffraction Analysis," X-Ray Spectrometry 3, 21-25 (1974).

23. J. J. Sahores, "New Improvements in Routine Quantitative Phase
     Analysis by X-ray Diffractometry," in L. S. Birks, C. S. Barrett,
     J. B. Newkirk, and C. O. Ruud, Editors, Advances in X-Ray Analysis,
     Vol. 16, pp. 186-197, Plenum Press (1973).

24. J. Leroux and C. A. Powers, "Direct X-ray Diffraction Quantitative
     Analysis of Quartz in Industrial Dust Films Deposited on Silver
     Membrane Filters," Occupational Health Review 21, 26-34 (1970).

25. H. E. Bumsted, "Determination of Alpha-Quartz in the Respirable
     Portion of Airborne Particulates by X-ray Diffraction," Amer. Ind.
     Hygiene Assoc. J. 34, 150-158 (1973).

26. J. V. Crable, "Quantitative Determination of Chrysotile, Amosite
     and Crocidolite by X-ray Diffraction," Amer. Ind. Hygiene Assoc. J.
     27, 293-298 (1966).

27. A. L. Rickards, "Estimation of Trace Amounts of Chrysotile Asbestos
     by X-ray Diffraction," Anal. Chem. 44, 1872-1873 (1972).

28. R. E. Ferrell, Jr., "Applicability of Energy-Dispersive X-ray Powder
     Diffractometry to Determinative Mineralogy," Amer. Mineral. 56,
     1822-1831 (1971).

29. F. A. Mauer, private communication.

30. A. P. Voskamp, "High-Speed Retained Austenite Analysis with an
     Energy Dispersive X-ray Diffraction Technique," in C. L. Grant,
     C. S. Barrett, J. B. Newkirk, and C. O. Ruud, Editors, Advances
     in X-Ray Analysis, Vol. 17, pp. 124-138, Plenum Press (1974).

31. M. Richesson, L. Morrison, J. B. Cohen, and K. Paavola, "An
     Inexpensive Computer Control for an X-ray Diffraction Laboratory,"
     J. Appl. Cryst. 4, 524-527 (1971).

32. R. Jenkins, D. J. Haas, and F. R. Paolini, "A New Concept in
     Automated X-ray Powder Diffractometry," Norelco Reporter 18, No. 2,
     12-27 (1971).

33. M. Slaughter, "A Modular Automatic X-ray Analysis System," in
     K. F. J. Heinrich, C. S. Barrett, J. B. Newkirk, and C. O. Ruud,
     Editors, Advances in X-Ray Analysis, Vol. 15, pp. 135-147, Plenum
     Press (1972).

34. A. Segmüller, "Automated X-ray Diffraction Laboratory System," in
     K. F. J. Heinrich, C. S. Barrett, J. B. Newkirk, and C. O. Ruud,
     Editors, Advances in X-Ray Analysis, Vol. 15, pp. 114-122, Plenum
     Press (1972).

35. P. J. King and W. L. Smith, "A Computer-Controlled X-ray Powder
     Diffractometer," J. Appl. Cryst. 7, 603-608 (1974).

36. R. Jenkins, "Quantitative Analysis with the Automatic Powder
     Diffractometer," Norelco Reporter 22, No. 1, 7-12 (1975).

37. R. H. Bragg, "Quantitative Analysis by Powder Diffraction," in
     E. Kaelble, Editor, Handbook of X-Rays, Chap. 12, McGraw-Hill,
     New York (1967).

38. R. Jenkins, "Provision, Suitability and Stability of Standards for
     Quantitative Powder Diffractometry," in C. L. Grant, C. S. Barrett,
     J. B. Newkirk, and C. O. Ruud, Editors, Advances in X-Ray Analysis,
     Vol. 17, pp. 32-43, Plenum Press (1974).

39. H. P. Klug and L. E. Alexander, X-Ray Diffraction Procedures,
     2nd Ed., pp. 202-206, 364-376, 540-544, Wiley, New York (1974).

40. J. Leroux, "Preparation of Thin Dust Coatings for Their Analysis by
     X-ray Emission and Diffraction," Occupational Health Review 21,
     19-25 (1970).

41. B. Post, "Laboratory Hints for Crystallographers," Norelco Reporter
     20, No. 1, 8 (1973).

QUANTITATIVE MATCHING OF POWDER DIFFRACTION PATTERNS

L. K. Frevel

1205 W. Park Drive

Midland, MI 48640

ABSTRACT

An updated d,I-MATCH procedure [Frevel and Adams, Anal. Chem., 40, 1335-1340 (1968)] is described which permits analysts to compare diverse powder-patterns in a quantitative,objective manner. Criteria are established for rating the quality of a diffraction standard. The d,I-MATCH program has been combined with an efficient SEARCH program to facilitate the quantitative matching of multiphase powder patterns.

INTRODUCTION

Comparing the powder diffraction pattern of an unknown sample with published standard patterns is not a straightforward procedure because of the wide variation in the quality of the published data (1). The simplest way to identify a crystalline phase is to compare visually its powder diffraction film with films of certified standards exposed in the same camera under nearly identical conditions. Most analytical laboratories, however, do not have a comprehensive catalogued file of standard films and thus must rely on literature data. The various methods of comparing powder patterns are conveniently characterized in Table 1, listing the types of comparisons and their principal advantages and disadvantages. Although counter techniques have certain quantitative advantages over photographic registration of powder patterns, the film methods (especially with focusing cameras utilizing $CuK\alpha_1$ or $FeK\alpha_1$ radiation) have maintained their demonstrated utility for identifying unknown phases. Nonetheless, it is the advent of automated recording of digitized powder data that has made practical the objective quantitative comparisons of powder diffraction patterns.

MATCHING OF INTERPLANAR SPACINGS

If $\{d_\nu, I_\nu\}$ represents the digitized data (2) of a powder pattern to be compared with the standard pattern $\{d_s,(\frac{I}{I_1})_s\}$, then each pattern is assigned a merit rating for the interplanar spacings and a separate rating for the corresponding relative intensities. The criterion for a valid match of d_s with d_ν is given by Expression (1),

TABLE 1. Methods of Comparing Powder Diffraction Patterns

Form of Diffraction Data		Type of Comparison		Advantages	Disadvantages
Unknown	Standard	d,Å	I_p or I_i		
film	film	visual	visual	Simplicity	Subjectiveness; Non-transferability of subjective data
film	trace	numeric	visual	Indexability (hkℓ), Information on line profiles	Film shrinkage; Bulky storage for traces
film	$\{d_s, (\frac{I}{I_i})_s\}$	numeric	numeric	Low capital cost	Eye fatigue from prolonged use of intensity scale
film	$\{d_s\}$ calc.	numeric	---	Utilization of unit cell and space group data	Lack of intensity data
film	$\{d_s, I_s\}$ calc.	numeric	numeric	Detection of preferred orientation	High labor cost for manual operation
trace	trace	visual/numeric	visual/numeric	Directness of comparison	Cumbersomeness
trace	$\{d_s, (\frac{I}{I_i})_s\}$	numeric	numeric	Quantitativeness	High labor cost for manual operation
trace	$\{d_s\}$ calc.	numeric	---	Usefulness in absence of powder data	Lack of intensity data
trace	$\{d_s, I_s\}$ calc.	numeric	numeric	Detection of preferred orientation	Human errors in manual operations
trace	trace (calc.)	visual/numeric	visual/numeric	High resolution capability	High cost
$\{d_v, I_v\}$	$\{d_s, (\frac{I}{I_i})_s\}$	numeric	numeric	Adaptability to computer processing, Automated matching of powder patterns, Nonsubjectiveness	High initial investment in reliable automatic digitization of powder diffraction technique
(machine-readable)	(machine-readable)	(print-out)	(print-out)		

d = interplanar spacing, Å

I_p = peak intensity of diffraction line in arbitrary units

I_i = integrated intensity in arbitrary units

Subscript s denotes Standard

$$d_{\nu} - \Delta d_{\nu} - \Delta d_{s,\sigma} \leqslant d_{s,\sigma} \leqslant d_{\nu} + \Delta d_{\nu} + \Delta d_{s,\sigma} \qquad (1)$$

where $d_{s,\sigma}$ is the σth d-spacing for standard s. The inaccuracy, Δd, for any interplanar spacing d depends not only on the error in the measurement of the Bragg angle, but also markedly on the particular value of d. This functional dependence is depicted graphically in Figure 1 and pertains to CuKα_1 radiation. The inaccuracy in measuring the Bragg angle is dependent on the particular camera or diffractometer employed and on the calibration of the effective camera diameter. If $\Delta\theta_r$ is 0.0000 radian for all measured reflections, then the merit rating for that particular pattern is 0 (no error). In like manner, if $10^4\Delta\theta_r$ = 1,2,4,6,8,10,12,15, and 20 radian, then the corresponding d-ratings are 1,2,3,4,5,6,7,8 and 9. Although the relative error in measuring θ has been stated as low as 0.00001 radian, the absolute error is usually considerably larger. The following d-ratings have been observed by the author: NBS-precision data (3,4) rate 2; patterns taken with a 114 mm Guinier camera (CuKα_1 radiation and internal standard), 3 to 5; 17 cm. powder diffractometer (CuKα with internal standard), 4 to 6; 17 cm. diffractometer without internal standard, 5 to 7; 5.73 cm. Debye-Scherrer camera (CuKα, no internal standard), 7 to 9. Most PDF patterns rate between 5 and 8. An instructive comparison between two patterns with different d-ratings is shown in Table 2 (columns 1 and 4) where a BaSO$_4$ pattern, taken with filtered MoKα (5), is matched with a calculated pattern of BaSO$_4$. Smith's (6) program was used to compute the interplanar spacings and corresponding integrated intensities which were converted to peak intensities by the empirical relationship published by the National Bureau of Standards (4). Required input data for BaSO$_4$ included unit cell dimensions (7) (a = 7.1565Å, b = 8.8811Å, c = 5.4541Å), space group, atomic coordinates (8), temperature factors of B = 2.5 for barium and sulfur and 3.0 for oxygen, and the atomic scattering factors (9). The d-rating for the calculated BaSO$_4$-pattern was assigned the value 2; whereas that for the DOW pattern 116 was 9. The d,I-MATCH program (1) encountered no difficulty in pairing all observed d-values for pattern 116. When a d-rating of 4 was used for pattern 116, the 101 reflection (4.36Å) and the 022 reflection (2.315Å) were not matched. The poor resolution of the old Debye-Scherrer pattern is evident for the following d-spacings: 4.36Å (101 and 020); 2.73Å (002 and 130); 2.315Å (022 and 310); 2.104Å (311, 140, 212, and 231). Nevertheless, by assigning a d-rating of 6, one could utilize this pattern for identification purposes if no better standard were in the Powder Diffraction File. Likewise this example shows that a precision pattern can profitably replace a standard pattern of lesser resolution for the purpose of matching interplanar spacings of an unknown sample. Another example is shown in Table 2, columns 7 and 9, where the most recent pattern for BaSO$_4$ with a d-rating of 2 is compared with the calculated BaSO$_4$-pattern. One finds that the first 19 reflections are correctly matched and that, for the unresolved reflections 311 and 140, the d,I-MATCH program selects the more closely fitting d-spacing to match the 2.121Å spacing. The same situation obtains for the unresolved reflections 212 and 231.

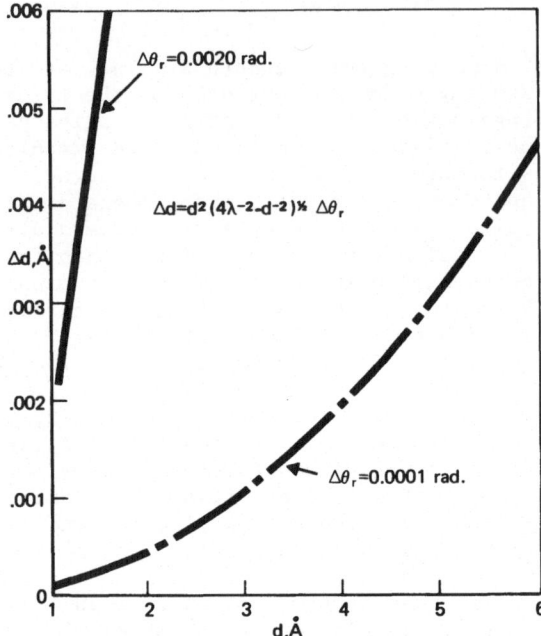

Figure 1.   Terminal curves displaying the error in d
as a function of d, λ, $\Delta\theta$.

TABLE 2.   Matching of $BaSO_4$ Patterns

Pattern 116 Dow 1938			d,I-Match			PDF 24-1035 NBS 1972		d,I-Match		
d,Å	I	hkℓ	d_s,Å	I_c	$\frac{\|I-I_c\|}{I_c}$	d,Å	I	d_s,Å	I_c	$\frac{\|I-I_c\|}{I_c}$
		110	5.572	1.8	1.00−	5.58	2	5.572	3.0	0.33−
		020	4.441	10.0	1.00−	4.440	16	4.441	16.8	0.05
4.36	20	101	4.338	21.8	0.08	4.339	30	4.338	36.6	0.18−
3.90	25	111	3.898	31.2	0.20	3.899	50	3.898	52.5	0.05
		120	3.773	7.1	1.00−	3.773	12	3.773	11.9	0.01
3.58	10	200	3.578	15.9	0.37	3.577	30	3.578	26.7	0.12−
3.45	63	021	3.444	58.8	0.07	3.445	100	3.444	99.0	0.01
3.32	35	210	3.319	38.8	0.10	3.319	70	3.319	65.3	0.07
3.11	63	121	3.103	57.6	0.09	3.103	95	3.103	97.0	0.02
2.84	40	211	2.835	27.6	0.45	2.836	50	2.835	46.5	0.07
		130	2.736	5.3	1.00−	2.735	15	2.736	8.9	0.68−
2.730	45	002	2.727	25.9	0.74	2.729	45	2.727	43.6	0.03
2.475	15	221	2.481	7.6	0.96	2.482	13	2.481	12.9	0.01
		131	2.445	0.6	1.00	2.447	2	2.445	1.0	1.02−
2.315	10	022	2.324	7.6	0.31	2.325	14	2.324	12.9	0.09
		310	2.304	2.4	1.00−	2.305	6	2.304	4.0	0.52−
		230	2.281	3.5	1.00−	2.282	8	2.281	5.9	0.35−
2.204	15	122	2.210	12.9	0.16	2.211	25	2.210	21.8	0.15−
		202	2.169	1.8	1.00−	2.169	3	2.169	3.0	0.01
		311	2.122	22.3	1.00−			2.122	37.6	1.00−
		140	2.121	17.1	1.00−	2.121	80	2.121	28.7	1.79−
2.104	100	212	2.107	22.9	3.36*	2.106	75	2.107	38.6	0.94−
		231	2.104	18.8	1.00−			2.104	31.7	1.00−

MATCHING OF INTENSITIES

After all valid matches of $\{d_\nu\}$ with $\{d_s\}$ have been carried out, the program computes the corresponding intensities from Expression (2)

$$I_c = \left(\frac{I}{I_1}\right)_{s,\sigma} \cdot \frac{\sum\limits_{\nu} I_\nu}{\sum\limits_{\sigma}\left(\frac{I}{I_1}\right)_{s,\sigma}} \qquad (2)$$

where the summations over ν and σ are restricted to paired matched lines. The weighting of lines according to their relative intensities reduces the undue influence of weak reflections in computing I_c. Errors in the automated measurement of peak intensities can be ascertained with reasonable accuracy so that a meaningful value of ΔI can be assigned to each I_ν. Unfortunately most published intensity data lack an explicit statement of the allowable error in the reported I_s. To circumvent this deficiency one can calculate ΔI from the empirical graph shown in Figure 2. The variable τ is defined in the article on automated measurement of powder diffraction patterns (10). Appropriate values for $\Delta\tau_r$ are selected according to the I-ratings as follows: take ratings $0,1,2,3,4,5,6,7,8,9$ as corresponding respectively to $10^2\Delta\tau_r = 0.0, 0.5, 1.0, 2.0, 4.0, 6.0, 8.0, 10.0, 15.0, 25.0$. In general the intensity data for most powder patterns are rather poor and are rated by the author between 5 and 9. The NBS intensity data are rated 3 rather than 2, because preferred orientation is difficult to eliminate completely from flat powder specimens. Since the history and preparation of a powder specimen play such a controlling role in affecting the observed intensities, it is not surprising that one seldom deals with ideal mosaic crystallites randomly distributed. Calculated patterns (6,11) therefore serve to reveal discrepancies arising from preferred orientation, primary extinction or microabsorption. In general one should not expect the observed intensities to match the calculated intensities with less than 20% error. An intensity match between I_ν and I_c is judged valid if Criterion (3) is met. The $BaSO_4$ patterns of Table 2 illustrate the manner in which

$$I_\nu - \Delta I_\nu - \Delta I_c \leqslant I_c \leqslant I_\nu + \Delta I_\nu + \Delta I_c \qquad (3)$$

Criterion (3) and the specification of a minimum intensity for an observable reflection denote invalid intensity-matches. Assigning an I-rating of 9 to Pattern 116 and a rating of 2 to the calculated $BaSO_4$-pattern, one observes from columns 2, 5, and 6 that the unobserved reflections (110, 020, 120, 130, 310, 220, and 202) should have been observed inasmuch as a minimum intensity of 1 was specified. A minus sign in column 6 denotes an invalid match; an asterisk warns of a possible superposition whenever $|I-I_c|/I_c$ is greater than or equal to 2 for a valid intensity match. If I (minimum) had been entered as 5, then the d,I-MATCH program would not have rated the absence of 110, 310, 230, and 202 as invalid. The asterisk for the 212 reflection, denoting pronounced superposition, is readily explained by the failure of the old G.E. camera to resolve four closely spaced strong reflections; namely 2.122Å, 2.121Å, 2.107Å, and 2.104Å. This coincidence accounts for the fact that "the" strongest line of Pattern 116 was recorded as 2.10 kX

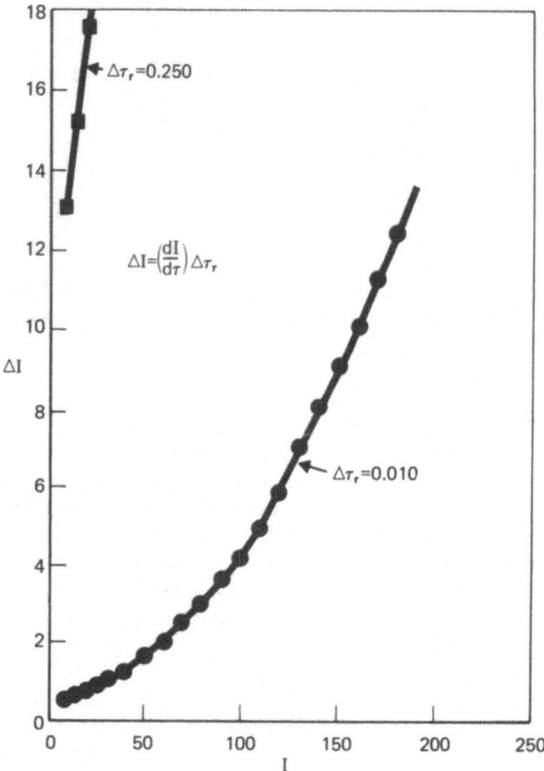

Figure 2.   Terminal curves displaying the error in I
as a function of I and Δτ.

rather than 3.44 kX, the 021 reflection. Poor resolution also accounts
for the partial overlap of reflections 020 and 101. By weighting these
two reflections (4.441Å and 4.338Å) according to their respective intensities, one computes an unresolved peak at 4.37Å. Applying the same
procedure to the unresolved reflections (130 + 002), (221 + 131), (022 +
310 + 230), and (122 + 202), one obtains the respective d-spacings of
2.729Å, 2.478Å, 2.309Å, and 2.205Å. This example clearly points up the
advantage of having an accurate calculated pattern for quantitative comparison. When an I-rating of 4 is assigned to Pattern 116, only three
valid matches are found: namely, for spacings 3.45Å, 3.32Å, and 3.11Å.

     For the second d,I-MATCH of Table 2, an I-rating of 3 was entered
for PDF 24-1035 and a rating of 2 for the calculated $BaSO_4$-pattern.
Eight out of 21 intensity matches were found to be invalid according to
Criterion (3). Superpositions are evident for reflections 311 and 140
as well as for 212 and 231. Because of the closeness of the superpositions, one is justified in adding the appropriate calculated intensities and in matching I = 80 with I_c = 66.3 and 75 with 70.3. An
I-rating of 2 for the calculated $BaSO_4$-pattern is probably too low

because of errors in the atomic positions determined from crystal structure analysis. For the tetrahedral sulfate ion in $BaSO_4$, the experimentally determined S-O distances vary between 1.476Å and 1.515Å. This difference can be ascribed to the errors in the nine spatial parameters required to fix the coordinates for sulfur and oxygen atoms.

MATCHING OF MULTIPHASE PATTERNS

When a pattern to be matched pertains to a multiphase powder, complications may occur because of fortuitous superpositions of two or more diffraction lines from two or more phases. To compensate for this contingency an iterative correction is applied. If the difference, $|I_v-I_c|$, is greater than a specified lower bound for ΔI (superposition), then I_v for the suspect reflection is replaced by its corresponding I_c in Expression (2). Three such iterations are sufficient to eliminate all significant superpositions. After all matches with stipulated standards have been corrected for superposed intensities, then the matched patterns are arranged in ascending order of their over-all strength; i.e., according to the increasing value of the sum of their computed peak intensities. The complete matching process is then repeated by matching the weakest pattern first; subtracting its pattern to obtain a difference pattern, $\{d_v, I_v-I_c\}$; matching this difference pattern with the second weakest pattern and again obtaining a difference pattern for subsequent exhaustive matching. The final accounting of all matched phases is then printed in tabular form as in Table 3. The data therein pertain to Mineral Sample B received from Mr. Ron Jenkins. As Chairman of the Computer Subcommittee of The Joint Committee on Powder Diffraction Standards, he sent five powder mixtures to twelve participants in a round-robin test to determine how efficiently computer search/matching can be, relative to manual searching. Mr. L. R. Ruhberg of The Dow Chemical Company ran Mineral B on a Norelco diffractometer: $CuK\alpha$ (32 KV, 20 ma), scanning rate of $1°/min.$, one degree divergence and receiving slits in conjunction with a focusing monochromator. The sample was first ground in an agate mortar, pressed into a sample holder with a glass slide, and revolved during exposure. Qualitative elemental data were obtained by energy dispersive spectroscopy: Ba, Cu, Fe, S, Al, and Si. With the aid of the ZRD-'74 SEARCH-MATCH program (12) he identified the sample as a mixture of barite, $BaSO_4$; chalcopyrite, $CuFeS_2$; gibbsite, $Al(OH)_3$; and possibly some kaolinite, $Al_2Si_2O_5(OH)_4$. It was learned later that the actual sample was made up of equal parts by weight of the first three minerals. In discussing these findings with Mr. Ruhberg, the author noted that the quantitative matching of the kaolinite phase appeared questionable. To obtain confirmative data, the $\{d_v, I_v\}$ values were requested from Prof. J. D. Hanawalt who had used a Guinier camera with $CuK\alpha_1$ radiation to obtain his powder pattern of Mineral B. His pattern of 63 lines had better resolution than Ruhberg's diffractometer trace; e.g., the splitting of the following pairs of d-spacings: 4.38, 4.34Å; 3.35, 3.32Å; 1.856, 1.836Å; 1.761, 1.753Å; 1.536, 1.527Å; and 1.428, 1.424Å. However, some weak reflections were not recorded: 7.06Å, 4.72Å, 4.16Å, 3.53Å, 3.18Å, 2.79Å, 2.565Å, 2.305Å, 1.994Å, 1.951Å. Entering Hanawalt's data into the ZRD-'74 program, one matches not only the three expected phases: $BaSO_4$, $CuFeS_2$, and $Al(OH)_3$; but also α-quartz and pyrrhotite, $Fe_{1-x}S$. To obtain comparable resolution, L. R. Ruhberg ran Mineral B in a 100 mm Hägg camera with $CuK\alpha_1$ radiation and observed

TABLE 3.   Completed Print-out of Decomposed Match of Mineral B

ν	SAMPLE d_ν,Å	I_ν	Pyrrhotite d_s,Å	I_c	α-Quartz d_s,Å	I_c	Amesite d_s,Å	I_c	Gibbsite d_s,Å	I_c	Chalcopyrite d_s,Å	I_c	Barite d_s,Å	I_c	$I_\nu - \sum_p (I_c)_p$
1	7.064	7.0					7.060	9.9							-2.9
2	4.844	44.0							4.850	37.8					6.2
3	4.720	4.2									4.710	8.7			-4.5
4	4.440	9.1											4.440	8.4	0.7
5	4.364	10.4							4.360	17.0					-6.6
6	4.336	16.7							4.320	3.8			4.340	17.8	-4.9
7	4.268	4.2			4.260	8.2									-4.0
8	4.166	1.9													1.9
9	3.898	24.7											3.900	28.2	-3.5
10	3.775	5.0											3.770	5.9	-0.9
11	3.576	39.6											3.576	15.4	24.2
12	3.534	11.1					3.520	9.9							1.2
13	3.443	74.6											3.442	49.6	25.0
14	3.342	21.3			3.343	23.5			3.355	2.3					(-2.2)-4.5
15	3.320	49.4							3.310	3.4			3.317	33.2	12.8
16	3.189	3.2							3.180	2.6					0.6
17	3.102	54.5							3.090	2.3			3.101	48.1	4.2
18	3.039	100.0									3.025	43.5			56.5
19	2.985	3.2	2.980	3.7											-0.5
20	2.837	26.2											2.834	26.3	-0.1
21	2.791	14.4													14.4
22	2.727	27.3											2.726	23.3	4.0
23	2.646	6.0	2.645	5.1							2.635	4.4			-3.5
24	2.610	1.0					2.610	2.0							-1.0
25	2.565	1.0													1.0
26	2.483	6.9					2.480	6.0					2.481	6.9	-6.0
27	2.454	5.1			2.450	2.8			2.448	5.7					-3.4
28	2.425	3.8							2.420	1.9					1.9
29	2.387	3.3							2.380	7.6					-4.3
30	2.326	9.0					2.320	5.0					2.322	7.4	(4.0)-3.4
31	2.305	6.9											2.303	3.0	3.9
32	2.281	4.9			2.282	2.8			2.283	1.9			2.281	3.5	-3.3
33	2.212	13.7											2.209	13.4	(13.7)0.3
34	2.167	2.8							2.159	3.0					-0.2
35	2.121	47.6											2.120	39.6	8.0
36	2.108	40.2					2.110	3.0					2.104	37.7	(37.2)-0.5
37	2.068	6.5	2.065	7.3											-0.8
38	2.057	12.8											2.056	11.4	1.4
39	1.994	2.0							1.987	3.8					-1.8
40	1.951	1.9							1.955	0.8					1.1
41	1.932	3.9					1.925	7.0					1.930	3.5	(-3.1)-6.6
42	1.918	2.8							1.910	3.8					-1.0
43	1.870	17.0									1.865	10.9			6.1
44	1.857	39.9									1.852	21.8	1.857	7.9	10.2
45	1.845	4.5													4.5
46	1.818	2.1			1.817	4.0									-1.9
47	1.790	5.1							1.795	3.8			1.787	1.5	(1.3)-0.2
48	1.762	4.1											1.760	4.5	(4.1)-0.4
49	1.754	11.2					1.749	5.0					1.754	4.5	1.8
50	1.724	4.1	1.720	3.7									1.723	3.0	-2.5
51	1.683	7.5							1.680	3.8			1.681	3.5	(3.7)0.2
52	1.674	10.4			1.672	1.6							1.673	6.9	1.8
53	1.660	3.7			1.659	0.7			1.652	1.5			1.669	4.9	(3.0)-1.9
54	1.635	5.3							1.635	0.8			1.636	4.0	0.6
55	1.594	19.5					1.596	6.0			1.590	17.4	1.593	4.0	(-3.9)-7.9
56	1.576	7.8							1.573	1.1	1.575	8.7			-2.0
57	1.564	2.8													2.8
58	1.535	8.1					1.531	6.0					1.534	8.9	-6.8
59	1.496	2.0					1.495	2.0					1.495	1.5	(0.0)-1.5
60	1.481	2.8													2.8
61	1.476	5.2											1.474	5.0	(5.2)0.2
62	1.458	3.5					1.461	6.0	1.455	3.8			1.457	1.5	(-2.5)-7.8
63	1.440	1.9	1.443	0.7											1.2
64	1.426	8.4											1.426	4.0	+7.9 (4.4)-3.5
65	1.407	4.7							1.410	3.8			1.406	3.5	(0.9)-2.6
66	1.402	6.1					1.401	6.0					1.401	5.0	(0.1)-4.9
67	1.385	3.6			1.382	1.6							1.384	3.0	(3.6)-1.0
68	1.363	4.2											1.363	3.0	1.2
69	1.349	2.3					1.342	5.0					1.349	2.5	(-0.2)-5.2
70	1.323	4.2	1.325	1.5							1.320	6.5			-3.8
71	1.263	4.5													4.5

SAMPLE: MINERAL B   DATE RECD'   ORIGIN: Ron Jenkins

SAMPLE WAS EXAMINED BY X-RAY POWDER DIFFRACTION AND PROCESSED BY COMPUTER PROGRAM ZRD -'74 .
FOLLOWING CRYSTALLINE PHASES WERE MATCHED: Barite, $BaSO_4$; Chalcopyrite, $CuFeS_2$; Gibbsite, $Al(OH)_3$; Amesite, $(Mg_{1.6}Al_{1.0}Fe_{0.4})(SiAlO_5)(OH)_4$; α-Quartz, SiO_2; Pyrrhotite, $Fe_{1-x}S$.

ANALYST: L.K. Frevel   DATE'   REPORT'   CHARGE'

COMMENTS: Unassigned lines 8, 21, 25, 45, and 60 imply at least one unidentified phase.

the following spacings which were not resolved by routine diffractometry: 4.364Å, 3.342Å, 1.845Å, 1.762Å, 1.660Å, and 1.564Å. It was also noted that the innermost reflection (7.06Å) was very weak and that the 3.534Å spacing as well as some other weak reflections were not observed on the Guinier pattern. Evidently the diffractometer sample-mounting induced pronounced preferred orientation of an unsuspected phase. Columns 2 and 3 of Table 3 give the averaged data from the diffractometer and the focusing camera. Barium sulfate was used as an internal standard to correct the diffractometer measurements of 2θ; aluminum was the internal standard for the Hägg camera.

Pairing the innermost unidentified spacing, 7.064Å, with 3.534Å (probably second order) and searching in an ordered file of clay minerals, one readily locates amesite as a likely candidate. Entering its powder pattern, PDF 9-493, as an external standard and applying the ZRD-'74 program, one obtains the quantitative matching of the six phases listed in Table 3. Inspecting the intensity matches for the strongest pattern, one finds only one serious discrepancy, namely the low I_c for the 3.576Å spacing. All other cases give valid intensity matches with an I-rating of 8. Although there are 28 cases of superposed lines, including four instances of triple superpositions, the iterative correction procedure is effective in resolving these complications.

The ZRD-'74 program does not exclude a matching candidate because of invalid intensity matches. It is up to the diffractionist to examine the quantitative print-out of each match and decide when to exclude a pseudomatch from the final tabulation. For Mineral B such a pseudomatch was found for rhombohedral AlF_3 after the major phases were correctly identified. Table 4 reproduces the data of the print-out for AlF_3. In column 3 are listed the residual intensities after subtracting from I_ν the calculated I_c's from prior matches. Column 4 contains the d-spacings for standard, PDF 9-138. The absence of the fourth strongest line of AlF_3 is a signal to suspect this fortuitous match. Subsequent assignment of the 3.534Å spacing to the amesite pattern confirmed the suspicion of a pseudomatch. Great caution has to be exercised in the valid matching of minor phases. The successful identification of minor phases in the presence of three or more prominent phases is critically dependent on: the resolution of closely spaced d-values, the detection of weak reflections, and the quantitative matching of intensities with appropriate adjustments for superpositions.

CONCLUSION

The quantitative matching of powder patterns by the ZRD-'74 program is recommended to all analysts interested in extracting the maximum information from precise digitized powder data. Regardless of the methods employed to unravel an unknown powder pattern, a final accounting of the total diffraction data is essential for a dependable analysis. In combination with calculated powder patterns, the d,I-MATCH procedure can assist in assigning a meaningful merit rating for the interplanar spacings of a standard and a separate rating for the corresponding relative intensities.

TABLE 4.   Print-out of Match with Rhombohedral AlF_3

ν	d_ν,Å	I_ν'	d_s,Å	I_c	$\dfrac{\|I_\nu' - I_c\|}{I_c}$
12	3.534	11.1	3.520	12.8	0.13
			2.510	0.5	1.00
35	2.121	3.3	2.119	2.6	0.27
37	2.068	6.5	2.074	0.5	11.74*
			2.019	0.3	1.00
48	1.762	4.1	1.759	3.2	0.29
			1.600	0.5	1.00
			1.587	2.0	1.00–
57	1.564	2.4	1.560	1.0	1.35
62	1.458	0.0	1.460	0.3	1.00
			1.431	0.5	1.00
			.	.	.
			.	.	.
			.	.	.
			0.800	0.1	1.00

$$\Sigma I_c = 30.9$$

$I_\nu' = $ residual intensity

ACKNOWLEDGMENTS

The author wishes to thank Mr. L. R. Ruhberg of The Dow Chemical Company and Prof. J. D. Hanawalt of The University of Michigan for their powder data cited in the text. He is also grateful to Dr. June W. Turley for carefully reviewing the manuscript.

REFERENCES

(1) L. K. Frevel and C. E. Adams, ANAL. CHEM., 40, 1335–1340 (1968).

(2) L. K. Frevel, ANAL. CHEM., 37, 471–482 (1965).

(3) H. E. Swanson and R. K. Fuyat, National Bureau of Standards Circular 539, Vol. 2, U.S. Government Printing Office, Washington, D.C. (1953).

(4) H. E. Swanson, M. C. Morris, R. P. Strinchfield, and E. H. Evans, National Bureau of Standards Monograph 25, Section 1, U.S. Government Printing Office, Washington, D.C. (1962).

(5)   J. D. Hanawalt, H. W. Rinn, and L. K. Frevel, Ind. Eng. Chem.,
      Anal. Ed. <u>10</u>, 457-512 (1938).

(6)   D. K. Smith, UCRL-50264, Lawrence Radiation Laboratory, Livermore,
      Calif. (1967).

(7)   H. E. Swanson, H. F. McMurdie, M. C. Morris, E. H. Evans, and
      B. Paretzkin, National Bureau of Standards Monograph 25, Section
      10, page 12, U.S. Government Printing Office, Washington, D.C.
      (1972).

(8)   R. W. G. Wyckoff, "Crystal Structures", 2nd ed., Vol. 3, Inter-
      science, New York (1965) p. 46.

(9)   "International Tables for X-Ray Crystallography" Vol. 3, The Kynoch
      Press, Birmingham, England (1962) p. 201.

(10) L. K. Frevel, ANAL. CHEM., <u>38</u>, 1914-1920 (1966).

(11) C. R. Hubbard, E. H. Evans, and D. K. Smith, J. Appl. Cryst. <u>9</u>,
      169-174 (1976).

(12) L. K. Frevel, C. E. Adams, and L. R. Ruhberg, J. Appl. Cryst.
      <u>9</u>, 199-204 (1976).

EXPERIMENTAL AND CALCULATED STANDARDS FOR QUANTITATIVE ANALYSIS BY

POWDER DIFFRACTION

C. R. Hubbard

Institute for Materials Research

National Bureau of Standards

Washington, D. C. 20234

and

D. K. Smith

Department of Geosciences

Pennsylvania State University

University Park, PA 16802

ABSTRACT

Quantitative analysis by x-ray powder diffraction methods has become increasingly important in recent years with the availability of computer-controlled automatic powder diffractometers. All data gathering techniques require suitable reference standards to scale the measured data properly. One means of achieving this scaling is through the *reference intensity ratio* which is defined as the intensity ratio of the strongest diffraction maximum of a substance to the strongest maximum of a reference material in a 1:1 mixture by weight. These ratios may be measured or they may be calculated if the crystal structures of the materials are accurately known.

Although the ratio for a 1:1 mixture of a substance to a specific reference material (currently corundum) is a good basis for tabulating the *reference intensity ratios*, the actual measurement is often facilitated by using diffraction peaks other than the strongest ones, mixtures other than 1:1 or different reference materials. These measurements are appropriately corrected for tabulation. The tabulated ratios for the suggested set of experimental conditions -- a diffractometer without any monochromators and employing CuKα radiation -- can be modified for different conditions using appropriate relationships. Carefully determined *reference intensity ratios* will prove invaluable in x-ray quantitative analysis either indirectly as a check on experimental standards or directly in the analytical equations.

INTRODUCTION

From the early days of x-ray powder diffraction as a tool in analytical chemistry, it was recognized that the method had potential both in quantitative analysis as well as qualitative analysis. In their important 1938 paper Hanawalt, Rinn and Frevel (1) indicated the need to place reference patterns on an absolute scale. They scaled their intensities by referencing them against NaCl. The first patterns issued under the auspices of the American Society for Testing and Materials (ASTM)* contained these scaling factors. Subsequent patterns did not follow this format.

The need for scaling reference patterns for quantitative applications has become more important in recent years with the extensive use of computer-controlled diffractometers and computer data processing. The *reference intensity ratio* is a means of placing every pattern on a common scale. It is defined as the ratio of the intensity of the strongest line in the pattern of a substance to the strongest line of a reference material in a 1:1 mixture by weight. Even though a material may be initially related to any of several reference materials it is an easy matter to relate it to another standard when the *reference intensity ratios* among the standards are known.

Around the middle 1960's, the *reference intensity ratio* was reintroduced in the Powder Diffraction File, PDF, at the suggestion of deWolff and Visser (personal communication). The reference standard was chosen to be α-Al_2O_3 in a microcrystalline state (0.3 μm average particle size). Since this time JCPDS sponsored projects, including the Associateship at the National Bureau of Standards, have routinely measured these ratios. Generally the ratio of peak heights has been used as an approximation to the ratio of integrated intensities.

The definition and use of the *reference intensity ratio* was discussed in Hubbard, Evans and Smith (HES) (2), and the details will not be repeated here. It is the purpose of this paper to discuss the methods of obtaining the *reference intensity ratio* and the sources of error in each method. This work is part of a study to locate additional suitable reference materials for powder diffraction standards to be issued through the Standard Reference Materials program at the National Bureau of Standards.

PROBLEMS AFFECTING INTENSITY MEASUREMENT

One of the most difficult aspects of powder diffraction techniques is the reproducibility of intensity values. In single crystal intensity measurements, data sets accurate to around 1% have been obtained, but in powder diffraction it is difficult to obtain 5% accuracy (3). More commonly data are only good to 10% accuracy.

---

*The publication of these and subsequent patterns in the Powder Diffraction File is now the responsibility of the Joint Committee on Powder Diffraction Standards, 1601 Park Lane, Swarthmore, Pennsylvania, 19081, USA.

The problems which limit the accuracy of intensity measurement may be classed into the two categories, instrumental and sample. Even when counter diffractometers are used, the instrumental problems are minimal but not negligible. The sample though is still a significant problem.

Instrumental Problems

With the precision of modern equipment, most of the instrumental problems result from some aspect of the counting procedure. With high resolution counters with negligible coincidence loses, the principal problem is due to counting statistics, and this problem is not really separable from the sample itself. Statistical accuracy of counting is determined by the number of counts accumulated, $\sigma(N) = N^{\frac{1}{2}}$. To obtain 1% accuracy, one must accumulate a minimum of 10,000 counts in the peak profile after the background has been subtracted.

In most quantitative powder diffraction work, integrated intensities are used, and the counts are accumulated over the whole profile. The total number of counts accumulated in a scan depends on the diffracting power of the sample and the scan rate. In most work one or more of the stronger diffraction maxima of a given phase are usually employed; thus normal scan times ($1/4°$ 2θ/min.) are usually adequate to accumulate sufficient counts for the desired accuracy. Because the total counts are obtained by integration, and because each point on a profile is not independent of adjacent points, it is not necessary to have a common accuracy at every point on the profile.

Another instrumental factor affecting intensity is the polarization, p. This factor should be well understood in order to compare experimental methods such as the diffractometer and Guinier camera. The Guinier camera has an incident beam monochromator where as the diffractometer may, or may not, have a diffracted beam monochromator. The latter case - diffractometer with no monochromator - is the condition generally assumed when reference powder patterns are calculated. Possible polarization factors (4) are

$$p \quad = \quad [1+\cos^2(2\theta)]/2$$

$$p_{DBM} \quad = \quad [1+\cos^2(2\theta)\cos^2(2\theta_m)]/2$$

$$p_{IBM} \quad = \quad [1+\cos^2(2\theta)\cos^2(2\theta_m)]/[1+\cos^2(2\theta_m)]$$

where the subscripts DBM and IBM indicate diffracted and incident beam monochromators, respectively and $2\theta_m$ and 2θ are the Bragg angles of the monochromator and sample, respectively. Note that p_{DBM} and p_{IBM} differ only in their denominators which are constant in each case. The ratio of p to p_{DBM} is a slowly varing function of angle. Taking $2\theta_m$=26.6° (graphite, CuKα) the ratio as a function of 2θ is:

2θ	5.0°	20.0	30.0	40.0	50.0	60.0	70.0	80.0	90.0
p/p_{DBM}	1.11	1.11	1.10	1.09	1.05	1.04	1.02	1.01	1.00

This ratio, symmetrical about 2θ=90.0°, is the factor that data recorded using a monochromator should be multiplied by, before a comparison is made with data recorded without a monochromator. Programs to simulate powder patterns such as POWD5 (5) generally assume that no monochromator is used. Such a small systematic effect will introduce no problem with search procedures for qualitative analysis. However, this difference in polarization is particularly troublesome in efforts to tabulate the *reference intensity ratio*. Currently, no standard condition has been established. Because various monochromators and wavelengths may be employed, it appears reasonable to specify that the standard of tabulation be data measured without any monochromator. Experimental data collected using a monochromator can be easily converted to this standard.

Sample Problems

The sample problems which affect intensities are numerous and may be subdivided into two categories, those which are inherent to the particles themselves and those which are due to the interrelations of the particles.

In the former category we find extinction effects, which have been shown by James (6) to be non-trivial even for very small particle sizes of such simple materials as NaCl (halite) and diamond. This diminishing of intensity is especially important in quantitative analysis where it is the common practice to utilize the strongest maximum in any given pattern. For example, if one compares the diffraction pattern of crushed optical grade Brazilian quartz with theoretical calculations the observed intensity of the 101 line is 30% lower than the calculated value. Similar extinction effects may be observed for other common reference materials such as $CaCO_3$(calcite). However, for commercial α-Al_2O_3($\approx$0.3μm) no extinction effects have been observed.

Also under particle effects, we would classify particle size. Profile broadening from small crystallite sizes (<0.2 μm) is well known. This broadening does not affect integrated intensity measurements, but it does markedly affect peak intensities and this is the single most important reason why peak intensities should not be used in quantitative studies. Less well recognized but equally well documented are problems resulting from too coarse a particle size. The theoretical number of particles necessary to achieve total randomization is infinite. The number of particles is still large when the finite divergence of the x-ray beam in the modern diffractometer is considered. To have one particle in a random sample oriented with its diffraction vector in any area 0.01°2θ square requires 6×10^6 particles. To achieve any sort of counting statistics at least 100 particles must be so oriented. Klug and Alexander (7) have shown that to achieve reasonable sample statistics particles no larger than 1μm must be used. The volume of sample seen by the x-ray beam is not large. The half depth of penetration, the depth above which one half the scattered intensity originates, is $\ln 2/(2\mu\cos\theta)$. For Al_2O_3 and $CuK\alpha$, the half depth of penetration is around 50μm. For UO_2 it is around 1μm. The greater the absorption, the smaller the sample volume contributing to the diffracted intensity and the finer the particle size necessary to achieve adequate particle statistics. Many diffractionists do not recognize the importance of obtaining an adequate number of particles to achieve true randomness. Samples are usually

prepared by seiving through a 400 mesh seive (37μm) which is obviously
still too coarse to yield the desired number of particles. Before one
blames preferred orientation for the lack of reproducibility of diffraction intensities from sample to sample, one should be sure the particle
size is in the proper range.

Under the category of particle interrelations we consider microabsorption and preferred orientation. Microabsorption was discussed in
HES and will not be discussed further here. Preferred orientation has
been discussed in almost every paper on quantitative analysis because it
is probably the single most important factor limiting the accuracy of
the method. Randomization of samples with small equant particles can be
nearly achieved through careful sample preparation. With less symmetrically shaped particles the problem is more difficult. Fiberous particles
are more difficult to randomize than platy ones. Methods which have
been used in many laboratories have been described in the literature (8)
and other methods have been learned by experience or personal communications. The most common method is to drift the powder into a cavity and
to carefully smooth the surface seen by the x-ray beam. The cavity must
be deep enough to allow for the full depth of penetration of the x-ray
beam. The sample surface should not be tamped by a polished surface
such as a microscope slide because this process would enhance any shape-
induced orientation in the critical near surface layers. Instead, the
surface should be carefully cut flat with a sharp edge or razor blade.
Some users prefer to pack the sample from the side or back of the specimen holder against a flat rough surface which is later removed. Although
it is desirable to achieve a high density of particle packing, care must
be taken not to enhance orientation by compaction. A compacted sample
may prove useful if the surface layers are removed before x-raying.

Dilution techniques are often used to achieve randomness but care
must be taken not to reduce the number of particles significantly.
Diluents should be x-ray amorphous and x-ray transparent which restricts
their choice to organic compounds or light element glasses. Cellulose
base cements, particularly collodion, are often used. This method
produces a permanent sample which can be stored for later reference.
Samples diluted with low melting glasses can often be fired and re-
crushed. In all cases the user must remember that the diluent absorption affects the bulk absorption coefficient of the sample as a whole.

EXPERIMENTAL REFERENCE STANDARDS

The most general approach to quantitative analysis by x-ray powder
diffraction is to use an internal standard. With an internal standard
the ratio of the weight fraction of the sample, X, to that of the standard, Xs, is proportional(9) to the ratio of intensities of a line for
each phase:

$$\frac{X}{X_s} = \frac{1}{k} \frac{I}{I_s} \qquad\qquad (1)$$

When the standard is corundum and the strongest line of the sample and of
corundum are measured, then the proportionality constant k is the *reference intensity ratio*, I/Ic. Tabulated ratios can be used even if an

internal standard s (instead of corundum) is added to the sample. In
such a case k is simply the ratio of I/Ic to Is/Ic. Diffraction lines
other than the strongest can readily be used in Equation 1 by introducing
a factor dependent only on the relative intensities of the lines used.
In practice, the ratios reported are rarely of sufficient accuracy to
allow complete dependence on them. They work better for materials with
nearly the same absorption than for those whose linear absorption coefficients differ by more than an order of magnitude. In theory they show
much potential which requires further investigation, as accurate values
would allow one to calculate concentrations from any well prepared
pattern without the need to prepare experimental reference mixtures.

     The choice of α-Al$_2$O$_3$ as the common reference material was one of
considerable thought, but it was not a difficult selection to make when
one considers the criteria. A good reference material should be readily
available in a pure, reproducible, finely powdered form. It should not
show marked profile broadening, and it should have a relatively simple
pattern with maxima scattered uniformly over the 2θ scale. Intensities
should not suffer extinction effects, and the material should not be
prone to orientation. Corundum, α-Al$_2$O$_3$, meets all of these criteria
and few other materials appear to be as suitable.

     The measurement of the *reference intensity ratio*, I/Ic, has been
accomplished by carefully preparing a 1:1 mixture by weight of the
sample and corundum, determining the intensity of the strongest line of
each phase and calculating the ratio of intensities. As a measure of
the intensity ratios, the ratio of peak heights has been routinely used.
This measure is not the best choice because it is beset with many problems as outlined above. It is only valid if the full width at half
maximum, FWHM, is identical for both materials. The ease of peak-height
measurement in contrast to the time-consuming measurement of integrated
intensities is insufficient justification to validate this procedure.

     There are several additional shortcomings to the use of I/Ic from a
1:1 mixture as a basis for experimental measurement, although it is an
appropriate basis for tabulation of the *reference intensity ratio*. To
properly determine the effective ratio it is statistically more valid to
measure the intensities of several maxima of both the sample and the
reference material. All measurements could be weighted and an I/Ic
value calculated considering all of them. This procedure would permit
measurement of I/Ic even when considerable overlap occurs with the (113)
corundum peak. Two other shortcomings are the restriction to a 1:1
mixture and to the use of corundum rather than some other reference
material. Experience has indicated that the best values of I/Ic result
when the measured intensities are nearly the same. By changing the
weight fractions or by using a different reference material, already
calibrated against corundum, this more optimum condition can be met.

     A new experimental procedure has been adopted at NBS which elimin-
ates some of the shortcomings. This procedure involves: 1) obtaining
the relative intensities, I^{rel}; 2) estimating I/Ic; 3) preparing a
mixture which will yield intensities of approximately the same magnitude
for both the sample and corundum; 4) measuring integrated intensities of
three or more reflections from both the sample and corundum; and 5)
repeating steps 3 and 4 until a satisfactorily consistent ratio is
obtained. Other reference materials can be used as soon as they are
available.

To obtain the I^{rel} values a thick, drifted, single phase sample is used. The I^{rel} values are obtained from the unscaled intensity measurements. To estimate I/Ic the unscaled $I(\underline{h}_0)$ and Ic(113) intensity values measured on the same diffractometer are used in

$$I/Ic(est.) = \frac{\rho_c}{\rho} \quad \frac{\mu}{\mu_c} \quad \frac{I(\underline{h}_0)}{Ic(113)} \qquad (2)$$

where $I(\underline{h}_0)$ is the intensity of the strongest line, $\underline{h}_0$, of the sample, and Ic(113) is the intensity of the 113(hex) line of corundum. The results using this equation have proven to be only approximate. However, the I/Ic estimate is useful as a guide for mixing the sample and corundum in reasonable proportions, because the statistical uncertainty in the *reference intensity ratio* is a minimum when $I(\underline{h}_0)$ and Ic(113) are equal.

More precise values of $I(\underline{h}_0)/Ic(113)$ are determined from the mixture of the sample and corundum. If the weight fractions of sample and corundum are respectively X_s and X_c ($X_s = 1 - X_c$), the ratio is given by

$$I(\underline{h}_0)/Ic(113) = \frac{X_c}{X_s} \quad \frac{Ic^{rel}(\underline{k})}{I^{rel}(\underline{h})} \quad \frac{I(\underline{h})}{Ic(\underline{k})} \qquad (3)$$

Several sets of reflections $\underline{h}$ and $\underline{k}$, usually adjacent reflections should be measured to provide data on possible preferred orientation, extinction or other systematic errors. Occasionally it may be more convenient to pair one corundum reflection with two sample reflections or vice versa, or, due to overlap, some reflections may not be usable at all. Alternatively, each measured line of the sample could be paired with each measured line of corundum to obtain an overall average. Table 1 lists the results of applying this procedure for ZnO. Measurement of another mixture yielded similar results. The average value for I/Ic is 5.2 and the standard deviation is 0.1. The consistency between the I/Ic values suggests that the sample is free of effects of extinction and preferred orientation. Also, no systematic error as a function of diffraction angle is apparent. Such experimental results can be evaluated by theoretical calculation of the *reference intensity ratio* from knowledge of the crystal structure.

Table 1
Data for Obtaining I/Ic for ZnO

$[I/Ic (est.) = 6.2; X_c/X = 4.57]$

ZnO			α-Al$_2$O$_3$			
2θ	I^{rel}	I	2θ	I^{rel}	I	$\dfrac{I(101)}{Ic(113)}$
31.78°	59	249	25.58°	61	235	4.99
34.44	46	191	35.16	92	333.5	5.24
36.26	100	413	35.16	92	333.5	5.21
56.62	31	128	57.50	83	296	5.29

CALCULATED REFERENCE INTENSITY RATIOS

Computer codes have been widely used for simulating powder patterns from the crystal lattice and atomic parameter information. In the code POWD5 (5) intensities are first calculated on the 'absolute/relative' scale:

$$\frac{I^{abs}(\underline{h})}{K} = \frac{M(\underline{h})Lp(\underline{h})|F_T(\underline{h})|^2}{2\mu V^2} \qquad (4)$$

where $M(\underline{h})$=multiplicity, $Lp(\underline{h})$=Lorentz-polarization factor for a diffrac-tometer without monochromator, $F_T(\underline{h})$=structure factor including thermal effects for reflection $\underline{h}$, μ=linear absorption coefficient and V=volume of the unit cell. The constant K is independent of the sample and diffraction geometry; the value $I^{abs}(\underline{h})/K$ is the intensity on the absolute/relative scale.

When the largest integrated intensity is assigned a value of 100 and all other reflection intensities are scaled relative to it, the integrated intensities are said to be placed on the relative scale (I^{rel}). A scale factor γ (2) is defined to convert any relative intensity to the absolute/relative scale. That is,

$$I^{abs/rel}(\underline{h}) = \frac{I^{abs}(\underline{h})}{K} = \gamma\, I^{rel}(\underline{h}) \qquad (5)$$

where

$$\gamma = \frac{M(\underline{h})\, Lp(\underline{h})|F_T(\underline{h})|^2}{2\mu V^2\, I^{rel}(\underline{h})} \qquad (6)$$

Equation 6 permits calculation of γ from any reflection $\underline{h}$. Other scale factors, which have been reported for simulated powder patterns, are related to γ in the paper HES (2).

In HES we derived the following expression for the *reference inten-sity ratio:*

$$\frac{I(\underline{h}_0)}{Ic(113)} = \frac{\mu\gamma/\rho}{\mu_c\gamma_c/\rho_c} \qquad (7)$$

where the physical constants μ, γ and ρ are the values for the pure
phases. The subscript c refers to the reference material corundum, and
the indices 113 refer to the hexagonal indexing. In some cases, another
reflection $\underline{h}$ may overlap with reflection $\underline{h}_0$. In such a case the table
of calculated I^{rel} values can be used to calculate $I(\underline{h}, \underline{h}_0)/Ic$ from
$I(\underline{h}_0)/Ic$.

$$\frac{I(\underline{h}, \underline{h}_0)}{Ic(113)} = \frac{\mu\gamma/\rho}{\mu_c \gamma_c/\rho_c} \cdot \frac{I^{rel}(\underline{h}) + I^{rel}(\underline{h}_0)}{I^{rel}(\underline{h}_0)} \qquad (8)$$

This calculated intensity ratio should agree with experimental measure-
ments which integrate over $\underline{h}$ and $\underline{h}_0$. For tabulation purposes if $\underline{h}_0$ and
$\underline{h}$ have identical d-spacings, the adjustment (Equation 8) should clearly
be made. But if $d(\underline{h}_0) \neq d(\underline{h})$, then it is probably preferable to tabulate
$I(\underline{h}_0)/Ic(113)$ because some instrumentation and/or some laboratories may
resolve the pair while others may not. The ratio of intensities
$I(\underline{h})/Ic(113)$ for any line of the sample can easily be obtained from the
reference intensity ratio, $I(\underline{h}_0)/Ic$, and the table of relative intensities.

$$I(\underline{h})/Ic(113) = \frac{I(\underline{h}_0)}{Ic(113)} \cdot \frac{I^{rel}(\underline{h})}{100} \qquad (9)$$

In practice, overlap involving the strongest line of the sample or
the standard often occurs and the equation above must be used. Thus, it
is important to have available both $I(\underline{h}_0)/Ic$ and the d-I^{rel} table of
integrated intensities. Such tables are available in the NBS Monograph
25 (10) and were published on PDF cards of calculated patterns up through
set 24. Currently the JCPDS publishes only the d-I table of simulated
peak height intensities. While these values are excellent for identification, the primary purpose of the PDF, their use can introduce unneces-
sary errors in quantitative analysis applications. Perhaps a limited
table of only the largest integrated I^{rel}'s should be added to the PDF
card of calculated patterns, or a book with integrated intensities
should be published.

The discussion up to this point has assumed that the same radiation
will be used experimentally as was used for the theoretical calculations
(usually CuKα). If a different radiation is desired, it is not difficult
to convert the given *reference intensity ratio* to one appropriate to the
wavelength used. Equation 6 shows that γ is dependent on the linear
absorption coefficient and the Lorentz-polarization factor which is a
function of the scattering angle 2θ. (We discussed above the changes in
the polarization factor for different experimental conditions). The
value of $\gamma|_{\lambda_2}$ can be calculated from $\gamma|_{\lambda_1}$, and from the diffraction
angle for reflection $\underline{h}$ by

$$\gamma|_{\lambda_2} = \frac{\mu|_{\lambda_1}}{\mu|_{\lambda_2}} \cdot \frac{Lp(\underline{h})|_{\lambda_2}}{Lp(\underline{h})|_{\lambda_1}} \cdot \gamma|_{\lambda_1} \qquad (10)$$

where $\big|_{\lambda_1}$ signifies evaluation at wavelength λ_1. Substituting this expression into Equation 7 yields

$$\frac{I(\underline{h}_0)}{Ic(113)}\bigg|_{\lambda_2} = \frac{Lp(\underline{h}_0)|_{\lambda_2}}{Lp(\underline{h}_0)|_{\lambda_1}} \cdot \left[\frac{Lp_c(113)|_{\lambda_1}}{Lp_c(113)|_{\lambda_2}}\right] \cdot \frac{I(\underline{h}_0)}{I_c(113)}\bigg|_{\lambda_1} \quad (11)$$

The factor in brackets is tabulated for a variety of wavelengths in Table 2 with λ_1 taken as CuKα.

If the sample exhibits anomalous scattering then Equation 11 is only approximate because $|F(\underline{h})|_{\lambda_2} \neq |F(\underline{h})|_{\lambda_1}$. As an example, the calculated I(104)/Ic(113) for Cr_2O_3 considering anamolous scattering is 2.156 for CuKα radiation. From I/Ic for CuKα radiation equation 11 would predict that for CrKα radiation I(104)/Ic(113) would be 2.240. However, the actual value is 1.849 due to the fact that $|F(104)|$ for CuKα is 154.5 and $|F(104)|$ for CrKα is 140.3. Thus, when anomalous scattering is a significant contribution to $|F(\underline{h})|$, $I/Ic|_{\lambda_2}$ should not be derived from $I/Ic|_{\lambda_1}$. Instead, the *reference intensity ratio* should be recalculated directly from Equation 7.

Another source of uncertainty in the calculated I/Ic values is the accuracy in the atomic parameters themselves. The structure factors are less sensitive to the thermal parameters than to the atomic coordinates. Table 3 lists the relative intensities of five selected lines from the Cr_2O_3 simulated patterns for various deviations from the atomic parameters of Newnham and deHaan (11). Changes in isotropic thermal parameters from B(Cr) = 0.14 and B(0) = $0.22A^2$ to B = 1.0 (these values are choosen only to provide an example) resulted in a 6% change in I(104)/Ic(113). The changes in relative intensities are more important at higher 2θ's than at low 2θ's. Changing the z atomic coordinate of Cr by approximately 4σ (assuming σ = 0.0003) again gives errors in I(104)/Ic(113) less than 6% and somewhat larger changes in the I^{rel} values. We have found in practice that even with limited errors in atomic parameters, the simulated pattern is adequate for identification of an unknown. However, only patterns calculated from well determined atomic positions and thermal parameters should be used in quantitative analysis so that errors in calculated I/Ic values may be kept to a minimum.

Table 2

Ratio of Lp factors for the 113 reflection of α-Al_2O_3

	$\mu_c (cm^{-1})$	γ_c	$\dfrac{Lp(113)|CuK\alpha}{Lp(113)|_\lambda}$
CuKα	124.14	0.5129×10^{-3}	1.0
CrKα	389.43	0.6233×10^{-4}	2.623
FeKα	240.18	0.1525×10^{-3}	1.739
CoKα	191.24	0.2325×10^{-3}	1.432
NiKα	153.47	0.3484×10^{-3}	1.191
MoKα	13.35	0.2621×10^{-6}	0.1820
AgKα	6.976	0.8188×10^{-1}	0.1115

Table 3

Variation in I^{rel} and $I(104)/Ic(113)$ due to shifts in atomic parameters of Cr_2O_3

$z(Cr^{3+})$	0.3475[+]	0.3475	0.3475	<u>0.3481</u>*	<u>0.3487</u>	0.3475
$x(O^{2-})$	0.306	0.306	0.306	0.306	0.306	<u>0.310</u>
$B(Cr^{3+})$	0.14	<u>0.5</u>	<u>1.0</u>	0.14	0.14	0.14
$B(O^{2-})$	0.22	<u>0.5</u>	<u>1.0</u>	0.22	0.22	0.22
$I(012)$	69	70	71	65	61	65
$I(104)$	100	100	100	100	100	100
$I(110)$	92	91	91	89	87	89
$I(116)$	97	94	89	92	88	95
$I(300)$	44	41	38	42	42	43
$I(104)/Ic(113)$	2.156	2.101	2.028	2.185	2.271	2.204

+ Positional parameters from Newnham and deHaan (11).
* A change in atomic parameter is indicated by the underline.

Simulated powder patterns and calculated ratios are very useful for evaluating experimental results. Such errors as extinction, preferred orientation and microabsorption can all be readily detected by comparing the observed and calculated I^{rel}'s and I/Ic. At NBS, simulated patterns are being used in just this way to evaluate potential powder diffraction standard reference materials.

REFERENCE MATERIALS

The criteria for a suitable x-ray powder diffraction intensity reference material are many, and occasionally they are contradictory. The material must be stable, pure, non toxic, homogeneous, and readily available in a fine particle size ($\approx 1\mu m$). The diffraction lines should not exhibit marked profile broadening, and the pattern should have only a few lines with approximately equal maxima scattered uniformly over the 2θ range 20 to 60° (CuKα). Effects due to orientation and extinction must be absent. For quantitative analysis applications by the internal standard method the reference material should meet two further criteria. First, for minimal microabsorption effects the linear absorption coefficient should be similar to that of the phase of interest. Second, the lines of the reference material should not overlap with many lines from the mixture and vise versa.

Although α-Al_2O_3 is a satisfactory reference material, it will not necessarily be appropriate for analysis of a particular mixture because of microabsorption or overlap. At NBS, a program has begun to identify suitable reference materials to be issued as Standard Reference Materials (SRM). In addition to α-Al_2O_3 four other materials are currently under study. These materials are listed in Table 4 with their linear absorption coefficient and calculated I/Ic value.

Table 4

Possible Reference Materials

material	$\mu(cm^{-1})$ (CuKα)	I/Ic (calculated)
α-Al$_2$O$_3$	124.9	1.00
ZnO	288.0	5.43
TiO$_2$-rutile	549.5	3.44
Cr$_2$O$_3$	952.1	2.16
CeO$_2$	2082.9	14.1

Each selected material will be certified for the *reference intensity ratio* and I^{rel} values. The materials selected as intensity SRM's can be used in a variety of ways. The most important use will be as standards in quantitative analysis using tabulated *reference intensity ratios*. An intensity SRM could also be used to evaluate diffraction apparatus alignment, experimental methods and algorithms for analyzing the data. In some laboratories the materials might be used to provide accurate constants relating the response functions of various apparatus.

Diffractometers using theta compensating slits (12) record a pattern with relative intensities quite different than those recorded on diffractometers with fixed slits. Since the PDF contains patterns recorded using fixed slits, conversion factors must be applied before searching the file. In quantitative analysis using tabulated I/Ic values (fixed slits) the same conversion factors are also needed. While the conversion factors are known theoretically, the reference materials could be used for experimental confirmation.

SUMMARY

In previous papers (2,3) errors in quantitative analysis using the calculated and experimental *reference intensity ratios* have been shown to be as small as 1 wt.%. Yet, use of the *reference intensity ratio* in routine analysis is limited by the scarcity of accurate tabulated values for a wide variety of compounds. The ratios in existing tables (14) are primarily based on peak height ratios, and thus cannot be expected to always yield satisfactory accuracy in analysis. Hopefully, these peak height ratios can be replaced by integrated ratios using the methods discussed in this paper. New standards for tabulating $I(h_0)/Ic(113)$ values should be adopted. The standards proposed are 1) measurements of integrated intensities; 2) use of an experimental procedure, which verifies the absence of systematic errors; and 3) tabulation of the $I(h_0)/Ic(113)$ value for a standard experimental arrangement - diffractometer, no monochromating crystal and CuKα radiation. Such carefully determined *reference intensity ratios* will prove invaluable in quantitative x-ray analysis applications.

REFERENCES

1.   J. D. Hanawalt, H. W. Rinn and L. K. Frevel, "Chemical Analysis by
      X-ray Diffraction," Ind. Eng. Chem., Anal. Ed., 10, 457-467 (1938).

2.   C. R. Hubbard, E. H. Evans, and D. K. Smith, "The Reference Intensity
      Ratio, I/Ic, for Computer Simulated Powder Patterns," J. Appl.
      Cryst. 9, 169-174 (1976).

3.   L. D. Jennings "Current Status of the I.U.Cr. Powder Intensity
      Project," Acta Cryst. A25, 217-222 (1969).

4.   L. V. Azaroff, "Polarization Correction for Crystal-Monochromatized
      X-radiation," Acta Cryst. 8, 701-704 (1955).

5.   C. M. Clark, D. K. Smith and G. G. Johnson, Jr., "A FORTRAN IV
      Program for Calculating X-ray Powder Diffraction Patterns - Version
      5," Department of Geosciences, Pennsylvania State Univ., University
      Park, PA. 16802.

6.   R. W. James, The Optical Principles of the Diffraction of X-rays
      p. 61-62 G. Bell and Sons (1950).

7.   H. P. Klug and L. E. Alexander, X-ray Diffraction Procedures, second
      edition, p. 365-368. John Wiley and Sons, (1974).

8.   ibid p. 368-376

9.   ibid. p. 537

10.   "Standard X-ray Diffraction Patterns," Monograph 25, Sections 7-13,
      Crystallography Section, National Bureau of Standards, Washington,
      D. C. 20234.

11.   R. E. Newnham and Y. M. deHaan "Refinement of the α-Al_2O_3, Ti_2O_3, V_2O_3
      and Cr_2O_3 structures," Zeit. Krist. 117, 235-237 (1962).

12.   R. Jenkins and F. R. Paolini, "An Automatic Divergence Slit
      for the Powder Diffractometer," Norelco Reporter 21, No. 1, p. 9-14
      (1974).

13.   F. H. Chung, "Quantitative Interpretation of X-ray Diffraction Patterns
      of Mixtures," J. Appl. Cryst. 7, 519-531, (1974).

14.   JCPDS, "Search Manual - Inorganic Compounds/Alphabetical Listing
      SMA-25, P. 895-905 (1975). Joint Committee on Powder Diffraction
      Standards, 1601 Park Lane, Swarthmore, PA 19081.

X-RAY DIFFRACTION EXAMINATION OF THE PHASES IN EXPANSIVE CEMENTS

Katharine Mather

U.S. Army Engineer Waterways Experiment Station

Vicksburg, Mississippi 39180

ABSTRACT

Beginning in December 1968, Type K expansive cements of both the shrinkage-compensating and self-stressing types, particularly ChemStress (a self-stressing cement) have been tested, and examined by X-ray diffraction in the Concrete Laboratory of the U. S. Army Engineer Waterways Experiment Station. They are interesting subjects for the study of mixtures by X-ray diffraction because they may contain as many as 14 identifiable phases, including most of the phases in portland cement. Their examination by X-ray diffraction is greatly assisted by chemical treatments, first with maleic acid in methanol, followed by treatment with ammonium chloride solution to remove the calcium sulfates and leave a residue that consists of tetracalcium trialuminate sulfate, quartz, corundum, and a few very minor substances which have proved to be unidentifiable. The constituents that have been identified in Type K cements include alite and belite (the substituted forms of tricalcium silicate and dicalcium silicate found in portland cement), calcium aluminoferrites, MgO (periclase), tricalcium aluminate, gypsum, tetracalcium trialuminate sulfate ($C_4A_3\bar{S}$)*, free lime (CaO), orthorhombic anhydrite, plaster of paris, calcite, calcium hydroxide, 5-calcium disilicate monosulfate, quartz, and corundum. The cements examined in the greatest detail have contained less than 0.60 percent total alkali expressed as Na_2O and we have not recognized or specifically looked for the several alkali sulfates which could be present in the portland-cement portion of the system.

Comparisons of the phase compositions of some of the self-stressing cements with relevant aspects of their compositions will be discussed.

EXPANSIVE CEMENTS

Expansive cements have been studied for many years especially in France and the Soviet Union and later in the United States (1). There

---

*$Ca_4(Al_6O_{12})SO_4$.

have been three basic types, one (Type S) consisting of portland cement
with a large amount of tricalcium aluminate and calcium sulfate; a second
(Type M) that is a mixture of portland cement with high alumina cement and
calcium sulfate; and a third (Type K) consisting of portland cement, tetracalcium trialuminate sulfate $(Ca_4(Al_6O_{12})SO_4 = C_4A_3\bar{S})$, with CaO and
$CaSO_4$. This paper is concerned with Type K, the only type now being made
in the USA.

CHEMICAL, PHYSICAL, AND X-RAY DIFFRACTION ANALYSES OF CHEMSTRESS CEMENTS

The cement mill producing the highly expansive Type K cement used
in the study reported herein has made a number of different production
runs of this cement beginning in 1968 (2). The earlier productions, hereinafter referred to as ChemStress I, did not have as great an expansive
potential as did the second generation productions, hereinafter referred
to as ChemStress II. This upgrading of the expansive potential was
intentional. Field designations (NTS) were given (3) to each production
run through 1970 with the distinction that ChemStress I and ChemStress II
referred to cements that were moderately and highly expansive, respectively. Carrying these designations forward through 1973 results in the
following:

Production Run	WES Designation	Field Designation
1	RC-610	ChemStress I(68)
2	RC-645	ChemStress I(70)
3	RC-644	ChemStress II(70)
4	RC-653	ChemStress II(71)
5	RC-657	ChemStress II(71)A
6	RC-659	ChemStress II(72)
7	RC-660	ChemStress II(72)A
8	RC-664	ChemStress II(72)B
9	RC-671	ChemStress II(72)C
10	RC-678	ChemStress II(73)

The numbers in the parentheses indicate the year in which the cement
clinker was burned and ground. The alphabetical suffix indicates separate productions within a given year.

Summaries of the chemical, physical, and X-ray diffraction information for these cements are contained in Tables 1, 2, and 4, respectively.
For identification purposes in these tables, additional shipments of the
same cement for testing purposes were given suffix numbers, for example,
RC-644 was the original shipment for which no testing was done. Additional shipments of the same cement were called RC-644(2), denoting
second shipment, and RC-644(3), denoting third shipment. Testing was
done on these additional shipments.

Chemical and Physical Analysis of Cements

As of the date of evaluation of the last ChemStress cement received
(RC-678), no specification existed for expansive cements. In lieu of a

specific specification for this type of cement, the Federal Specification
for Portland Cement was used as the basis for the chemical and physical
analyses of the cements; results of those analyses are shown in Tables
1 and 2, respectively. Cube strength information on unrestrained cubes
of hardened ChemStress cement pastes is meaningless as these highly expansive cements usually expand themselves apart in a few days unless
adequately restrained; hence no strength data are presented.

Physical Properties of Mortar

In order to compare all the cements on a common basis, a control
mixture was adopted that would remain constant with regard to propor-
tioning but would have each of the various cements substituted in it for
evaluation. This mixture was a mortar mixture having a sand-cement ratio
of 2.75:1, a water-cement ratio of 0.50, and a ChemStress content of
25 percent (by volume) of the cement in the mixture. The remainder of
the cement is Type II laboratory stock cement.

The batch data for this mixture made with each cement and the values
for unit weight, air content, flow, and time of set (4) are contained in
Table 3.

Only restrained expansion determinations were made on the hardened
mortar. These results are also given in Table 3.

X-Ray Diffraction Analyses

Type K expansive cements have been examined by X-ray diffraction
from time to time beginning in December 1968. During this period,
changes have taken place in technique and equipment so that charts of
RC-610, -610(2), and -645 are not strictly comparable with RC-644 through
RC-678. Not all of the ChemStress cements were analyzed by XRD.

Test Procedures. Four samples (RC-610, -610(2), -645, and -644)
were examined using a GE XRD-5 diffractometer with a large-target copper
tube with the radiation filtered by nickel foil and a constant-potential
X-ray generator; pulse height selection was used. All of these were
examined in their as-received fineness, as powders tightly packed in a
3-in. aluminum holder, with the sample holder and its contents contained
in a sealed chamber flushed with dry nitrogen and closed with a slight
static pressure of nitrogen to prevent carbonation and hydration of the
sample during the scan, which takes about 5 hours at the scanning speed
of 0.2 degrees two-theta per minute used. All four samples were examined
on a logarithmic intensity scale in which full-scale deflection was 4000
counts per second.

The remaining samples, examined in the period 24 November 1971
through May 1973, were evaluated on an XRD-700 diffractometer at the same
scanning speed as before, with a similar copper tube, X-ray generation,
and filtration of the diffracted beam with nickel. Collimation was
standardized so that the region from 6 to 20 degrees two-theta was scanned
using a 1-degree beam slit, a 3-degree beam slit as a Soller slit, and a
0.2-degree detection slit. X-ray intensity was standardized using an
external standard. As a consequence, in the range 6 to 20 degrees

TABLE 1. Summary of Chemical Analysis Data, Percent

Constituents	ChemStress I					ChemStress II									
	RC-610	RC-610(2)	RC-645	RC-644(2)	RC-644(3)	RC-653	RC-657	RC-657(2)	RC-659	RC-659(2)	RC-660	RC-664	RC-671	RC-671(2)	RC-678
SiO_2	18.5	18.5	17.6	15.1	14.6	15.1	14.5	14.5	13.5	14.0	14.3	14.1	15.1	15.3	15.3
Al_2O_3	6.0	6.2	6.3	8.9	8.4	7.4	10.2	9.0	9.3	8.6	10.1	10.3	8.0	7.8	8.3
Fe_2O_3	1.8	1.8	1.7	1.6	1.6	1.7	1.9	1.9	1.4	1.5	1.7	1.5	1.4	1.4	1.5
CaO	61.5	61.9	61.5	57.9	57.6	60.3	55.2	56.4	56.0	56.2	57.6	57.8	58.6	59.3	57.5
MgO	3.0	2.8	2.4	2.8	2.7	2.8	2.7	2.7	2.1	2.2	2.3	2.3	1.7	1.9	1.6
SO_3	7.0	6.7	7.7	11.8	12.4	11.5	13.0	12.6	14.5	15.1	12.8	12.7	12.9	13.0	14.2
Ignition Loss	1.7	1.3	1.3	1.6	1.9	0.8	2.3	2.4	1.9	1.9	1.1	1.1	1.6	1.2	1.6
Insoluble Residue	0.69	0.55	0.34	1.19	1.09	0.31	1.26	0.75	1.24	2.16	0.73	0.43	0.47	0.49	0.67
Na_2O	0.23	0.25	0.21	0.16	0.15	0.20	0.14	0.13	0.18	0.19	0.19	0.17	0.17	0.18	0.14
K_2O	0.42	0.43	0.42	0.51	0.49	0.83	0.38	0.38	0.41	0.42	0.47	0.52	0.53	0.56	0.45
Total Alkalies as Na_2O	0.51	0.53	0.49	0.50	0.47	0.75	0.39	0.39	0.45	0.47	0.50	0.51	0.52	0.55	0.44

TABLE 2. Summary of Physical Properties

Property	ChemStress I								ChemStress II						
	RC-610	RC-610(2)	RC-645	RC-644(2)	RC-644(3)	RC-653	RC-657	RC-657(2)	RC-659	RC-659(2)	RC-660	RC-664	RC-671	RC-671(2)	RC-678
Specific Gravity	--	3.06	3.05	3.04	3.04	3.02	2.96	2.96	3.02	3.02	3.12	3.10	3.06	3.03	3.08
Surface Area, Air Permeability Fineness (Blaine), cm2/g	--	--	3745	4380	4240	5120	4750	4575	3750	3955	3730	3660	3510	3800	4175
Air Content of Mortar, %	--	--	7.6	8.1	9.9	6.2	6.0	6.2	9.1	10.9	--	--	8.1	7.7	8.9
Setting Time, Gillmore, Hours:Minutes															
Initial	--	--	0:35	0:25	0:25	0:35	0:25	0:30	0:35	0:30	0:30	0:30	0:45	0:35	0:20
Final	--	--	1:35	1:05	1:00	1:15	1:05	1:20	1:30	1:20	1:20	1:10	1:30	1:25	0:45
Heat of Hydration, cal/g															
1 Day	--	--	54	57	51	--	39	41	50	--	--	--	--	--	--
2 Days	--	--	63	66	--	--	47	48	59	--	--	--	--	--	--
3 Days	--	--	67	72	70	--	55	54	62	--	--	--	--	--	--
7 Days	--	--	83	82	79	--	--	--	--	--	--	--	--	--	--
28 Days	--	--	91	98	94	--	--	--	--	--	--	--	--	--	--

TABLE 3. Summary of Batch Data and Restrained Expansion Determinations

	RC-644 (CSII-70)	RC-645 (CSI-70)	RC-653 (CSII-71)	RC-657 (CSII-71A)	RC-657(2) (CSII-71A)	RC-659 (CSII-72)	RC-660 (CSII-72)	RC-664 (CSII-72B)	RC-671 (CSII-72C)	RC-671(2) (CSII-72C)
Age, Days					Restrained Expansion, Percent					
1	0.020	0.008	0.027	0.015	0.008	0.023	0.024	0.030	0.032	0.037
2	--	0.012	0.041	--	0.013	0.034	0.032	--	--	0.050
6	0.033	0.017	--	0.021	0.016	0.042	0.038	0.062	0.059	0.057
8	0.032	0.018	0.050	0.021	0.017	0.042	0.039	0.062	0.059	0.057
9	--	0.019	0.051	--	0.017	0.042	0.040	0.063	--	0.058
12	--	0.019	0.050	0.022	0.017	0.044	0.040	0.065	0.063	--
13	0.032	0.020	0.050	0.023	0.016	0.042	0.040	0.064	0.064	0.057
14	0.032	0.019	0.050	0.023	0.016	0.042	0.040	--	0.063	0.061
21	0.033	0.022	0.052	0.024	0.017	0.044	0.042	0.066	0.059	0.059
28	0.039	0.023	0.053	0.025	0.017	0.042	0.041	0.065	0.060	0.058
Time of Initial Bar Reading Hours:Minutes	6:50	6:00	6:20	6:50	5:55	6:55	6:23	5:21	6:00	6:45
					Batch Data					
Unit Weight, pcf	140.0	--	--	--	140.6	140.4	140.4	140.0	139.2	138.4
Air Content, %	3.1	--	--	--	3.3	3.7	3.5	3.2	3.0	4.4
Flow, % (Flow Table) 10 Drops	22.0	--	--	--	34.4	26.7	46.9	20.6	53.7	58.7
Flow, % (Flow Table) 25 Drops	45.4	--	--	--	73.0	66.0	88.9	43.0	94.3	98.3
Time of Set (Proctor) Hours:Minutes Initial	3:30	--	--	--	--	--	3:10	2:44	3:17	3:33
Time of Set (Proctor) Hours:Minutes Final	5:20	4:30	4:50	4:20	4:25	5:25	4:43	3:51	5:00	5:15

two-theta, with kilovoltage constant potential (KVCP) at 27, the milliamperage was adjusted in the range 39 to 40; in the range 20 to 65 two-
theta, with milliamperage at 20, KVCP was adjusted in the range 48 to 50.
Pulse height selection was used. The intensity scale of the charts is
nominally log 4000 but there is more magnification of low intensities or
a greater ability to detect trace amounts with this instrument than with
the XRD-5. However, the high sensitivity is accompanied by less stability
over long time periods.

In standard procedure, samples of the whole cement were run in their
as-received particle-size distribution because the increased exposure to
air in grinding and sieving has been regarded as allowing increased opportunity for dehydration of gypsum by CaO and more opportunity for aeration
and carbonation of the cement. In summary, grinding and sieving this
cement is a procedure with inherent risks of making the sample less representative, with dubious benefits, and requiring increased cost and time.
In cases in which the intensities of unground samples have unusual characteristics that suggest unusually intense diffraction by an oversize crystal or preferred orientation, a new portion of sample is ground but not
sieved; and the part of the scan showing unusual intensity is repeated.
This situation has not been encountered often and is usually confined to
regions 1 to 3 degrees two-theta in width.

Chemical Treatments of Samples. In February 1972 it became standard
procedure in X-ray diffraction examination of Type K expansive cements to
examine a second sample of the cement after treatment with a solution of
maleic acid in methanol to remove the calcium silicates and calcium oxide.
The ground sample of cement is stirred on a magnetic stirrer during this
treatment; if gelation begins, the addition of 10 to 20 ml of distilled
water usually stops it. After 10 minutes of stirring, following the addition of the maleic acid, the solution and residue are filtered with aspi-
ration over a No. 50 Whatman filter paper on a small scrap of nylon
stocking in a Buchner funnel. The residue is washed three times with
methanol and dried under a heat lamp until it is free flowing. It is then
examined by X-ray diffraction like the sample of the whole cement. The
maleic acid procedure used was developed by the Southwestern Portland
Cement Co.

Some use has been made of a procedure in which 2 grams of cement are
stirred for about 3 hr on a magnetic stirrer in 800 ml of distilled water
that is kept cold by adding ice cubes made from distilled water. The
suspension and solution are filtered, washed with methanol, and dried at
50°C. This rapidly removes water-soluble constituents (CaO, calcium
sulfate hemihydrate, gypsum) and makes the silicates more visible.

If it is desired to simplify the material insoluble in maleic acid,
the anhydrite and other calcium sulfates can be removed by stirring the
material in an excess of 10 percent NH_4Cl solution at room temperature
for about 2 hr with a magnetic stirrer, followed by filtering, washing,
and drying.

Results. The results of the examinations are shown in Table 4.
Because two diffractometers with different detection apparatus were used,
and because of the relatively long time between the first and last examinations, it did not seem justifiable to rank the abundance of constituents from sample to sample. Accordingly, abundance within samples is

TABLE 4. Composition by X-Ray Diffraction
Abundant, Common, minor, trace, and not detected are abbreviated A, C, M, tr, and nd, respectively.

Constituents	ChemStress I			ChemStress II									
	RC-610	RC-610(2)	RC-645	RC-644	RC-653	RC-657	RC-659	RC-659(2)	RC-660	RC-664	RC-671	RC-671(2)	RC-678
Alite	C	C+	C	M-	M-	M-	M-	M-	M-	M	M-	M-	M--
Belite	C+	C-	C+	A-	A-	A-	A-	A--	A-	A	A-	A	A
Aluminoferrite	M-	M-	M-	M-	M-	M-	M-	M-	M-	M--	M--	M--	M--
MgO	tr	tr	tr	tr	tr	tr	tr	tr	tr	tr	tr	tr	tr
C_3A	M-	M-	M-	nd	nd	nd	nd	nd	nd	nd	nd	nd	nd
Gypsum	M-	nd	nd	M-	nd	nd	M-	nd	tr+	nd	M-	nd	tr?
$C_4A_3\bar{S}$	C	C	C	C+	C+	C+	C	C+	C	C+	C	C	C
CaO	C	M-	C	C+	C+	M+[a]	C	C-	C	C	C	C	C
$CaSO_4$	C+	C	C+	C+	C+	C	C+	C	C	C-	C-	C-	C-
$CaSO_4\cdot1/2H_2O$	tr?	tr	tr?	tr	M-	M	M-	M	M-	M	tr+	tr+	M-
$CaCO_3$	nd	M-	tr	tr-	?	?	?	?	?	?	?	?	?
$Ca(OH)_2$	nd	tr?	tr?	tr?	tr?	tr+	tr	tr+	tr	M-	tr	tr	tr
$2C_2S\cdot C\bar{S}$	nd	nd	nd	nd	M-	M-	M-	M-	tr+	tr	tr	tr	?
Quartz	nd	nd	nd	nd	tr	tr	tr	tr?	tr	nd	tr	tr	tr
Sample Received	11-29-68	1-14-69	12-3-70	11-18-70	6-1-71	11-17-71	2-14-72	3-1-72	4-17-72	6-1-72	12-13-72	1-18-73	5-7-73
Sample Examined	12-18-68	12-3-70	12-17-70	12-4-70	1-4-72	11-24-71	2-17-72	3-2-72	4-18-72	6-16-72	1-8-73	1-20-73	5-9-73

[a] CaO depleted by dehydration of $CaSO_4\cdot2H_2O$ to $CaSO_4\cdot1/2H_2O$ and formation of $Ca(OH)_2$.

shown as A = abundant, C = common, M = minor, tr = trace, and each of
these categories is modified as necessary with plus (+) or minus (−)
signs.

The constituents of normal portland cements found in ChemStress I
were alite (C to C+), belite (C− to C+), aluminoferrite (M−), tricalcium aluminate (M−), and MgO (tr). Constituents of the expansive combination found in all of the ChemStress I samples were $C_4A_3\bar{S}$, CaO, anhydrite (5,6). Gypsum was present in one.

The constituents of normal portland cements found in ChemStress II
samples were, in order of abundance from most to least, belite, alite
and aluminoferrite, and MgO. Constituents of the expansive combination
found in all of the ChemStress II samples were $C_4A_3\bar{S}$, CaO, and anhydrite.
Gypsum was present in four, $CaSO_4 \cdot 1/2H_2O$ in all nine, and calcium
hydroxide in eight. Calcium silicosulfate was present in seven and
possibly in an eighth, and traces of quartz were present in seven and
probably in an eighth where it was not verified by chemical treatment.

Maleic acid as used in this program removed the following constituents:

> Calcium silicates: alite, belite, and calcium silicosulfate
> Calcium oxide

It did not remove:

> $C_4A_3\bar{S}$
> $CaSO_4$, $CaSO_4 \cdot 1/2H_2O$, $CaSO_4 \cdot 2H_2O$
> Aluminoferrites
> MgO
> Quartz

The presence of quartz was verified in the samples treated with maleic
acid and then with 10 percent NH_4Cl solution, which removed the calcium
sulfates and raised the concentration of quartz in the residue to a
level that revealed not only the strongest line at 3.34 A but the
second and third at 4.26 and 1.82 A, respectively.

Calcium silicosulfate forms in kilns in which the atmosphere is
high in SO_3 (8) and apparently persists in some cases in the final product in Type K self-stressing cements. The Powder Diffraction File (9)
contains a diffraction pattern (18-307) from Gutt and Smith (10);
another diffraction pattern was published by Nakamura, Sudoh, and
Akaiwa (11) which shows somewhat different spacings and intensities.
The identification of calcium silicosulfate in Type K self-stressing
cements appears to depend on several weak lines. The intensity indications in the pattern by Nakamura, Sudoh, and Akaiwa agreed better, but
not perfectly, with the relative frequency with which they were detected here than did the intensity indications in Card 18-307. In one
instance (RC-671(2)), only the lines at 4.10 and 3.95 A were detected.
The identification of calcium silicosulfate is discussed because it
was initially misidentified (I am obliged to Professor P. K. Mehta of
the University of California at Berkeley, California, for correcting

the error) and because strong lines of the compound are uniformly subject to interference in Type K self-stressing cements. In the minor to
trace amounts in which it was present in the cements examined, it is
probably initially a relatively inert diluent that may ultimately contribute some strength in hydration.

All of the chemical treatments mentioned are useful in the examination of Type K cements. The ammonium chloride procedure was developed
in this laboratory. The maleic acid in methanol method has been described (12). It is recommended that anyone who wishes to examine cements like these should use these or equivalent procedures because Type
K self-stressing cements are complex mixtures and difficult subjects
for identification by X-ray diffraction.

What can be said of the relations between the composition of these
cements as examined by X-ray diffraction and their physical behavior in
mixtures in which sand-to-cement was 2.75:1, water-cement ratio 0.5,
and ChemStress 25 percent by volume of the cement in the mixture? Comparing the expansions at 1, 8, 14, 21, and 28 days, listing them in
order from lowest to highest, is as follows:

	Restrained Expansion, % at Ages Shown					Percent Increase from 1-Day Reading at Ages Shown		
RC-	1	8	14	21	28	28	21	14
657(2)	0.008	0.017	0.016	0.017	0.017	112.5	112.5	(100)
657	0.015	0.021	0.023	0.024	0.025	66.6	(60.0)	(53.3)
644	0.020	0.032	0.032	0.033	0.039	95.0	(65.0)	(60.0)
660	0.024	0.039	0.040	0.042	0.041	(70.8)	75.0	(66.7)
659	0.023	0.042	0.042	0.044	0.042	(82.6)	91.0	(82.6)
653	0.027	0.050	0.050	0.052	0.053	96.3	(92.6)	96.0
671(2)	0.037	0.057	0.061	0.059	0.058	(56.8)	(84.4)	64.9
671	0.032	0.059	0.063	0.059	0.060	(87.5)	(84.4)	96.9
664	0.030	0.062	0.064[a]	0.066	0.065	(117.0)	120.0	(113.0)
Total maxima at age						4	4	3

[a]13-day age.

The expansions fall into three or four groups, depending on whether it
is believed that RC-653 should be grouped with RC-671(2) and RC-671.
It appears that in this group, RC-657(2), RC-659, RC-660, RC-671(2),
and RC-671 have achieved a plateau in expansion at 8 days and do not
maintain additional expansion gained thereafter until 28 days. On
the other hand, the other four cements continue to expand to 21 or 28
days.

RC-657 had less CaO than any of the others examined and the most

$CaSO_4 \cdot 1/2H_2O$, suggesting that more CaO was hydrated by dehydrating gypsum than in any other cement except the companion field sample RC-657(2). It is believed that the shortage of CaO unbalanced the expansive combination in these two cements which gave the lowest original and ultimate expansions.

RC-664 had the greatest increase in expansion (120 percent) after 1 days age. It contained the most alite (minor) and was one of the three with the most belite; its 1-day strength in this mixture should have been the highest of those with finenesses below 4000 cm^2/g. In the constituents entering the expansive combination, slightly low anhydrite was compensated for by abundant hemihydrate; and although part of the lime was hydrated as shown by the absence of gypsum and presence of calcium hydroxide, this cement had the shortest final Proctor setting time so that the maximum amount of ettringite formed contributed to useful expansion. It is not known whether the short final Proctor setting time or early strength gain, or both, maximized expansion; but RC-671 with the second highest expansion had a long final Proctor setting time (Table 3). RC-671(2) contained high belite, and the strength-expansion relation may have been well balanced. The three cements with the highest 28-day expansions, RC-664, -671, and -671(2), had in common lower aluminoferrite than any of the others; RC-664 and -671(2) had higher belite than the rest; and RC-664 had the most alite of any, although it was still minor.

The lower aluminoferrite probably means that more of the aluminum was located more usefully than in aluminoferrite, which is probably neutral at 28 days. It is also possibly significant that RC-664, -671, -671(2), and -678 were similar in the amounts of $C_4A_3\bar{S}$, CaO, and $CaSO_4$ they contained but RC-664 contained the most. Possibly the proportions of the expansive constituent expressed (Table 4) as $C_4A_3\bar{S}$, C or C+; CaO, C; $CaSO_4$, C- are significant; all three occur only in the three cements with the largest expansions, and RC-678.

The remaining four cements with 28-day restrained expansions from 0.039 to 0.053 (RC-644, -660, -659, -653) are all similar in alite and belite contents, have more aluminoferrite than the three with the highest expansions, and do not have the relative proportions of $C_4A_3\bar{S}$, CaO, and $CaSO_4$ found in the three cements with highest restrained expansion.

ACKNOWLEDGEMENTS

This paper is based on Appendix 1 by Katharine Mather of Technical Report C-74-6, by George Hoff, "Use of Self-Stressing Expansive Cements in Large Sections of Grout, Mortar, and Concrete, Report 1, Pumpable Mortar Studies," U. S. Army Engineer Waterways Experiment Station, CE, Vicksburg, Mississippi 39180, August 1974.

REFERENCES

1. Report by ACI Committee 223; "Expansive Cement Concretes--Present State of Knowledge"; Journal of the American Concrete Institute, Proceedings, August 1970, Vol. 67, No. 8, Pages 583-610; American Concrete Institute, Detroit, Michigan.

2. G. C. Hoff; "Investigation of Expanding Grout and Concrete; Summary of Field Mixture Test Results, July 1969 Through June 1970"; Miscellaneous Paper C-71-5, Report 1, June 1971; U. S. Army Engineer Waterways Experiment Station, CE, Vicksburg, Mississippi.

3. G. C. Hoff; "Investigation of Expanding Grout and Concrete; Summary of Field Mixture Test Results, July 1970 Through June 1971"; Miscellaneous Paper C-71-5, Report 2, January 1973; U. S. Army Engineer Waterways Experiment Station, CE, Vicksburg, Mississippi.

4. U. S. Army Engineer Waterways Experiment Station, CE; "Handbook for Concrete and Cement"; August 1949 (with quarterly supplements); Vicksburg, Mississippi.

5. P. K. Mehta; "Chemistry and Microstructure of Expansive Cements"; Conference on Expansive Cement Concrete, University of California, Berkeley, California, June 1972.

6. B. Mather, P. K. Mehta, and Committee; Discussion of the report by ACI Committee 223, "Expansive Cement Concretes--Present State of Knowledge"; Journal of the American Concrete Institute, Proceedings, April 1971, Vol. 68, No. 4, Pages 293-296; American Concrete Institute, Detroit, Michigan.

7. A. D. Ross; "Shape, Size, and Shrinkage"; Concrete and Constructional Engineering, August 1944, Vol. 39, No. 8, Pages 193-199; London, England.

8. M. A. Smith and W. Gutt; "Studies of the Mechanism of the Combined Cement/Sulphuric Acid Process"; Cement Technology, November-December 1971, Pages 167-177.

9. Joint Committee on Powder Diffraction Standards; Powder Diffraction File, 1973; JCPDS, Swarthmore, Pennsylvania.

10. W. Gutt and M. A. Smith; "A New Calcium Silicosulphate"; Nature, Vol. 210, No. 5034, 23 April 1966, Pages 408-409.

11. T. Nakamura, G. Sudoh, and S. Akaiwa; "Mineralogical Composition of Expansive Cement Clinker Rich in SiO_2 and its Expansibility"; Supplement Paper IV-74, December 1969, Vol. IV, Pages 351-365; Proceedings, Fifth International Symposium on the Chemistry of Cement, Tokyo, 1968.

12. J. E. Mander, L. D. Adams, and E. E. Larkin; "A Method for the Determination of Some Minor Compounds in Portland Cement and Clinker by X-Ray Diffraction"; Cement and Concrete Research, 1974, Vol. 4, Pages 533-544.

THE SINGLE CRYSTAL VS. THE POWDER METHOD FOR IDENTIFICATION OF
CRYSTALLINE MATERIALS

Alan D. Mighell

National Bureau of Standards

Washington, D.C. 20234

ABSTRACT

Single crystal X-ray diffraction methods for the study of crystalline materials, although reliable, have been mainly confined to the
academic laboratory because of the rather lengthy and complex procedure
necessary to determine the unit cell and the space group. The situation
has now changed. Several recent developments give single-crystal methods
considerable potential for routine industrial use. They include growth
of the data base, advances in lattice theory, and automation of the
single-crystal X-ray diffractometer. To identify an unknown, one can
start with a single crystal, mount it on the diffractometer, determine
a refined primitive cell, reduce the cell, and check against a file of
known reduced cells. The entire procedure can be automated. As a result,
the single-crystal X-ray diffraction method can now complement the powder
method for the routine analysis of crystalline materials.

INTRODUCTION

The powder X-ray diffraction method has been extensively used in
industry for the routine analysis of crystalline materials. The single-
crystal method, in which diffraction data are collected from one crystal,
has been mainly confined to the academic laboratory by lengthy and
expensive procedures. But the method has been improved and is now in
a position to play a more active role in industry. This change is due
to three factors: the development of a large file of single crystal data,
the availability of mathematical methods to characterize lattices, and
the application of semi-automated methods to determine unit cells. This
paper will discuss these factors, show how the single-crystal method can
be used to identify materials and show that the single-crystal method in
conjunction with the powder method offers an attractive combination for
the characterization of crystalline materials.

THE SINGLE-CRYSTAL DATA BASE

The single-crystal data base is large and rapidly expanding.

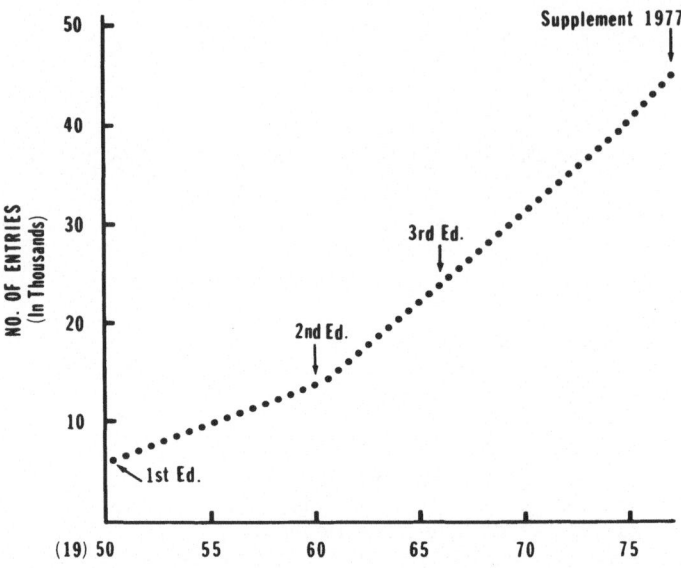

Fig. 1.   Growth of Single-Crystal File

Before a crystalline material can be included, its cell parameters must
be reported in the literature.   Cell parameters for materials in the
data base are determined in a number of ways.   Three common methods are
the precession, the powder, and the single-crystal diffractometer.   The
current rate of file expansion is more than 5000 entries a year and
because of equipment automation and of the seemingly unlimited number of
compounds, especially organic, the rate of expansion will probably
increase.

Figure 1 shows the growth of the single-crystal file.   In the first
edition of Crystal Data Determinative Tables (1), there were 6000
entries, in the second (2) 13,000, and in the third edition (3) 24,000.[1]
In the inorganic and organic supplements to the third edition of Crystal
Data, to be issued in 1977, data on about 24,000 more materials will be
reported which will bring the total to nearly 50,000 entries.   Another
30,000 entries still need to be extracted from the literature for a
fully up-to-date file.

---

[1]
 A tape has been prepared containing selected data for each entry in
  the 3rd edition.   For information concerning this tape contact Dr.
  S.A. Rossmassler, Office of Standard Reference Data, National Bureau
  of Standards, Washington, D.C.   20234.

There are now about 19,000 organic entries in the file. This number
covers data through 1974 and includes entries in the third edition (3) and
the compounds to appear in the forthcoming Organic Supplement.[2] New
organic entries are being added at about 2500 per year. Figure 2 shows
the distribution according to crystal system. Note that most fall into
the low-symmetry crystal systems. Consequently, these compounds would
have relatively complex powder patterns. In contrast, the inorganic
materials are skewed toward the high symmetry systems.

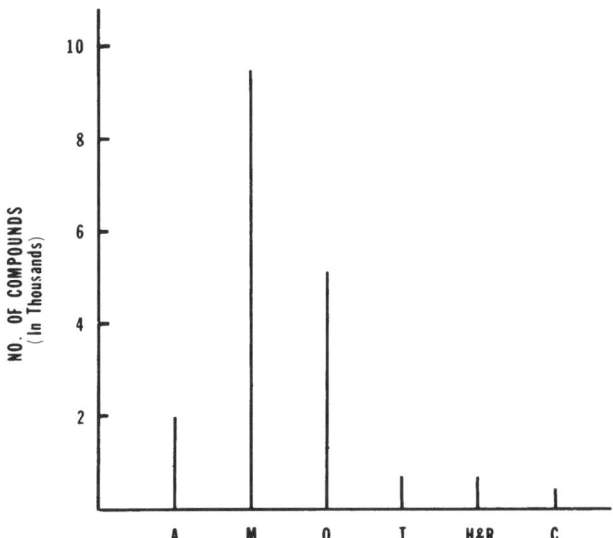

Fig. 2. Occurrence by Crystal System of Organic Compounds:
         A = Anorthic, M = Monoclinic, O = Orthorhombic, T = Tetragonal,
         H and R = Hexagonal and Rhombohedral, C = Cubic.

The size of the Powder Diffraction File[3] is somewhat smaller. This
file now contains nearly 30,000 entries. Many entries in the powder
file are not in the single-crystal file and vice versa. For example, the
powder file contains some entries for which the unit cell has not been
established. None of these would be in the single-crystal file. On
the other hand, many more organic entries are in the single-crystal
than in the powder file.

---

[2]The Single-Crystal file, for convenience, consists of two parts:
   inorganic and organic. The inorganic entries are being compiled by the
   Crystal Data Center at the National Bureau of Standards and the organic
   entries by the Crystallographic Data Centre at Cambridge, England.

[3]Powder Diffraction File. Issued by the Joint Committee on Powder
   Diffraction Standards, Swarthmore, Pennsylvania.

It is planned to correlate the single-crystal and the powder files
via the chemical registry number. [Drs. G.W. Milne and S. Heller, of
the National Institutes of Health (NIH) and Environmental Protection
Agency (EPA), respectively, are organizing a Chemical Information System
(CIS) composed of a number of data bases. The CIS System is accessible
by computer terminal and a principal use of the system will be the
identification of materials. The material in the respective data bases
will be uniquely represented by their chemical registry numbers.] In
the future, it is likely that both single-crystal and powder files will
be available to the user by on-line computer networks. With a file as
large as the single-crystal file, the information on each compound must
be compactly represented for efficient computer search and storage.
The reduced cell and the empirical formula represents one such compact
format.

THEORY

The lattice of a crystal can be uniquely represented by the reduced
cell. (The mathematical conditions for reduction are given in Table 1.)
Hence, this cell can be used to identify crystalline materials, to tell
whether two lattices are the same or not, and to determine symmetry. It
should be emphasized that the symmetry as determined from the reduced
cell is purely metric. The metric symmetry is equal to or higher than
the crystal symmetry. However, an analysis of the single-crystal file
indicates that in the vast majority of cases, the two are identical.
The conditions for reduction, procedures to obtain the reduced cell, and
procedures to determine the Bravais lattice from this cell have been
reported in the literature. A summary is given in the International
Tables for X-ray Crystallography (4). Errata or Addenda to the Tables
are reported in (5), (6), and (7). Efficient computer algorithms are
also available to go from any cell in the lattice to the reduced cell
(8) and from the reduced cell to the Bravais cell.

An undistorted representation of the reciprocal lattice of a crystalline material can be obtained with a precession camera. Precession
films can be used, therefore, to illustrate the procedure to obtain the
reduced cell. On these films, nodes of the reciprocal lattice are
recorded. A primitive reciprocal cell is selected. In choosing this
cell, nodes are not allowed on the faces or in the center of the cell.
This cell is then converted to a cell in direct space which is then
reduced. [The reduction should be carried out using the direct space
cell because a reduced cell in reciprocal spaces does not lead, in the
general case, directly to a reduced direct space cell.] To be reduced,
both the main and the special conditions for reduction must be satisfied
[Table 1]. The main conditions assure that the cell is based on the
three shortest non-coplanar vectors of the lattice. The special conditions take care of those cases in which there are two or more cells (not
symmetrically related) based on the three shortest vectors. The special
conditions are not uncommon and must be satisfied for the sake of unique-
ness. The most important experimental step in the reduction procedure
is the determination of a primitive cell of the lattice. Strategies for
checking whether a cell is primitive or for finding a primitive cell
from a supercell in reciprocal space are given in (9).

TABLE 1. Conditions for a Reduced Cell*

A. Positive Reduced form, Type I cell, all angles < 90°.

Main conditions: $a \cdot a \leq b \cdot b \leq c \cdot c$; $b \cdot c \leq \frac{1}{2} b \cdot b$; $a \cdot c \leq \frac{1}{2} a \cdot a$; $a \cdot b \leq \frac{1}{2} a \cdot a$ I

Special conditions:

(a)	if $a \cdot a = b \cdot b$	then	$b \cdot c \leq a \cdot c$	
(b)	if $b \cdot b = c \cdot c$	then	$a \cdot c \leq a \cdot b$	
(c)	if $b \cdot c = \frac{1}{2} b \cdot b$	then	$a \cdot b \leq 2a \cdot c$	II
(d)	if $a \cdot c = \frac{1}{2} a \cdot a$	then	$a \cdot b \leq 2b \cdot c$	
(e)	if $a \cdot b = \frac{1}{2} a \cdot a$	then	$a \cdot c \leq 2b \cdot c$	

B. Negative reduced form, Type II cell, all angles ≥ 90°.

Main conditions: (a) $a \cdot a \leq b \cdot b \leq c \cdot c$; $|b \cdot c| \leq \frac{1}{2} b \cdot b$; $|a \cdot c| \leq \frac{1}{2} a \cdot a$; $|a \cdot b| \leq \frac{1}{2} a \cdot a$ III

(b) $(|b \cdot c| + |a \cdot c| + |a \cdot b|) \leq \frac{1}{2}(a \cdot a + b \cdot b)$

Special conditions:

(a)	if $a \cdot a = b \cdot b$	then	$|b \cdot c| \leq |a \cdot c|$	
(b)	if $b \cdot b = c \cdot c$	then	$|a \cdot c| \leq |a \cdot b|$	
(c)	if $|b \cdot c| = \frac{1}{2} b \cdot b$	then	$a \cdot b = 0$	IV
(d)	if $|a \cdot c| = \frac{1}{2} a \cdot a$	then	$a \cdot b = 0$	
(e)	if $|a \cdot b| = \frac{1}{2} a \cdot a$	then	$a \cdot c = 0$	
(f)	if $(|b \cdot c| + |a \cdot c| + |a \cdot b|) = \frac{1}{2}(a \cdot a + b \cdot b)$ then $a \cdot a \leq 2|a \cdot c| + |a \cdot b|$			

*For further details, see the discussion in International Tables for X-ray Crystallography (4).

As the reduced cell is unique, one can immediately determine the symmetry of the lattice. In all there are 44 reduced cell types (4). Each reduced cell can therefore be related by a transformation matrix to the conventional Bravais cell. In summary the lattice symmetry can be determined from any three vectors which define a primitive cell. A discussion for the various ways to classify materials based on the reduced cell is given in (9).

The analyst often wants to determine the lattice symmetry of a material to better understand its physical properties. This can be done by determining a primitive cell of the lattice from a single-crystal or by deducing the cell from powder data alone. If the primitive cell is deduced from powder data the analyst must be cautious because the possibility for error occurs. It can be shown (10) that two different lattices (i.e. different reduced cells) with different volumes of the primitive cells can be consistent with exactly the same set of d-spacings. That is, one can go uniquely from the lattice to the powder pattern d-spacings but not always the reverse. The exact equivalence in certain cases of the d-spacings for two different lattices implies that in other cases the d-spacings would be almost equivalent. For example, it can be shown that subsets of d-spacings from two or more different lattices can be exactly equivalent. Therefore the prudent procedure would be to use a cell determined from a single crystal whenever possible to supplement powder indexing and powder identification procedures.

UNIT CELL PARAMETERS

The determination of the cell parameters and the space group have been major obstacles to the routine use of the single-crystal method in identifying crystalline materials. It has been much simpler to record a powder pattern. Now, however, data from a single crystal can be routinely recorded in a short period of time with a computer-controlled procedure on a single-crystal X-ray diffractometer. The procedure still requires limited operator interaction. However, a technician can be trained in a short time to carry out the procedure. There is no theoretical or practical reason that prevents the procedure from being completely automated. Three factors account for this:

1. The nodes in reciprocal space are well resolved and subject to exact mathematical treatment.

2. Derivative cell in reciprocal space can be systematically calculated (9, 11, 12, 13).

3. Automation capability is being improved and expanded.

If fully automated, the operator would need only to mount the crystal on the diffractometer, and to return in a short time to collect the refined cell along with the lattice symmetry. Even with present procedures, one can often determine a refined cell in ∿2 hours from a single crystal that is less than 0.1mm in diameter and that contains only light atoms.

An identification procedure, based on the primitive cell of the lattice, is immune to certain errors that plague other procedures. For example, identification based on the powder pattern uses both the peak position and the peak height. Normally, in the powder technique, many peaks are recorded and they are subject to a number of experimental problems which make routine reproducibility sometimes difficult. Also for identification, the storage of a large amount of data in the computer is required.

In contrast, to determine a unit cell suitable for identification from a single crystal, one is concerned only with the measurement of peak positions. Moreover, many resolved peaks can be measured. Thus a large monoclinic cell can readily be determined and refined. Normally one measures as many peaks as required to obtain a good least-squares refinement of the unit cell. These peak positions, then, are input data to a refinement procedure that gives a refined unit cell defined by a maximum of six independent parameters. This is a good basis for an analytical identification procedure. Compressed into a few refined parameters are many independent, precise experimental measurements. So now the identification scheme can be based on cells that can be refined, that contain few parameters, that can be reproducibly determined by different experimentors, and that can be transformed to the unique reduced cell.

This is ideal for computer storage, for computer searching, and for selectivity (i.e. when one checks an unknown against a file of known materials, in many cases, only one or several materials need be considered).

IDENTIFICATION

Identification by the single-crystal method can now play an active and useful role in the industrial laboratory. Figure 3 outlines how the reduced cell file of known compounds is prepared. Each crystal lattice is represented by its primitive reduced cell or by the reduced cell matrix. Such a file is being prepared by the Crystal Data Center at the National Bureau of Standards and should contain data on about 50,000 crystals by mid-1977.

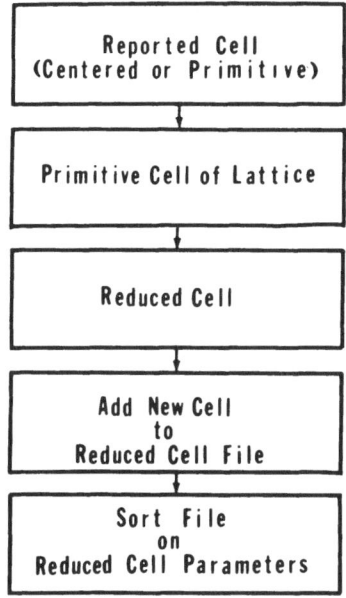

Fig. 3.   Preparation of Reduced Cell File

Figure 4 shows how the identification scheme works. The analyst selects a single crystal of an unknown, mounts it on the diffractometer, determines a refined primitive cell, reduces this cell, and then checks the cell against the reduced-cell file of known materials. If a match is not found, derivative cells are calculated and they in turn are checked against the file of knowns. All of the required derivative lattices can be systematically calculated via unique sets of upper triangular matrices. A discussion of the calculation of derivative lattices is given in (9, 11, 12, & 13).

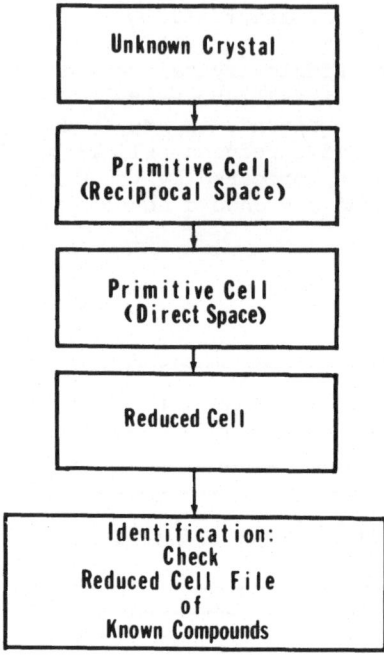

Fig. 4. Crystallochemical Identification Procedure Based on the Reduced
         Cell

     The entire process can be automated by combining into one instru-
ment the ability to determine a refined cell from an unknown with the
capability to search a large file of known crystals for a match.
This would be ideal not only for identification but also as the first
step of a complete structure determination. Thus, one could establish
whether a structure or a related structure has been already determined
before collecting a large set of diffraction data. Currently, the
Crystal Data Center at NBS calculates the reduced cell for all compounds
that will appear in the supplements to the third edition of Crystal Data
(3). The reduced cell is used routinely to identify duplicate and iso-
structural compounds, and to evaluate the correctness of the reported
symmetry of each material.

                              CONCLUSION

     Either the single-crystal or the powder method can be used indepen-
dently. However, when used in conjunction to study the crystalline state,
they complement and enhance each other. Often in the analysis of an
unknown sample, the analyst may wish to employ both techniques. When
both methods are used, the analyst can compare an unknown against a
greatly expanded number of known compounds.

     In analyzing a sample of crystals, one may wish to establish the
identity of a component crystal. Here the two methods reinforce each

other. The powder method can be used to check the overall quality of
the crystalline sample and its degree of purity. Often impurities
will be immediately revealed by the presence of "extra" lines. If such
lines appear, one can try to isolate the suspect crystal from the batch,
identify it with the single-crystal method, and decide whether it is
responsible for the lines.

The single-crystal method aids in studying the crystalline state in
the following three situations. First if the sample is limited and must
be conserved, second if the symmetry must be established, and third if it
is desired to routinely confirm or reject a postulated structure. The
first situation may arise if there is danger of losing a limited sample
by grinding, if the crystal transforms or decomposes on grinding, or if
the material has been grown under conditions that produced only one
crystal. In some of these cases one may use the Gandolfi camera which
gives the equivalent of a powder pattern from a single crystal. In the
second situation, the use of the single-crystal method to establish the
lattice and symmetry unambiguously is helpful to index a powder pattern,
to understand the physical properties of a material better, or to predict
the possibility of twinning or of coincidence site lattice formation. In
the third situation, the analyst can confirm a structure by determining
the lattice (via the reduced cell), space group, Z, and the density by
flotation. If the postulated structure is correct, then the X-ray
calculated density must agree with the experimental density measured by
flotation.

In summary, as both the single-crystal and powder methods depend on
the unit cell of the lattice, they can, and should, be used jointly for
identification and for the study of crystalline materials.

REFERENCES

1. J.D.H. Donnay and Werner Nowacki, Crystal Data, Classification of
   Substances by Space Groups and their Identification from Cell
   Dimensions, Geological Society of America Memoir 60, New York (1954).

2. J.D.H. Donnay, Gabrielle Donnay, E.G. Cox, Olga Kennard, and M.V.
   King, Crystal Data, Determinative Tables, 2nd Ed., American
   Crystallographic Association Monograph Number 5, (1963.)

3. J.D.H. Donnay and Helen M. Ondik, Crystal Data, Determinative Tables:
   Third Edition, Vol. 1 and 2, U.S. Department of Commerce, National
   Bureau of Standards, and the Joint Committee on Powder Diffraction
   Standards (1972, 1973).

4. A.D. Mighell, A. Santoro, and J.D.H. Donnay, "Reduced Cells",
   International Tables for X-ray Crystallography, Vol. 1, 530-535,
   Birmingham: Kynoch Press (1969).

5. A.D. Mighell, A. Santoro and J.D.H. Donnay, "Errata in International
   Tables for X-ray Crystallography," Acta Cryst. B27, 1837-1838 (1971).

6. E. Parthé and J. Hornstra, "Corrections to the Tables" in Chapter 5.1
   given in the 1969 Edition of Vol. 1 of International Tables, Acta Cryst.
   A29, 309 (1973).

7. A.D. Mighell, A. Santoro and J.D.H. Donnay, "Addenda to International Tables for X-ray Crystallography," Acta Cryst. B31, 2942 (1975).

8. J.M. Stewart, P.A. Machin, C.W. Dickinson, H.L. Ammon, H. Heck and H. Flack, The X-ray System, version of 1976, Technical Report TR-446, Computer Science Center, Univ. of Maryland, College Park, Maryland (1976).

9. A.D. Mighell, "The Reduced Cell: Its Use in the Identification of Crystalline Materials," Acta Cryst. In press (1976).

10. A.D. Mighell and A. Santoro, "Geometrical Ambiguities in the Indexing of Powder Patterns," J. Appl. Cryst. 8, 372-374 (1975).

11. A. Santoro and A.D. Mighell, "Properties of Crystal Lattices: The Derivative Lattices & Their Determination," Acta Cryst. A28, 284-287,(1972).

12. A. Santoro and A.D. Mighell, "Coincidence-Site Lattices," Acta Cryst. A29, 169-175 (See Appendix) (1973).

13. R. Bucksch, "A Theory of Sublattices and Superlattices," J. Appl. Cryst. 5, 96-101 (1972).

PHASE IDENTIFICATION BY X-RAY POWDER DIFFRACTION

EVALUATION OF VARIOUS TECHNIQUES

J.D. Hanawalt[*]

University of Michigan

Ann Arbor, Michigan 48109

*Chairman, Joint Committee on Powder Diffraction Standards

ABSTRACT

Three powder mixtures, each composed of four or more phases, were submitted for phase identification by x-ray diffraction. Laboratory technicians supplied tables of "d" values and of relative intensities as obtained separately and independently by use of the diffractometer, the Debye camera and the Guinier camera. These tables of diffraction data were "solved" by utilization of the Joint Committee search manuals and reference to the Joint Committee Powder Diffraction File (P.D.F.). The same tables of data were then submitted to the 2dTS:Diffraction Data Tele-Search for a computer printout of results. Experimental data are also presented which provide a quantitative comparison of the accuracy of measurement of "d" values and of the resolution of Debye cameras vs Guinier cameras, since this information is necessary for efficient search procedures whether by manual or computer methods.

The "solutions" of the three unknown mixtures confirm the general experience that only those diffraction data of the highest quality with respect to resolution of close lines and with respect to accuracy and intensity of "d" values are adequate for an easy and complete solution of complicated mixtures of phases. The Debye pattern does not entirely meet these requirements though for many simpler problems its great usefulness and especially its sensitivity to minute amounts of sample are well known. The diffractometer is a very versatile instrument which can provide high quality powder data and has as well the considerable advantage that it provides quantitative intensity values directly. Furthermore, units are now commercially available which automatically produce the diffraction data for immediate computer search and printout of results, thus almost completely eliminating manual labor in the determination of unknown phases in a powder mixture. The Guinier camera also provides powder diffraction data of the highest quality though for quantitative intensities the additional step of making a microphotometer trace of the film is necessary. However, for identification purposes relative intensities as can be obtained by the use of a Frevel type visual density gauge are shown to be sufficient.

With respect to the involved question of computer vs manual search methods much more experience is necessary. But the evidence from these three problems is that each method can determine the phases present (insofar as they are included in the P.D.F.) and that while the computer is probably faster the manual method is probably cheaper.

INTRODUCTION

It is almost 60 years since the "discovery" of powder diffraction by Debye and Scherrer and independently by Hull. The type of powder camera designed at that time is still in use today and in most x-ray laboratories is probably still the "work-horse" in the utilization of x-rays for phase identification. The diffractometer is a highly developed instrument likewise in use for the purpose of powder diffraction. So also is the Guinier design of camera which is finding considerably increasing popularity within the last decade.

X-ray diffraction phase identification became more broadly usable and important with the availability of a file of standard patterns and practical search systems. At present the JCPDS powder diffraction file (P.D.F.) contains about 34,000 patterns and is expanding at the rate of about 2000 patterns each year. Computer programs are now available for retrieval of unknown patterns.

Thus, the analyst in an x-ray laboratory has a considerable choice of techniques for obtaining the diffraction data of unknown phases and also a choice of methods of "solving" these data. The objective of the present paper is to make some comparison of the available techniques and methods and to illustrate with some examples how limitations in the "quality" of the data reflect in the degree of completeness and reliability of the identification of the unknowns.

DEBYE AND GUINIER DATA COMPARED

In order to determine the degree of uniformity to be expected in powder diffraction data from various types of x-ray units, x-ray patterns of certain standard materials were solicited from a score of x-ray laboratories in the U.S.A., Canada, and Europe. It would be impractical to reproduce the films in this paper but in Figure 1 is seen the microphotometer traces of two of the best films which were received. The traces in Figure 1 illustrate the greater dispersion, the sharper diffraction lines and the much lower background of the Guinier pattern and are typical of the differences between Guinier and Debye powder patterns.

Accuracy

Experience in measuring the positions of the lines in a Philips vernier scale viewing box indicates that on a Debye pattern this can be done to about ± 0.1 mm, while on a Guinier pattern this figure is about ± 0.05 mm. Since for the same diameter camera the dispersion of the Guinier unit is twice that of the Debye, the accuracy of measurement of "d" values for the Guinier film is four times better. Figure 2

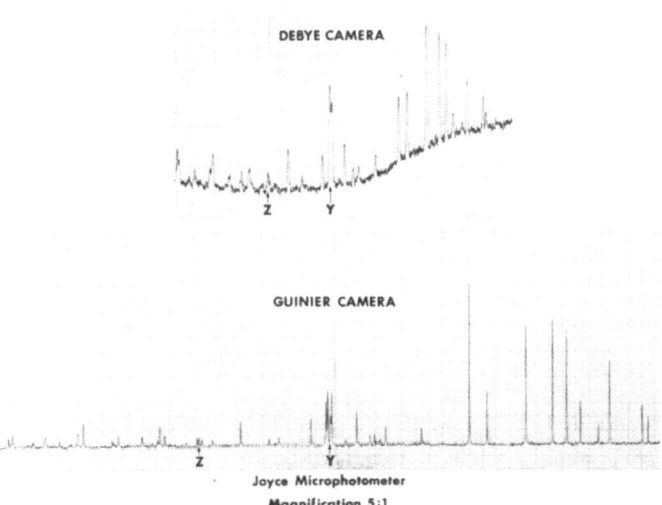

Figure 1.   Microphotometer traces of powder patterns of BaSO$_4$. Debye film from Lawrence Radiation Laboratory, courtesy of D.K. Smith. Guinier film from University of Paris, courtesy of A. Guinier.

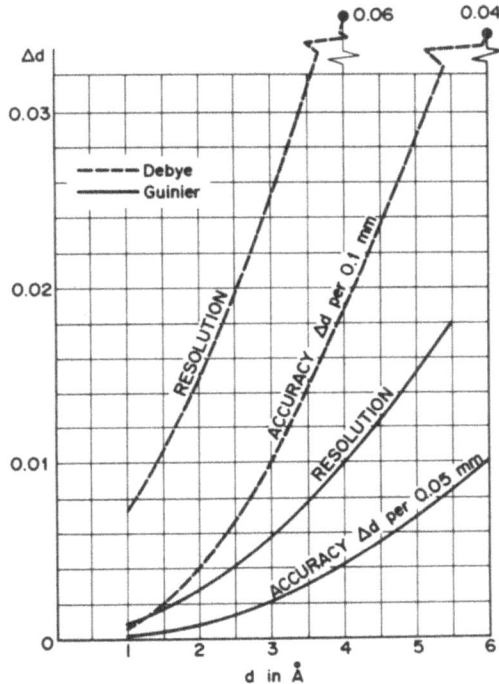

Figure 2.   Graph of values of experimentally observed accuracy and resolution for cameras of 114.6 mm diameter.

shows this plot of $\pm\Delta d$ as a function of d for 114.6 mm diameter cameras. Of course to obtain this degree of accuracy of absolute d values, the film must be calibrated by use of an internal standard. For the particular case of the three powder mixtures it can be seen by examining the tables of data that the actual accuracy of measurement for the Guinier films was considerably better than given by the curve of Figure 2. In the range up to about 3.5 Å most all of the d values of the stronger lines agreed with the Bureau of Standards determinations to within ±0.001 Å or less.

Resolution

The question of resolution also involves the sharpness of the lines and the background density and for practical purposes requires an experimentally determined answer. The curves drawn in Figure 2 represent the results of observations of films from a considerable number of different cameras. The Debye cameras were 57.3 mm, 114.6 mm and 140.0 mm in diameter. The Guinier camera diameters were 80.0 mm, 100.0 mm, 114.6 mm and 120.0 mm. Interestingly, while the 140.0 mm Debye did not make a noticeable improvement in the resolution of a 114.6 mm Debye, the 80.0 mm Guinier provided almost as good resolution as the 120.0 mm Guinier. Within the important and most commonly used range of d values for phase identification the resolution of the Guinier type camera is about five times greater than that of the Debye camera. This is illustrated in Figure 3 which compares Guinier and Debye patterns for the KNO_3 two lines at A, 3.78 and 3.73, and the three lines at B, 2.662, 2.647 and 2.632. The separations of 0.05 Å at A and 0.015 Å at B are unresolved in the Debye pattern but clearly resolved in the Guinier pattern.

Another illustration is given in Figure 4 which shows the Y and Z regions of Figure 1 but at a microphotometer magnification of 50 to 1. These close lines can be clearly seen by direct visual inspection of the original film. The four lines at Y, 2.118, 2.1151, 2.1048 and 2.1012 which are separated by Δd 0.0032 Å, 0.014 Å and 0.0036 Å respectively as seen in the Guinier pattern (these measurements are from Professor Andre Guinier's laboratory) are barely resolved into two lines in the Debye pattern. Likewise the three lines at Z, 1.7605, 1.7554, and 1.7509 which are separated by Δd 0.0051 Å and 0.0045 Å respectively are entirely unresolved in the Debye pattern. (None of the Debye patterns received showed better resolution than this pattern from the Lawrence Radiation Laboratory, courtesy of D.K. Smith.)

The curves of Figure 2 showing attainable resolutions of Debye and Guinier cameras have been generated from the above type of experimental observations of many films. It should be mentioned that the resolution of the lines at Y and Z in the $BaSO_4$ diffraction pattern is only possible if the $BaSO_4$ has been heated to above 800°C. However, with this qualification, these lines of $BaSO_4$ provide a quite practical and critical test of the excellence of adjustment of the diffraction camera or instrument. In Figure 5 is shown the resolution by step scanning with a diffractometer by M. Nichols at Sandia Laboratories, Livermore, California. The same resolution can also be observed with the diffractometer with a slow scan of about 1°2θ per minute. In Figure 6 is shown the resolution of these lines of $BaSO_4$ as demonstrated

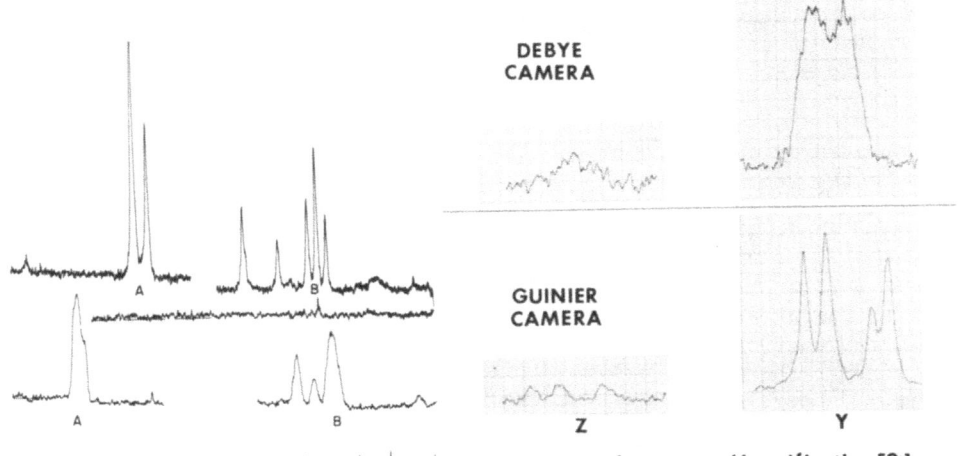

Figure 3.   Microphotometer trace
of KNO$_3$ powder patterns. Upper
traces Guinier, lower traces Debye.

Figure 4.   Regions Y and Z of Fig. 1
at higher magnification.

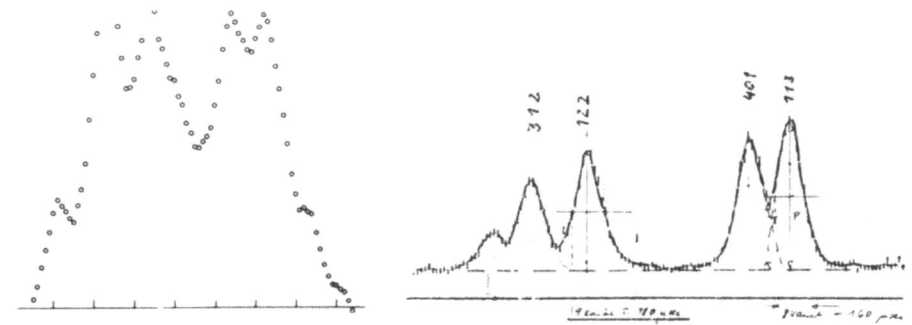

Figure 5.   Diffractometer step
scan of BaSO$_4$ at Y region.
Courtesy of M.C. Nichols.

Figure 6.   Neutron diffraction of
BaSO$_4$ at Y region. Courtesy of
E. Steichele.

with his neutron time-of-flight diffractometer by Dr. E. Steichele, at
the Techn. Univ. Munich. Dr. Steichele heated the BaSO$_4$ to 1000°C and
used 20 cc for his sample. The resolution by neutron diffraction illustrated in Figure 6 is superior to any of the x-ray diffraction techniques which we have observed. It is uncertain, however, why the Δds
of the neutron diffraction lines are not exactly the same as the Δds by
x-ray diffraction.

     Dr. Steichele has called our attention to another critical test of
resolution which is posed by the CaWO$_3$ pattern. The CaWO$_4$ (scheelite)
powder pattern has a strong line (312) at 1.5921 Å which is separated

from a weak line (303) by a Δd of 0.0045 Å. The 303 line is clearly
resolved by the Guinier camera in spite of the fact that the 312 line
has about 30-fold greater intensity. The 303 line is not resolved by
the Debye camera and can be detected with a diffractometer only by a
very slow scan.

In summary it can be concluded that the Guinier camera of the same
diameter as the Debye camera provides about four times the accuracy of
determination of d values and about five times the resolution. The
quantitative plots of accuracy and resolution as a function of d value
as given in Figure 2 are necessary for efficient search procedures
whether by manual or by computer methods. One needs to know for each
step of the search how much error to allow for or, in the case of computer search, what error "window" to use. The analyst must, however,
keep in mind if the search proves difficult that not all of the patterns
in the P.D.F. are high quality "starred" patterns.

ILLUSTRATIVE SOLUTIONS OF UNKNOWN POWDER PATTERNS

Tables 1, 2 and 3 list the diffraction data from unknown mixtures
supplied by Mr. Ron Jenkins of Philips Electronic Instruments, Inc.
These data were produced by a graduate student, Mr. Chi-Hung Leung, who
had no other connection with the project. The main steps in Mr. Leung's
work were to mix the unknown samples with Si powder as an internal
standard, make the 1-1/2 hour exposure to $CuK\alpha_1$ radiation in a 100 mm
diameter Hägg design XDC-700 Guinier camera, develop the film, use a
Philips vernier scale viewing box to make the linear measurements of
the diffraction lines, and type these measurements into a teletypewriter terminal of the University of Michigan computer in which had
been stored the LaGrangian program (15) using six lines of silicon
for calibrating the film. The "d" values as printed out by the computer are listed in Tables 1, 2 and 3. Mr. Leung also made microphotometer traces of the films and calculated the relative intensities
after measuring the peak height above background of each line. It is
interesting to note that equivalent "d" values can be obtained by less
sophisticated methods simply by visual reading on a Nies scale with
careful use of the silicon lines for calibration. Also, though somewhat tedious, the visual reading of intensities with a Frevel type
gauge (5) is satisfactory for identification purposes.

Manual Solutions

Beginning with these three tables of data, the author, J.D.
Hanawalt, "solved" the three unknown mixtures by manual search of the
J.C.P.D.S. Mineral File for Table 1 and search of the Mini File for
Tables 2 and 3. It was fortuitous that the Mini File did contain all
of the inorganic phases present. If any phase had been absent in the
Mini File, the search would have been continued using the Maxi File,
i.e. the total listing of inorganics in the P.D.F. (The Mini File is
limited to about 2000 of the more commonly occurring inorganics in the
P.D.F.).

The results of the manual search for each of the three unknown
mixtures are shown in the respective three tables. No other information than the diffraction data was used in the search. Inspection of

TABLE 1. Mineral #1

Guinier Data	9-423 CuFeS$_2$ Chalcopyrite Toronto Univ	5-448 BaSO$_4$ barite NBS	7-324 Al$_2$O$_3$·3H$_2$O Gibbsite Penn State	5-490 SiO$_2$ α-quartz NBS	20-535 FeS Pyrrhotite Tokyo U.
		5.58 2			
4.849 19			4.85 320		
4.450 10		4.44 16			
4.382 11			4.37 50		
4.339 14		4.34 30	4.32 23		
4.269 10				4.26 35	
3.899 35		3.90 50			
3.777 10		3.773 12			
3.581 21		3.577 30			
3.444 70		3.445 100			
3.346 21				3.343 100	
3.318 49		3.319 70	3.306 15		
3.102 66		3.103 95	3.187 12		
3.040 100	3.03 100				
2.989 6					2.98 40
2.835 34		2.836 50			
2.727 42		2.729 45			
2.645 7					2.65 60
	2.63 5				
2.606 4					
2.481 11		2.482 13			
2.451 5			2.454 23	2.458 12	
2.422 3			2.420 20		
2.385 5			2.388 27		
2.324 15		2.325 14			
2.280 10		2.282 8	2.285 5	2.282 12	
2.210 22		2.211 25	2.244 10		
2.166 3		2.169 3	2.168 7		
2.121 56		2.121 80		2.128 9	
2.104 48		2.106 75			
2.069 12					2.07 100
2.056 21		2.057 19	2.043 17		
1.930 10		1.932 7	1.993 11		
1.870 20	1.865 40		1.921 11		
1.856 47	1.854 80	1.857 18			
1.836 3					
1.819 7				1.817 17	
1.800 2			1.799 13		
1.790 2		1.789 4			
1.761 7		1.762 8			
		1.758 10			
1.753 7		1.754 8	1.750 16		
1.728 6		1.728 4			
1.721 6		1.724 5			1.72 20
1.684 10		1.682 8	1.689 13		
1.674 15		1.674 14			
		1.670 11			
1.660 3		1.660 2			
1.594 30	1.591 60	1.594 8			1.61 5
1.576 14	1.573 20				
1.536 14		1.535 15		1.541 15	

these three tables will show that in each case the identification of
the phases contained in the mixtures is positive and convincing. The
purpose in presenting these tables in detail is to permit the viewer
to check the exactness of the agreement of "d" values from the Guinier
pattern with the "d" values from the P.D.F. It will be seen that with
a few exceptions here and there, "d" values agree within $\pm$ 0.001 A or
less. By chance most of the standard patterns involved originated at
the U.S. National Bureau of Standards and were produced by their
exacting diffractometer techniques. The other standard patterns also
came from very reliable sources. It is thus demonstrated that Guinier
camera "d" values agree quite exactly with the best diffractometer data.
With regard to intensities the agreement is satisfactory considering
the many well known factors which may affect this quantity.

TABLE 2.   Inorganic #2

Guinier Data	21-1272 TiO₂ Anatase NBS		21-1276 TiO₂ Rutile NBS		25-1135 BaCl₂·2H₂O NBS		9-387 V₂O₅ NBS	
					5.96	7		
5.757 8					5.72	12	5.76	40
5.456 18					5.45	85		
4.939 10					4.94	60		
4.847 12					4.84	20		
4.500 12					4.50	55		
4.424 28					4.42	100		
4.374 15					4.33·	17	4.38	100
4.090 7							4.09	35
3.659 8					3.661	45		
3.617 3					3.662	17		
3.517 100	3.52	100			3.569	15	3.48	7
3.407 25							3.40	90
3.393 16					3.390	55		
3.364 5					3.359	25		
3.244 98			3.25	100	3.240	12		
2.950 5					2.948	50		
2.929 8					2.928	65		
2.908 22					2.908	90		
2.884 12							2.88	65
2.861 14					2.861	45		
2.770 6							2.76	35
2.714 12					2.712	50		
2.669 4					2.671	15	2.687	15
2.611 7							2.610	40
2.545 15					2.547	75		
2.487 45			2.487	50			2.492	7
2.431 10	2.431	10						
2.408 17					2.409	42	2.405	7
2.378 30	2.378	20						
2.366 5					2.365	19		
2.344 5					2.344	20		
2.334 5	2.332	10						
2.296 7			2.297	8				
2.257 10					2.256	25		
2.228 8					2.229	25		
					2.214	20		
2.210 7					2.209	30		
2.187 34			2.188	25			2.185	17
2.152 9					2.148	13	2.147	11
2.119 7					2.116	13		
2.087 12					2.085	50		
2.054 14			2.054	10				
2.025 5					2.023	12		
2.001 5					1.999	19		
1.992 5					1.989	12	1.992	17
1.892 46	1.892	35					1.919	25
1.782 10					1.783	3		
1.699 32	1.670	20						
1.687 70			1.687	3				
1.666 27	1.666	20						
1.623 21			1.624	20				
1.564 4							1.564	11

Computer Solutions

These same three tables of diffraction data for the three unknown mixtures were also submitted for computer solution via 2dTS.   2dTS: Diffraction Data Tele-Search is designed by the Joint Committee on Powder Diffraction Standards to make available to the user data retrieval and identification of multiphase mixtures via a remote computer facility.   By teletypewriter, access to the P.D.F. and utilization of the Johnson-Vand Search-Match program is achieved.   2dTS prints out the ten best choices along with a "scale factor" which is an indication of the relative amounts of the phases.   It is then necessary

TABLE 3.   Inorganic #3

Guinier Data		4-531 KBr NBS		5-378 BaCO$_3$ NBS		4-857 LiF NBS		11-652 Na$_5$P$_3$O$_{10}$ Delft		10-187 Na$_4$P$_2$O$_7$ Delft	
7.995	5							8.02	20	7.69	1
4.706	18							4.72	100	4.67	11
4.575	24			4.56	9			4.57	100		
4.471	10			4.45	4			4.47	30		
4.415	10									4.41	100
3.807	57	3.80	15							3.84	7
3.722	82			3.72	100			3.74	25		
3.662	36			3.68	53			3.68	25		
3.583	7							3.58	35		
3.371	3									3.37	35
3.297	100	3.292	100								
3.214	21			3.215	15			3.18	20		
3.011	11			3.025	4			3.01	50		
2.959	8							2.959	20		
2.733	5			2.749	3			2.732	7	2.734	60
2.694	20							2.694	85	2.698	70
2.683	5							2.686	35		
2.656	7			2.656	11						
2.648	5									2.648	15
2.628	43			2.628	24			2.627	65		
2.593	34			2.590	23			2.592	65		
2.503	4							2.503	17		
2.458	4							2.456	20		
2.421	8							2.421	30		
2.411	3									2.413	7
2.333	100	2.333	57					2.357	13	2.336	25
2.325	50					2.325	95				
2.283	7			2.281	6						
2.149	29			2.150	28						
2.104	12			2.104	12						
2.048	12			2.048	10						
2.018	33			2.019	21			2.019	20		
2.014	54					2.013	100				
1.990	37	1.989	7								
1.941	23			1.940	15						
1.932	13									1.921	20
1.905	64	1.905	16								
1.862	7			1.859	3						
1.790	3									1.767	7
1.751	5			1.737	2					1.752	5
1.716	3			1.706	1						
1.676	7			1.677	5						
1.650	46	1.649	10	1.649	4						
1.530	5			1.521	4						
1.514	21	1.514	2								
1.475	72	1.475	17								
1.424	25					1.424	48				
1.375	8			1.375	6						
1.367	4			1.366	4						
1.347	57	1.346	8	1.348	4						

for the analyst to manually inspect these ten answers and decide which
and how many of the ten are valid and correct answers to the problem.
The 2dTS search for problem 1 was carried out on the Mineral File and
also on the Maxi File; for problems 2 and 3, it was carried out on the
Mini File and the Maxi File. For these three problems 2dTS provided
the same answers whether the search was made on the Maxi File or on the
smaller files. For each of the three problems the computer gave the
same principal solutions as obtained by the manual search plus extra
answers which must be inspected as has been mentioned.

Table 4 shows some comparisons of the times and costs of the solutions by manual vs computer methods. The manual times listed give the
total time, i.e. the time for search plus the time for matching.
Generally the time required for manual matching of all of the lines of
the standard pattern given in the P.D.F. is considerably more than the

the search time required for locating the possible answer based on the
eight lines in the Search Manual.

The 2dTS figures in Table 4 show the seconds and dollar costs for
central computer time and the time required for punching the diffraction data on tape to feed the teletypewriter terminal. The figures do
not include the elapsed time of a few minutes to get the answer back
(or possibly more minutes if access to the computer is not immediate)
nor the time which may be required to inspect the ten most probable
answers printed out by the computer. When the computer search is made
on the Maxi File rather than on the smaller specialized files the search
is equally successful for these three problems, but the times and costs
for the computer central processing unit are considerably greater.
Also, as to be expected, the manual search method is considerably more
tedious when the much larger Maxi Search Manual must be used.

TABLE 4. Time and Costs for Solution of Unknowns

| | Manual Method | 2dTS Computer Method | | |
Problem	Total Time	Entry of Data Time	Central Time	Computer Cost
Mineral 1	2 hr 45 min	15 min	67 sec	$42.18
Inorganic 2	1 hr 17 min	15 min	26 sec	$19.04
Inorganic 3	2 hr 47 min	15 min	24 sec	$17.96

The diffraction data for these three problems were also processed
by Mr. M.C. Nichols with the Maxi File using his program on a CDC 6600
computer at Sandia Laboratories in Livermore, California. Mr. Nichols'
program was also successful in obtaining the correct solutions to the
problems. The figures are not at hand for the times and costs of Mr.
Nichols' solutions.

Solutions with Debye Camera Data

To save space the Debye and the diffractometer data of the three
problems are not shown. The diffractometer data essentially match the
Guinier data. The Debye data, however, are not equivalent, some lines
being missing and others having shifted "d" values in spite of the fact
that the Debye films were carefully calibrated with internal standards.
These diffraction data provided very unsatisfactory results in the 2dTS
computer printout of answers. For problem 1, only three of the ten
listed phases agreed with the five correct answers given by the Guinier
data. For problem 2, only two of the four correct answers were given.
For problem 3 two of the five correct answers were given in the final
printout though two more were included among the 50 phases with highest
reliability ratings. It should be mentioned that if information
regarding the chemical elements present or absent had been included in
the computer input, more complete answers might have been obtained.
It should also be mentioned that experience shows that for many simpler
multiphase unknowns in which super-positions and interference of lines
are absent the Debye pattern will be fully adequate for a complete
solution.

SUMMARY AND CONCLUSIONS

Instruments and methods in the field of phase identification by x-ray powder diffraction have been evolving over the past 60 years (1-10) and have reached a high state of automation. The advent of computer programming has contributed greatly and has become quite indispensable to the continuing development of data banks and data retrieval (11-17).

The evidence submitted in the present paper is that with the superior quality of data provided by the Guinier camera, manual methods can also be very successful. It can be concluded that the small x-ray laboratory with limited equipment or the laboratory so located in the world as not to have access to large computers can with adequate personnel take full advantage of the inherent power of x-ray powder diffraction phase identification.

REFERENCES

1. P. Debye and P. Scherrer, Phys. Z. 17, 277 (1916); 18, 291 (1917).
2. A. W. Hull, Phys. Rev. 10, 661 (1917); J. Am. Chem. Soc. 41, 1169 (1919).
3. A. N. Winchell, Am. Min. 12, 261 (1927).
4. A. Guinier, C. R. Acad. Sci. 204, 1115 (1937); Ann. Phys. 12, 161 (1939).
5. J. D. Hanawalt, H. W. Rinn, and L. K. Frevel, Ind. Eng. Chem., Anal. Ed. 10, 457 (1938).
6. A. K. Boldyrev, V. I. Mikheev, V. N. Dubinina and G. A. Kovalev, Ann. Inst. Mines Leningrad 11, 1 (1938).
7. W. P. Davey, J. App. Phys. 10, 820 (1939).
8. P. M. DeWolff, Acta Cryst. 1, 207 (1948).
9. W. Parrish, Am. Min. 33, 770 (1948); X-ray Analysis Papers, Centrex Publ. Co., Eindhoven 1965.
10. W. C. Bigelow and J. V. Smith, ASTM, S.T.P. 372, 54 (1965).
11. D. K. Smith, Law. Rad. Lab. UCRL-7196 (1963); ibid. UCRL-50264 (1967).
12. M. C. Nichols, UCRL-70078, 19 (1966).
13. G. G. Johnson and V. Vand, Ind. Eng. Chem. 59, 19 (1967).
14. L. K. Frevel and C. E. Adams, Anal. Chem. 40, 1335 (1968).
15. Lagrangian Interpolation, Fortran IV, Ingeniorsfirman Instrumenttjanst, Sundbyberg, Sweden (1970).
16. G. G. Johnson, Data Base and Search Programs, J.C.P.D.S. (1974).
17. The Joint Committee on Powder Diffraction Standards, 1601 Park Lane, Swarthmore, Pennsylvania 19081.

CHEMICAL IDENTIFICATION AND PHASE ANALYSIS OF TRANSPLUTONIUM ELEMENTS AND COMPOUNDS VIA X-RAY POWDER DIFFRACTION[*]

J. R. Peterson

Department of Chemistry, University of Tennessee, Knoxville, Tennessee 37916 and Transuranium Research Laboratory, Oak Ridge National Laboratory, Oak Ridge, Tennessee 37830

ABSTRACT

X-Ray powder diffraction (XRPD) techniques are compatible with studies of the transplutonium elements and their compounds. Only limited amounts of these materials are available for study, and microchemical techniques exist for the preparation of samples suitable for X-ray analysis. The microscale syntheses, systems for annealing/quenching to bring about phase transformations, and a modified powder diffraction camera are described. Examples of recent and current work are presented to illustrate some specific applications of XRPD to basic, transplutonium element research.

INTRODUCTION

Most of the experimental data available on the physico-chemical properties of the transplutonium elements and their compounds has been obtained through the use of specialized microchemical techniques. The radioactivity of these elements has often precluded otherwise routine manipulations, and the small sample sizes available have inspired the development of new techniques for determination of a particular property of interest. One especially useful technique is X-ray powder diffraction (XRPD), which is ideally suited to small quantities of materials. XRPD actually made possible the chemical studies of transplutonium elements and compounds on the microgram and submicrogram scale, because of its capability to provide identification of such crystalline materials. More often than not, XRPD has allowed the researcher to identify and characterize actinide samples by comparison to isomorphous lanthanide materials of well-known stoichiometry. This approach has played a major role in the identification of many actinide materials where quantities are insufficient for elemental analysis.

_____

[*] Research sponsored by the U. S. Energy Research and Development Administration under contracts with the University of Tennessee (Knoxville) and Union Carbide Corporation.

Historically, the discovery of a new transplutonium element (Z>94) has resulted from its separation on a cation-exchange column, followed by detection of its characteristic nuclear radiation. Since the "discovery isotope" is usually produced by charged-particle bombardment in an accelerator, the quantity actually separated for study is not weighable and is therefore suitable only for tracer-level experimentation. Weighable amounts of the transplutonium elements through einsteinium (Z=99) are produced via long-term, neutron irradiation of plutonium (Z=94) in a nuclear reactor. With the early generation reactors only multimicrogram quantities were available for study, but at the present time, milligram or greater amounts of americium (Z=95), curium (Z=96), berkelium (Z=97), and californium (Z=98) are being produced. The short half-life of einsteinium-253, about 20 days, precludes its production in more than multimicrogram amounts. Weighable amounts of the elements with $Z \geq 100$ are unavailable.

Traditionally, the study of bulk physico-chemical properties of a new transplutonium element has begun with the microscale preparation and crystallographic characterization (usually structure type and lattice parameters) of its oxides and halides and then progressed to study of the elemental state. The fundamental analytical tool supporting the study of transplutonics in the solid state is XRPD.

It is the scope of this paper to summarize several of the historical and modern-day microchemical techniques used to prepare samples of these actinide materials suitable for XRPD, to indicate the variety of studies possible via XRPD analysis using only slightly modified, standard powder diffraction equipment, and to illustrate some of these applications with examples of work carried out at the Oak Ridge National Laboratory. No effort is made to be exhaustive; instead, emphasis is placed on indicating the scope of studies made possible by XRPD being able to provide sample characterization on microamounts of material.

EXPERIMENTAL

Sample Preparation on the Microscale

One of the more formidable problems encountered in working on the microscale is that of maintaining a high degree of sample purity through a series of chemical and mechanical manipulations. This is due in part to the large increase of the surface-to-volume ratio on the micro-scale. An effective method of dealing with this problem is to use the single ion-exchange resin bead technique pioneered by Cunningham and Wallmann (1). A carefully purified resin bead, of appropriate size to contain from nanogram to multimicrogram amounts of the trivalent actinide ions, is saturated with the desired actinide by equilibration in solution. The loaded and washed bead, representing about 2 M concentration of actinide, is an ideal, clean container for the storage and manipulation (via quartz fibers or tungsten needles) of microamounts of highly purified actinides. Conversion to the oxide is effected by calcining the loaded bead in air or oxygen. This results in a spherically shaped oxide sample which can be transferred to a quartz X-ray capillary drawn on the end of a standard taper joint (to facilitate attachment

to a preparation vacuum line). The oxide can be sealed in vacuum (or in
any appropriate gas) and subjected to analysis as is, or it can be
chemically converted in situ to another compound prior to X-ray analysis.
In fact, a series of compounds using the same sample can be prepared
and characterized by XRPD through the use of a long capillary, mounted
backwards in a Debye-Scherrer powder camera provided with an extension
of the cover and a capillary support. Following X-ray analysis the
capillary is cracked open and resealed onto its original standard taper
joint. After the appropriate chemical treatment, the capillary is again
sealed off for X-ray analysis. This process for the preparation of
successive compounds (2) is limited only by the length of the initial
capillary and the skill of the user! A useful extension of this technique
was developed by Copeland and Cunningham (3), who added a twist stopcock
to the capillary containing the sample, thus providing an easy means
for the sequential sealing of the desired sample, analyzing by XRPD,
and converting to another compound, etc.

In general the chemical treatments of the initial sample involve
oxygen and hydrogen to produce the various oxides, anhydrous hydrogen
halides to produce the various trihalides, mixtures of hydrogen
halides and water vapor to produce the various oxyhalides, etc. All
of the above reactions can be carried out in a quartz capillary with
the exception of those involving HF (4). For detailed accounts of
the experimental conditions used to synthesize these various compounds,
the reader is referred to Keller's book (5) and the references therein.

Preparation and XRPD study of the transplutonium metals have also
been accomplished on the microgram scale. One method, producing a bulk
metal sample suitable for analysis by X rays, is based on the lithium
metal vapor reduction of the transplutonium element trifluoride (AnF$_3$)
in accord with the equation AnF$_3$ + 3Li $\rightarrow$ An + 3LiF. The single bead
technique is employed to produce the trifluoride by treatment of the
oxide with anhydrous hydrogen fluoride. The trifluoride is supported
by a tungsten-wire spiral inside a tantalum crucible, which also contains the lithium reductant. The crucible is heated to bring about the
reduction, and the metal sample is removed from the supporting spiral
(often by just pulling the wire straight) and sealed in a quartz
capillary for analysis by X rays. Details are available elsewhere (6,7).
Following study of the metal samples, they can be used as starting
material for the preparation of hydrides, nitrides, carbides, sulfides,
etc. (e.g., 8,9).

Annealing/Quenching Systems for Phase Studies

Several heating systems have been designed for specific applications
in the analyses of polymorphic actinide materials. Common to all the
systems is a minimal total heat capacity to allow both rapid attainment
of thermal equilibrium (to minimize loss of volatile samples or to
mimimize interaction between the quartz capillary and the sample) and
rapid quenching of samples (to study high-temperature phases at room
temperature). The heating element in each case is a wire, formed either
into a coil to support the sample-containing capillary or into a spiral
to support and contain the sample itself. In the latter case, compatibility between the sample and the spiral is important. These systems
are easily mounted in a controlled atmosphere enclosure.

Perhaps the most sophisticated system consists of a wire coil
formed from 5-mil Pt and Pt-10% Rh wires such that the two types of
wire are joined in the center turn of the coil to form a- thermocouple
junction. The coil is heated by resistance using a 3 kHz ac variable
power supply. Sample temperature can be monitored continuously by
measuring the dc output of the thermocouple. Thermal equilibrium
is attained in less than 10 sec. Quenching to room temperature can
be achieved in a few seconds by just turning off the power to the coil,
or more rapidly, by simultaneously turning off the power and submerging
the coil and sealed capillary in a cool liquid. A photomicrograph of
this coil system in use is given elsewhere (10).

It is possible to use a tungsten or iridium wire spiral to contain
and heat/quench samples of some transplutonium metals and oxides. Heat
is transferred to the sample by thermal contact and by radiation as the
spiral is heated electrically. Temperature is monitored by use of a
micro-optical pyrometer and/or by reference to a calibration curve
of input power versus sample temperature. The environment of the
spiral is controlled to prevent undesired chemical reactions from
taking place during the heat treatment and to effect rapid quenching.
Following heat treatment the sample is transferred to and sealed in a
capillary for X-ray analysis.

X-Ray Diffraction Equipment and Data Analysis

Standard X-ray generators (CuK X radiation) are employed in this
work. Placement of a Ni foil filter between the diffracted X rays and
the film serves to reduce the film darkening caused by the sample's
radioactivity. Indeed, unique to the study of transplutonics is the
oftimes necessity of adjusting the time of exposure to X rays to optimize
the contrast between the diffraction lines and the isotropic, radiation-
produced film darkening.

Standard, 57.3-mm diameter Debye-Scherrer powder cameras are each
equipped with a beryllium cup enclosing the sample area to prevent interference with the fragile capillary (11). One camera is also equipped
with a heating coil of nichrome wire, surrounding the capillary on both
sides of the X-ray beam, to allow studies of thermal properties at
temperatures up to about 800°C (10). A brass microcollimator which slips
over the end of the standard collimator is available for use when an
X-ray beam of only 275-μ diameter is desired. This equipment has provided
us with sufficient XRPD data to identify the structure types and determine the lattice parameters of a variety of transplutonium elements and
compounds, to study the polymorphism in many of these materials, and
to determine some lattice parameter expansion coefficients.

Analysis of the XRPD data follows standard procedures. Experimental patterns are compared to those calculated by the POWD program
(12) using the best available atomic coordinate data. Use is made of
a CALCOMP subroutine (9) to plot the calculated powder patterns for
a convenient and efficient visual display. Examples are shown in
several recent publications (9,13-15).

Lattice parameters are refined by the method of least squares
using the LCR-2 program (16) and including appropriate line weights

to prevent overweighting of an experimental line considered to be composed of two or more theoretically possible lines.

RESULTS AND DISCUSSION

Presented here are examples of work done at the Oak Ridge National Laboratory to illustrate some of the applications of XRPD to the elucidation of the solid-state behavior of the transplutonics. Because a complete discussion of the results obtained is unwarranted here, reference is made to the literature wherever possible.

Determination of the accessibility and stability of the various chemical oxidation states of an element is an important early goal in the understanding of its chemistry. In the late sixties and early seventies a number of papers appeared offering evidence from solution studies for the existence of Cf(II). At the same time a lesser number of papers were published refuting the evidence or the interpretation of the data as indicating Cf(II). It was felt by this author that identification of Cf(II) could best be obtained by the analysis of XRPD data from a compound of divalent californium. First $CfBr_3$ was synthesized on the microgram scale by treatment of Cf_2O_3 with anhydrous HBr. The product was identified on the basis of its XRPD pattern in comparison to those obtained by two independent research groups. [Actually, because these two groups were unable to identify the structure type from their data, the present author grew a single crystal for a complete structure analysis (14)]. A product resulting from the treatment of light-green $CfBr_3$ with hydrogen gas at elevated temperature was amber-colored, and it produced a diffraction pattern consistent with the tetragonal structure characteristic of $EuBr_2$ and $SrBr_2$. The measured lattice parameters led to a calculated Cf(II) ionic radius which was smaller than that of Eu(II) by about 0.01 Å, a result in agreement with the relative magnitudes of the respective trivalent ionic radii. Thus, XRPD provided the key for the positive identification of Cf(II) in the solid state and for the establishment of the preparation of the first salt-like, divalent transuranium element compound (17).

Study of the chemistry of einsteinium is greatly hampered by its intense radioactivity. In fact, the alpha particle decay energy amounts to some 3600 kilocalories per mole per minute, far in excess of normal chemical bond energies. Although a number of attempts to obtain XRPD data from simple compounds of einsteinium have failed, limited success has been achieved for $EsCl_3$ (18), EsOCl (18), $EsBr_3$ (19), and EsI_3 (20). Two factors seem to yield success: one is to have a sample size and geometry that maximizes escape of the alpha particles, thus minimizing deposition of the decay energy within the sample; the second is to provide for a continual resynthesis system in which the chemical effects of self-irradiation can be reversed, and a steady-state concentration of the desired compound maintained. One endeavors to find the conditions where the crystallinity of the sample is sufficient to produce a diffraction pattern in a time period shorter than that dictated by the radiation-produced darkening of the film. [The potential application of energy dispersive X-ray diffraction analysis to obtain sufficient data for determination of the structure type and lattice parameters of compounds of einsteinium is now being considered].

Using the powder camera equipped with a heating coil and sealing the einsteinium trihalide samples in the corresponding hydrogen halide gas, we have been successful in obtaining adequate diffraction data. One approach is to keep the sample between 400 and 450°C during the exposure to X rays, while the other is to anneal the sample in place at about 425°C just prior to X-ray exposure at ambient temperature. We have not been very successful in obtaining diffraction data from the products resulting from the treatment of $EsBr_3$ and EsI_3 with hydrogen or from EsOBr and EsOI. Indeed, it is this failure to achieve sufficient characterization of einsteinium compounds via XRPD that has led us to develop a microscope-spectrophotometer capable of recording the absorption (or luminescence) spectrum of the same sample of compound used for XRPD and held at the same temperature (15). Applying these two complementary techniques to the same sample of actinide material has greatly expanded our control and understanding of the chemistry (21). It is the spectral data that support the chemical identification of a compound from which insufficient XRPD data are obtained. Wherever possible data from both techniques are obtained and correlated, so that experience is built up to justify the analysis of a compound on the basis of spectral data alone.

In 1973 it was reported (22) that the available evidence suggested the existence of $EsBr_2$ and EsI_2 to the exclusion of the corresponding trihalides, perhaps as a result of the intense self-irradiation. Our complementary XRPD and absorption spectrophotometric studies of $EsBr_3$ (19) and EsI_3 (20) have shown conclusively the stability of these trihalides.

Many of the transplutonium metals and compounds exhibit polymorphism. Identifying the various phases in a given actinide material, determining their temperature relationship, and correlating the range of stabilities of the various structure types across the series of elements is an integral part of actinide research. Structure types identified in phase studies that were carried out with use of the minimal heat capacity annealing/quenching systems or the coil-equipped XRPD camera are summarized in Table 1.

Table 1. Polymorphism in Some Transplutonic Materials

Material	Structure Types Low Temperature → High Temperature			Reference
Cm, Bk	dhcp	→	fcc	9, 6
Cf	dhcp → fcc-β	→	fcc-α	7
Am_2O_3, Cm_2O_3 Bk_2O_3, Cf_2O_3	bcc → monoclinic	→	hexagonal	23
BkF_3, CfF_3	orthorhombic	→	trigonal	4,13
$BkCl_3$, $CfCl_3$	hexagonal	→	orthorhombic	24,25
$BkBr_3$	orthorhombic→(rhombohedral)?→monoclinic			14
$CfBr_3$	rhombohedral	→	monoclinic	14

dhcp = double hexagonal closest packed   fcc = face-centered cubic
bcc = body-centered cubic

The behavior of elemental californium is unusual. By comparison of the metallic radii calculated for its three forms with those of the preceding elements, it was suggested (7) that the two, lower temperature modifications represent californium with a metallic valence of three, while the highest temperature form represents a metallic valence of two. Zachariasen (26) has interpreted these same data to indicate metallic valences of four, three, and two for the three forms, respectively. We are extending our capabilities for phase analysis via XRPD to include pressure effects for further study of californium metal.

CONCLUSION

XRPD analysis has played a dominant role in the study of the transplutonium elements in the solid state, by allowing the determination of bulk physico-chemical properties on even submicrogram-sized samples. By analysis and comparison of powder data obtained on these materials with that known for the lanthanides, structure types, lattice parameters, and chemical oxidation states have been identified. XRPD studies performed either at high temperature or at room temperature after quenching have resulted in the determination of phase relationships, thermal decomposition products, and lattice parameter expansion coefficients. Even as larger samples of the transplutonics are synthesized for the determination of other physico-chemical properties, e.g., heats of solution and sublimation, and magnetic susceptibility, XRPD will still provide the data necessary to confirm preparation of the material under study.

ACKNOWLEDGEMENTS

The author is pleased to acknowledge the special contributions of Dr. James N. Stevenson, who expended considerable effort in the development of the annealing/quenching systems, the coil-equipped camera, and the CALCOMP plotting subroutine. Dr. John H. Burns of the Oak Ridge National Laboratory is thanked for his helpful discussions concerning this manuscript.

REFERENCES

1. B. B. Cunningham, "Submicrogram Methods Used in Studies of the Synthetic Elements," Microchem. J. Symp. Ser., 69-93 (1961).

2. J. R. Peterson, "The Solution Absorption Spectrum of Bk^{3+} and the Crystallography of Berkelium Dioxide, Sesquioxide, Trichloride, Oxychloride, and Trifluoride," Ph.D. Thesis, University of California, Berkeley, October 1967 (UCRL-17875), p. 43.

3. J. C. Copeland, "Preparation and Crystallographic Analysis of Californium Sesquioxide and Californium Oxychloride," M.S. Thesis, University of California, Berkeley, August 1967 (UCRL-17718), p. 8.

4. J. R. Peterson and B. B. Cunningham, "Crystal Structures and Lattice Parameters of the Compounds of Berkelium. IV. Berkelium Trifluoride," J. Inorg. Nucl. Chem. 30, 1775-1784 (1968).

5. C. Keller, <u>The Chemistry of the Transuranium Elements</u>, Verlag
    Chemie GmbH (1971).

6. J. R. Peterson, J. A. Fahey and R. D. Baybarz, "The Crystal
    Structures and Lattice Parameters of Berkelium Metal," J. Inorg.
    Nucl. Chem. <u>33</u>, 3345-3351 (1971).

7. M. Noé and J. R. Peterson, "Preparation and Study of Elemental
    Californium-249," in W. Müller and R. Lindner, Editors,
    <u>Transplutonium 1975</u> (Proceedings, 4th Intern. Transplutonium
    Element Symp., Baden-Baden, Germany, Sept. 13-17, 1975), p. 69-77,
    North-Holland/American Elsevier (1976).

8. J. A. Fahey, J. R. Peterson and R. D. Baybarz, "Some Properties
    of Berkelium Metal and the Apparent Trend Toward Divalent
    Character in the Transcurium Actinide Metals," Inorg. Nucl.
    Chem. Letters <u>8</u>, 101-107 (1972).

9. J. N. Stevenson, "The Microchemical Preparations of Metallic
    Curium-248 and Californium-249 by the Reduction of Their Tri-
    fluorides, and Related Structural Studies," Ph.D. Thesis,
    University of Tennessee, Knoxville, August 1973 (ORO-4447-004).

10. J. N. Stevenson and J. R. Peterson, "Some New Microchemical
    Techniques Used in the Preparation and Study of Transplutonium
    Elements and Compounds," Microchem. J. <u>20</u>, 213-220 (1975).

11. R. L. Sherman and O. L. Keller, "Modified Debye-Scherrer X-Ray
    Diffraction Camera for Radioactive Compounds," Rev. Sci. Instrum.
    <u>37</u>, 240 (1966).

12. D. K. Smith, "A Fortran Program for Calculating X-Ray Powder
    Diffraction Patterns," University of California Report No.
    UCRL-7196 (1963).

13. J. N. Stevenson and J. R. Peterson, "The Trigonal and Ortho-
    rhombic Crystal Structures of CfF_3 and Their Temperature Relation-
    ship," J. Inorg. Nucl. Chem. <u>35</u>, 3481-3486 (1973).

14. J. H. Burns, J. R. Peterson and J. N. Stevenson, "Crystallo-
    graphic Studies of Some Transuranic Trihalides: $^{239}PuCl_3$,
    $^{244}CmBr_3$, $^{249}BkBr_3$ and $^{249}CfBr_3$," J. Inorg. Nucl. Chem.
    <u>37</u>, 743-749 (1975).

15. J. P. Young, K. L. Vander Sluis, G. K. Werner, J. R. Peterson and
    M. Noé, "High Temperature Spectroscopic and X-Ray Diffraction
    Studies of Californium Tribromide: Proof of Thermal Reduction
    to Californium (II)," J. Inorg. Nucl. Chem. <u>37</u>, 2497-2501 (1975).

16. D. E. Williams, "LCR-2, A Fortran Lattice Constant Refinement
    Program," Iowa State University Report No. IS-1052 (1964).

17. J. R. Peterson and R. D. Baybarz, "The Stabilization of Divalent
    Californium in the Solid State: Californium Dibromide,"
    Inorg. Nucl. Chem. Letters <u>8</u>, 423-431 (1972) and the references
    therein.

18. D. K. Fujita, B. B. Cunningham and T. C. Parsons, "Crystal Structures and Lattice Parameters of Einsteinium Trichloride and Einsteinium Oxychloride," Inorg. Nucl. Chem. Letters 5, 307–313 (1969).

19. R. L. Fellows, J. R. Peterson, M. Noé, J. P. Young and R. G. Haire, "X-Ray Diffraction and Spectroscopic Studies of Crystalline Einsteinium (III) Bromide, $^{253}EsBr_3$," Inorg. Nucl. Chem. Letters 11, 737–742 (1975).

20. R. G. Haire, J. R. Peterson, J. P. Young and R. L. Fellows, Transuranium Research Laboratory, Oak Ridge National Laboratory, unpublished results (1976).

21. J. P. Young, R. G. Haire, R. L. Fellows, M. Noé and J. R. Peterson, "Spectroscopic and X-Ray Diffraction Studies of the Bromides of Californium-249 and Einsteinium-253," in W. Müller and R. Lindner, Editors, Transplutonium 1975 (Proceedings, 4th Intern. Transplutonium Element Symp., Baden-Baden, Germany, Sept. 13–17, 1975), p. 227–234, North-Holland/American Elsevier (1976).

22. J. R. Peterson, "Compounds of Divalent Lanthanides and Actinides," in C. J. Kevane and T. Moeller, Editors, Proceedings of the Tenth Rare Earth Research Conference, Vol. 1, p. 4–14, U.S. AEC Tech. Info. Center (1973).

23. R. D. Baybarz, "High-Temperature Phases, Crystal Structures and the Melting Points for Several of the Transplutonium Sesquioxides," J. Inorg. Nucl. Chem. 35, 4149–4158 (1973).

24. J. R. Peterson, Transuranium Research Laboratory, Oak Ridge National Laboratory, unpublished results (1976).

25. J. H. Burns, J. R. Peterson, and R. D. Baybarz, "Hexagonal and Orthorhombic Crystal Structures of Californium Trichloride," J. Inorg. Nucl. Chem. 35, 1171–1177 (1973).

26. W. H. Zachariasen, comment on paper listed here as reference 7, p. 76.

X-RAY DIFFRACTION EXAMINATION OF COAL COMBUSTION PRODUCTS RELATED TO BOILER TUBE FOULING AND SLAGGING

M. H. Mazza and J. S. Wilson

Energy Research and Development Administration

Morgantown Energy Research Center

Morgantown, West Virginia 26505

ABSTRACT

The Morgantown Energy Research Center of ERDA is studying fouling and slagging of pulverized coal fired boiler tubes resulting from mineral matter contained in the coal. The research is dependent on the development and application of reliable analytical techniques which can be applied to the coal and ash products to determine the mineral interactions responsible for fouling and slagging. X-ray powder diffraction is one of the techniques used to characterize the feed coals and their combustion products. A summary of the methods and techniques applied at MERC is the subject of this paper.

BACKGROUND

The specific fouling and slagging problems arising in coal combustion are dependent on the characteristics of the non-hydrocarbon fractions of the fuel. These coal impurities can occur in two forms. They are found in the carbon matrix of the fuel as organically bound sulfur, or as organometallic alkali, alkaline earth, or heavy metals. In addition to organic impurities, minerals are present which precipitated within the organic material during coal formation (inherent mineral matter) or may have been mined along with the coal (extraneous mineral matter). For purposes of this paper, coal minerals have been grouped into four categories: aluminosilicates, carbonates, sulfides, and silica, and are summarized in Table 1. Crushing and flotation can remove some of these minerals, but a portion is finely distributed throughout the carbonaceous matrix and remains to be released during coal conversion or combustion. Interest in these materials has increased dramatically since they are responsible for many of the detrimental effects associated with coal processes, among them catalyst poisoning, pollution, and fouling and corrosion of metal surfaces which come in contact with the ash.

The problems of ash fouling and fireside corrosion in coal-fired boilers have been investigated for years, but they continue to affect

Table 1

Characteristic Minerals in Coal		
Aluminosilicates	Kaolinite	$Al_2Si_2O_5(OH)_4$
	Illite	$K(Si_3 \cdot Al)Al_2O_{10}(OH)_4$
Carbonates	Siderite	$FeCO_3$
	Calcite	$CaCO_3$
	Dolomite	$CaMg(CO_3)_2$
Sulfides	Pyrite and Marcasite	FeS_2
Silica	Quartz	SiO_2
Accessory Minerals	Feldspars	$(K,Na)\ AlSi_3O_8$
		$CaAl_2Si_2O_8$
	Gypsum	$CaSO_4 \cdot 2H_2O$

design and operation of boilers (1)[1]. Design steam temperatures have been limited to about 1000°F (575°C) because of such problems, thus limiting plant efficiency gains which would be possible at higher temperatures. Existing plants must often be derated from design capacity to accommodate slagging and corrosion problems. Most research in the past has sought practical cures rather than attempting to obtain general understanding of the mechanisms involved. General agreement does exist that liquid phase reactions lead to the accelerated corrosion rates observed in boilers, with sulfur and alkali metals being important contributors to the problem. Complex alkali-iron-trisulfates ($K_3Fe(SO_4)_3$ and $Na_3Fe(SO_4)_3$) have been identified as the reaction products destroying metal surfaces. Unfortunately, correlations developed thus far are unsatisfactory for predictive purposes and expensive trials must still be carried out with new power plants and coal feed sources. "Solutions" have typically been compromises to maintain operation rather than real changes resulting from an understanding of the problem.

Previous studies have been made on coal mineral matter, and on the behavior of minerals when heated (2-12). Usually the latter studies were made by heating the minerals to prescribed temperatures in a crucible. Although providing insight into the chemical behavior of the mineral matter, these studies do not simulate the conditions encountered in pulverized coal combustion. Recently a report appeared on the behavior of ash under simulated combustion conditions (13). This study used a small experimental furnace, feeding coal at the rate of 0.1g/min in an air stream of 2 cc/min.

---

[1]Numbers in parenthesis refer to references listed at the end of the manuscript.

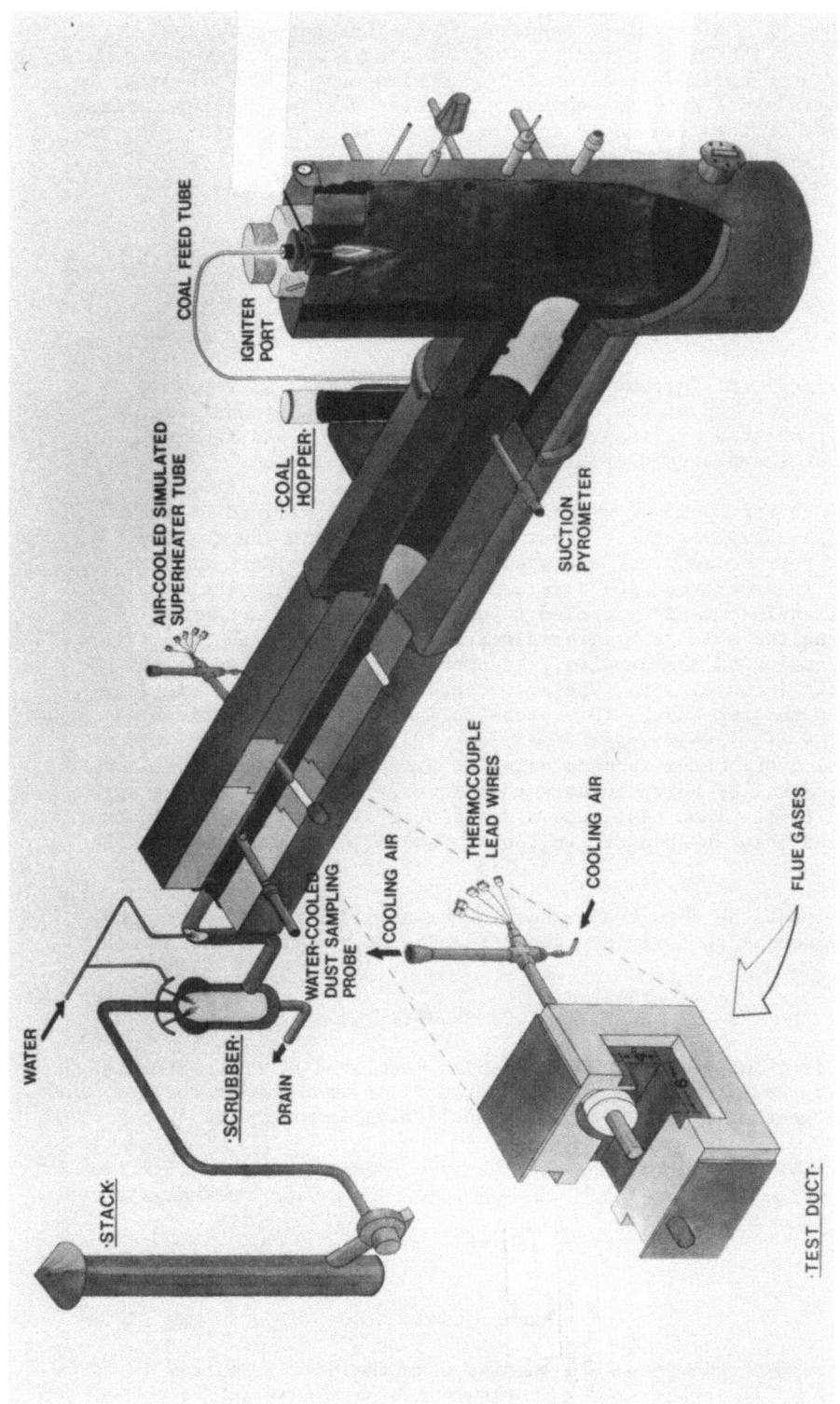

Figure 1. Coal-Fired Combustor.

In an attempt to establish more fundamental relationships about fireside fouling and slagging, research at the Morgantown Energy Research Center (MERC) is aimed at studying alkali sulfate formation and behavior in a pulverized-coal combustion environment (14). In order to examine mineral interactions, emphasis has been placed on x-ray diffraction analysis of the feed coals and their combustion products. The experimental program is carried out on a pulverized-coal test combustor, large enough to simulate conditions that one might expect in a coal-fired boiler yet small enough to be easily modified.

COMBUSTION TESTS

Figure 1 depicts the combustion furnace and sampling systems.

Pulverized coal (ground to 70% through a 200 mesh sieve, 74μ) is injected into the top of the combustor by a screw feeder where it is mixed with preheated air (approximately 500°F, 300°C) and ignited. Coal feed rates average 10 lb/hr.

The reactions occurring within the furnace are examined by two methods. (a) Sampling the combustion products within the furnace wherein a water-cooled probe equipped with a filter holder attachment is inserted in a port and the sample removed by suction. The sampled gases are therefore rapidly cooled to quench any occurring chemical reactions and the solid ash materials are then deposited on the filter. During collection of the samples, the probe nozzle is placed at the center of the furnace. (b) The second probe is designed to simulate a boiler superheater tube. This probe is inserted in the horizontal section for a predetermined exposure time in the ash-laden flue gas stream. The probe metal is maintained at a set temperature by an automatic controller which adjusts the flow of compressed cooling air through the tube. In a test run, a deposit is allowed to build up on the probe and then brushed off following removal of the probe from the combustor.

Analysis of the feed coals and mineral products collected in the furnace are needed to formulate the mechanics of the reactions.

ANALYTICAL PROCEDURES

A standard analytical scheme has been developed at MERC which is applied to x-ray diffraction analysis of samples from the combustion tests. The general sequence of analytical steps is as follows:

```
Prelim → LTA → Grind → Slurry → Dry → Inspect → Qual Scan → Quantitative
Grind           (-325)  & Pack           & Load    Overnight    Analysis
                    ↑_____|
                    If Necessary
```

Sample Preparation

Separation of the inorganic species from the organic matrix is accomplished by electronic low temperature ashing. ASTM ashing (800°C)

is unsatisfactory since some of the minerals in coal decompose at this
temperature or are oxidized along with the organic material. Commercial
low temperature ashers produce an oxygen plasma by passing commercial
grade oxygen through a radio frequency field (13.56MHz) at reduced pressure. The activated oxygen passes over the sample, removing the organic
material as oxides of carbon, nitrogen, etc. Temperatures can be maintained at 150°C or less by controlling the oxygen flow and RF power.
Gluskoter (7) has investigated the effects of electronic low temperature ashing on a variety of minerals found in coal. Calcite, pyrite,
kaolinite, and illite exhibit no discernible structural changes.
Gypsum does undergo partial or complete dehydration depending on the
temperature and pressure in the ashing chamber.

Prior to low temperature ashing, samples are ground for five minutes
in a Spex Mixermill[2] to increase the surface area exposed to the oxygen
plasma. Samples are stirred twice during the ashing period. Ashing is
complete when a constant weight is obtained. The resulting low temperature ashes will generally be light grey in color.

The operating parameters are summarized below.

| Low Temperature Ashing Analysis LFE Model 504 | | | | |
Sample	Quantity	Time Required	New Power Output	Pressure
Coal Bituminous	2 1-gram samples	30 hours maximum	110 watts	1 torr
Combustion Products	1 1-gram sample	15 hours maximum	110 watts	1 torr

Following low temperature ashing, samples are further ground to
pass 325 mesh (44 microns) and packed into sample holders as a slurry
with amyl acetate (15). The slurry is made by placing a sufficient
quantity of sample ($\approx$1 gram) on a watch glass and adding just enough
amyl acetate to compact the powder. Keeping the slurry as dry as possible minimizes cracking and sinking of the sample in the holder. Data are
collected after the samples have dried and their surfaces inspected to
insure they are uniform, smooth, and level with the sample holder.

Diffraction Analysis

Qualitative and quantitative intensity measurements are made on a
Philips APD-3500 automated powder diffractometer equipped with a copper
target x-ray tube and stepping motor goniometer (16). This unit employs a
theta compensating slit which functions as an automatic divergence slit
and soller slit. The area of the irradiated sample remains nearly constant within the scan range. A curved graphite monochromator is used to
diffract the copper radiation and direct it to a scintillation detector.

---

[2]Instruments and equipment brand names are mentioned for information
only and do not represent endorsement by the Federal Government.

Table 2

Instrument Settings for
Qualitative Analysis

Radiation: Copper target, $\lambda = 1.5418\text{Å}$
                            40 KV 20 MA

Scan: $4°2\theta$ to $72°2\theta$

  Stepping Increment $0.02°2\theta$
  Count Time   0.50 sec/increment

Detection: Scintillation Detector
                            Focusing graphite monochromator
                            0.2° receiving slit

The principles of pulse height selection are employed. All intensity
measurements are made with a 0.2° receiving slit.

Qualitative Analysis: The automatic capabilities of the diffractometer are utilized by collecting the data for qualitative analysis
overnight. Scans are made from $4°2\theta$ to $72°2\theta$ in stepping increments
of $0.02°2\theta$, and counting for 0.5 second per increment. With these
parameters, approximately ten scans can be made in 16 hours. Determination of peak height, and conversion from 2θ angle to d-spacing (Bragg's
Law, $n\lambda = 2d\sin\theta$) is done by the APD controller during data collection.
Automatic calibration on an α-quartz standard is made before and after
a series of samples. These and other operating parameters are summarized in Table 2. The d-spacings used in identification of the different
phases are summarized in Table 3.

Identification of the coal combustion products is generally straight
forward and unambiguous. They are identified by the following d-spacings:

Quartz: Quartz gives two strong peaks at 3.34Å(100) and 4.26Å(35). The
4.26Å peak is used for positive identification. Generally the entire
quartz pattern is well defined in the region scanned.

Mullite: Mullite can be identified by its 5.39Å(50) peak if present at
greater than two percent concentration. The 3.43Å(100) and 3.39Å(95
peaks appear as leading edges on the 3.34Å peak of quartz at low concentrations, but become distinct as the mullite concentration increases.

Hematite: Hematite is identified by the 2.69Å(100) and 2.51Å(50) peaks.
The mullite pattern also contains a peak at 2.69Å(60).

Magnetite: Magnetite has a peak at 2.53Å(100) and a broad, ill-defined
peak at 2.97Å(50). Interferences result from the 2.53Å peak of magnetite and the 2.51Å peak of hematite.

Anhydrite: Anhydrite is identified by its 3.49Å(100) peak and confirmed
by the 2.85Å(35) peak.

Lime: Generally only two peaks appear for lime, the 2.40Å(100) and the
2.78Å(40) peaks.

Table 3

Principal X-Ray Diffraction Spacings of Coal
Minerals and Coal Combustion Products

Coal Combustion Products

Mineral	Diffraction Spacing (Å)
Mullite	5.39(50), 3.43(95), 3.39(100)
Quartz	4.26(35), 3.34(100)
Anhydrite	3.49(100), 2.85(35)
Magnetite	2.97(70), 2.53(100), 2.10(70)
Hematite	2.69(100), 2.51(50), 1.69(60)
Lime	2.78(34), 2.40(100), 1.70(45)
Feldspars	3.18-3.24(100)

Coal Minerals

Mineral	Diffraction Spacing (Å)
Illite	10.1(100), 4.98(60), 3.32(100)
Gypsum	7.56(100), 4.27(50), 3.06(55)
Kaolinite	7.15(100), 3.57(80), 2.38(25)
Quartz	4.26(35), 3.34(100), 1.82(17)
Feldspars	3.18-3.24(100)
Pyrite	3.13(35), 2.71(85), 2.42(65), 2.21(50)
Calcite	3.04(100), 2.29(18), 2.10(18)

Relative intensities are shown in parenthesis.

Feldspars: Feldspars show peaks between 3.18 and 3.24Å.

Coal minerals are identified by the following d-spacings.

Illite: Illite is usually identified by broad peaks at 10Å(100), and at 5Å(40).

Kaolinite: Kaolinite is most reliably identified by its peaks at 3.57Å and 7.15Å.

Sulfides: Both polymorphs of FeS_2, pyrite and marcasite, show strong reflections at 2.71Å(100) and are distinguished by their lower order reflections.

Carbonates: The carbonate minerals, calcite, dolomite, and siderite are identified by their (100) reflections at 3.04, 2.88, and 2.97Å respectively.

Gypsum: Gypsum is identified by its peak at 7.56Å(100).

Quantitative Analysis. After the qualitative scans have been made and all phases identified, the samples are analyzed quantitatively (17). Concentration of the different phases is determined through the use of external calibration curves developed from standards prepared in our laboratory. These curves relate the net integrated intensity of a suitable diffraction peak to the phase concentration, and can be expressed as $C = I(K_1 + K_2 I) + B$, where C = concentration of phase and I = net integrated intensity of the peak. The values of K_1 (slope), K_2 (correction), and B (background) are determined from the standard curves, either graphically (Figure 2) or with regression analysis.

Coal combustion products typically contain amorphous material in addition to the crystalline phases. For this reason standards are prepared by mixing appropriate pure materials and then diluting them with ground (-325 mesh) glass. This glass, obtained from a local glass factory, is scanned to verify that it is completely amorphous. In addition to providing an amorphous matrix for the standards, this method also controls absorption effects by introducing the same elements into the matrix of each standard. After the standards have been prepared, they are qualitatively scanned to confirm they are similar in appearance to the samples. Intensity data are then collected for each phase in the standards, and the coefficients K_1, K_2, and B are determined. The results are checked through a back calculation on the standards. When satisfactory results are obtained, this information is then summarized, and filed with the qualitative scans and intensity measurements for future reference. An example of a representative standard is shown in Figure 3. Additional standards are made by replacing magnetite with either anhydrite or lime.

Measurement of the net integrated intensity is controlled by the APD controller according to specified parameters. These parameters include the upper and lower 2θ limits of the diffraction peak, and stepping increment between the two angular limits, and the count time per increment. The controller also calculates the phase concentration of unknown samples, given the values of K_1, K_2, and B for the corresponding phase.

The standard curves have been calibrated from zero to twenty-five weight percent. The values of K_1, K_2, and B have been obtained through linear regression analysis, and have yielded linear correlation coefficients above 0.97 for all six phases. A comparison of the known and calculated values of a synthetic mixture series is shown in Table 4. This mixture was prepared following the procedures outlined above but diluted to different concentration values. The average percentage difference for all six phases is below 5%. The accuracy of the calculated values is dependent on the errors normally associated with x-ray diffraction analysis, homogeneity, preferred orientation, matrix effects, and counting statistics. No corrections have been made for these effects in this study. However, there is acceptable agreement between the known and calculated values to apply these techniques to the analysis of combustion products.

RESULTS

Some test results obtained by these techniques will be briefly discussed as applications of the analytical procedures.

Two coals have been recently compared in combustion tests: Pittsburgh and Waynesburg seam high volatile bituminous coals. The studies have emphasized the interactions of the minerals containing the alkali (sodium and potassium) and alkaline earth (calcium, magnesium) elements, and sulfur. Results from the Pittsburgh coal tests have been useful in determining an overall view of the possible interactions of

Estimate Value of "B" from the Curve

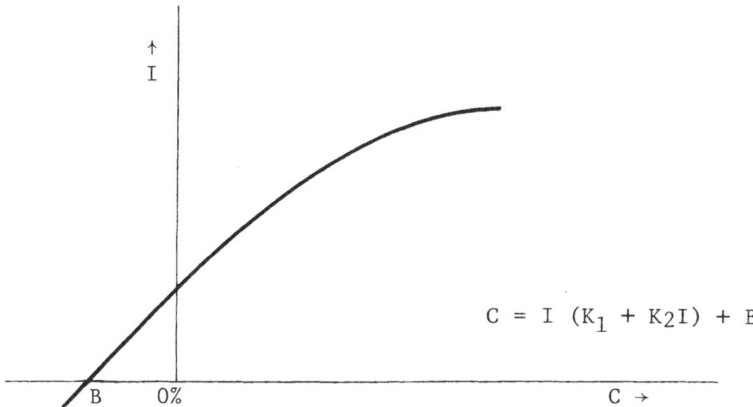

$$C = I (K_1 + K_2I) + B$$

Derive Values of K_1 and K_2 from the Curve

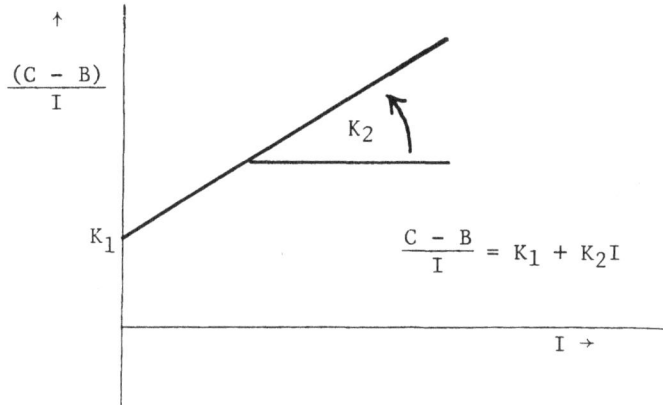

$$\frac{C - B}{I} = K_1 + K_2I$$

Figure 2.   Determination of Constants Used in the
Concentration Algorithm

Standard	Composition	Weight %
D-1	100% Std D 0% glass	25.00 mullite 25.00 quartz 25.00 magnitite 25.00 hematite
D-2	64% Std D 36% glass	16.00 mullite 16.00 quartz 16.00 magnitite 16.00 hematite
D-3	50% Std D 50% glass	12.50 mullite 12.50 quartz 12.50 magnitite 12.50 hematite
D-4	32% Std D 68% glass	8.00 mullite 8.00 quartz 8.00 magnitite 8.00 hematite
D-5	16% Std D 84% glass	4.00 mullite 4.00 quartz 4.00 magnitite 4.00 hematite
D-6	0% Std D 100% glass	0.00 mullite 0.00 quartz 0.00 magnitite 0.00 hematite

	d(Å)	2θ(Cu)	$(2\theta_1-2\theta_2)$ (approx.)	Slope	Bkg.	Correction
mullite	5.39(50)	16.45	16.15-16.75	2.99	-1.60	0
quartz	4.26(35)	20.85	20.50-21.10	1.30	-0.14	0
magnetite	2.97(70)	30.09	29.80-30.60	1.66	-1.59	0
hematite	3.65(25)	24.40	24.00-24.80	0.30	-4.80	0

Figure 3. X-Ray Diffraction Standards

Table 4

Phase Determination[a] in X-Ray Powder Diffraction Standards

Standard	Mullite (5.39Å)[b]	Quartz (4.26Å)	Hematite (3.65Å)	Magnetite (2.97Å)	Anhydrite (3.49Å)	Lime (2.40Å)
Weight%						
7.50	7.14(4.80)[c]	7.17(4.51)	7.63(1.73)	7.78(3.73)	7.77(4.58)	7.82(4.30)
12.50	11.78(5.76)	12.79(2.30)	12.15(2.80)	11.83(5.36)	12.25(2.00)	13.02(4.80)
17.50	18.22(4.1)	17.10(2.86)	18.56(6.06)	17.13(2.11)	17.90(2.28)	18.12(3.54)
22.50	23.20(3.11)	22.14(1.60)	21.30(5.33)	21.43(4.76)	22.10(1.78)	21.20(5.77)
Avg. % Error	4.44	2.82	4.68	3.99	2.66	4.60

[a] expressed as weight percent
[b] peak used for intensity measurements
[c] numbers in parenthesis are percentage errors

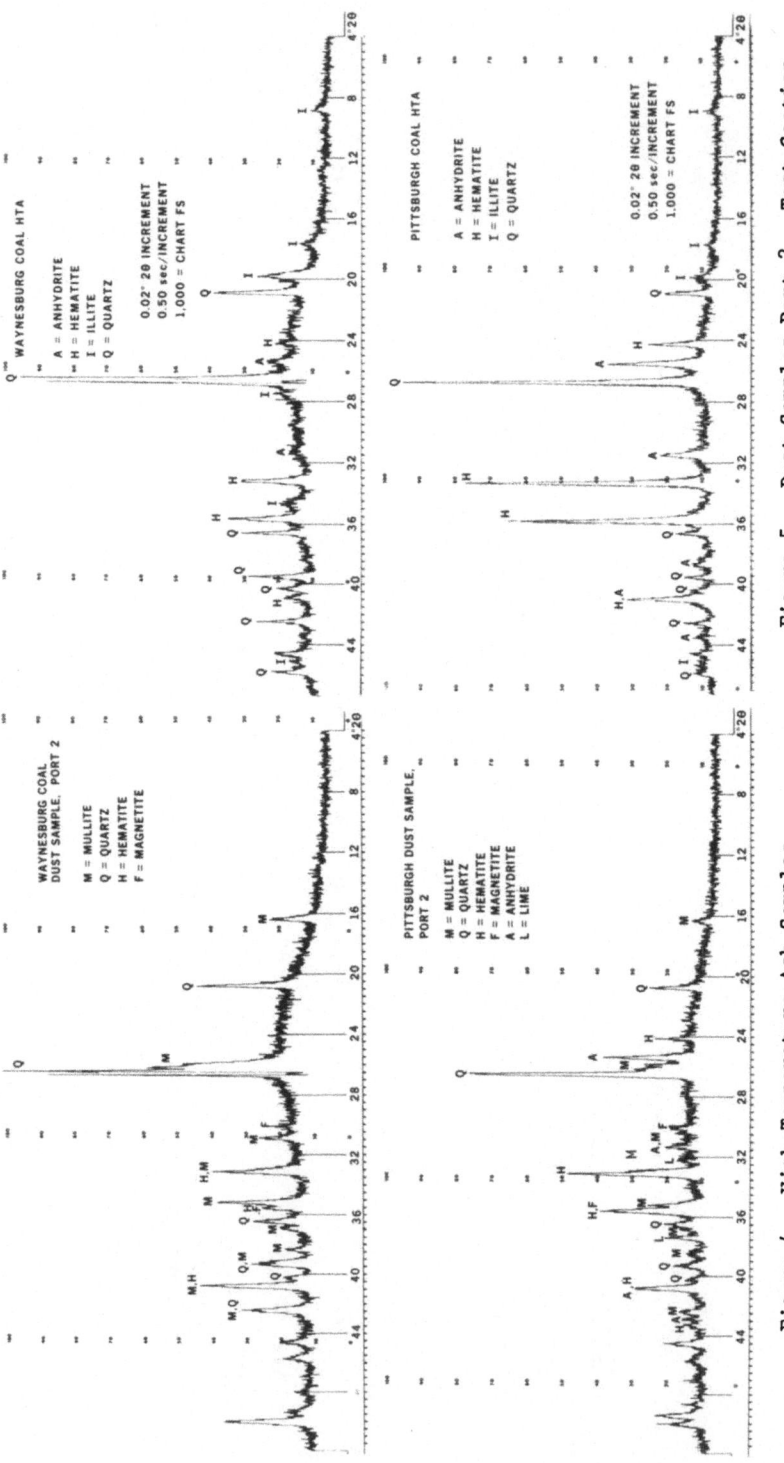

Figure 5.   Dust Samples Port 2, Test Section

Figure 4.   High Temperature Ash Samples

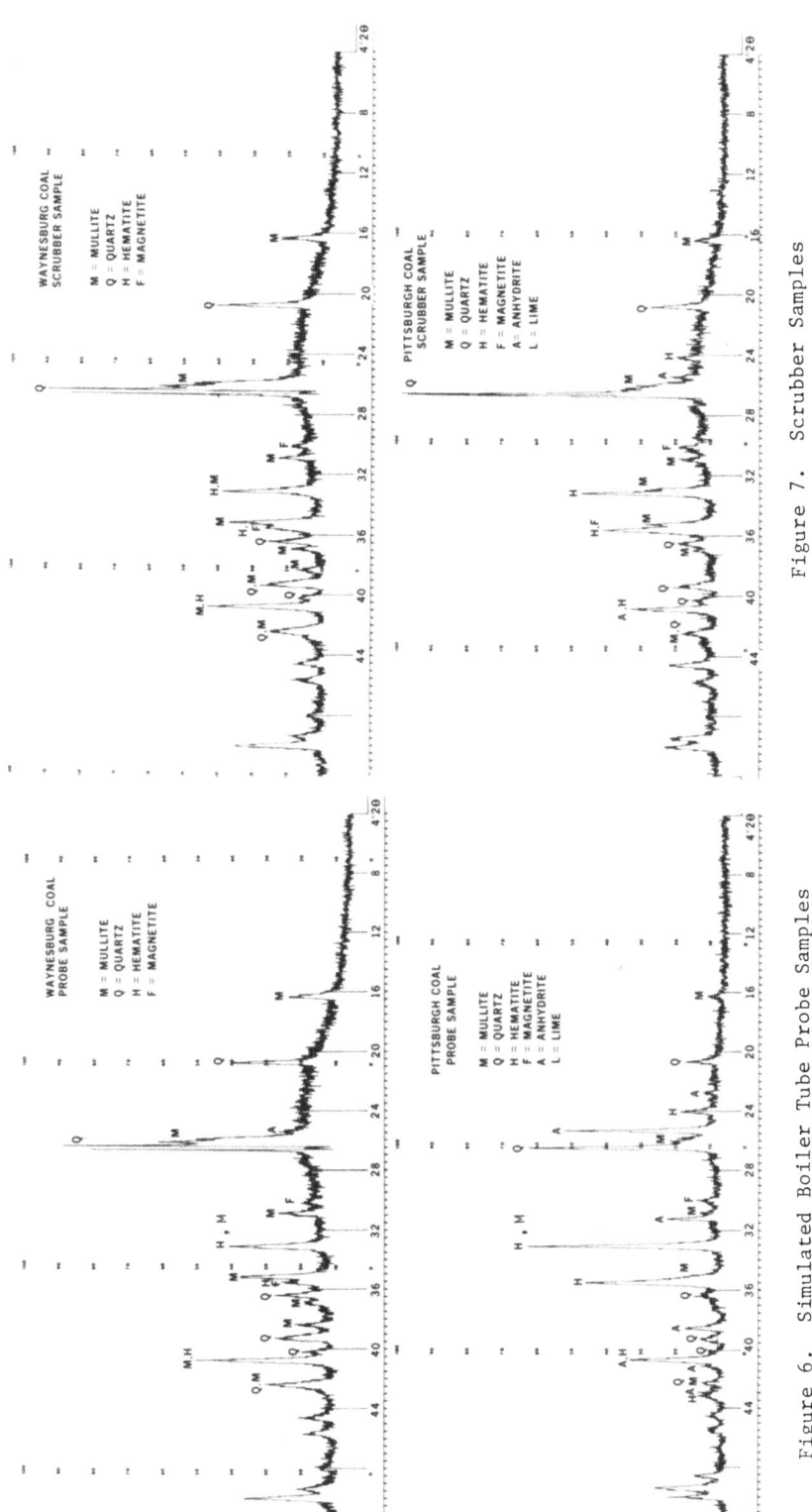

Figure 7. Scrubber Samples

Figure 6. Simulated Boiler Tube Probe Samples

the major minerals, and the location of the products of the combustor.
Combustion of the Waynesburg coal results in a simpler matrix in which
additives can be assessed.

The low temperature ashes of the two coals were scanned to examine
the unaltered minerals. Both coals contain four major minerals, kaolinite, illite, quartz, and pyrite. Calcite is present in the Pittsburgh
coal but was not detected in the Waynesburg coal. Comparison of the two
powder patterns shows that the Pittsburgh coal contains a higher concentration of pyrite (and of course calcite), while the Waynesburg coal contains a higher concentration of the refractory materials, kaolinite,
quartz, and illite.

Samples of the two coals then were ashed according to ASTM procedures (18). These ashes were scanned to examine the effect of heating
the minerals collectively under controlled conditions. The powder
patterns for the two ashes are shown in Figure 4. A comparison of the
two coals is summarized below.

High Temperature Ash Summary

Minerals Sources	HTA Species	Relative Concentrations
Kaolinite $SiO_2 \cdot Al_2O_3(OH)_4$	Metakaolinite (amorphous)	Waynesburg>Pittsburgh
Illite	Illite, glass	Waynesburg>Pittsburgh
Quartz $(\alpha-SiO_2)$	Quartz $(\alpha-SiO_2)$	Waynesburg>Pittsburgh
Pyrite (FeS_2)	Hematite (Fe_2O_3)	Pittsburgh>Waynesburg
Calcite $(CaCO_3)$	Lime (CaO) +SO_3→Anhydrite $(CaSO_4)$	Pittsburgh>Waynesburg (none det.) Pittsburgh>Waynesburg (trace det.)

The low calcium content of the Waynesburg coal is reflected in the
HTA products. Little or none of the calcium-containing species lime and
anhydrite was detected in this pattern. In addition, the Waynesburg coal
diffraction pattern has a high background from the formation of the
metakaolinite.

Combustion Analysis

Samples for both coals were selected from the dust from port 2 in
the test section, the simulated boiler tube probe, and the scrubber.
Powder patterns of the different samples are shown in figures 5, 6, and
7, respectively. The species identified are listed below, and the quantitative results of the analysis are summarized in Table 5.

Combustion Species

Mineral Source	
Kaolinite $SiO_2 \cdot Al_2O_3(OH)_4$	Mullite $(2SiO_2 \cdot 3Al_2O_3)$
Illite	Glass
Quartz $(\alpha-SiO_2)$	Quartz $(\alpha-SiO_2)$
Pyrite (FeS_2)	Hematite, Magnetite (Fe_2O_3) (Fe_3O_4)
Calcite $(CaCO_3)$	Lime (CaO) $+SO_3 \rightarrow$ Anhydrite $(CaSO_4)$

Calcium feldspars are detected in bottom ash sample.

The six major phases detected in the Pittsburgh coal are mullite, quartz, hematite, magnetite, anhydrite, and lime. The diffraction data from this coal have revealed an inverse relationship between the two calcium phases, lime and anhydrite. Lime is predominant in the dust, but decreases in concentration with increasing distance from the combustion zone. A corresponding increase in anhydrite concentration is also seen, with the major change occurring at the first port in the test section. However, the probe samples show anhydrite as the major calcium containing phase, with little or no lime. Neither lime nor anhydrite appear in any appreciable amount in the scrubber sample.

The Pittsburgh coal data also show that the refractory materials, mullite and quartz, preferentially remain in the dust and are carried through the test section to the scrubber. The iron tends to increase in

Table 5

	Mullite	Quartz	Hematite	Magnetite	Anhydrite	Lime
Waynesburg Coal						
Dust, Port 2 Test Section	28	27	present	present	N.D.	N.D.
Probe (1200°F)	25	25	present	1	present	N.D.
Scrubber	26	32	present	1	N.D.	N.D.
Pittsburgh Coal						
Dust, Port 2 Test Section	10	13	present	6	6	3
Probe (1200°F)	5	8	9	7	9	1
Scrubber	8	14	8	6	N.D.	N.D.

N.D. = Not Detected

oxidation state with increasing distance from the combustion zone. Both magnetite and hematite concentrate on the probe. The feldspars formed predominate in the ash sample removed from the bottom of the vertical section at the end of a run.

In contrast, the Waynesburg coal forms only three major phases, mullite, quartz, and iron oxide. The same trends prevail: mullite and quartz are more concentrated in the dust and are carried to the scrubber. Anhydrite is present in small quantity and is only detected on the probe. The behavior of iron is not clear due to interferences from mullite.

To date a distinct alkali phase has not been detected from feeding these two coals. One test run was made with Pittsburgh coal plus 7% sodium acetate to encourage the formation of sodium sulfate. The phase detected has been identified as sodium sulfate (form III).

Concluding Remarks

X-ray powder diffraction has proven to be a valuable technique to examine the mineral composition of the coal combustion products. Qualatative analysis of the unaltered minerals and their combustion products can be used to examine the thermal behavior of the minerals. Although the techniques outlined in the this paper suffer from limited accuracy (4-10%), they are still sensitive enough to obtain a quantitative distribution of the products in the combustor. It is believed that the accuracy of the analysis, particularly in the lower concentration ranges, can be improved by applying absorption corrections to the standards.

ACKNOWLEDGEMENTS

The author wishes to express her appreciation to Michael W. Paris for performing the analyses, and to Jeanne M. Turner for her patience and skill in typing the manuscript. She would also like to thank John J. Kovach for his helpful comments on the manuscript.

REFERENCES

1. W. T. Reid, External Corrosion and Deposits--Boilers and Gas Turbines, 199 pp., American Elsevier Publishing Co., New York (1971).

2. W. A. Selvig and F. H. Gibson, "Analysis of Ash from United States Coals," Bulletin 567, U. S. Bureau of Mines, 33 pp. (1956).

3. F. Rost and P. Ney, "Study Concerning Boiler Cinders," Brennstoff Chemie, 37, 201-210 (1956).

4. M. R. Brown and D. J. Swaine, "Inorganic Constituents of Australian Coals," J. Inst. Fuel, XXXVII, 422-440 (1964).

5. M. Kemezys and G. H. Taylor, "Occurrence and Distribution of Minerals in Some Australian Coals," J. Inst. Fuel, XXXVII, 389-396 (1964).

6. N. A. Sinarskii, et al., "X-ray Analysis of Fuel Ash," Teploenergetika, 12, 65-70 (1964).

7. H. J. Gluskoter, "Electronic Low Temperature Ashing of Bituminous Coal," Fuel (London), 44, 285-291 (1965).

8. R. F. Littlejohn, "Mineral Matter and Ash Distribution in 'As-Fired' Samples of Pulverized Fuels," J. Inst. Fuel, XXXIX, 59-67 (1966).

9. H. J. Gluskoter, "Clay Minerals in Illinois Coals," J. Sed. Pet., 37, 205-214 (1967).

10. J. V. O'Gorman and P. L. Walker, "Mineral Matter Characteristics of Some American Coals," Fuel (London), 50, 136-151 (1971).

11. J. V. O'Gorman and P. L. Walker, "Thermal Behavior of Mineral Fractions Separated from Selected American Coals," Fuel (London), 52, 71-91 (1973).

12. H. J. Gluskoter and R. S. Mitchell, "Mineralogy of Ash of Some American Coals: Variations with Temperatures and Source," Fuels (London) 55, 2, (1976).

13. A. S. Padia, et al., "The Behavior of Ash in Pulberized Coal under Simulated Combustion Conditions," presented at Combustion Institute's Central States Meeting April 5-6, 1976.

14. J. S. Wilson, "Radioisotope Tracer Analysis of Gas Phase Reaction of Alkali Metals with SO_2 in a Combustion Environment," Unpublished PhD Dissertation, West Virginia University, 1975.

15. H. P. Klug and L. E. Alexander, X-Ray Diffraction Procedures, John Wiley and Sons, Inc., New York, 1962, 716 pp.

16. R. Jenkins, et al., "A New Concept in Automated X-Ray Powder Diffractometry," Norelco Reporter, 18, 1-16 (1971).

17. R. Jenkins, "Quantitative Analysis with the Automatic Powder Diffractometer," Norelco Reporter, 22, 8-12 (1975).

18. 1970 Annual Book of ASTM Standards, Part 19, "Gaseous Fuels; Coal and Coke," p. 19-20 (1970).

COMPUTER IDENTIFICATION TECHNIQUES FOR CRYSTALLINE COMPOUNDS
USING THE JCPDS POWDER DIFFRACTION FILE AS A DATA REFERENCE

George VanTrump, Jr., and Phoebe L. Hauff

U.S. Geological Survey

Box 25046, Denver Federal Center

Denver, Colorado 80225

ABSTRACT

The mineralogy laboratory of the U.S. Geological Survey in Denver has developed a series of time-sharing oriented computer programs which aid in the identification of crystalline compounds from chemical and X-ray diffraction data. These programs operate on a data base compiled primarily from the Powder Diffraction File of the Joint Committee on Powder Diffraction Standards (JCPDS). Diagrammatic X-ray diffraction patterns and various search tables are products of these programs. Additional programs can retrieve information from the data base by chemical formula components or Powder Diffraction File number, and can search and match reflections of an unknown against reference patterns.

INTRODUCTION

Identification of crystalline compounds from X-ray diffraction data can be a tedious process when dealing with multiple phases or a complete unknown. Since the procedures involved in the initial identification stages are essentially number matching, this can be done more thoroughly and rapidly by use of the computer. With this as a premise, a series of time-sharing oriented computer programs has been written.

The examples used in this report are mineralogical because the author's primary concern is mineral identification. However, these programs and their products may be applied to chemical and X-ray diffraction data of any crystalline material.

At this time the compound identification scheme, includes the main data base (MINBAS); its subsidiary file (XRDFIL); the three major programs, XRDPLT, XRDSCH, and 2THETA; and their nonterminal products, the sieve and search tables.

This system has been designed to take full advantage of the interactive capabilities of a time-sharing computer. The programs can be run by the most casual user of a computer. Each program prompts and guides the user with a series of questions for data entry and manipulation. The output or answers are returned to the user at his terminal. These programs are written in standard FORTRAN and are as machine independent as possible. Currently, they are running on the U.S. Geological Survey DEC System 10 computer, located in Denver. Conversion to other computing systems should not be difficult.

THE MAIN DATA BASE

The foundation of the system is the main data base MINBAS, which is described in detail by VanTrump and Hauff(1). It is a composite of various data sources, the primary one being the file of the Joint Committee on Powder Diffraction Standards (JCPDS) (2). This has been supplemented with data from Geological Society of America Memoirs (3,4), data from the U.S. Geological Survey and other scientists, and general literature references. Because the JCPDS considers parts of its data proprietary information, the U.S. Geological Survey is not allowed to distribute MINBAS directly. However, once the potential user has acquired the JCPDS data tape, the authors can assist in the creation of a MINBAS-type file. Similar data bases can be constructed from either the entire JCPDS Inorganic or Organic files or from parts of them.

Contents of MINBAS

MINBAS contains: mineral name; chemical formula (or at least elemental constituents of the compound); a mineral group code, which is explained in VanTrump and Hauff (1), p. 6; the six digit PDF (Powder Diffraction File) Number or IDNO (identification number for non-JCPDS data); crystallographic parameters consisting of unit-cell data, crystal system and hkl values when available; some optical data; powder diffraction data stored as "d" values, intensities, and target material wavelengths if source data was given as 2-theta; references for all data sources; and a code for data reliability and source. All analytical data must have a valid reference before it is accepted for entry into the main data base.

At present, there are approximately 3,000 mineral records stored in MINBAS. Most of them contain data in all the categories listed. The minimum information required for an entry to be useful are name, formula, group, IDNO, powder diffraction data, and some type of data source reference.

MINBAS is dynamic in that data can be added to it at any time. In the near future, the Crystal Data File (5), distributed by the National Bureau of Standards and based on the Crystal Data Determinative Tables of Donnay and Ondik (6) may be added to MINBAS to further supplement the cystallographic information.

THE Program 2THETA and Its Products

The program 2THETA, operating directly on MINBAS, produces, on the line printer, a series of search tables expressed in degrees 2-theta for any type of target material. (Some of these tables have been loosely patterned after the format used by the JCPDS for their "Search" tables.) Figure 1 is a page from the search tables based on the Hanawalt method (7). In this method the entries between 0 and 180 degrees 2-theta have been divided into groups, each of which has a specific range of 2-theta values. The first column of 2-theta data gives the group values, and the range of those values is listed at the top of the page and between groups. Within each group, the minerals are ranked in ascending numerical order by the second-column 2-theta values. Each entry appears three times in this table, sorted into different groups by its three strongest lines. Eight additional lines of each pattern are also included. The intensities appear after the 2-theta values, set off by a hyphen. The ranges of 2-theta values overlap about 25 percent. For further verification of the identification, the individual table which has been created for each entry in the data file may be consulted. This table contains the powder diffraction data expressed in "d", 2-theta copper values, and 2-theta iron values with intensities and hkl's when available.

The search tables also include an alphabetical list by mineral name and a listing sorted by the strongest line of each reference pattern. Both of these tables include the five strongest lines of each reference pattern, and a version of the chemical formula which may not always be the complete stoichiometric expression. The examples in this report have been generated using Cu $K\alpha_1$ radiation. The program can produce tables for any standard wavelength.

The published manual on the data file (1) also contains indexes to the file. These indexes are sorted alphabetically, by identification number, and also by mineral groups.

THE SUBFILE XRDFIL

MINBAS is the nucleus of the entire system. Its data potential has only been superficially utilized with the limited number of programs now operational. MINBAS occupies a fairly large area in the U.S. Geological Survey's DEC-10 computer, necessitating its storage on an off-line private disk pack. It was not designed to accommodate any specific program, but rather has been built with a flexible format so that more data categories can be added in the future. It would be too time consuming for a smaller computer to search a data base as large as MINBAS when only parts of the information it contains are needed to solve a particular problem. Consequently, XRDFIL, a smaller, more efficient file, with a selected data content, has been created for the programs XRDPLT and XRDSCH. This was accomplished by stripping the nonessential items from MINBAS, rearranging the data for the discreet applications of these programs, and using more efficient data storage compression techniques. XRDFIL contains only the mineral name, formula, identification number, mineral group code, "D" spacing and intensities. Its size is an order of magnitude less than that of MINRAS.

The table below is printed rotated 90° on the page. Column group headers: **2-THETA VALUES** (with range sub-labels **0.00 – 8.84** and **8.54 – 10.36**) and **CU**. Each entry is given as a 2-theta value followed by a relative intensity (X = 10).

IDNO	MINERAL NAME	2-THETA VALUES									CU
130413	CHLOROTILE	7.30-X	36.49-9	19.80-8	24.92-8	30.27-8	27.33-5	33.15-5	34.88-5	62.96-4	21.24-3
130414	MIXITE	7.36-X	36.49-9	24.92-8	30.27-7	33.15-6	34.88-6	19.85-5	21.14-5	21.14-5	27.25-5
150781	GLUCINE	8.18-X	37.28-X	67.30-X	46.58-8	60.50-8	77.62-7	79.94-7	33.53-6	28.40-5	42.19-5
080243	SAUCONITE (AIR DRIED)	5.73-X	59.43-8	11.19-6	19.49-6	33.53-6	28.22-5	70.54-4	22.55-1	31.59-1	73.13-1
130305	GRIFFITHITE	5.73-X	59.98-7	11.19-5	19.28-5	28.49-5	33.82-5	35.02-5	70.78-3	16.78-1	22.61-1
060002	SAPONITE	4.70-X	60.19-7	34.33-6	9.71-5	19.49-5	24.64-5	29.65-4	52.68-4	71.34-4	39.85-2
130086	SAPONITE	6.22-X	60.45-9	24.23-8	19.41-5	17.87-4	71.40-4	49.49-3	34.74-2	37.12-2	39.13-2
160613	VERMICULITE	6.22-X	60.54-7	19.41-6	34.26-5	34.88-5	35.52-5	37.77-4	38.01-4	31.36-3	24.99-3
110056	SAPONITE	5.32-X	60.89-9	19.67-8	34.74-5	72.03-5	24.03-4	24.03-4	122.16-4	53.21-3	102.16-3
140331	CACOXENITE	9.71-1	3.82-X	7.42-X	28.04-1	12.80-1	6.35-1	9.11-6	11.05-1	18.09-1	26.75-1
090469	OKENITE	10.04-8	4.20-X	24.99-8	30.48-8	29.06-6	29.26-6	50.67-6	29.96-5	32.29-5	11.95-4
190229	TRUSCOTTITE	9.32-X	4.65-X	28.40-X	31.47-X	33.33-8	18.78-4	23.14-4	23.64-4	25.43-4	49.55-4
210289	STRASHIMIRITE	9.85-9	4.72-X	31.25-0	28.49-9	9.34-8	18.51-8	21.08-8	26.59-8	36.25-7	33.80-6
200659	COPIAPITE-AL	9.61-X	4.88-8	15.87-8	14.34-7	24.85-5	25.45-5	16.65-3	18.95-3	21.14-3	22.15-3
110146	LISKEARDITE	10.22-8	5.02-X	26.75-X	11.16-7	7.24-6	11.88-6	36.93-6	20.88-4	22.49-4	27.68-4
120228	FAUJASITE	10.10-8	6.18-X	15.51-X	11.77-8	20.21-8	23.46-8	31.17-6	18.55-4	22.60-4	26.82-4
110191	SHERWOODITE	8.84-X	7.18-X	9.50-8	34.33-7	43.00-7	19.07-5	11.33-4	12.46-4	17.04-4	27.42-4
110191	SHERWOODITE	9.50-X	7.18-X	8.84-X	34.74-5	43.04-7	19.07-5	11.33-4	12.46-4	17.04-4	27.42-4
210150	TERUGGITE	8.85-2	7.30-X	32.11-3	10.56-2	19.07-2	24.85-2	19.41-2	23.08-2	27.42-2	16.97-2
200679	SLAVIKITE	9.78-X	7.55-8	15.18-8	16.37-8	21.08-8	25.65-8	30.30-8	33.14-8	8.75-6	19.49-6
170158	AMARANTITE	10.17-X	8.26-X	24.92-8	29.26-8	21.68-6	17.80-4	17.80-4	26.11-4	34.17-4	13.61-3
140600	WIGHTMANITE	9.74-X	8.26-X	29.45-3	33.55-3	18.70-2	36.71-2	38.27-2	42.61-2	53.17-2	16.16-1
180032	TUCANITE	10.21-X	9.01-X	20.49-8	18.05-6	23.77-6	15.73-4	22.32-4	36.34-4	36.65-4	7.89-2
200703	SYNADELPHITE	9.47-7	10.09-X	33.97-X	16.81-6	22.61-5	19.03-4	19.45-4	31.07-4	57.59-4	59.13-4
180032	TUCANITE	9.01-8	10.21-X	20.49-8	18.05-6	23.77-6	15.73-4	15.73-4	36.34-4	36.65-4	7.89-2
130191	FERRIMOLYBDITE	8.95-X	10.52-6	13.14-2	28.96-2	25.43-1	17.17-0	17.86-0	11.48-0	14.98-0	15.18-0
150289	FERRIMOLYBDENITE	8.84-X	10.65-X	13.10-0	26.42-8	27.68-7	29.16-6	33.90-5	25.29-5	32.05-5	38.44-5
080142	CHALCOALUMITE	9.91-X	10.66-X	20.93-X	20.35-9	45.07-9	35.45-8	38.96-8	28.96-7	52.78-7	18.05-6
200742	METAVOLTINE	9.75-X	10.70-9	27.16-8	29.06-7	23.77-5	31.03-5	11.66-3	32.17-3	14.32-3	39.31-3
120528	GOWERITE	9.61-3	10.78-X	27.95-5	32.78-3	21.71-2	22.55-1	23.14-1	26.67-1	30.17-1	31.70-1
130419	KIVUITE	8.58-X	11.11-9	28.96-8	31.14-8	15.05-6	20.03-6	22.55-6	23.02-6	17.10-4	26.03-4
170155	HOHMANNITE	10.17-8	11.16-X	8.49-6	25.75-6	16.68-4	28.59-4	22.49-3	27.33-2	27.77-2	21.24-1
190571	HOWIEITE	9.63-X	11.18-8	27.42-7	34.19-6	33.41-5	32.17-4	29.16-4	40.83-4	24.50-2	40.49-2
080055	ALUMINITE	9.82-X	11.33-X	23.90-X	13.98-8	18.86-8	16.40-7	17.87-7	21.14-7	26.11-7	29.16-6
181225	HUEMULITE	8.66-5	11.62-X	8.33-9	9.70-5	10.78-4	31.00-4	29.16-3	23.44-2	24.85-2	31.44-2
080443	URANOPILITE	9.63-8	12.42-X	20.74-8	24.37-5	16.07-4	26.91-4	29.86-3	30.81-3	41.18-3	49.90-3
120462	WEEKSITE	9.84-8	12.44-X	15.90-9	25.06-7	27.00-7	30.70-6	27.86-5	36.93-5	39.490-5	19.36-4
140685	BISBEEITE	8.66-X	12.71-5	26.83-5	18.20-4	23.15-4	21.87-2	22.61-2	30.17-2	38.61-2	28.04-1
110296	LIEBIGITE	10.18-9	12.99-X	16.40-9	19.49-6	28.77-6	26.75-5	26.91-5	45.35-5	53.34-5	41.99-4
080172	SYMPLESITE	9.85-2	13.04-X	11.79-2	17.63-1	21.82-1	22.17-1	23.75-1	26.18-1	28.06-1	30.06-1
170159	KORNELITE	8.84-X	13.32-8	18.99-8	12.56-6	20.64-6	28.40-6	20.21-4	25.43-4	17.69-3	21.71-3
170171	SIGLOITE	9.12-X	13.70-9	18.24-9	27.59-7	21.70-6	35.10-5	35.01-5	22.72-4	28.87-4	18.51-3
140147	PARAVAUXITE	9.00-X	13.87-9	21.14-9	28.04-8	31.36-6	34.74-5	18.05-4	28.96-4	22.72-3	33.15-2
150529	UNNAMED MINERAL	10.27-X	14.56-8	20.64-7	16.28-6	26.42-5	28.87-5	30.70-4	30.70-3	36.96-3	35.45-3

Figure 1.   Example of Hanawalt type 2-theta search table.

THE PROGRAM XRDPLT

The Plot Option of the Program XRDPLT

The program XRDPLT, Mineral X-ray Diffraction Data Retrieval/Plot Computer Program, (8) was initially created to plot diagrammatic X-ray diffraction patterns (see Figure 2). It is capable of varying the horizontal scale, which is expressed in degrees 2-theta, and the vertical scale, which contains the intensities; both scales are expressed to the computer in units per inch. The illustration used in this paper was plotted to the scale of the manuscript.

XRDPLT is also able to plot powder film patterns. There is an optional angstrom scale which can be included at the top and bottom of each plot. These plots are termed "sieves". When they are produced on mylar and used with a light table, they can be overlain on the chart or film of the unknown, and the reference patterns which do not match the line of the unknown can be eliminated or "sieved out" until a possible match is obtained. A series of sieves for the more commonly encountered mineral groups has been made up for use in the X-ray laboratories of the U.S. Geological Survey. These have proven to be invaluable time savers, especially when the user has an idea of the approximate identity or mineral group of the unknown. These plots can also illustrate structural similarities within mineral groups, and they can be made for specific projects where the mineral suite is known.

XRDPLT can also work from a special purpose data file rather than the main file. Substitution in the alunite/natroalunite group has been studied by Charles Cunningham et al. (9). By plotting the X-ray data, shifts in peak positions and intensities as a function of potassium content become very apparent. This plot can be seen in the above publication (9). These examples represent only a few of the applications this program could have. Others are given in the XRDPLT operations manual by Hauff and VanTrump (10).

The plotting instructions for XRDPLT are relatively simple. Once the program starts into execution, the user simply answers questions posed to him through the terminal. The actual plotting is done by a Gerber flat bed plotter from instructions created by the user. A complete explanation of these instructions and examples of the interactive questions are given in the manual for the program (10).

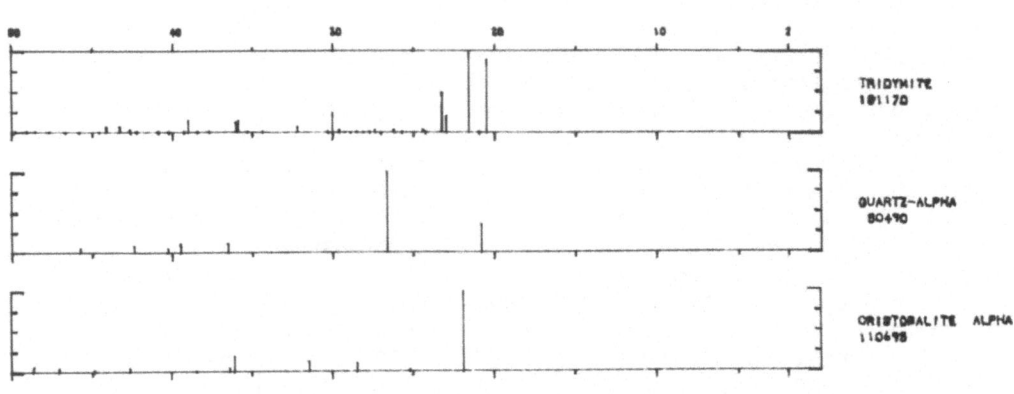

Figure 2.   Example of a plot produced by the program XRDPLT.

The Retrieval Option of the Program XRDPLT

To plot information it must first be retrieved from the file.
However, retrieval can be made without plotting. XRDPLT will retrieve
on the IDNO or PDF number. It can also retrieve by mineral group code.
Using the codes given on page 7 of the MINBAS manual (1), a user can
create a list of minerals which compose a designated group, such as
feldspars, potassium fledspars, amphiboles, garnets, pyroxenes, and/or
carbonates. In this listing the chemical formula is also given.

The program can also retrieve by various formula components, such
as: all silica minerals; minerals containing only silica and oxygen;
minerals containing copper and arsenic but not oxygen. This is perhaps
the most versatile feature of the XRDPLT program. For the formula
components retrieval the program employs "AND" logic symbolized by
a comma (,); "OR" logic symbolized by a slash (/); "NOT" logic, using
a minus sign (-); and "ONLY" logic, indicated by a plus sign (+). A
question mark (?) indicates a chemical subscript. An "exclusive"
 retrieval, using "ONLY" logic, is indicated by a plus sign (+) at
the start of the search string. Only the specified components will
be considered to satisfy the search.

An inclusive retrieval may also be performed based on formula
components. In this retrieval, additional formula constituents, other
than those listed in the search string, would also be acceptable.

Space limitations of this publication prevent us from illustrating these logic permutations so they are more understandable. However, a detailed explanation is given in the manual on the XRDPLT program (10).

THE PROGRAM XRDSCH

The object of the program XRDSCH is to provide the user with a list of possible identifications for an unknown compound. It accomplishes this by a search-match of the powder diffraction lines and intensities of an unknown against those of reference patterns

---

Figure 3   Input Interactive Questions for the Program XRDSCH

.RUN XRDSCH

ENTER DATA FILE TO BE SEARCHED: XRDFIL

ENTER SCRATCH FILE NAME IF TO BE SAVED: DUMMY

TITLE : MINERAL UNKNOWN

NO. OF "BEST FIT" PATTERNS SAVED : 10

IS DATA 2-THETA? YES

TYPE OF RADIATION : CU

MAXIMUM NO. OF ITERATIONS : 5

ENTER ANY KNOWN CHEMICAL ELEMENTS :

DO YOU WISH TO SUBTRACT AKNOWN MINERAL BEFORE
          SEARCH IS STARTED? YES

ENTER IDNO :   50490

ERROR WINDOW :
ERROR LIMITS ON D WILL BE DETERMINED INTERNALLY
D RANGE TO BE CALCULATED INTERNALLY

IS INPUT DATA IN A DISK FILE? NO

DO YOU WANT TO SAVE D-1 PAIRS AFTER ENTRY? YES
IN WHAT FILE : MINUK

ENTER 2-THETA & INTENSITY VALUES.
        (SEPARATE BY A COMMA)
NO. 1 : 8.95,25
NO. 2 : 11.2,80
NO. 3 : 14.05,30   (ETC.)

INTENSITY CORRELATION WILL BE USED TO CALCULATE MATCH

DO YOU WANT TO WANT TO SEE ERROR WINDOW CALCULATIONS? NO

---

stored in the data base. XRDSCH is based on a program developed by
M. C. Nichols of Sandia Laboratories, Livermore, California (11,12)
and L. K. Frevel of the Dow Chemical Company, Midland, Michigan (13).
The X-ray diffraction data, which consists of either "d" or 2-theta
values with intensities, are entered into this program, and are compared against a file of reference patterns, and the best matches or
identifications are listed for the user. The program is capable
of solving single-or multiple-phase unknowns. Figure 3 is a listing
of the interactive input questions for the program.

There are various ways for refining the search parameters of the
program by chemistry, by changing the error window, by pre-disposing
the program towards a specific pattern, or by subtracting out a
known substance before processing begins. Figure 4 shows part of the
terminal output of the program: a brief listing giving the potential
matches and their statistical reliability. The authors' version of
this program is still in the experimental stages. The subtraction
processes do not perform satisfactorily as yet, nor are the interactive
conversational formats finalized.

Figure 4   Output from the Program XRDSCH

NUMBER OF MERIT CALCULATIONS = 570
NUMBER OF PATTERNS SURVIVING MERIT CALCULATION = 290
NUMBER OF FIGURES OF MERIT GREATER THAN 0.01 = 32

CALCITE/QTZ/GOLD/ SYN MIXTURE

LIST OF POSSIBLE SUBSTANCES

PDFND	RHD	UFIT	MINERAL NAME	FORMULA
40783	0.58	0.29	SILVER	AG
40784	0.51	0.29	GOLD	AU
110065	0.23	0.29	CARLSBERGITE	CR N
210574	0.19	0.65	ILSEMANNITE	MO3 08 .N H20
70271	0.19	0.43	SODA NITER	NA N 03
120730	0.18	0.39	IKUNOLITE	B14 (S, SE)3
230599	0.18	0.30	SUKULAITE	SN2 TA2 07
30219	0.08	0.28	CHRYSOCOLLA	CU SI 03 . 2 H20
10830	0.06	0.29	STISTAITE	SB SN
10386	0.01	0.61	LAUTARITE	CA (I 03)2

ENTER IDNO TO BE SUBTRACTED

40784

These programs are a tool for the user. Their value corresponds
to the intelligence with which they are used. The computer eliminates
routine steps and speeds up the identification process, saving the user
of X-ray diffraction techniques much time and effort. However, it is
still the user who makes the ultimate determinations.

ACKNOWLEDGEMENTS

The authors wish to express appreciation to Mr. Arthur J. Gude,
3rd, and Mr. Theodore Botinelly of the U.S. Geological Survey for their
many contributions in support of this project.

REFERENCES

1. George VanTrump, Jr. and P. L. Hauff, "The Mineral X-Ray Diffraction
    Data File; Description and Indices", U.S. Geol. Survey Open-file
    Report 76-406, 145 p. (1976).

2. Joint Committee on Powder Diffraction Standards, "The Powder
    Diffraction File, sets 1-25," available from the Joint Committee on
    Powder Diffraction Standards, 1601 Park Lane, Swarthmore, Pa. 19081
    (1975).

3. I. Y. Borg and D. K. Smith, "Calculated X-Ray Powder Patterns for
    Silicate Minerals," Geol. Soc. Amer. Memoir 122, 896 p. (1969).

4. L. G. Berry and R. M. Thompson, "X-Ray Powder Data for Ore Minerals
    --The Peacock Atlas," Geol. Soc. Amer. Memoir 85, 281 p. (1962).

5. H. M. Ondik and A. D. Mighell, "Crystal Data Tape, Derived from the
    3rd Edition of Crystal Data Determinative Tables, National Bureau
    of Standards," available from the National Technical Information
    Service, U.S. Department of Commerce, 5285 Port Royal Road, Spring-
    field, Va. 22151 (1976).

6. J. D. H. Donnay and Helen M. Ondik, "Crystal Data Determinative
    Tables, 3rd Edition: U.S. Bureau of Standards and the Joint
    Committee on Powder Diffraction Standards," available from JCPDS,
    1601 Park Lane, Swarthmore, Pa. 19081 (1973).

7. Joint Committee on Powder Diffraction Standards, "Powder Diffraction
    File Search Manual--Hanawalt Method--Inorganic, sets 1-25," avail-
    able from the Joint Committee on Powder Diffraction Standards, 1601
    Park Lane, Swarthmore, Pa. 19081, 994 p. (1975).

8. George VanTrump, Jr. and P. L. Hauff, "Mineral X-Ray Diffraction
    Data Retrieval/Plot Computer Program--Program Listing," U.S. Geol.
    Survey Open-file Report 76-404, 24 p. (1976).

9. Charles G. Cunningham, Jr., Robert B. Hall, Phoebe L. Hauff, and
    George VanTrump, Jr., "Alunite-Natroalunite Identification Using
    Field Tests and a Computer Plotted Overlay for X-Ray Diffraction
    Charts," U.S. Geol. Survey Open-file Report 76-403, 9 p. (1976).

10.   P. L. Hauff and George VanTrump, Jr., "Mineral X-Ray Diffraction
      Data Retrieval/Plot Computer Program--Laboratory Manual," U.S.
      Geol. Survey Open-file Report 76-407, 55 p. (1976).

11.   Monte C. Nichols, "A Fortran II Program for the Identification of
      X-Ray Powder Diffraction Patterns, UCRL-70078," Lawrence Radiation
      Laboratory, University of California, Livermore, California, 119
      p. (1966).

12.   Monte C. Nichols, Quinton C. Johnson, and M. Eileen Kahara,
      "Characteristics of an Effective Computer Based Powder Search
      System," American Crystallographic Association, Twenty-Fifth
      Anniversary Meeting--Program and Abstracts (1975).

13.   Ludo K. Frevel, "Computational Aids for Identifying Crystalline
      Phases by Powder Diffraction," Analytical Chemistry, Vol. 37
      p. 471-482 (1965).

COMPUTER SEARCHING OF THE JCPDS POWDER DIFFRACTION FILE

Paden F. Dismore

E. I. du Pont de Nemours & Company

Wilmington, Delaware 19898

ABSTRACT

Jackson Laboratory is using computer searching of the powder diffraction file on an "in-house" IBM 360/65 computer for about 1000 searches per year. This paper will discuss our experiences with the search program.

A Philips Automatic Powder Diffractometer (APD), modified to communicate with an XDS Sigma 5 computer collects the data. A program in the Sigma 5 provides both hard copy output and a computer deck of punched cards for input to the JCPDS search program.

The JCPDS search program is indispensable for our large volume of samples. The majority of these samples are corrosion products which generally have broad peaks, and frequently, preferred orientation.

The use of the search program (choice of parameters and interpretation of output) for this type of sample is discussed. Some suggestions for improving the program are made.

INTRODUCTION

Jackson Laboratory is the principal research laboratory of the Organic Chemicals Department of the Du Pont Company. About 1500 requests for x-ray analyses are received each year from the research work as well as from manufacturing and sales service activities. On these samples we perform about 2000 x-ray fluorescence deter-

minations and about 1000 diffraction scans. The samples
vary widely and very few are pure materials. Most
samples are inorganic, and many are corrosion products.
The typical request is for a qualitative fluorescence
scan to determine what elements are present and a diffraction scan to determine what specific compounds are present.

We find that computer searching of the JCPDS powder
diffraction file is quite useful for identifying this
large number of complex samples. However, discussions
with x-ray diffractionists both in and outside of the
Du Pont Company suggest that computer searching is not
very popular. This paper will discuss use of the computer
search system at Jackson Laboratory, and the procedures
used to meet the particular requirements of our organization.

DATA COLLECTION

With a large sample load, efficient collection of the
raw data was handled by equipping our diffractometer with
an automatic sample changer and computer control. We
chose the Philips Automatic Powder Diffractometer (APD
3500) with a Texas Instruments Model 733 ASR "Silent 700"
terminal, modified by Philips. The automatic operation
of this instrument, allowing unattended overnight processing of samples, has completely eliminated the backlog
of samples.

The large sample load increases the clerical work to
identify the diffraction scan and to get the raw data into
the correct format for searching the diffraction file. The
APD, used in the peak search mode, has helped solve this
problem by: (1) locating peak positions and converting
2-theta values to d-spacings and (2) normalizing intensities. Thus, the data are immediately available for manual
searching of the diffraction file.

The small size (4 K of core) of the computer in the
APD, however, does create clerical problems for us. Since
only one line is provided for sample identification, if
there are several samples in a run, each sample is identified only by its position in the sample changer. The
operating conditions for the APD are printed only at the
start of the run, not with each sample. At Jackson Laboratory, the output is filed under the name of the chemist
submitting the sample. Thus the identification and operating conditions must be manually copied onto the data
sheet for each sample.

This manual copying was eliminated by using a large computer (Xerox "Sigma 5") that was available in the laboratory. Sample identification and instrument operating conditions (voltage, current, slits, monochromatization, etc.) are recorded in the "Sigma 5", followed by the APD operating conditions and the output data.

Several hardware modifications were made to the APD and to the "Silent 700" to allow this communication to the large computer. The teletype interface board in the "Silent 700" was modified to communicate with either the APD or an acoustic coupler. This allowed recording on the magnetic tape cassette followed by transmission of the data over a telephone line at the end of a run. A second modification was made to the APD so that it transmits data to the "Sigma 5" as the data is collected, with the magnetic tape cassette used only as a backup. A third modification increased the speed of transmission to 30 characters per second. The details of these hardware modifications can be obtained from the author.

A batch program was written for the "Sigma 5" to provide hard copy output with the sample description and instrument operating conditions printed along with the APD data. The program also punches a complete set of cards for using the JCPDS search program. Because of the peculiarities of our particular computer system, this program would probably have to be greatly modified to work on another system.

We find it absolutely essential to have knowledge of the elemental composition of a sample for the JCPDS search to be effective. Thus we always make a qualitative elemental analysis by x-ray fluorescence.

JCPDS COMPUTER SEARCH

The JCPDS computer search of the diffraction file is used routinely for all samples. The remainder of this paper will describe our experience with the program and our solutions to problems that were encountered with the hope that these remarks will make it easier for others to use the program. I want to emphasize that these results should be interpreted as suggestions for improving a very good system.

I suspect that computer searching is not widely used because it has not been given a fair trial. In order to perform the very complicated task of matching an unknown diffractogram with a very few of the 25,000-odd patterns in the file, the program must be complex. Thirty-five options in the program must be carefully chosen for each run. Although default values for these options are provided, these are frequently incorrect for a particular job.

A thorough knowledge of how the program works is absolutely essential for satisfactory results. This requires thorough study of the available documentation, thorough study of the complete output (not just the summary page), and continued practice. This knowledge can only be acquired after much use of the program. I would suggest that JCPDS provide more detailed user documentation in a single source. At present several references (1, 2, 3) must be studied and there is variation in nomenclature among these references which can cause confusion. The use of some elegant programming techniques causes some intermediate results of the program to be slightly different from those calculated manually. This also is confusing until the explanation is found.

Another possible deterrent to general use of the search program may be the computer costs for running the program. Large differences in costs are found between various computer organizations depending upon the type of computer used and the accounting practices of the particular computer group. With our particular computer group the best way to reduce computer costs is to run several searches in one batch. Thus, it is more economical to use the program for every sample than to use it only occasionally.

USE OF SEARCH PROGRAM

The program uses "packed spacings and intensities" (2) because: (1) they have an essentially constant error window, and (2) they are integers which are much easier for the computer to handle. Since this usage can cause some confusion, a brief explanation is in order. Packed spacing (PS) is defined as

$$PS = 1000/d = d' \qquad (1)$$

where d is the interplanar spacing. Since d ranges from about 200Å to about 0.6 Å, PS is an integer between 5 and 1666. Packed intensity (I_1) is defined as

$$I_1 = 5* \log_{10} I_3 \qquad (2)$$

where I_3 is the intensity normalized on a scale of 0-100. This I_1 is a single digit integer.

A minor source of confusion is that the program does not calculate I_1 from equation (2) but instead uses Table I which gives slightly different answers.

TABLE I

CONVERSION OF INTENSITY TO PACKED INTENSITY

I_3	0-1	2	3	4-6	7-10	11-15	16-25	26-39	40-63	64-100
I_1	0	1	2	3	4	5	6	7	8	9

For example $5* \log_{10} 10 = 5$ but $I_1 = 4$ in the table.

As mentioned earlier about thirty-five parameters must be specified. Although default values are provided for all of these parameters, the default values are generally not the best for each particular sample. Our choices for some of these parameters are discussed later in this paper.

A very important parameter is the error window, IW, which compensates for errors both in the experimental measurement and errors in the data file. This is defined as (3):

$$IW = \left(\frac{1}{d} - \frac{1}{d + \Delta d}\right) *1000 \qquad (3)$$

As a rule of thumb, for copper radiation, the error window is about ten times the error in two-theta. Reference 3 suggests that each laboratory determine its own error window while Reference 1 suggests that the error window be varied depending on the line width of each sample spectrum. A test on a diffraction diagram of a 50-50 mixture of Si (NBS standard reference material 640) and KCl (analytical reagent) with the default value of IW = 2 found both Si and KCl. Thus, a window of 2 was excellent for pure compounds. However, IW = 2 retrieved no chemically correct matches for a typical Jackson Laboratory

sample. This was a corrosion product from a process
pipeline and contained Fe, Ni, S, F, Cl. Manual identification showed iron and nickel chlorides and fluorides.
The pattern was searched using the parameter ICYCLE which
allows repeated searches of the same pattern with different parameter values. The instructions (1) recommend
not using ICYCLE but we have found it very helpful. However, only one parameter (e.g., error window) should be
changed. Thus care must be taken to prevent other
parameters from changing (1).

Only the match of the experimental pattern with NiF_2
(JCPDS No. 24-792) will be discussed. Table II gives the
values of d and I from JCPDS 24-792 along with the error
window for each observed line with IW = 2 and 5.

TABLE II

Effect of Error Window, IW

JCPDS 24-792		Experimental	
d $\overset{\circ}{A}$	I	IW = 2 $\overset{\circ}{A}$	IW = 5 $\overset{\circ}{A}$
3.2877	100	3.2733-3.3167	3.2415-3.3501
2.5701	49	(2.5740-2.6008)	2.5543-2.6212
2.3259	5		
2.2489	20	(2.2497-2.2701)	2.2346-2.2857
2.0796	6		
1.7245	50	(1.7256-1.7376)	1.7167-1.7467
1.6445	15	(1.6434-1.6543)	1.6353-1.6625
1.5420	7	(1.5468-1.5564)	1.5396-1.5637
1.4709	6	1.4631-1.4717	1.4567-1.4782
1.3960	8	1.3937-1.4015	1.3879-1.4075
1.3850	15	(1.3860-1.3937)	1.3803-1.3996
1.3274	1	(1.3149-1.3219)	1.3098-1.3271
1.2850	1	1.2829-1.2895	1.2780-1.2945

Those lines in parenthesis are outside of the error
window with IW = 2. Since two of the three strongest
lines of the pattern are outside of the error window, the
program rejects the match. With IW = 5, all of the lines
except the very weak one at 1.3274Å are within the error
window and the computer calls this a good match. This observed shifting of lines from that of NiF_2 is not surprising since the sample was a corrosion product and probably was a solid solution of NiF_2 containing iron and

chlorine. Since most of our samples are corrosion products, as a result of this and similar experiments, we routinely use an error window, IW, of 5.

An insignificant detail that can cause trouble is illustrated in Table III which compares the first six lines of our experimentally observed spacings and intensities with those found on card 24-792 and those returned by the program.

TABLE III

NiF2 Spacings and Intensities

Observed		Card 24-792		Program	
d (Å)	I	d (Å)	I	d (Å)	I
3.300	100	3.2877	100	3.284	100-63
2.586	48	2.5701	49	2.567	63-39
		2.3259	5	2.328	6-3
2.262	31	2.2489	20	2.250	25-15
		2.0796	6	2.081	6-3
1.733	88	1.7245	50	1.726	63-39

A comparison of the data from card 24-792 with those returned by the program shows slight differences which are caused by conversion of the card data to packed intensities and spacings and then reconversion. The intensity is printed as a range, indicating that the conversions occurred. However, the conversion is not obvious in the d-spacing data and caused confusion until the reason was explained.

Table IV gives the output of the program for the match with NiF2. We find this very useful and always have the "OUTPUT" section printed.

TABLE IV

Output Summary, NiF_2

	IW = 5	IW = 8
Number of matches	9	9
Per cent of matches	82	82
Minimum intensity of lines compared	2	2
Log intensity match	25	25
F (Del D)	0.578	0.736
F (Del I)	0.897	0.897
Reliability factor (raw)	42	54
Reliability factor (corrected)	31	39

In Table IV the minimum intensity of lines compared is 2. The last two lines of Table II show an intensity of 1. Thus these lines are ignored and there are eleven (not 13) lines in the range measured of which matches were found of 9 (row 1 of Table IV) or 82% (row 2 of Table IV).

The minimum number of line matches is one of the parameters in the program (NUMBER) with a default value of 3. This is adequate for most materials but if pure metals are suspected to be present (e.g. Cu, Pt) which have very few lines in the measured range, we lower this parameter to NUMBER = 2.

Log intensity match is described in Reference 2 as Davey Maximum concentration (DMC). This is the percentage of the matched phase in the unknown assuming that both the experimental pattern and the JCPDS data are correct and on an absolute scale. It is defined as

$$DMC = (I_{JCPDS} - I_{unknown})_{max}$$

$$= (5 \log_{10} I_{3_{JCPDS}} - 5 \log_{10} I_{3_{unknown}})_{max} \quad (4)$$

The program converts DMC back into a ratio of the relative intensity of the unknown to that of the standard or

$$\text{log int match} = \frac{I_{3_{unknown}}}{I_{3_{JCPDS}}} = \text{antilog } (-DMC/5) \quad (6)$$

Table V shows the first six lines of the NiF_2 data
with intensities converted to packed intensities. The
last column shows the difference in packed intensities
($I_{JCPDS} - I_{unknown}$) and if negative is called zero. The
largest difference is 3 for each of the missing lines.
Hence log intensity match is antilog $(-3/5) = 0.25$ or 25%.
Again, the program uses packed intensities, which are
single digit integers, and Table I to convert the percentages, so the printed results are limited to the values
in Table I and may not be exactly as calculated by the
above formulae.

TABLE V

Calculation of Log Intensity Match

Sample			JCPDS 24-792			Differences
d	I	Packed	d	I	Packed	Packed
Å		I	Å		I	I
3.300	100	9	3.2877	100	9	0
2.586	48	8	2.5701	49	8	0
			2.3259	5	3	3
2.262	31	7	2.2489	20	6	0
			2.0796	6	3	3
1.733	88	9	1.7245	50	8	0

In Table II, the lines at d = 2.5701 and 1.7245,
each with a packed intensity of 8 were outside the error
window of IW = 2. Thus DMC = 8 in this case or log int
match = 2.5% which is a packed intensity of 1 (Table I).
This is below the default value of 2 for the parameter,
KUTOFF. Thus an error window of IW = 2 actually discarded
the match because of a poor log int match. The important
point here is that if only one strong line (relative
intensity greater than 40) is absent, no match will be
made.

If the minor option is chosen, the log intensity
match is not used. We occasionally use this option if a
satisfactory identification is not obtained. In these
cases, a strong line on the JCPDS pattern is generally
absent in the unknown. Of course a penalty is paid in
that a very large number of matches are reported.

F (Del D) in Table III is a measure of the goodness of
fit of the d spacings. On p. 10 of Reference 3 it is defined as

$$F\ (Del\ D) = 1 - \frac{\sum\limits^{N} |d'_{JCPDS} - d'_{unknown}|}{\sum\limits_{N} number\ of\ matches\ band\ pass}$$

$$\qquad(6)$$

$$= 1 - \frac{\sum\limits^{N} |d'_{JCPDS} - d'_{unknown}|}{IW * number\ of\ lines\ matched}$$

Thus if all of the packed spacings of the JCPDS card
exactly match the packed spacings of the unknown, the
second term is zero and F (Del D) = 1. We like F (Del D)
to be greater than 0.5, an arbitrary choice that seems to
work. In my mind, there is a question of the logic of
Equation 6 in that the error window, IW, is in the denomina-
tor of the second term. Thus, other things being equal,
an increase of the error window causes an increase in the
goodness of fit of the spacings as seen in Table IV. I
would think that a larger error window should cause a
poorer fit.

F (Del I) is a measure of the error in matching intensities and is defined as (3)

$$F\ (Del\ I) = 1 - \frac{\sum\limits^{N} |I_{1\ JCPDS} - I_{1\ Unknown}|}{\sum I_{1\ JCPDS}} \qquad(7)$$

if $I_{1\ JCPDS} < I_{1\ Unknown}$, $|I_{1\ JCPDS} - I_{1\ Unknown}| = 0$

In this case, the standard must first be scaled by the log
intensity match parameter. Since our match of intensities
is generally poor, our practice is to ignore this parameter.

Reliability factor is defined as (3):

Rel factor = % lines matched *F (Del D) * F (Del I) (8)

Reference 3 suggests a value of greater than 33 is indicative of the presence of a particular phase. As mentioned
earlier, our matches of intensities are not very good and
thus we check the individual components of equation (8)
instead of using the reliability factor.

The program <u>reduces</u> the reliability factor if chemical
information is supplied. In the case of NiF_2, Ni was
said to be a major element and F an undetermined element.
Thus the reliability factor was recalculated to be only 77%
of that calculated from Equation (8). This is shown in
Table IV as the corrected factor. I would expect that the
reliability factor should be increased if chemical information is available.

This reduction of reliability factor for chemistry
caused us some trouble in an unknown where the only element
detected (x-ray fluorescence) was Cl which was listed as
a major element (an element known to be present). H, 0,
C, N, F, Na and Mg were listed as undetermined (elements
which were suspected as possibly being present). With the
usual error window of 5 and the major option, 1089 matches
were found of which only one, NaCl, was chemically correct
and had a reliability factor of only 8. In the list
labelled "with question mark", the highest reliability
factor was only 10 for $Na_2CO_3 \cdot H_2O$. In the "OUTPUT" section,
NaCl had a reliability factor of only 12 before rescaling
for chemistry while $Na_2CO_3 \cdot H_2O$ was 58 (29 or 94% lines
matched; F (Del D) = 0.738; F (Del I) = 0.838) which is
very good. To avoid this problem we now call all elements
either major elements present or absent and do not use the
minor, trace or undetermined categories.

As described above, the use of the JCPDS search program at Jackson Laboratory greatly relies on chemical information. Thus, for our purposes, it would be desirable
not to print any chemically incorrect matches. We also
use the "OUTPUT" section very extensively and it would be
desirable to have those data included in the summary
report. A third suggestion is to have an option in the
program to delete any phases that are not stable at normal
laboratory conditions (e.g. H_2O, SO_2). Finally, I would
suggest that the instruction for use of the program be
expanded and clarified.

<u>REFERENCES</u>

1. Johnson, G. G., Jr., <u>User Guide Data Base and Search
    Program, JCPDS</u>, Swarthmore, Pa. (1975).

2. Johnson, G. G., Jr., Vand, V; <u>Ind. Eng. Chem.</u>, <u>59</u>, 19
    (1967).

3. Johnson, G. G., Jr., Fortran IV Programs Version XII,
    JCPDS, Swarthmore, Pa. (1970).

A ROUND ROBIN TEST TO EVALUATE COMPUTER SEARCH/MATCH METHODS FOR

QUALITATIVE POWDER DIFFRACTOMETRY.

Ron Jenkins*

Philips Electronic Instruments, Inc.

Mount Vernon, New York   10550

*Chairman, Computer Sub-Committee, Joint Committee on Powder
Diffraction Standards.

ABSTRACT

A preliminary round robin test to evaluate available computer
search/match has been carried out and comparisons made with hand-
searching techniques. Participants in the test were all experienced
diffractionists and a variety of data collection instrumentation was
employed. Hand-searching and computer searching were found to be
equally efficient, about 70-80%, for the mineral and inorganic samples
provided that the quality of the "d" values was good (1 in 1000).
Although the computer methods were much quicker, they were found to
be much more sensitive to the quality of the "d" values. Where elemental data was used to assist in computer searching, much poorer "d"
values (3 in 1000) were acceptable. Hand-searching was found to be
vastly superior to computer searching for the organic specimen.

INTRODUCTION

X-ray powder diffractometry has been employed for many years for
the routine qualitative analysis of multi-phase polycrystalline
materials. Qualitative phase identification is carried out almost
exclusively by matching the recorded pattern of the "unknown" material
to standard patterns of pure phases, using a search/match procedure
based on successive matching of lines in the unknown pattern with a
standard pattern, subtraction of the standard pattern from the original,
and repeating the process until the residue from the original pattern
is at an acceptable level. Manual search/match procedures such as
those described by Hanawalt[1], Fink[2], etc., are well documented and
have been in use for many years, but the success with which they may
be applied in practice, varies from being extra-ordinarily good to
frustatingly poor, the degree of success being directly correlatable
with the experience of the user, the quality of his experimental data
and the complexity of the problem. Hand searching may also be very
time consuming unless the diffractionist can make a series of inspired

guesses based on prior knowledge of the specimen history, or on additional analytical information such as a semi-quantitative elemental
analysis.

The largest commercially available index of standard patterns is
that maintained by the Joint Committee on Powder Diffraction Standards[3]
(J.C.P.D.S.) and this index currently contains around 25,000 patterns.
With data manipulation problems of this size, the use of the computer
is an obvious candidate to assist in the search/match procedure and
indeed several programs for this purpose have been generated[4-6]. The
JCPDS has traditionally offered the Johnson/Vand program[7] for use with
their data base and a Computer Subcommittee of the JCPDS was established
two years ago to ensure compatibility of the data base with existing
and potential computer based search/match procedures. This present
communication describes the format and results of a round robin test
inspired by the Computer Subcommittee of the JCPDS, this test being
designed as a first attempt to quantitate the success of computer search/
match techniques in the analysis of unknown specimens under routine
laboratory conditions.

Scheme of the Round Robin Test

A series of five specimens were made up, two inorganic mixtures, two
mineral mixtures and one organic mixture, and these specimens submitted
to a number of different diffractionists with the request to identify
the phases present, treating the specimens as routine analytical problems. Thirteen different scientists were involved representing a
variety of data collection techniques, including Guinier and Debye-
Scherrer Cameras, plus manual and automated powder diffractometer instrumentation. Table 1 lists the participants in the Round Robin and
Table 2 lists the data collection and search/match procedures used.

It will be seen that these represent both hand-searching
methods and computer search methods, this latter category including three different procedures, namely those by Johnson, Nicholls
and Frevel. Each participant was simple told that a given mixture
was inorganic, mineral or organic. For the purpose of reporting
data, each participant was assigned a code, which was established by
taking a random selection of the order shown in Table 1.

Table 3 lists the composition of the actual mixtures. These
were carefully selected to yield specific types of data, rather than
to reflect a typical analytical situation. For example, the inorganic
and mineral specimens were devised to include two readily identifiable
phases, one rather difficult phase, plus one additional phase selected
because it presents some special problem such as preferred orientation
or hygroscopy or, as in the case of montmorillonite, because it gives
major lines in a difficult portion of the angular range of the instrument. One specimen also included $\alpha-SiO2$ which was added as a reference standard (the participants were unaware of this) to guage a
measure of the quality of the recorded "d" values. Mixtures were made
from pure phases, wherever possible, and although these ware partially
pre-ground and sieved, no specific attempt was made to inhibit potential problems due to grain size or density variations.

TABLE 1

Participants in the Round Robin

Dr. J. D. Hanawalt	University of Michigan
Ms. C. Foris	E. I. duPont de Nemours, Wilmington, Delaware
Dr. C. Freiburg	Kernforschungsanlage G.mb H., Juelich, Germany
Dr. J. A. Huston	Sun Oil, Marcus Hook, Pennsylvania
Dr. E. J. Graeber	Sandia, Albuquerque, New Mexico
Dr. Monte Nichols	Sandia, Livermore, California
Dr. Gerry Johnson	Penn State, Materials Research Laboratory
Dr. Greg McCarthy	Penn State, Materials Research Laboratory
Prof. Ben Post	Polytechnic Institute of New York
Dr. Jan Visser	Technisch Physiche Dienst, TNO, Delft, The Netherlands
Dr. Dean Smith	Penn State, Dept. of Geology
Dr. Ven Ruhberg	Dow Chemical
Dr. Ludo Frevel	Dow Chemical

Note: The participant codes were established by taking a random selection of the above participant list. The Round Robin Coordinator was Ron Jenkins, Philips Electronic Instruments, Mt. Vernon, NY.

TABLE 2

Participant Code	Data Acquired With				Search Procedure			
	Diffracto-meter	Guinier	Debye-Scherrer	Elemental Data Used	Hand	J/V#17	MN	FR
A		X			X			
B		X				X		
C	X					X		
D	X			X		X		
E	X			X		X		
F	X				X			
G			X				X	
H	X						X	
I	X					X		
J	X				X			
K	X			X		X		
L	X							X
M		X				X		
TOTALS	9	3	1	3	3	7	2	1

NOTE: J/V #17   Johnson/Vand
       MN   Monte Nichols
       FR   Ludo Frevel

TABLE 3

Sample Compositions Used in Round Robin Tests

Sample	Composition	PDF No.*
#1 (Mineral A)	25% Montmorillonite	?
	25% Calcite	5-586
	25% α-Quartz	5-490
	25% Kaolinite	14-164
#2 (Mineral B)	33% Chalkopyrite	9-423
	33% Gibsite	7-324
	33% Baryte	5-448
#3 (Inorganic C)	35% Rutile	21-1276
	35% Anatase	21-1272
	15% Vanadium Pentoxide	9-387
	15% Barium Chloride Dihydrate	11-137
#4 (Inorganic D)	30% Sodium triphosphate	11-652
	30% Lithium Fluoride	4-857
	30% Barium Carbonate	5-378
	10% Potassium Bromide	4-531
#5 (Organic E)	50% Pentaerythritol	3-214
	50% Lauric Acid	8-528

*Due to the continuing effort to annually update the Joint Committee
PDF file, some of these numbers may have been withdrawn from the most
recent versions of the PDF file. In most instances, the withdrawn pattern will have been replaced by an updated version.

Results reported by participants

Tables 4 through 8 summarise the reported analyses on the five
specimens. In the table a "Y" under a given phase indicates that
the phase was identified and an "X" indicates that the phase was
not identified. Also shown in the tables is any additional information that was reported. Study of this latter data suggests that
some of the samples may have been cross contaminated during preparation. As an example, several participants reported the presence of
α-SiO2, Calcite and Kaolinite in Mineral B, and their submitted data
indeed showed lines from these phases. Careful preparation of
Mineral B on the other hand, showed no indication of the presence of
these additional phases. It will be noted, however, that the three
additional phases are all present in Mineral A.

TABLE 4

Sample 1, Mineral A

Participant	Correct Identification				Additional
	Calcite	α-SiO$_2$	Kaolinite	Montmorillonite	
A	y	y	x	x	−
B	y	y	x	x	Al$_2$Si$_2$O$_5$(OH)$_4$
C	y	y	x	x	⊢
D	y	y	x	x	Ca(Al$_2$Si$_2$O$_8$)·4H$_2$O
E	y	y	y	x	Plagioclase-Feldspar
F	y	y	x	y	+ unidentified phase
G	y	y	x	x	+ unidentified phase
H	y	y	y	x	muscovite
I	y	y	x	x	various
J	y	y	y	x	
K	y	y	x	x	VO(OH)$_2$
L	y	y	y	x	probably a bentonite clay
M	y	y	x	x	+ some clay mineral

TABLE 5

Sample 2, Mineral B

Correct Identification

Participant	Chalkopyrite	Gibsite	Baryte	Additional
A	y	y	y	quartz, pyrrhotite
B	y	y	y	
C	y	y	y	
D	y	x	y	FeS_2, $FeSO_4$, H_2O
E	y	y	y	quartz, chamosite
F	y	x	y	pyrite, mullite
G	y	y	y	
H	y	y	y	$\alpha-SiO_2$
I	x	x	y	various
J	x	x	y	
K	y	x	y	various
L	y	y	y	probably kaolinite
M	x	y	y	calcite,trace of anatase

TABLE 6

Sample 3 - Inorganic C

Correct Identification

Participant	Rutile	Anatase	V_2O_5	$BaCl_2,2H_2O$	Additional
A	y	y	y	y	–
B	y	y	y	y	–
C	y	y	y	y	–
D	y	x	y	y	–
E	y	y	y	y	NiO ?
F	y	y	y	x	–
G	y	y	y	x	–
H	y	y	y	x	–
I	y	y	x	x	various
J	y	y	y	x	
K	y	y	x	x	various
L	y	y	y	x	possibly gibsite
M	y	y	x	y	–

TABLE 7

Sample 4, Inorganic D

Correct Identification

Participant	LiF	$BaCO_3$	K Br	$Na_5P_3O_{10}$	Additional
A	y	y	y	y	$Na_4P_2O_7$
B	y	y	y	y	–
C	y	y	y	y	–
D	y	y	x	y	–
E	y	y	y	y	–
F	x	y	y	x	aluminum phosphate dihydrate
G	x	y	x	y	
H	x	y	y	y	plus a cubic material
I	y	y	y	y	plus various others
J	y	y	x	y	
K	x	y	x	x	various
L	y	y	y	y	–
M	y	y	y	y	–

TABLE 8

Sample 5, Organic E

Correct Identification

Participant	Pentaerythritol	Lauric Acid	Additional
A	y	x	–
B	x	x	–
C	x	x	–
D	x	x	various
E	y	x	D(or L)-alanine
F	y	y	+ unidentified phase
G		no data	
H		no data	
I	y	x	+ various others
J		no data	
K		no data	
L	y	–	carnuba wax
M	y	y	+ dioctadecyl adipate

One disquieting feature about some of the reported data was the failure on the part of the participant to draw conclusions beyond the list of phases found by the computer program. In a few instances, the "analytical result" comprised a computer print-out of several hundred possibilities with little or no additional information as to which of the phases were likely or unlikely.

Rating and summary of results

In order to place some quantitative significance to the success or failure of the search matching, a rating system was designed as illustrated in Table 9. The maximum possible rating was 20 for the two Inorganics plus Mineral A, 15 for Mineral B and 10 for the Organic. All scores were also normalised and expressed as a "percentage success".

TABLE 9

System of Rating

For correct phase identification

Positive	+5
Possible	+3
Near Miss	+2
Unidentified Phase	+1

For incorrect phase identification

Positive	-3
Possible	-1

Table 10 lists the overall ratings plus the success percentages. Of the thirteen reporters, only nine included results on the Organic specimen. Table 11 shows the individual average success percentage ratings for the Mineral and Organic pairs.

Table 12 shows the summary ratings for the five specimens based on the search method, i.e., Hand Search, Computer Search without elemental data, and Computer Search with elemental data. The Computer Search is broken down further into three individual procedures.

In the interpretation of the results, it should be noted that only one user employed the Frevel program, two used the Nichols program and seven used the Johnson program. Also in the case of the Frevel program, only a limited portion of the data file was searched.

TABLE 10

Overall Ratings

Participant Code	Sample #1 (Mineral A)	Sample #2 (Mineral B)	Sample #3 (Inorganic C)	Sample #4 (Inorganic C)	Sample #5 (Organic E)
A	10	13	20	19	5
B	12	15	20	20	0
C	10	15	20	20	0
D	12	8	15	15	-3
E	17	13	19	20	2
F	16	8	15	7	9
G	11	15	15	10	-
H	17	14	15	17	-
I	6	2	7	17	2
J	15	3	15	15	-
K	9	7	7	3	-
L	17	12	14	20	4
M	12	8	15	20	7
Average	12.62	10.23	15.15	15.62	2.89
Max possible	20	15	20	20	10
% success	63	68	76	78	30

TABLE 11

Individual % Success Ratings

	Sample	
Participant Code	Minerals A&B	Inorganics C&D
A	66	99
B	77	100
C	71	100
D	57	75
E	86	98
F	69	55
G	74	63
H	89	80
I	23	60
J	48	75
K	46	25
L	83	85
M	57	88

Results and Conclusions

Even though only a relatively small number of data sets are available, several interesting conclusions can be drawn:

a) It appears that "hand-searching" and computer-searching" are equally efficient for inorganic and mineral type analysis.

b) Hand-searching is vastly superior to computer-searching for organics.

c) Where no elemental data is included in the input to the computer, the computer search/match is extremely sensitive to the quality of the "d" values. 1 part in 1000* seems highly desirable. This is apparently less so in hand-searching.

d) Where elemental data is included, an accuracy of 2 parts in 1000* for the "d" values is probably sufficient.

e) The overall quality of reported "d" values was rather poor, most data laying in the range 1 in 1000* to 3 in 1000*.

f) Reported data indicates that in some instances there was cross contamination in the samples as run.

g) Once a diffractogram (i.e., a chart of 2θ vs "I") has been obtained, the elapsed time required for interpretation of the diffractogram is about 3 times less for computer-searching compared to hand searching.

Once a set of d's and I's have been obtained, computer-searching is almost an order of magnitude faster than hand-searching.

* At a "d" value of 1 to 2Å.

TABLE 12

Summary Ratings Based on Search Method (%)

Search Method	Sample #1 (Mineral A)	Sample #2 (Mineral B)	Sample #3 Inorganic (C)	Sample #4 Inorganic (D)	Sample #5 Organic (E)
Hand Search	68	40	83	68	70
Computer Search (no elemental data)	61	77	76	89	17
Computer Search (with elemental data)	63	62	68	63	--
Individual Breakdown					
Johnson/Vand	56	65	74	82	--
Frevel	85	80	70	100	--
Nichols	70	97	75	68	--

Point (e) deserves some further clarification. As was mentioned previously, α-SiO$_2$ was used as an unknown phase in Mineral A in order to roughly establish the quality of the "d" values being used. As an example, Table 13 lists the reported d value of the (101) reflection along with the deviation from the mean. These data give a standard deviation of 0.004A or about 1 in 1000. There appears to be a definite correlation between the groups giving a significant error in the reported "d" and the percent success in search matching. Comparison of the data in Tables 11 and 13 confirms this, even though at first sight a couple of data sets (D and E) appear to be out of line. Note, however, that these are the data sets where elemental analysis was also used in the search matching.

TABLE 13

Measured Value of the (101) α-SiO$_2$ Line in Mineral A

Participant Code	Reported d (101) Å	Deviation from Mean
A	3.340	− 0.003
B	3.344	0
C	3.346	+ 0.002
D	3.340	− 0.004
E	3.350	+ 0.006
I	3.347	+ 0.003
J	3.342	− 0.002
K	3.336	− 0.008
L	3.345	+ 0.001
M	3.346	+ 0.002

Mean Value = 3.344Å
JCPDS Card #5-0490 = 3.343Å
No data available from groups F, G or H.

___

One last point should be made, this being that the scientists involved in this round robin could all be classified as the "upper crust" of the powder diffraction world. It is likely that the quality of the data being obtained in the average routine X-ray laboratory is significantly worse than reported in this paper.

Further Study

The above data were presented to a meeting of the Joint Committee on Powder Diffration Standards in February of this year. One of the conclusions of the JCPDS meeting was that a second round robin test should now be designed in which 3 mixtures (1 each inorganic, mineral and organic) plus 3 carefully recorded d/I data sets (1 each inorganic, mineral and organic) be submitted to as wide a range of Powder Diffractionists as possible. The unknowns plus the data sets for this test are now being prepared and should be ready for circulation in the Fall.

References

1. Hanawalt, J.D., Rinn, H.W. and Frevel, L.K., Ind. Eng. Chem. Anal. Ed, 10 (1938) 457.

2. The Fink Numerical Index to the Powder Diffraction File, Joint Committee on Powder Diffraction Standards.

3. The Joint Committee on Powder Diffraction Standards, 1601 Park Lane, Swarthmore, Pernsylvania 19081.

4. Johnson, G.G., and VAnd, V., Ind. Eng. Chem., 59 (1967) 19.

5. Frevel, L.K., and Adams, C.E., Anal. Chem., 40 (1968) 1335.

6. Nichols, M., and Bideaux, R.A., 24th Pittsburgh Diffraction Conference (1966) Paper No. B-3.

7. User Guide, Data Base and Search Programs, by G.G. Johnson, JCPDS (1974).

DIRECT QUANTITATIVE DETERMINATION OF SILICA BY X-RAY DIFFRACTION ON PVC MEMBRANE FILTERS

W. W. Henslee and R. E. Guerra

Central Laboratory, Dow Chemical U.S.A.

Freeport, Texas 77541

ABSTRACT

Exposure of industrial workers to respirable crystalline silica is receiving considerable attention at present from industrial laboratories and regulatory agencies. Analytical methodology for rapid and accurate determinations of microgram quantities of silica is necessary. The X-ray diffraction method which has been recommended by NIOSH requires lengthy sample preparation.

An X-ray diffraction method has been developed for the quantitative analysis of α-quartz directly on polyvinyl chloride filters in the microgram range. The minimum detectable concentration of α-quartz is approximately 2 $\mu g/cm^2$. This represents a sensitivity comparable to that claimed using the Bumsted method. The necessity of ashing the filter and depositing the residue on silver membrane filters, as well as the addition of a standard, is avoided. This minimizes handling and alteration of the sample. Determinations are made via a calibration curve based on the deposition of known quantities of SiO_2. Matrix effects of iron oxides, present as 20% of the total sample weight, were found to be negligible in the range investigated (<77 $\mu g/cm^2$).

INTRODUCTION

The Target Health Hazards Program was initiated by the Occupational Safety and Health Adminstration (OSHA) in 1972. Respirable silica, especially in the crystalline form, was listed as one of the most severe hazards to the American industrial worker (1). Quartz, cristobalite and tridymite are the most common forms of crystalline silica with quartz the most frequently encountered. The National Institute of Occupational Safety and Health (NIOSH) has proposed that no worker be exposed to a time-weighted average (TWA) concentration of free silica greater than 50 $\mu g/m^3$ as determined by a full-shift sample for up to a 10 hour workday, 40-hour workweek (2). In addition, a second level of 25 $\mu g/m^3$ free silica would be considered "an exposure". The method of analysis previously recommended by NIOSH is an X-ray diffraction method developed primarily by Bumsted (3,4) who cites a detection limit of 5 $\mu g/cm^2$. The method involves ashing the polyvinyl chloride (PVC) membranes on which the samples have been collected and

redepositing the residue on a low background silver membrane filter
together with an internal standard such as CaF_2.   This method suffers
several disadvantages. Specifically, the sample must be transferred
and significant time and expertise must be invested in sample preparation (5).   Several laboratories, including our own, have been working
on methods to analyze the silica directly on the PVC filters by X-ray
diffraction (6,7).

EQUIPMENT

A Philips Electronic Instrument vertical diffractometer which is
equipped with a graphite monochromator, enclosed beam path, $CuK\alpha$, theta
compensating slit and flat sample spinner was utilized for these
studies. The beam path was modified to permit a helium purge. Two
copper tubes, the Philips fine focus and the new long fine focus, were
tried, as well as two detectors, a flow proportional counter and a
scintillation counter. A 0.01 mm receiving slit (but no incident beam
filter) was used. The generator is a full wave rectified Philips type
12045. A Perkin-Elmer Autobalance AM-2 was used in the preparation
of standards. This electromagnetic microbalance is readable to 1 µg in
the 2 mg range and has a 3g capacity using counterweights.

PVC membrane filters used were the MSA type FWS-B, 5 µm and the
Millipore, 0.6 µm filter. Silver membrane filters were the Selas Flo-
tronics FM-25, 0.45 µm. All discussions of PVC filters refer to the
MSA filter unless specifically noted otherwise.

Scanning electron microscopy was performed on a Cambridge Stereo-
scan Mark II. Particle size distributions were determined with a
Reichert Zetopan research microscope and a JEOL 100C transmission
electron microscope.

DISCUSSION

Preparation of Standards

The silica standards were prepared in the following way. The
ground quartz powder that passed through a 400 mesh screen (<37 µm)
was suspended in ethanol (0.5 g in 400 ml). The material in suspension
was decanted off and discarded after about two hours. The remaining
SiO_2 was suspended in a fresh aliquot of ethanol. After two hours, the
liquid was again decanted and the residue dried. Approximately 99% of
these particles had a maximum size <30 µm, median value ∿5 µm. About
1 gram of this material was suspended in 400 ml of filtered, deionized
H_2O (0.2 µm filter) and mixed vigorously. After 12 minutes, the liquid
was decanted and the water evaporated. In this residue ∿90% of the
particles have a size distribution between 1 and 10 µm and 99% of the
are <19 µm. The median value of this size distribution is 4.8 µm.
Size distributions were obtained by electron microscopy. A stock solution was made by adding 100 mg of this residue to 500 ml of filtered
water along with a few drops of a surfactant solution (Dowfax 9-N-15).
A dilute stock solution was made by pipetting 5 ml of the above solution into a 250 ml volumetric flask to make up a solution which contained 4 µg/ml.

The PVC filters were dried in a vacuum oven at 50°C, ∿200 mm for ∿24 hrs. and then weighed. The desired quantity of freshly shaken stock solution (4 µg/ml) was pipeted into a 100 ml beaker with enough filtered H_2O to yield a total of ∿80 ml. After ultrasonic treatment for 15 minutes the solution was deposited on the MSA filters, washed to ensure complete transfer and dried in the vacuum oven overnight. The filters were reweighed and then scanned. A calibration curve was constructed from this data (Figure 1).

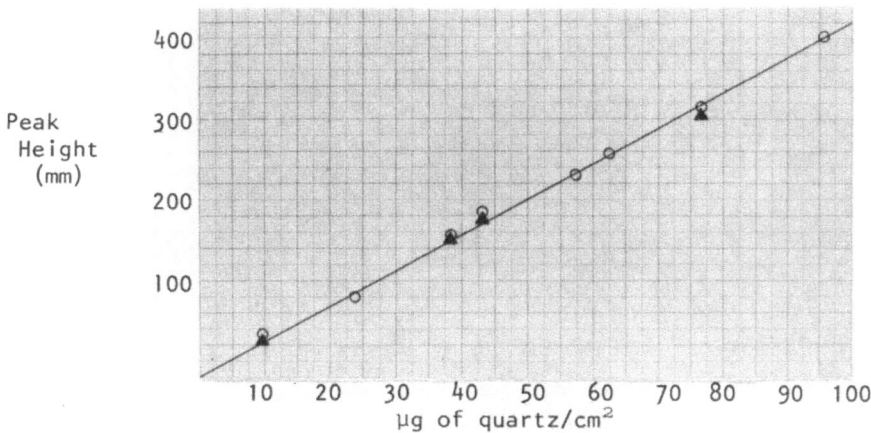

Figure 1. Calibration curve of quartz standards prepared by deposition of known quantities of quartz aqueous suspension on MSA 5 µm filters. Encircled points are pure quartz; points enclosed in triangles are a mixture of quartz and iron oxides in the weight ratio 4:1, respectively. (see Matrix Effects)

Methodology

MSA filters used in the preparation of wet standards were weighed before and after deposition with a microbalance. Several blanks were included which received the identical treatment given to the standards. It was found that on 84% of the 30 filters weighed, microbalance data agreed with dilution data to within ±12 µg. Occasionally, however, large discrepancies were encountered. Further, plots of peak intensity versus weight added, as indicated by microbalance data, did not yield a calibration curve with less scatter than that based on dilution data. It thus seems that, while microbalance data can usually provide a check for quantity of silica deposited, it may be more valuable to invest time in multiple determinations based solely on dilution data.

An interior section with a diameter of 25 mm was cut out of field samples collected on 37 mm MSA filters. This center section was mounted and scanned at 0.25°/min. A standard filter is scanned periodically and used to correct for variations in instrumental effects. A baseline is drawn through the middle of the background noise and the peak height is measured. The quantity of quartz is then determined from the calibration curve.

As is seen from the scan of a filter with 9.6 µg of quartz per cm^2 (Figure 2) the peak-to-peak noise is ∿ 0.1 times the signal. Using the criterion that a peak must be twice peak-to-peak noise in order to be considered detectable this yields a threshhold value of 1.9 µg/cm² for the method. The same criterion applied to the quartz peak at 4.26Å leads to a threshold value of ∿15 µg/cm².

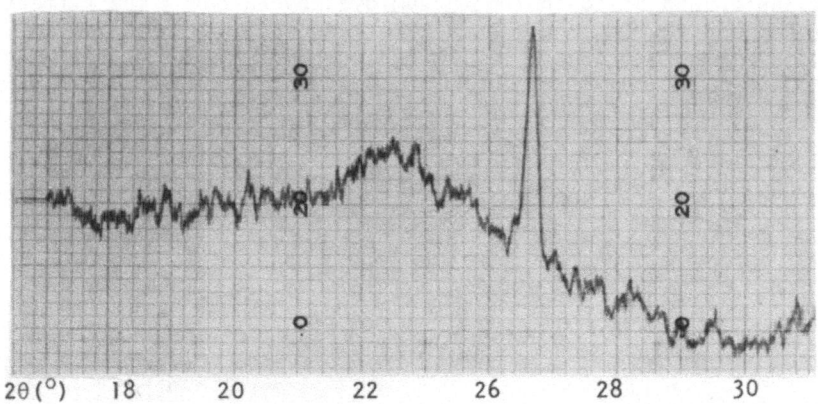

Figure 2.   Diffractometer scan of 9.6 µg of quartz per cm² on a 5 µm MSA filter deposited from a suspension of quartz.   CuKα 101.

For the proposed limit of 50 µg/m³ of air the following is valid:

$$50 \text{ µg/m}^3 \times R \text{ ℓ/min} \times 60 \text{ min/hr} \times T \text{ hrs} \times 10^{-3} \text{ m}^3/\ell$$
$$= 3RT \text{ µg on the filter}$$

where
  R = flow rate in ℓ/min
  T = sampling time in hrs

It therefore follows that at the suggested 1.7 ℓ/min and a sampling period of 8 hrs., 41 µg of SiO₂ would be deposited on a given filter. If the filter area is small, 2-3 cm², a detection of this amount or even half as much is easily achieved.   Even a conventional 37 mm filter would yield 4.8 µg/cm² for the above conditions.

Several laboratories are currently investigating the use of a cassette modified to accept a 25 mm filter (7).   This would produce a significant concentration effect at a given flow rate.

Additional concentration of the sample may be achieved by increasing the flow rate. Coulter counter data from our Midland laboratories indicate that the size distribution of quartz collected by a 10 mm nylon cyclone is virtually identical at 1.7 ℓ/min. and 2.1 ℓ/min. However, data taken at flow rates of 1.3 ℓ/min. and 2.5 ℓ/min. show significant shifts to a higher and lower particle size distribution, respectively (8).

Collection of field samples directly on silver membrane filters is currently practiced by some (9,10) and is under consideration by others. With the use of silver membrane filters directly, or indirectly via the Bumsted method, difficulties might arise if the particle size

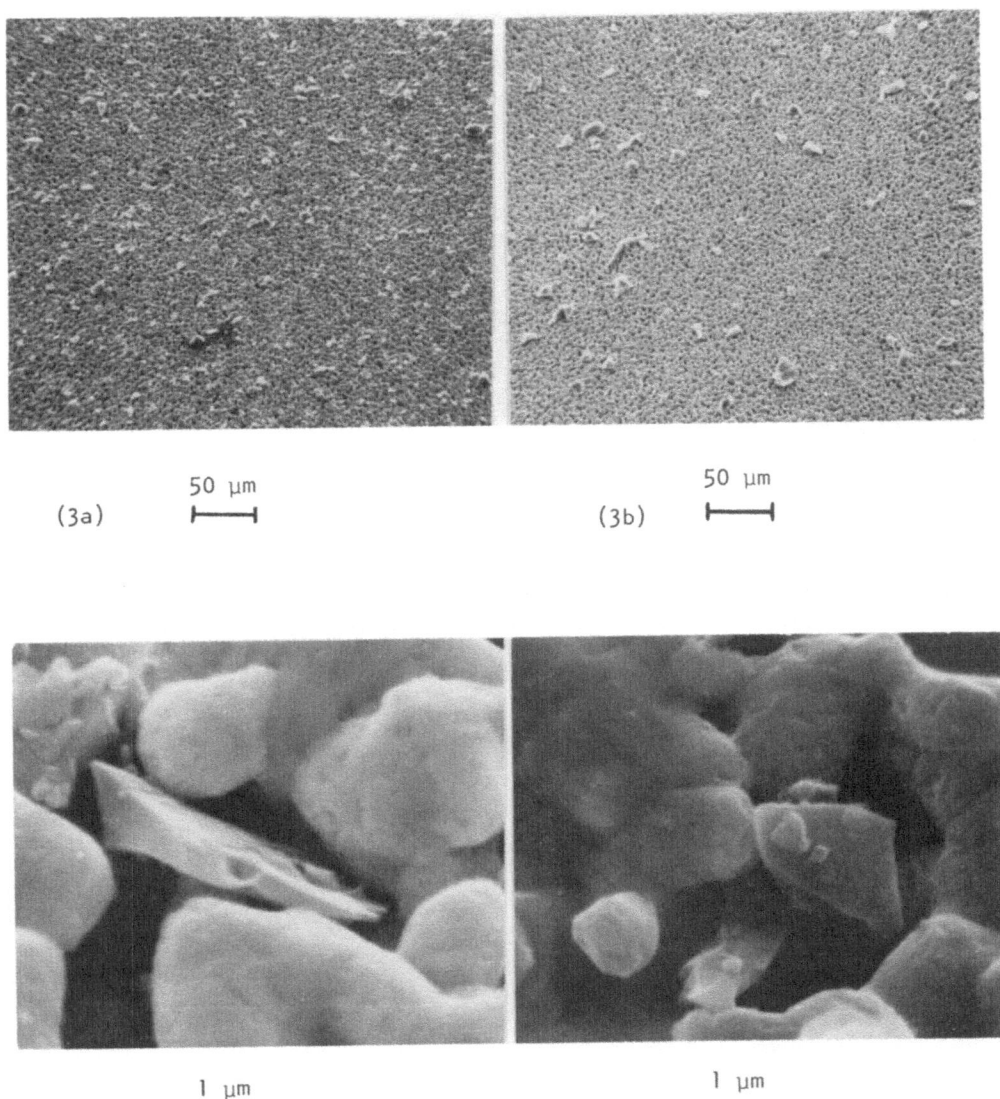

(3a)   50 μm
              ├──────┤

(3b)   50 μm
              ├──────┤

(3c)   1 μm
              ├──────┤

(3c)   1 μm
              ├──────┤

Figure 3.   Electron micrographs of quartz standards deposited from
            aqueous suspension on 0.45 μm silver membrane filters:
            (3a) 38 μg of Minusil*5 μm quartz per cm^2; (3b) 38 μg of
            quartz per cm^2 (99% of the particles with the maximum
            size >1 μm but <19 μm); (3c) closeup view of Minusil 5 SiO_2
            particles in which shadowing by the surrounding silver
            would occur.
            *Trade name for SiO_2 Standard made by Pennsylvania Glass
             Sand Corporation, 3 Penn Center, Pittsburgh PA 15235

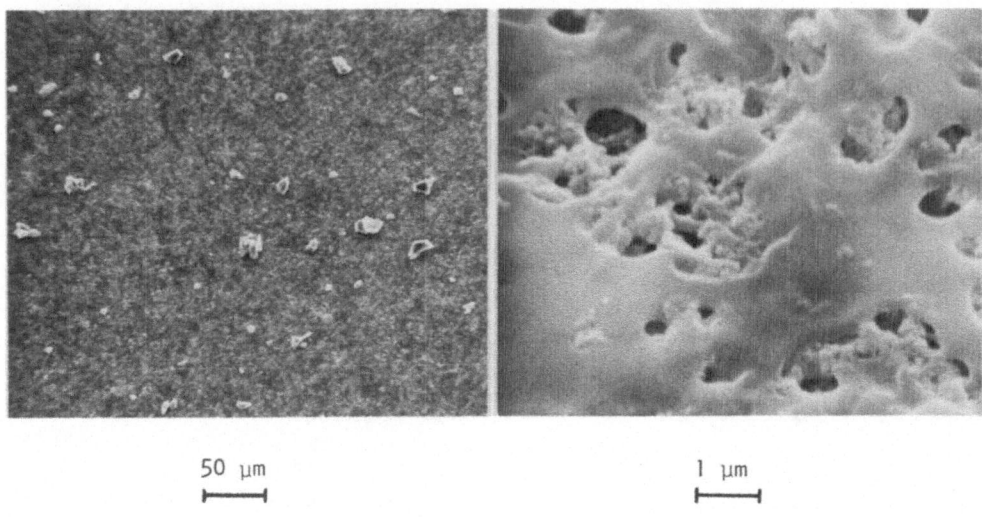

50 µm

1 µm

Figure 4.   Electron micrograph of a quartz standard (38 µg/cm²) deposited
from aqueous suspension on a 0.6 µm Millipore PVC filter.
Material in large pores   is low molecular weight polymer.

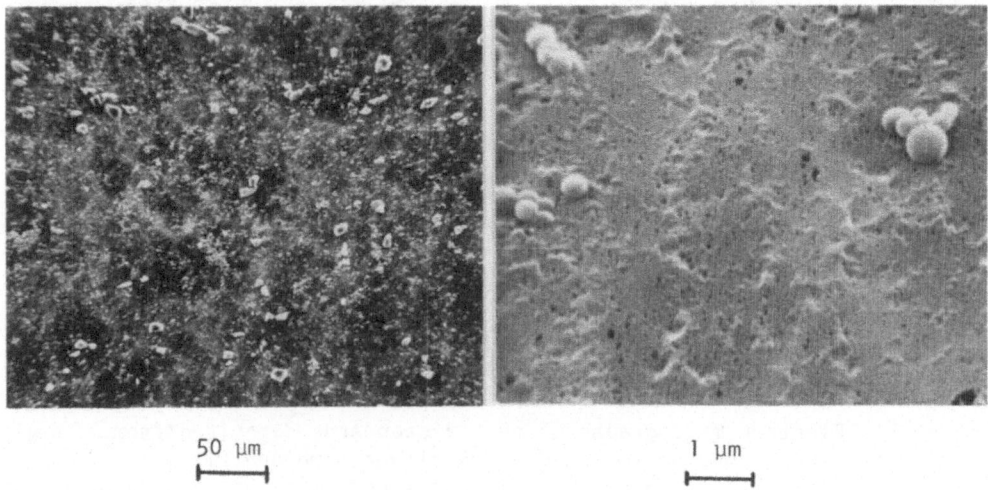

50 µm

1 µm

Figure 5.   Electron micrograph of a quartz standard (38 µg/cm²) deposited
from aqueous suspension on the smooth side of a 5 µm MSA
filter. Spheres of low molecular weight polymer are visible
in the high magnification picture.

distribution of the field sample is very different from that of the
standard. This is particularly important because of the shielding
effect of the heavy metal. Knight et al (10) stated that, for very low
masses of dust, only 55 to 60% of the quartz was detected using a 5 μm
silver membrane filter. We note a decrease of ∿25% in signal for a
given amount of quartz deposited from our Minusil 5 μm standard versus
our quartz standard in which most of the material <1 μm has been removed (see Preparation of Standards). The smaller effect may be due
to the difference in pore size of the silver filter used, 5 μm versus
0.45 μm. Electron microscopy reveals that some particles, but not the
majority, are deposited in such a way that shadowing by the silver
would occur at low angles (Figures 3a, b, and c). The peaks determined
from our quartz standard on 0.45 μm silver filters fall on a straight
line but have an intensity ∿15% lower than the line determined by the
same standard on MSA filters. No significant differences were observed
when the front and back side of a blank silver membrane were examined
by electron microscopy.

Another filter successfully used for direct determinations (7) is
the 0.6 μm Millipore PVC filter. Electron micrographs of the front side of
this filter reveal a surface which is somewhat rougher than the smooth
side of the MSA filters (Figures 4 and 5). Scans of our standard on
the 0.6 μm Millipore filters (top side as received) reveal intensities
which are ∿15% lower than those observed using MSA filters. A more
dramatic decrease (∿50%) is observed when Minusil 5 is used with Millipore filters. Calibration curves for our standard and Minusil 5 on
MSA, Millipore and silver membrane filters, as well as investigations
using a standard with 90% of the quartz particles between 1 μm and
6 μm, will be reported in the near future.

Scans of blank MSA, Millipore and Selas silver membrane filters
reveal little difference in peak-to-peak noise of our conditions of
operation (Figure 6). The Millipore filter gives rise to a baseline
with more slope at 26.7° than the other filters.

(a) Millipore
    (0.6 μm)

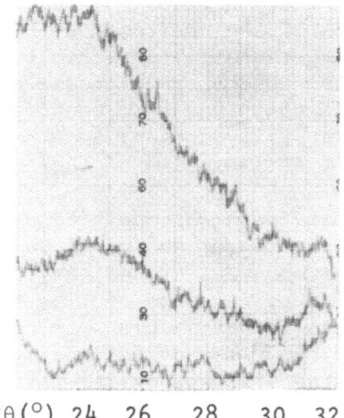

(b) MSA (5 μm)

(c) Selas Silver
     Membrane (0.45 μm)

2θ(°) 24   26   28   30   32

Figure 6. Diffractometer scans of blank filters; (6a), 0.6 μm Millipore
filter; (6b) 5 μm MSA filter; (6c) 0.45 μm Selas silver membrane filter.

Experimental Techniques

As discused below, a number of factors were found to critically influence the analysis.

1. Rotation of the sample is preferred in order to achieve reproducible results. The larger the size of the silica used in the preparation of standards the more important this variable will be, especially with light loadings. Differences of several hundred per cent were noted with changes in the orientation of the filter using a stock solution (see preparation of standards) in which ∿99% of the SiO_2 crystallites exhibit a maximum dimension of <30 μm with a medium value of 5 μm. These same filters, however, produced a reproducible, straight-line calibration curve when a sample spinner was employed. Using a stock solution which more closely approximates the "respirable" distribution, as described under preparation of standards, the effect was greatly diminished, but not eliminated. Differences as large as ± 13% of the mean intensity with a loading of 77 μg/cm^2 and ± 10% with a loading of 43 μg/cm^2 were observed. As noted by Bumsted (4), rotation of the sample would also average out the effects of uneven distribution on the filter.

2. Various ways to mount the filter were tried. The method initially adopted was to attach the filter to a machined aluminum disk with a small quantity of a hydrocarbon grease around the edge of the filter. This disk was designed to exactly fit the Philips sample spinner. The aluminum alloy used proved to have a diffraction peak close to that of quartz. Accordingly, various other materials were inserted between the PVC filters and the aluminum sample holder. These included a glass microscope cover glass, a silver membrane filter and thin metal sheets of silver or gold. The heavy metal foils gave the best background; however, they suffer the slight disadvantage that they must be periodically cleaned due to chloride formation via the breakdown of PVC by the X-ray beam. The filters are currently attached by a thin brass ring at the outer edge. Hydrocarbon grease also alters the weight of the filter and contributes to background. If used, the grease must be carefully chosen. Some have SiO_2 impurities and some yield diffraction peaks of their own (notably, Vaseline).

3. The enclosed path was modified at the specimen chamber to permit introduction of helium. A polyethylene bag was placed over the entire open end around the scintillation counter and monochromator. A range of flow rates was found in which the peak intensity was stable with respect to large alterations in helium flow. The difference in signal with and without a helium purge is substantial (∿30%) while the random noise, primarily the result of the filter, is unchanged.

4. When the first set of standards on MSA filters were scanned directly, a reproducible straight line calibration could not be obtained. This was ultimately traced to the fact that these filters exhibit a smooth side and a coarse side

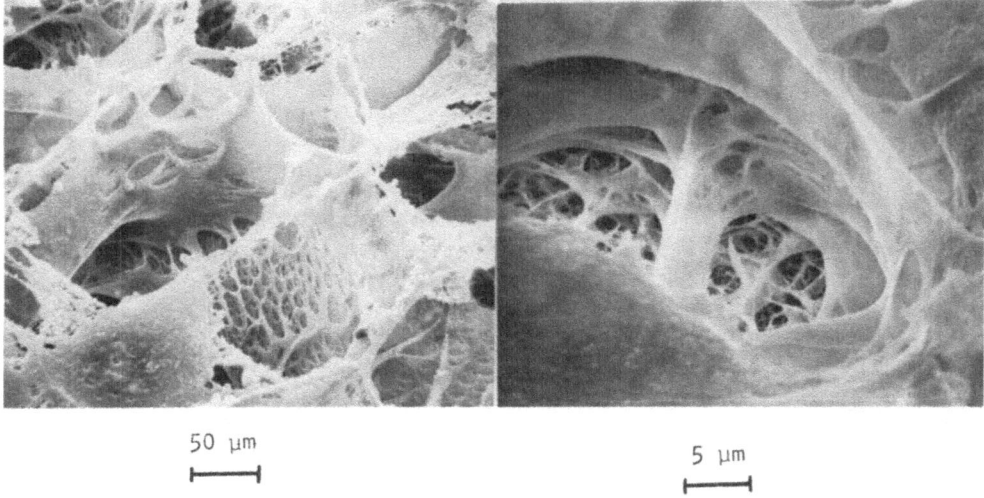

50 μm
├───┤

5 μm
├──┤

Figure 7.   Electron micrograph of a quartz standard (38 μg/cm^2) deposited from aqueous suspension on the coarse side of a 5 μm MSA filter.

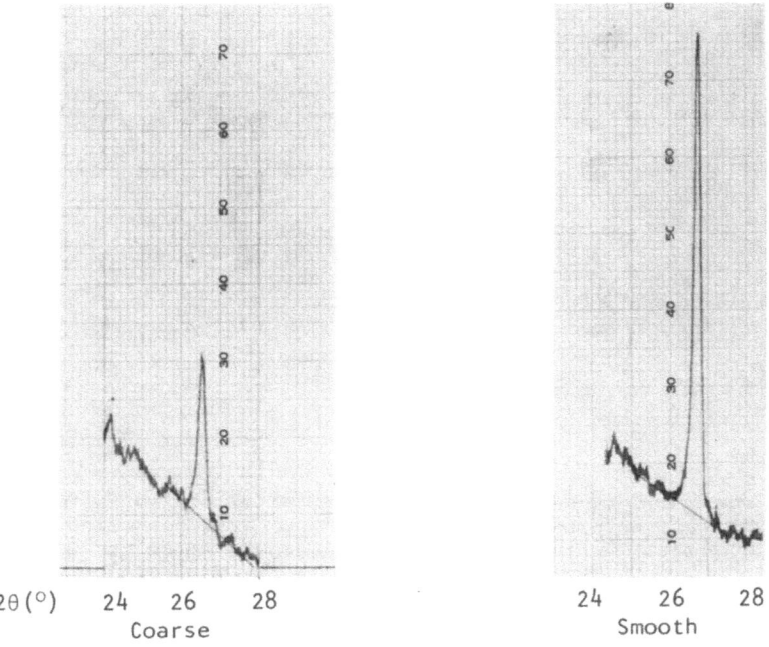

2θ(°)   24   26   28
Coarse

24   26   28
Smooth

Figure 8.   Diffractometer scans of 38 μg of quartz per cm^2 deposited from an aqueous suspension on the smooth and the coarse side of a 5 μm MSA filter.

(Figure 5 and 7). The difference in the surface structure of
the two sides is striking. The rough side exhibits very large
pores >100 μm while the smooth side has few pores >1 μm.
Differences in peak heights as large as 100% for a given loading of silica were observed for standards filtered with the
coarse side up versus the smooth side up (Figure 8). The
coarse side filters also exhibit peak broadening due to
deposition of the silica at different levels on the coarse
side. A serious question which remains unanswered to our
knowledge is whether or not the side presented to the air flow
in a cyclone affects the collection efficiency and/or the size
distribution collected. The 0.6 μm Millipore filter also
exhibits a coarse side but the difference is not as pronounced
as it is with the MSA (Figure 9).

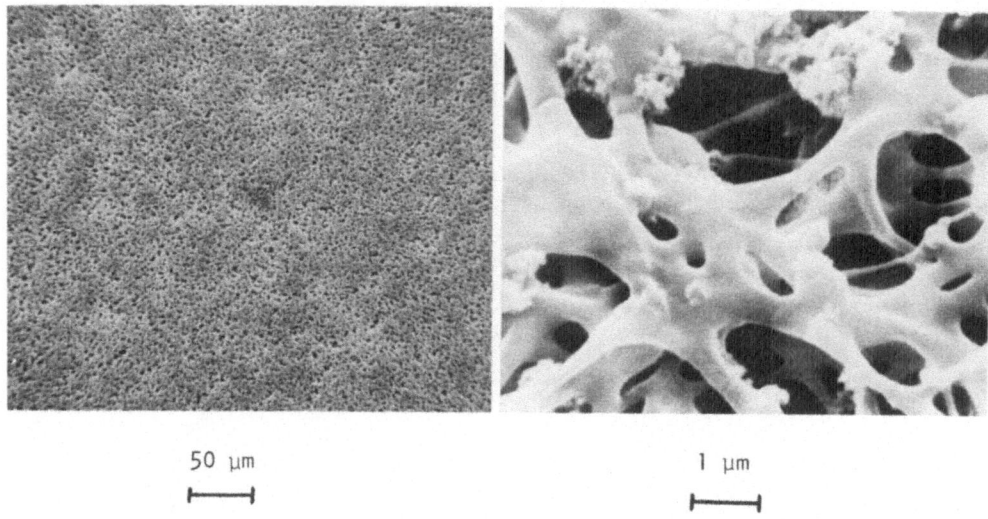

50 μm
├──────┤

1 μm
├──────┤

Figure 9. Electron micrograph of the back (coarse) side of a 0.6 μm
Millipore filter.

5. As mentioned under equipment, two X-ray source-detection
systems were used in this work. The first consisted of a
Philips fine focus copper tube operated at 40 KV, 20 ma and
a flow proportional counter. The present system consists of
a Philips long fine focus copper tube operated at 50 KV, 30 ma
and a Philips scintillation detector. While a definite improvement in signal-to-noise ratio has been noted with the
second system, satisfactory detection limits ($\sim$5 μg/cm^2) were
easily achieved with the earlier equipment.

6. As stated under equipment the diffractometer is equipped with
    the Philips theta compensating slit (11). The purpose of
    this device is to automatically open the divergence slit
    adjacent to the source to provide higher intensity as 2θ in-
    creases at the expense of resolution. The effect of this slit
    is large in the region of interest when compared to 0.5°
    divergence slit but is negligible when compared to a 1° di-
    vergence slit. Accordingly, the absence of a compensating
    slit would not be a serious limitation.

7. When comparison of field samples to filtered standards are
    made, one of three conditions should be fulfilled. The effec-
    tive filter area can lie completely within the beam rectangle.
    This case has the disadvantages that part of the beam is
    irradiating filter which has no quartz deposited on it and the
    filter area would be quite small. Alternately, the filter area
    might be larger than the diagonal of the beam spot. In this
    case some quartz will be outside the area of irradiation.
    A convenient way to measure this diagonal is to expose a
    piece of film protected by a thin metal foil while spinning
    the sample holder. In the two cases cited above, the area
    of the wet standard would not necessarily have to equal that
    of the field sample as long as both were known. If the filter
    area does not completely inscribe the rectangle irradiated
    by the beam, or vice versa, the area of the standard should
    equal that of the field sample or appropriate correction
    factors must be applied. Centering of the filter area will
    also become critical for this last case.

Matrix Effects

The possibility of matrix effects due to the presence of iron
oxides was investigated. Iron compounds are a likely contaminant in
some field samples and fluorescence of iron presents a particularly
unfavorable situation when Cu radiation is used. Bradley (12) investigated this effect using equal weights of quartz and Fe_2O_3 (hematite) in the range from 50 $\mu g/cm^2$ to 350 $\mu g/cm^2$ of each. The curve
for the mixture deviated slightly from that of pure quartz at the high
end (>200 $\mu g/cm^2$). The 200 $\mu g/cm^2$ figure corresponds to $\sim$140 $\mu g/cm^2$,
as iron. However, Knight et al (10) estimates a 25% suppression of
the quartz signal with the presence of 150 $\mu g/cm^2$ of iron, present as
pyrite.

For our investigation a 1:1 mixture by weight of Fe_2O_3 and Fe_3O_4
was obtained in which >90% of the particles were between 5 μm and 15 μm.
A mixed suspension was prepared with the iron oxides and quartz in a
1:4 ratio. Standards of this suspension were prepared in the manner
previously described with the following concentrations of quartz:
9.6, 38, 43 and 77 $\mu g/cm^2$. As indicated in Figure 1, the filters
containing iron oxides fall within the experimental error of the calibration curve determined by filters with only quartz present. Examination of the filter containing 38 μg of quartz and 9.5 μg of iron
oxides per cm^2 by scanning electron microscopy showed that most
particles were well separated. The absence of an effect is therefore
not surprising (Figure 10a). One sample was prepared using 38 μg of

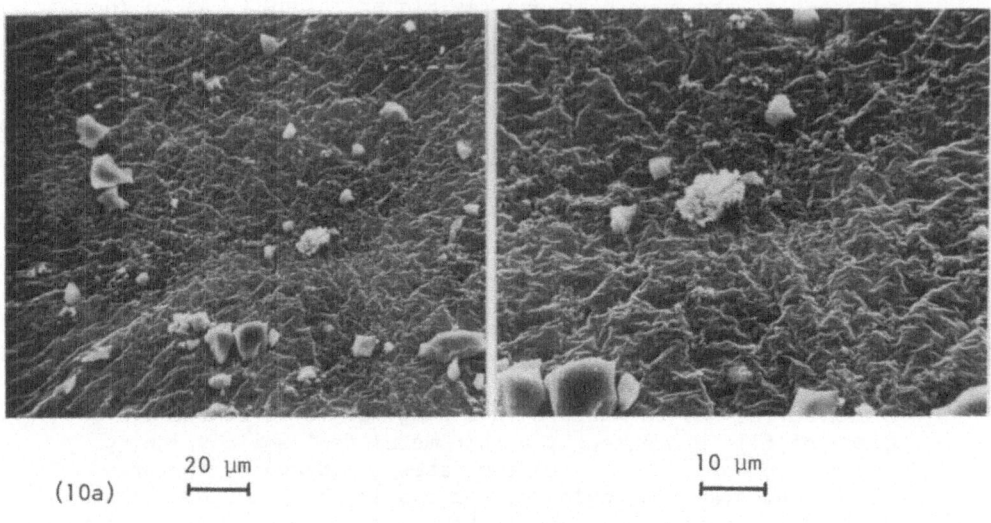

20 µm

10 µm

(10a)

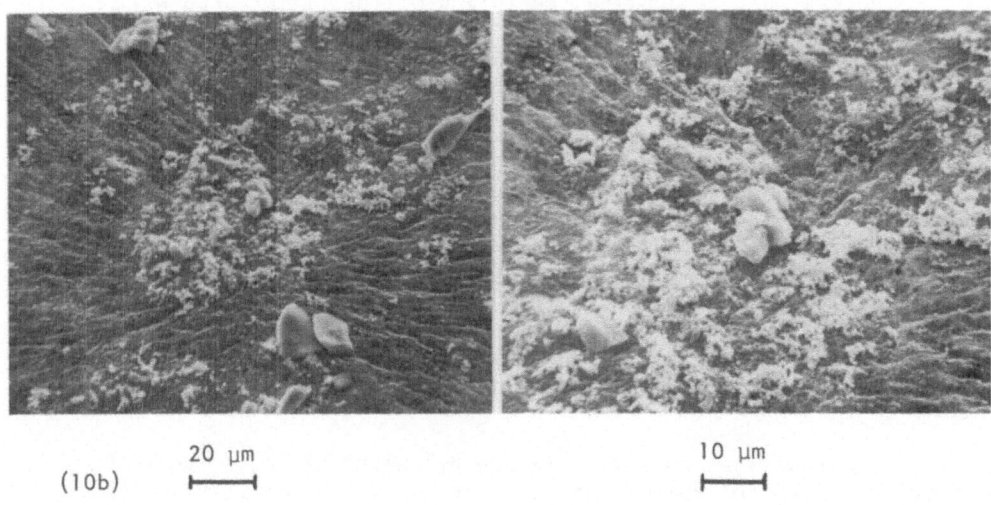

20 µm

10 µm

(10b)

Figure 10.   Matrix effects of iron oxides: (a) electron micrographs of
            38 µg of quartz and 9.6 µg of a 1:1 Fe_3O_4/Fe_2O_3 mixture
            per cm^2 on a 5 µm MSA filter; (b) micrograph of 38 µg of
            quartz and 38 µg of the iron oxide mixture per cm^2 on a
            5 µm MSA filter.

quartz and 38 μg of the iron oxide mixture per cm^2. A decrease in intensity of ∿35% was observed. Scanning microscopy shows intimate contact of the SiO_2 with the iron oxides (Figure 10b).

A study of the matrix effects of calcite ($CaCO_3$), as well as those of feldspar ($KAlSi_3O_8$), in the range from zero up to ∿25 μg of quartz per cm^2 has been reported (7). Mixtures by weight of 27-84% calcite,were found to lie on the same calibration curve as that of pure quartz. The same result was observed for binary mixtures of quartz and feldspar which were 36.5% and 59% feldspar by weight.

CONCLUSIONS

A method suitable for the determination of microgram quantities of quartz directly on the PVC filters used to monitor workers has been demonstrated. The method is much more rapid than that of Bumsted and is superior to it in sensitivity and reproducibility. Important questions concerning choice of filter, flow rate and suitable standards, as well as the problem of matrix effects, remain only partially answered. The value of electron microscopy to characterize field samples and standards has been shown.

ACKNOWLEDGEMENTS

The authors gratefully acknowledge the work of L. D. Martin, M. R. McCullough and D. W. Krieg on the scanning electron microscope and D. C. Benefiel on the transmission electron microscope and for the review of the manuscript. In addition, conversations with J. W. Edmonds provided much useful feedback concerning this work.

REFERENCES

1. C. J. Williams and R. E. Hawley, "An Industrial Hazard Silica Dust", Amer. Laboratory, 17-27 (July, 1975).

2. NIOSH Criteria Document, "Recommendations for an Occupational Exposure Standard for Crystalline Silica", Federal Register, Vol. 39, No. 250 (December, 1974).

3. Ibid, Appendix II.

4. H. E. Bumsted, "Determination of alpha-Quartz in the Respirable Portion of Airborne Particulates by X-ray Diffraction", Amer. Ind. Hyg. Assoc. U., p. 150-158 (April, 1973).

5. J. W. Edmonds, Dow Chemical Company, private communication (1975).

6. S. S. Pollack, "X-Ray Diffraction from Micrograms of Quartz on Filters Using a Rotating Anode", Amer. Ind. Hyg. Assoc. J., 73-75 (January, 1975).

7. R. F. Gebhardt, "Respirable Dust Regulations and Measurement", paper presented to the Sept. 1975 meeting of the Portland Cement Assoc. at Tuscon, Arizona.

8. G. McGowan, Dow Chemical Company, private communication (1975).

9. J. Leroux and C. A. Powers, "Direct Xay Diffraction Quantitative
    Analysis of Quartz in Industrial Dust Films Deposited on Silver
    Membrane Filters", Staub-Reinhalt Luft, Vol. 29, No. 5, 26-31
    (1969).

10. G. Knight, W. Stefanich, and G. Ireland, "Laboratory Calibration of
    a Technique for Determination of Respirable Quartz in Mine Air",
    Amer. Ind. Hyg. Assoc. J., 469-475 (July, 1972).

11. R. Jenkins and F. R. Paolini, "An Automatic Divergence Slit for
    the Powder Diffractometer", Norelco Reporter, Vol. 21, No. 1,
    9-14 (1974).

12. A. A. Bradley, "The Determination of Quartz in Small Samples
    by an X-Ray Technique", J. Sci. Instru., Vol. 44, 287-288 (1967).

INTERNAL STANDARD AND DILUTION ANALYSES APPLIED TO THE KINETICS OF TiB

FORMATION

G. C. Walther and R. W. Gould

Materials Science and Engineering

University of Florida

Gainesville, Florida 32611

ABSTRACT

The reaction kinetics of forming TiB in a mixed powder compact were analyzed using internal standard and dilution methods. The dilution method was not successful due to insufficient dilution, while the internal standard method was satisfactory to confirm that diffusion is the rate controlling mechanism. Difficulties analyzing metal powders and diffusional film products in mixed powders are discussed.

INTRODUCTION

The resistance of TiB_2 to acid attack and oxidation and its high temperature strength are promising characteristics for applications in refractories, cermets, and for dispersion strengthening of metals. In contact with Ti, however, it transforms to TiB:

$$Ti + TiB_2 \rightarrow 2TiB$$

The rate of this transformation was not known (1) so it was felt the kinetics of this degradation merited further attention.

EXPERIMENTAL PROCEDURE

The particle size distributions of the starting powders were found to be in the 20-80 μm range for titanium, and the 20-30 μm range for TiB_2. These were mixed in equimolar amounts, pressed into pellets, and reacted in a high purity argon atmosphere. Reaction times were zero to 120 minutes at 100°C intervals in the temperature range 1100-1500°C. The resulting pellets were broken into nuggets, some of which were mounted and polished for metallographic examination while the rest were pulverized to pass a 325 mesh sieve. Ductile titanium particles greater than 325 mesh that did not grind were re-added to the sieved material.

Quantitative analysis was attempted using both the internal standard (2) and dilution methods (3). A Philips diffracted beam monochromated diffractometer equipped with theta compensating slits and a rotating sample holder was employed for the x-ray measurements. The sample was rotated at 20 rpm to reduce the effects of particles size, while the goniometer speed was 0.5° of 2θ per minute. The integrated intensity was measured by use of a timer–scaler operating during the duration of the peak and subtracting the background intensity from it. An earlier attempt to load samples using the McCreery technique (4) to improve sample packing was made (on a non–rotating sample holder), but the small improvement obtained for these powders did not justify the extra effort and the rotating holder was retained. Cu k-alpha radiation was used.

Sodium chloride was selected as the internal standard and dilutant material because it was the best pattern found in the JCPDS powder files with strong, sharp peaks that did not interfere with those of the titanium and boride phases. It also grinds easily, has a low absorption coefficient ($\mu/\rho = 76$ cm^2/gm), is stable and readily available. The calibration standards were made by mixing reactant and product powders in the appropriate amounts to give nominal volume fractions of 0, 10, 25, 50, 70, 90, and 100 percent TiB. The reactant powder was obtained by mixing equimolar amounts of –325 mesh titanium and TiB$_2$ powders with –325 mesh NaCl. Since commercial TiB was not available, it was made by reacting several pressed pellets of mixed titanium and TiB$_2$ at 1500°C for 2 hours. These conditions were chosen to avoid the formation of eutectic liquid and because longer times did not provide a greater yield of TiB. The product mixture for standards was made by pulverizing this to pass 325 mesh and mixing it with NaCl.

Two levels of dilution were used: standards and samples were run with ten weight percent NaCl and then further diluted to 30 weight percent and run again. The standard calibration data was obtained by averaging ten runs on each standard but only four runs were averaged for the reacted samples. Two TiB peaks (d = 2.140 Å and d = 2.161 Å) were combined to increase their intensity and avoid overlap problems. Volume fractions for each sample constituent were found from the calibration curves. The extent of reaction was calculated as

$$x = \frac{V^{TiB}}{V^{TiB} + V^{TiB_2}}$$

and used to evaluate the kinetics of the reaction in terms of several spherical shell models (5).

RESULTS and DISCUSSION

Metallographic examination and the results of several additional experiments (6) showed that the reaction initiates at the contact points between titanium and TiB$_2$ particles and proceeds rapidly due to surface diffusion of titanium over the TiB$_2$ particles. As smaller particles of TiB$_2$ are completely converted to TiB, larger particles develop a coating of TiB product through which the titanium diffuses to continue reacting with the TiB$_2$ core.

The results of x-ray error determinations show that rotating the sample significantly reduced the experimental variability compared to the non-rotating case. The machine error (run ten times on a single sample) was about 0.5 percent while the particle orientation, sample packing, and mixing error (with samples remounted after each run) was between 1-2%.

The internal standard calibration curves (Figures 1 and 2) are approximately linear, as expected. The reaction to form TiB material for standards did not go to completion in a reasonable length of time and residual amounts of Ti and TiB_2 contaminated the nominal volume fraction quantities. Attempts to remove this residue by dissolving in several different acids did not prove satisfactory and were abandoned. Utilizing the method of additions (inset, Figures 1 and 2) the residual amounts were determined to be four volume percent titanium and six volume percent TiB_2, so the nominal volume fractions were corrected by these amounts. The intensity ratios for the 30 weight percent NaCl standards are about half of that for the ten weight percent case. However, the direct intensities for both cases are of comparable magnitude (Figures 3 and 4). This may be attributed to the lower mass absorption coefficient of NaCl and the ability of finely ground NaCl particles to pack between the larger titanium and boride phases. Hence the quantity of material exposed on the surface to the x-ray beam and their relative intensities are not significantly reduced over this range of dilution. This factor also makes the dilution standards unreliable (3). The dilution standard data had a large amount of scatter, including negative numbers, so analysis using the dilution method was not possible.

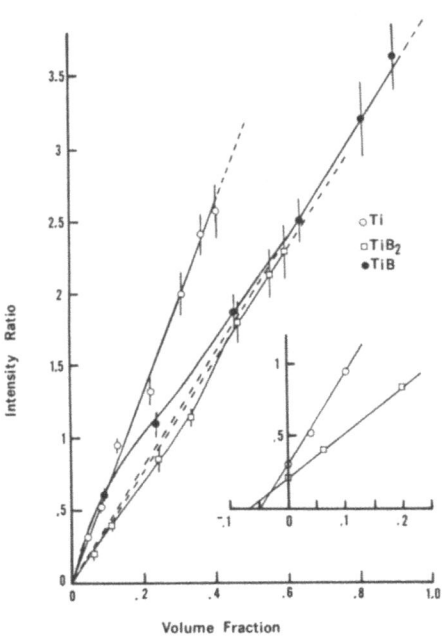
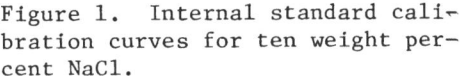
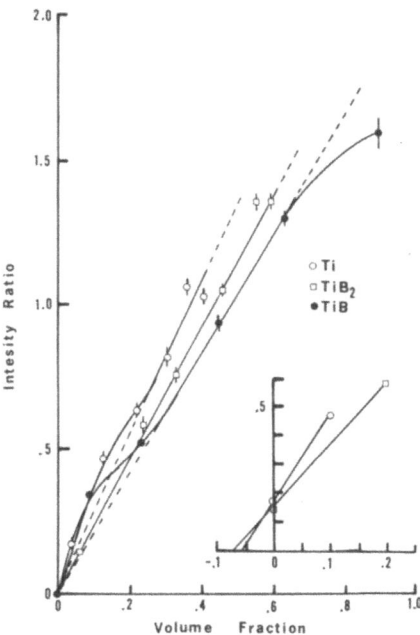

Figure 1.   Internal standard calibration curves for ten weight percent NaCl.

Figure 2.   Internal standard calibration curves for 30 weight percent NaCl.

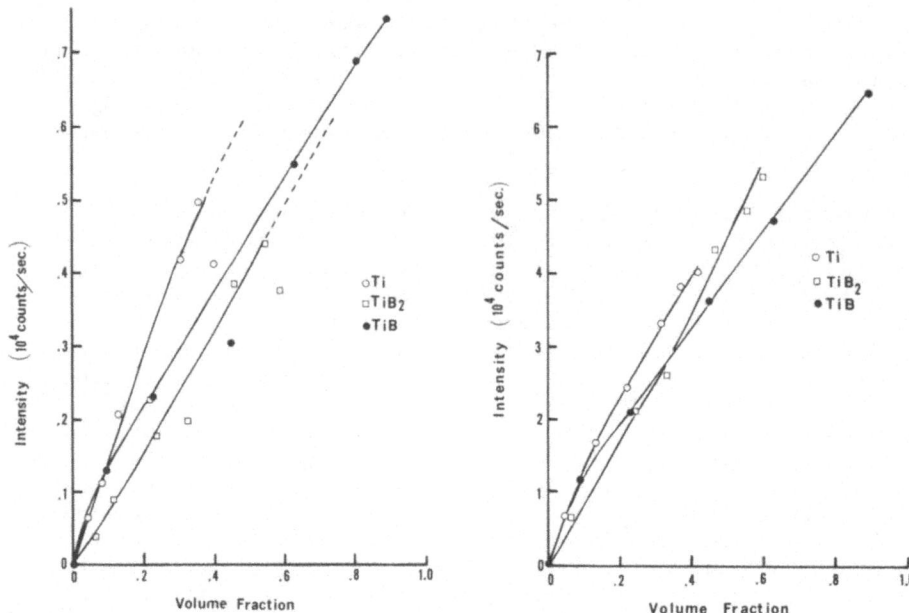

Figure 3. Direct intensity of ten
weight percent NaCl standards.

Figure 4. Direct intensity of 30
weight percent NaCl standards.

        Using the standard curves and the resulting sample data, volume
fractions as a function of time were determined. The resulting data for
the 1200°C reaction sequence with both ten and 30 weight percent NaCl
are given in Figures 5 and 6 respectively. The solid lines were obtained

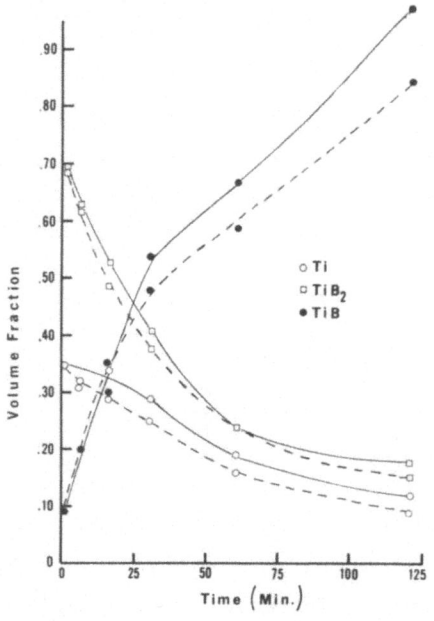

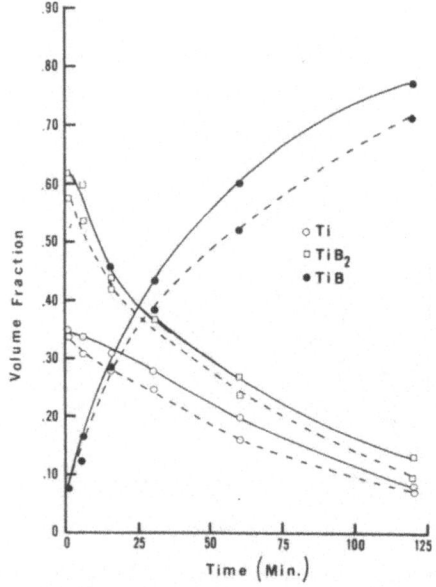

Figure 5. Volume fraction results
for 1200°C using ten weight percent NaCl dilutant.

Figure 6. Volume fraction results
for 1200°C using 30 weight percent
NaCl dilutant.

using the internal standard curves of Figures 1 or 2 and the dashed
lines resulted from using the direct intensity curves of Figures 3 or 4.
Agreement between the internal standard and direct intensity methods is
good and the results are also similar for both dilution levels. However,
the initial titanium volume fractions are lower than expected (∿40 percent) and the TiB_2 fractions are higher than they should be (∿60 percent).
The total of these volume fractions is ∿110-130 percent. Similar results
apply to other temperatures also. There are several differences between
samples and standards that may account for this.

1. The ductile titanium grains in the samples that did not pass the
325 mesh sieve were re-added to the already sieved material while -325
mesh titanium powder was used in mixing the standards. Thus, the larger
titanium particles from the samples would be under-exposed on the surface
layer in comparison to the titanium powder used in the standards.

2. A similar effect is caused by pulverizing the sample, where the
harder boride phases (and NaCl) become embedded in the ductile titanium
and/or form an adherent coating of fine particles (Figure 7), which decreases titanium intensity.

3. The standards were made from mixed, separate powder phases,
while in the samples, the TiB and TiB_2 phases were present in the same
particles. During pulverizing, the TiB could "flake off" but still
leave a coating that would diminish TiB_2 intensity and hence distort the
boride volume determinations. Calculation of the mass absorption coefficients and the penetration depth (∿11μm) show some absorption effects
could occur. However, the coating before grinding is only ∿10 μm, so if
it flakes off an even thinner layer should remain. Further, metallographic examination of a pulverized sample showed most fractures to be
across the particles and not just flaking, so each boride (TiB and TiB_2)
should have been exposed equally to the x-ray beam.

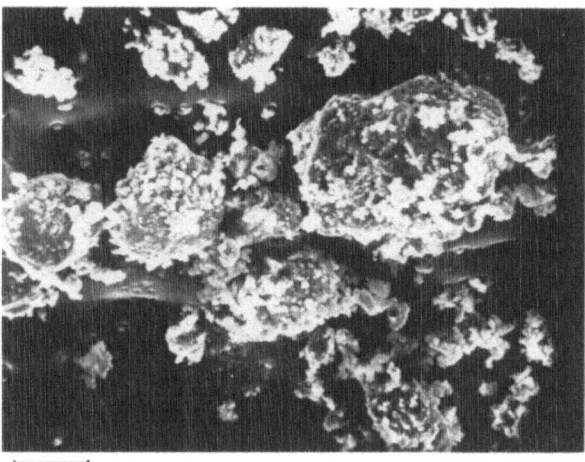

10 μm

Figure 7. Pulverized sample with smaller grains
coating larger ones.

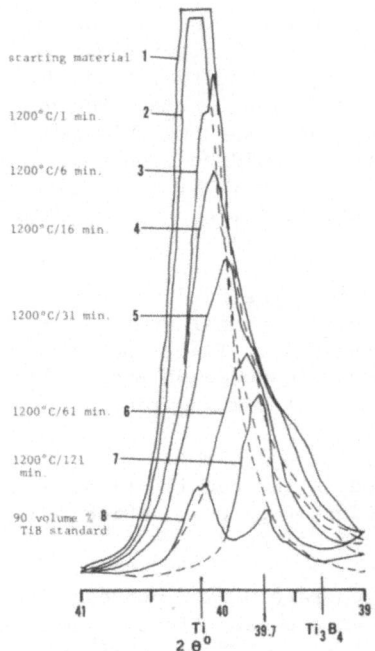

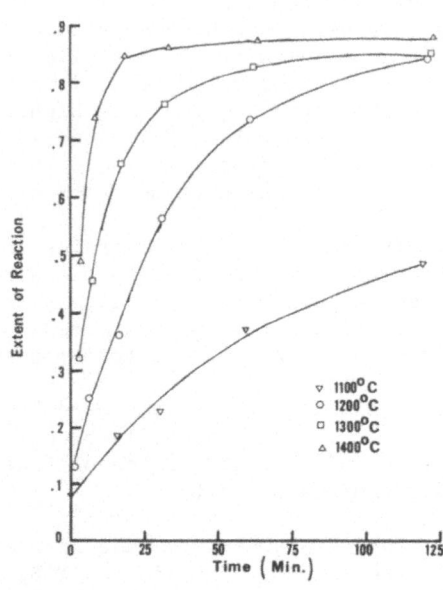

Figure 8. Peak shift for titanium   Figure 9. Extent of reaction for
(011) line.   ten weight percent NaCl dilution.

4. Another possiblity is peak overlap between titanium (d = 2.244
Å) and TiB (d = 2.346 Å) combined with an observed shift in titanium
peak with reaction time. This effect is shown schematically in Figure
8. The standards made with unreacted titanium showed the peak in its
normal position (curve 1), while for the samples, it has shifted (curves
2-7) toward its neighboring Ti-TiB peak (d = 2.342 Å and 2.346 Å, respectively) during the reaction. Thus, the number of titanium counts
made for the sample would be lower than the corresponding number obtained
from the standard and the volume fraction of titanium in the sample would
be underestimated. This is further supported by the results for the 90
volume percent standard (curve 8, Figure 8), which shows peaks from both
residual titanium in the TiB and four volume percent unreacted titanium
added to make the standard. This curve indicates a definite separation
of ∿0.45° 2θ. This shift could be due to formation of Ti_3B_4 or to solid
solution of B or some contaminant, but the final peak location is not
shifted sufficiently to support the conclusion that Ti_3B_4 has formed.
Examination of alternative major Ti_3B_4 peaks (d = 2.533 Å or 2.117 Å)
is not possible due to overlap with titanium and TiB peaks. Clark (7)
has shown the influence of oxygen and nitrogen on the titanium lattice
constants, but the results obtained here suggest the maximum amount of
these contaminants is already present in the starting titanium powder
(99.7 percent purity). The influence of boron solubility on the titanium
lattice could not be found (8). The exact cause of the shift is not
clear but Figure 8 shows it is real.

From the above discussion, titanium is the phase with the least
reliable volume measurement. Since the parameter of interest for the
kinetic analysis is the extent of reaction, x, or relative amounts of the
borides, this error is not serious if the borides are measured in the
proper proportions.

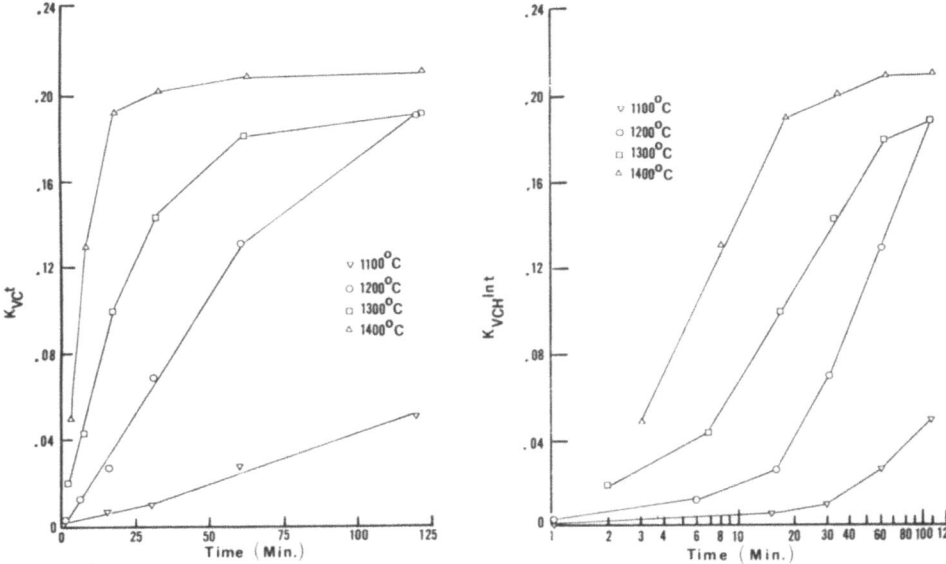

Figure 10. Valensi-Carter model results.

Figure 11. Hulbert modified model results.

The extent of reaction determined for samples containing ten weight percent dilutant is shown in Figure 9, and it increases with time and temperature, as expected. Figures 10 and 11 show the results for the Valensi-Carter and Hulbert modified models (5) of diffusion controlled reactions based on spherical particle geometries. A straight line indicates the data agree with the modeled mechanism. The good agreement seen in the early stages of reaction in Figure 11 is fortuitous since the actual microstructure does not fit the geometry assumed for the model. The agreement between the later stage data and the Hulbert modified model is more tenable. Hulbert incorporated the effect of a diffusion coefficient with an inverse (Tammann) time dependence into the Valensi-Carter geometry. This result is reasonable since the same effect is obtained during the later stage if the titanium available for reaction is reduced due to prior reaction, greater surface diffusion distances, and increased TiB product thickness through which volume diffusion must occur. This is in agreement with an independent analysis of this reaction by Quantitative Microscopy (6).

CONCLUSIONS

The internal standard analysis proved satisfactory (at two levels of dilution) to confirm that the reaction forming TiB is diffusion controlled. The attempt to use the dilution method did not work because of insufficient dilution for the second run, giving rise to unreliable intensity differences. The titanium measurements were compromised by the difficulty of grinding ductile metal powder and by the peak shift that occurred during the reaction. The results were also influenced by particle geometry and packing and by the presence of product phase coating reactant particles.

ACKNOWLEDGEMENTS

The authors would like to thank the Pratt and Witney Aircraft Co.
(Contract No. F33615-76-C-5136) and the Department of Materials Science
and Engineering of the University of Florida for partial financial
support. Thanks are also due to R. E. Loehman for helpful discussions
during the course of this work.

REFERENCES

1. L. V. Strashinskaya and A. N. Stepanchuk, "Contract Interaction of
    Titanium Diboride with Titanium, Zirconium, and Vanadium," Soviet
    Materials Science, 6, 722-25 (1970).

2. L. Alexander and H. P. Klug, "Basic Aspects of X-ray Absorption,"
    Anal. Chem., 20, 886-89 (1948).

3. N. H. Clark and R. J. Preston, "Dilution Methods in Quantitative
    X-Ray Diffraction Analysis," X-ray Spectrometry, 3, 21-25 (1974).

4. G. L. McCreery, "Improved Mount for Powdered Specimens Used on the
    Geiger-Counter X-ray Spectrometer," J. Am. Ceram. Soc., 32, 141-46
    (1949).

5. S. F. Hulbert, "Models for Solid-State Reactions in Powdered Com-
    pacts: A Review," J. Brit. Cer. Soc., 6, 11-20 (1969).

6. G. C. Walther and R. E. Loehman, "Kinetics of TiB Formation," to be
    published in the proceedings of the 8th International Symposium on
    the Reactivity of Solids, Gothenburg, Sweden, (1976).

7. H. T. Clark, "The Lattice Parameters of High Purity Alpha Titanium;
    and the Effects of Oxygen and Nitrogen on Them," Trans. AIME, 185,
    588-89 (1949).

8. W. B. Pearson, A Handbook of Lattice Spacings and Structure of Metals
    and Alloys, Pergamon Press, New York, Vol. 1, 1958; Vol. 2, 1967.

THE EFFECT OF THE K_α DOUBLET DIFFRACTED PEAK POSITION ON THE PRECISION OF THE LATTICE CONSTANT

C. P. Gazzara

Army Materials and Mechanics Research Center

Watertown, Massachusetts 02172

ABSTRACT

The effect of the position of the X-ray unresolved K_{α_1} - K_{α_2} line doublets, diffracted from powder specimens, on the precision of the calculated lattice constant has been determined using a least-squares analysis. An analytical procedure to synthesize CuK_α doublet X-ray diffraction peaks with X-ray characteristic lines (half widths ranging from 0.1° to 0.4° 2θ) has been applied in correcting the weighted wavelength of the doublet peak position. A series of correction curves was established from which the true 2θ peak position of the weighted K_α wavelength could be determined from the measured 2θ peak position.

Data taken from silicon powder and an yttrium aluminum garnet (YAG) powder standard to calculate the lattice constants indicate:

a. A loss in precision of approximately one part in 25,000 in determining the lattice constant is obtained when calculated using doublet 2θ peak positions applied in the conventional manner (λ weighted = 1.54178 Å, at peak, for copper).

b. An increase in precision is obtained by using low as well as high 2θ angle diffraction peaks for calculating the lattice constant.

The effect of the measured characteristic line width, simulated by modifying the slit system, on the calculated lattice constant was determined in the case of the YAG powder specimen.

Suggestions to increase the precision of the lattice constants determined from X-ray diffraction powder samples are presented.

INTRODUCTION

The determination of the lattice constant a_0 of a unit cell, using the positions of characteristic X-ray lines diffracted from a powder sample, has been treated by many investigators. Calculations of the lattice constant made by extrapolating d values against various functions of the Bragg angle θ have been made using a least-squares analysis. A summary

of the systematic error functions associated with an X-ray diffractometer
has been presented by Vassamillet and King (1), who found that the best
precision to be expected using a commercial diffractometer is approximately 1:20,000.

A comprehensive procedure that has been developed by Mueller, Heaton
and Miller (2) (MHM method) involves varying the systematic error functions and computing the change in d for each systematic error function,
or combination of functions, until the error in the computed lattice
constant has been reduced to an acceptable level.

In a study (3) of the systematic errors involved in the computation
of a_0 for yttrium aluminum garnet powder $Y_3Al_5O_{12}$ (YAG), it was found
that the precision of the computed lattice constant was a function of
the following.

(1) The slope of the least-squares straight line of the a_i values,
where a_i is the lattice constant computed for each diffraction line, versus $\cot\theta_i$. A slope value of O indicated that the systematic errors had
been adequately corrected for and yielded the most precise a_0 values.

(2) A statistical error term, i.e., standard error, absolute average deviation, rms deviation of functions of a_i and $\sin\theta_i$ from their
computed functions or other error terms. A comparison of these error
functions ultimately yield the same lattice constant. Whereas a difference was seen between error functions, their relative values were useful
in establishing the same most precise a_0 values.

(3) The number and selection of the diffraction peaks. Using all
35 diffraction peaks (unresolved K_{α_1} - K_{α_2} doublets and high-angle resolved peaks) yielded the highest error. Restricting the analysis to
the 12 highest 2θ angle peaks resulted in a reduction of 8% in error.
Taking the 20 highest 2θ angle peaks increased the precision, reducing
the error 28%. The fourth case gave the highest precision (42% reduction in error term), and occurred when 23 high- and low-2θ angle peaks
were chosen, eliminating those doublet peaks in mid-2θ range, along with
the diffracted peaks of low intensity.

Recently, in an attempt to relate the peak height to the integrated
intensity of X-ray diffracted K_α doublets by convoluting the characteristic peaks, it was found that the weighted K_α wavelength approximation was
inaccurate except at very low Bragg angles (4). It therefore seemed appropriate to determine what the effect of this doublet peak position
error would be in computing the lattice constant of a material whose lattice constant was known and to afford a means for correcting for the doublet position.

PROCEDURE

A doublet peak position correction curve could be generated for
CuK_α radiation from the curve shown in Figure 1, relating the value of
the positional parameter X_1 to a. X_1 is the position of the K_α doublet
peak, as determined from the analytical expressions describing the K_{α_1}
and K_{α_2} lines (4), relative to the K_{α_1} line. The parameter a is defined

as the K_{α_1} - K_{α_2} peak separation in terms of the wavelength dispersion.
Note that it is only in the linear portion of the curve, that the
weighted K_α peak position approximation is valid (i.e., $X_1 = a/3$).

The solid curve in Figure 1 was generated assuming a Gaussian dis-
tribution of the K_{α_1} and K_{α_2} diffraction peaks as a function of θ. The
dashed curve was computed by fitting a computed characteristic K_{α_1} peak
to an experimental CuK_{α_1} (531) Si diffraction peak, where the upper half
was assumed to be Gaussian and the lower half to have a Cauchy distribution (Gaussian and Cauchy).

It can be shown that the equation relating the change in wavelength
to a or X_1a is:

$$\Delta\lambda = \lambda a\omega_{h_{\frac{1}{2}}}/4.72 \ \tan\theta$$

where $\omega_{h_{\frac{1}{2}}}$ is the characteristic peak width at half peak height. Values
of $\omega_{h_{\frac{1}{2}}}$ can be obtained from Figure 2 from the measured peak width at half
height of the unresolved K_α peak doublet $\omega_{T_{\frac{1}{2}}}$. A series of curves was
generated, see Figure 3, relating the 2θ doublet peak position to the
t correction term whereby

$$2\theta_{corrected} = 2\theta_{measured} + t.$$

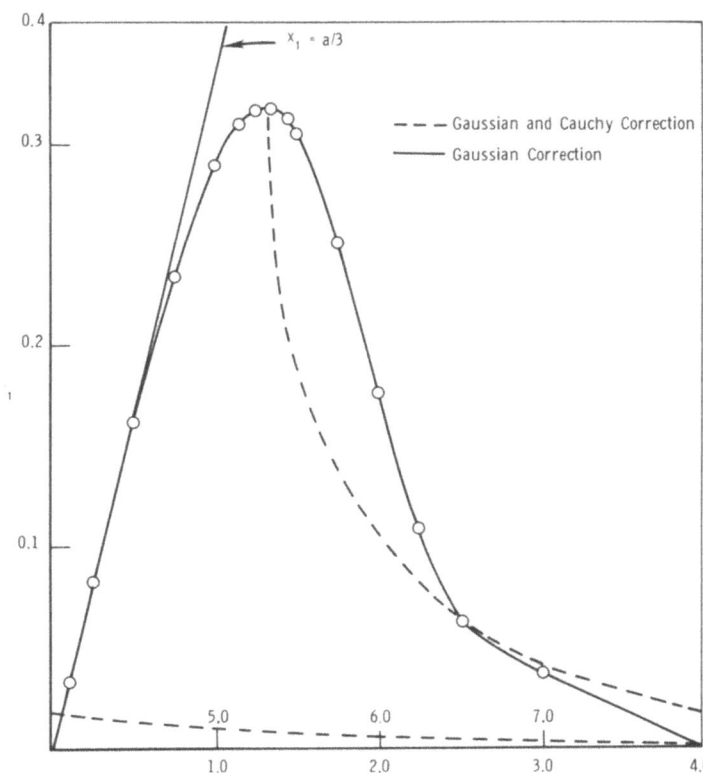

Figure 1. Plot of posi-
tional parameter X_1
versus CuK_{α_1} - CuK_{α_2}
diffracted peak separa-
tion a, assuming Gaus-
sian and Gaussian plus
Cauchy distribution.

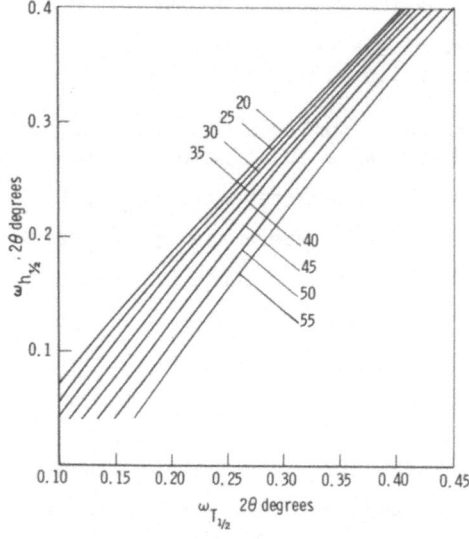

Figure 2. $\omega_{h_{1/2}}$ versus $\omega_{T_{1/2}}$ curves calculated for various levels of 2θ from 20° to 55° assuming Gaussian and Cauchy peak distributions.

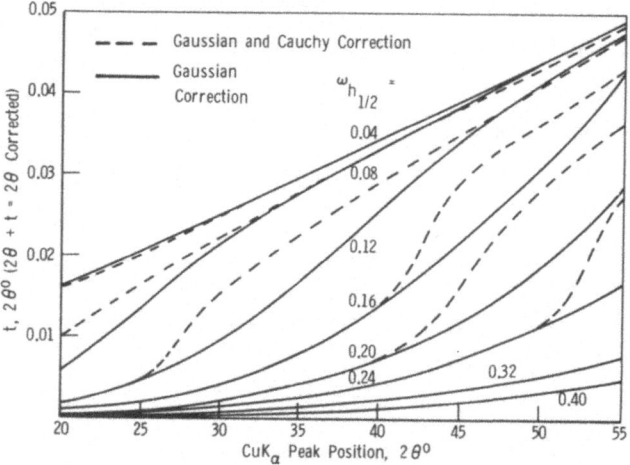

Figure 3. Plots of peak correction term t versus observed diffracted 2θ peak position of CuK_{α} doublets for constant values of $\omega_{h_{1/2}}$.

Figure 3 is also made up of solid lines relating to Gaussian diffraction peaks while the dashed curves were calculated assuming a Gaussian and Cauchy characteristic diffraction line.

To determine the effect such a t correction would have on the lattice constant precision, CuK_{α} diffraction peak positions of silicon powder were measured. A CuK_{α} powder pattern of silicon powder with a particle size of less than 30 microns was taken with a 0.05° receiving slit. A scanning speed of 0.10° (2θ)/min and a 1° (2θ) beam slit was used on a G.E. XRD-5 X-ray diffractometer. For the first three diffraction peaks an $\omega_{h_{1/2}}$ value of approximately 0.13° 2θ was determined from Figure 2 and applied in deriving the t correction in Figure 3. A least-squares extrapolation procedure utilizing a systematic error term, $\cot\theta$, and a weighting factor, $\tan\theta$, was employed.

The plane of the Si powder, characterized as a thin specimen (less than 0.002" thick), was adjusted with respect to the distance perpendicular to the plane of the specimen. The purpose for this operation was to minimize the slope (slope = $\Delta a_i / \Delta \cot \theta_i$), and thereby increase the precision of the a_0 measurement. Values of a_i are shown plotted versus θ in Figure 4 for three specimen settings. For position A, which represents the ideal alignment for an infinitely thick specimen, the extrapolated value of a_0 at $\theta = 90°$ is too high. The plotted values and computed curves applying the t correction are shown below the uncorrected curves. The specimen was moved inward 0.005" to position B, and the effects of the least-squares computation are shown in Figure 4. Next, the calculated slope values for positions A and B were plotted versus a_0 (see Figure 5) and, following a known procedure (3), the correct a_0 value was determined for slope = 0. Using this value of a_0, position C was found by varying the specimen position until the (111) CuK_α peak was maximized at the proper 2θ value for slope = 0. Plotting values of a_i for position C are shown in Figure 4 and the respective calculated values of a_0 are shown in Figure 5. The computed results of this procedure are summarized in Table 1.

Since a prior study of lattice constant measurements was performed with YAG powder, it was next deemed appropriate to ascertain the effects of the t correction on a YAG powder, where:

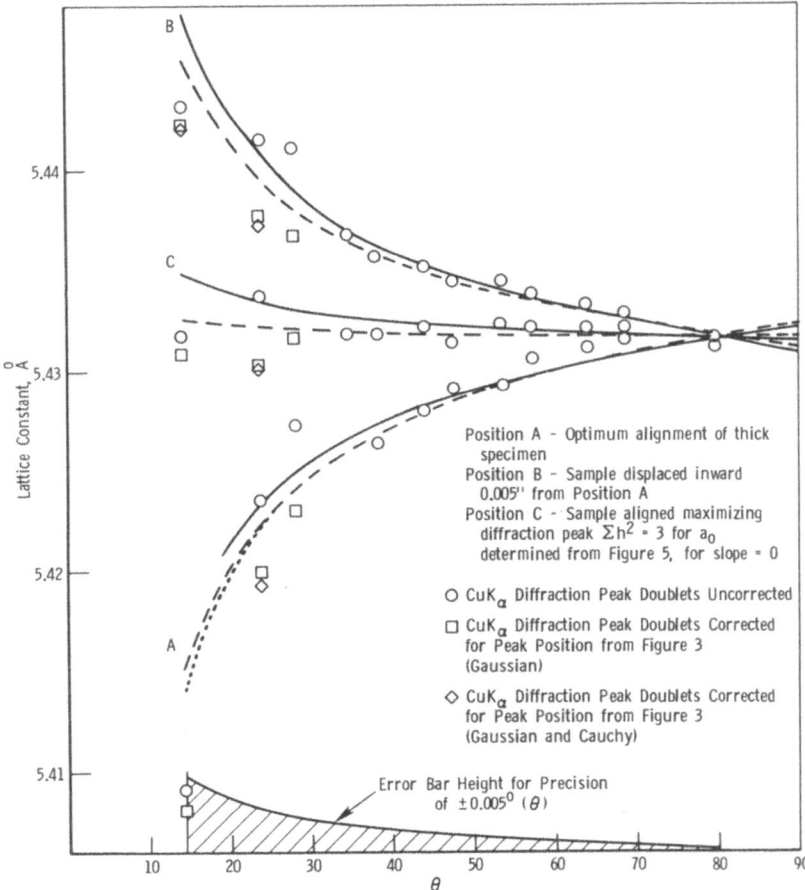

Figure 4. Values of lattice constant a_i calculated from diffracted peak position of family of planes $\{hkl\}_i$ versus Bragg angle θ for a thin silicon powder specimen.

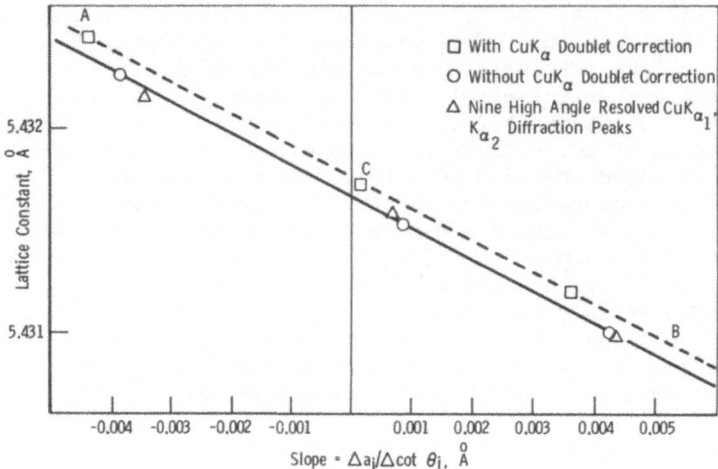

Figure 5. Lattice constant a_0 versus slope term for thin silicon powder specimen showing effect of t correction and the use of high angle diffraction peaks only.

Table 1. SILICON LATTICE CONSTANT VALUES MEASURED FROM THIN SPECIMEN AS A FUNCTION OF SPECIMEN DISPLACEMENT

(Powder Size Less Than 30 µM, 0.05° Receiving Slit, 1° Beam Slit, 0.10°/Min Scan Speed)

Specimen Position	Number of Diffracted Peaks	t Correction	a_0, Å	Slope	Weighted RMS Deviation
A (aligned for thick specimen)	12	None	5.4323	-0.0038	0.00130
	12	Gaussian	5.4325	-0.0044	.00120
	12	Gaussian & Cauchy	5.4325	-0.0045	.00124
	9 High Angle	-	5.4322	-0.0035	.00083
B (specimen displaced 0.005" inwards)	12		5.4310	0.0043	0.00083
	12		5.4312	.0036	.00070
	12		5.4312	.0036	.00073
	9 High Angle		5.4310	.0043	.00028
C (specimen in optimum alignment)	12		5.4315	0.0008	0.00072
	12		5.4317	.0002	.00049
	12		5.4317	.0002	.00051
	9 High Angle		5.4316	.0007	.00034

    a.   many more diffraction peaks are available (12 Si → 35 YAG);

    b.   an accurate value of the lattice constant was available for the YAG powder;

    c.   the effect of $\omega_{h_{1/2}}$ and the lattice constant could be found by varying the receiving slit width, from 0.05° 2θ to 0.10° 2θ to 0.20° 2θ;

    d.   the case of using only the high angle diffraction peaks could be examined; and

    e.   unresolved K_α doublets beyond the 2θ correction range could be eliminated (this case denoted as m). For a 0.05° receiving slit, peak $\Sigma h^2 = 52$ was eliminated, where $\Sigma h^2 = (h^2 + k^2 + l^2)$; for a 0.10° slit, peaks $\Sigma h^2 = 52, 54, 56$ were removed; and for a 0.20° slit peaks $\Sigma h^2 = 52$ to 88 were taken out of the least-squares analysis. The results of the analysis of the thin YAG powder specimen are summarized in Table 2.

RESULTS

The effect of the error in assigning a CuK_α wavelength of 1.54178 Å to the unresolved characteristic line doublet peak position, on the calculated lattice constant of either silicon or YAG always lowers the true value of a_0 by approximately one part in 25,000 (see Figure 5). It can also be seen from Table 1 and Figure 4 that the t correction is practically the same whether the K_{α_1} and K_{α_2} lines are considered to be Gaussian or Gaussian and Cauchy. It is also evident that with a t correction, the error terms decrease and the slope decreases when the systematic error function is properly treated. Table 1 and Figure 5 show the effect on the analysis resulting from eliminating the first three CuK_α silicon powder diffraction peaks and using the remaining nine higher angle peaks. The data points in Figure 5 are approximately the same as those obtained using all the diffraction peaks, without the t correction, in agreement with earlier work (3).

From Figure 4, notice the effect of improperly treating the systematic error in extrapolating to a_0. In this case an error of one part in 5000 in a_0 is possible, indicative of the importance of considering the slope term, unless a procedure such as the MHM method is employed.

The lattice constants computed from the experimental data obtained from the YAG powder may be seen in Table 2 and in the plot of Figure 6. The value of the lattice constant at slope = 0 was determined from an earlier MHM analysis to be 12.0107 Å (3).

In the case of YAG, it can be seen that the most precise values of a_0 are obtained with a 0.05° receiving slit, progressing from using the uncorrected K_α doublet peaks, to employing the t correction, to using the high angle peaks alone. These results represent the only conditions

Table 2. YAG LATTICE CONSTANT VALUES MEASURED
AS A FUNCTION OF RECEIVING SLIT SIZE

(Powder Size Less Than 30 μM, 1° Beam Slit,
0.10°/Min Scan Speed)

Receiving Slit (2θ°)	Number of Diffracted Peaks	t Correction	a_0, Å	Slope	Weighted RMS Deviation
0.05	35	None	12.0068	0.0063	0.00347
	34(m)	None	12.0069	.0060	.00330
	35	Gaussian	12.0071	.0056	.00329
	34(m)	Gaussian	12.0072	.0053	.00305
	23 High Angle		12.0075	.0047	.00195
0.10	35		12.0052	0.0081	0.00380
	32(m)		12.0054	.0076	.00357
	35		12.0055	.0075	.00401
	32(m)		12.0057	.0070	.00371
	21 High Angle		12.0050	.0085	.00233
0.20	35		12.0064	0.0126	0.00417
	26(m)		12.0068	.0113	.00277
	35		12.0066	.0123	.00428
	26(m)		12.0070	.0108	.00280
	15 High Angle		12.0068	.0113	.00187

where the high angle data gave the best results, in conflict with earlier work (3). The lowest lattice constant values are obtained with the
intermediate size receiving slit (0.10°) when only the high angle peaks
are utilized. For both cases, however, the change in lattice constant
with slope moves along the same line shown in Figure 6. For the 0.10°
receiving slit, when eight additional mid-2θ angle diffraction peaks are
removed from the analysis, the change in lattice constant with slope de-
parts from the curve. For the 0.20° receiving slit, the slope values
increase, indicating a further change in the magnitude of the systematic
errors, with a displacement of the points from the curve. The error
terms shown in Table 2 indicate no unusual behavior between cases except
that in treating the 0.20° receiving slit case, the sensitivity of removing the mid-range diffraction peaks seems to be more severe. This is
expected when one considers that the doublet correction t has not been
taken into account, since these corrections occur beyond the angular
range considered in Figure 3.

CONCLUSIONS

1. A correction for the 2θ X-ray diffracted peak position of CuK_α
unresolved doublets results in an increase in lattice constants, calculated from a least-squares extrapolation procedure, of one part in 25,000
for silicon and YAG powder.

2. The same peak position of the unresolved K_α doublet results from
assuming either a Gaussian or a Gaussian plus Cauchy distribution of the
CuK_{α_1} and CuK_{α_2} X-ray diffraction lines.

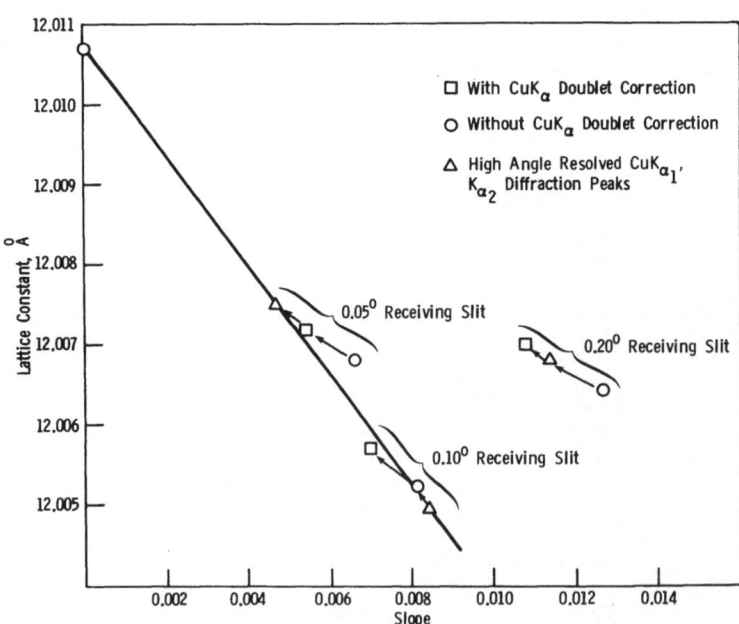

Figure 6. Lattice constant a_0 versus slope term for thin YAG powder specimen
showing the effect of t correction and the use of high angle diffraction peaks only.

3. In the absence of an extensive procedure such as that of MHM (2), an extrapolation procedure utilizing a plot of a_0 versus slope can be applied to minimize the effect of systematic errors and increase the precision in a_0.

4. The most precise values of a_0 are usually obtained when high- and low-2θ angle diffraction peaks are employed in a least-squares analysis. This is particularly true when few diffraction peaks are available, i.e., silicon powder with CuK_α. High precision can also be achieved when many diffraction peaks are available, i.e., YAG case, utilizing only the high 2θ angle diffraction peaks.

ACKNOWLEDGEMENT

The author wishes to thank Mr. T. Sheridan for assisting with the operation of the X-ray diffractometer.

REFERENCES

1. L. F. Vassamillet and H. W. King, "Precision X-Ray Diffractometry Using Powder Specimens," in W. M. Mueller et al., Editors, Advances in X-Ray Analysis, Vol. 6, p. 142-157, Plenum Press (1963).

2. M. H. Mueller, L. Heaton and K. T. Miller, "Determination of Lattice Parameters with the Aid of a Computer," Acta Cryst., 13, 828-829 (1960).

3. C. P. Gazzara, "The Effect of Systematic Errors on the Measurement of Lattice Constants," Army Materials and Mechanics Research Center, AMMRC TR 70-29 (September 1970).

4. C. P. Gazzara, "Peak Height Approximation for X-Ray Diffracted Integrated Intensity," in R. W. Gould, C. S. Barrett, J. B. Newkirk and C. O. Ruud, Editors, Advances in X-Ray Analysis, Vol. 19, p. 735-748, Kendall Hunt (1976).

ENERGY DISPERSIVE X-RAY DIFFRACTOMETRY

Michael Mantler and William Parrish

IBM Research Laboratory

San Jose, California 95193

ABSTRACT

This paper describes the principles, methods, instrumentation and results of EDXRD and a computer method of profile fitting to obtain corrected intensities and peak energies from isolated and overlapping reflections. The profile, P, of a diffraction peak is a convolution of the incident X-ray spectrum, X, the geometrical aberrations, T, the contribution from the specimen, S, and the detector resolution function, D:

$$P(E,\theta) = [X(E) \cdot T(E,\theta)] * S(E) * D(E)$$

where (E,θ) indicates dependence on energy and angle of incidence, respectively. T is the specimen transparency (which is the main source of profile asymmetry and is strongly dependent on E and the absorption coefficient) and the divergence of the beam along the irradiated specimen length. S is negligible for a well crystallized specimen but must be taken into account when the profile is broadened by small particle size, strain, etc. X(E) was determined from reflections at various incident angles of silicon single crystal plates cut parallel to (111) and (110) as well as tungsten powder specimens, and their known relative intensities. T was obtained from a theoretical model and D was measured using fluorescence spectra of several elements. A computer program was developed to calculate the peak intensities and energies by varying the parameters in the above expression until the best fit was obtained from the experimental profiles. A Rigaku horizontal diffractometer was modified to mount a solid state detector and still allow use of the wavelength dispersive scintillation counter geometry to make accurate comparisons. The system is controlled and the data stored in an IBM System 7 computer to allow fully automatic operation and the data are transferred to an IBM 360/195 computer for calculations and storage.

INTRODUCTION

This paper presents the results of a theoretical and experimental study of the use of the energy dispersive method for X-ray powder diffraction (EDXRD). Since the first description of the method by Giessen and Gordon in 1968 (1) a number of papers have been published on the instrumentation and applied to very simple powder patterns (2-5). Wilson (6) has described the geometrical and physical aberrations. In this paper a profile fitting method is described for better analysis of the data and for extending the method to more complicated patterns and overlapped peaks. The method is equally applicable to single crystal diffraction analysis (7).

The main problem of the energy dispersive method is that the peaks are more than an order of magnitude wider than in conventional diffractometry. The principal sources of the breadth are the limited energy resolution of the solid state detector which causes a symmetrical broadening of the diffracted peaks and specimen transparency which gives rise to asymmetric broadening. Both factors increase with increasing X-ray photon energies and may shift the peaks and limit the precision unless corrected. In our attempt to solve this problem we predict the shape of a diffraction peak, as it is recorded in the multichannel analyzer, from a theoretical model which takes into account the physical, geometrical and electronic conditions. The number, heights, and positions of the theoretical peaks are varied until optimum congruence with the experimental peaks is obtained.

Another major difficulty arises from the anomalous situation that although a recognizable powder pattern can be obtained very rapidly, in say less than a minute, it takes much longer times than in conventional scanning diffractometry to obtain patterns with good counting statistics. The reason is that the total input count rate of all the photons entering the detector must not exceed approximately 5000 to 10000 cps to minimize the dependence of resolution and peak shift on count rate, because of limitations in the present-day electronics. These photons are distributed among the many peaks of the pattern. To increase the statistical accuracy may require overnight runs as is often necessary to obtain data for profile fitting procedures.

It will also become evident that a considerable computer facility is required to apply the methods.

Energy dispersive X-ray diffraction is a technique related to the conventional angular dispersive technique as the Laue method is to the Bragg method. The latter employs monochromatic radiation and the reflections appear at angles determined by the lattice spacings as shown by the Bragg equation

$$d \sin \theta = \lambda/2 = \text{constant.} \tag{1}$$

In the energy dispersive case the reflections are generated by X-rays from the continuous radiation at a selected 2θ angle and satisfy the Bragg law written in the form

$$d\ E = h\ c/2\ \sin\ \theta = 12.398\ (keV\ \text{Å})/2\ \sin\ \theta = constant. \qquad (2)$$

The symbols have their usual meanings. In all cases we use a $\theta:2\theta$ setting of the specimen and detector so that θ is the angle of incidence of the primary beam to the specimen surface and 2θ is the angle between the primary and diffracted beams.

FACTORS EFFECTING FORM OF POWDER PATTERNS

The patterns of a tungsten powder specimen using both conventional scanning and energy dispersive methods, Figure 1, illustrate the

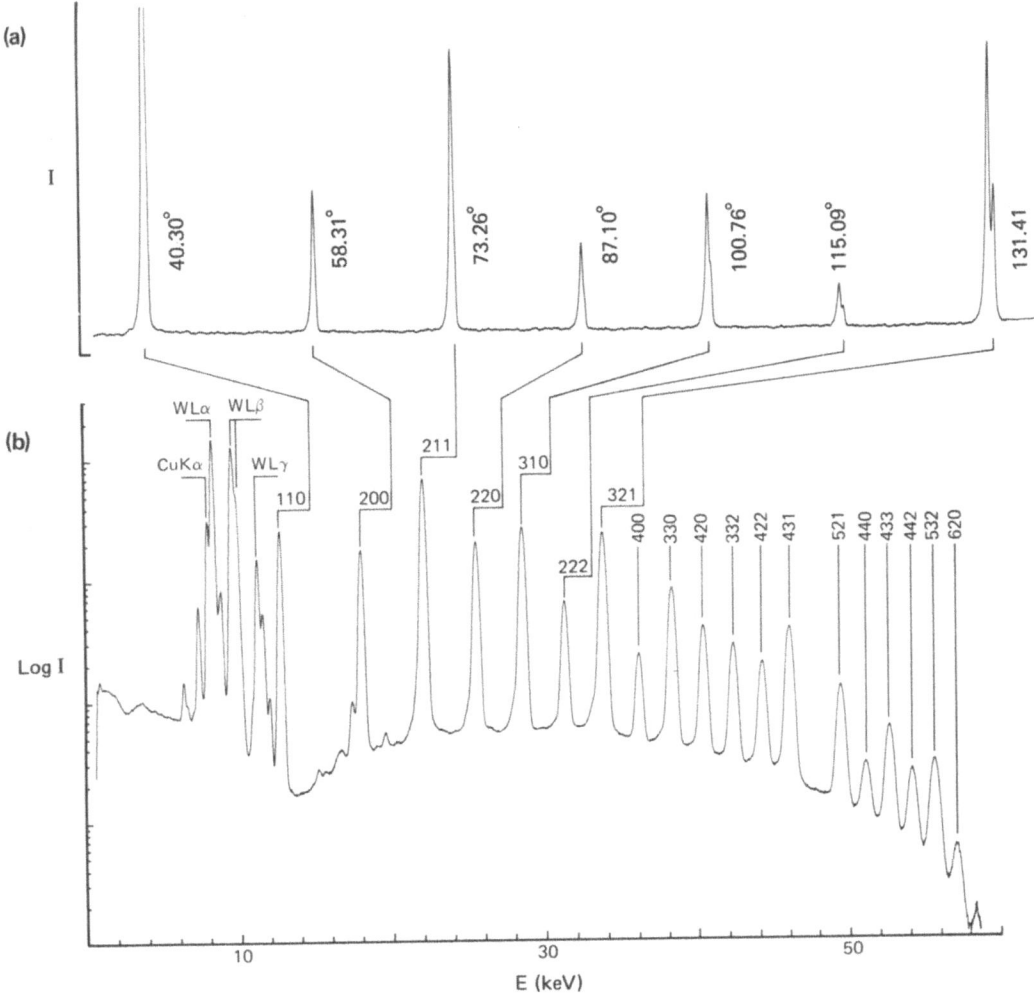

Figure 1. (a) Diffraction pattern of tungsten powder sample obtained with conventional diffractometer using $CuK\alpha_{1,2}$ radiation. (b) Angular dispersive pattern using Cu tube operated at 60 kV, $2\theta = 25°$.

principal differences. The resolution in terms of peak widths and
their separation in the EDXRD pattern is clearly poorer than the
conventional pattern. Each reflection arises from a different band
of X-ray energies but no overlapping occurs in this case because
tungsten is BCC with a small unit cell and therefore has a very simple
pattern. The EDXRD pattern contains additional peaks in the high
energy portion of the pattern which do not occur in the CuKα pattern.
It is easy to shift and contract or expand the EDXRD pattern simply
by changing 2θ. This is illustrated graphically in Figure 2 for a wide
range of 2θ's; the horizontal line is drawn for $2\theta=25°$, the angle at
which the pattern in Figure 1 was obtained.

The effect of decreasing 2θ on the form of the EDXRD pattern is
illustrated in Figure 3 for the far more complex pattern of the mineral
topaz, $Al_2SiO_4(F,OH)_2$, which is orthorhombic with a moderate size unit
cell. The separation in the cluster of four reflections [(120), (022),
(121), (112)] indicated by the black peaks to the right of the CuKα
peak improves as 2θ is decreased to 15° and then becomes poorer with

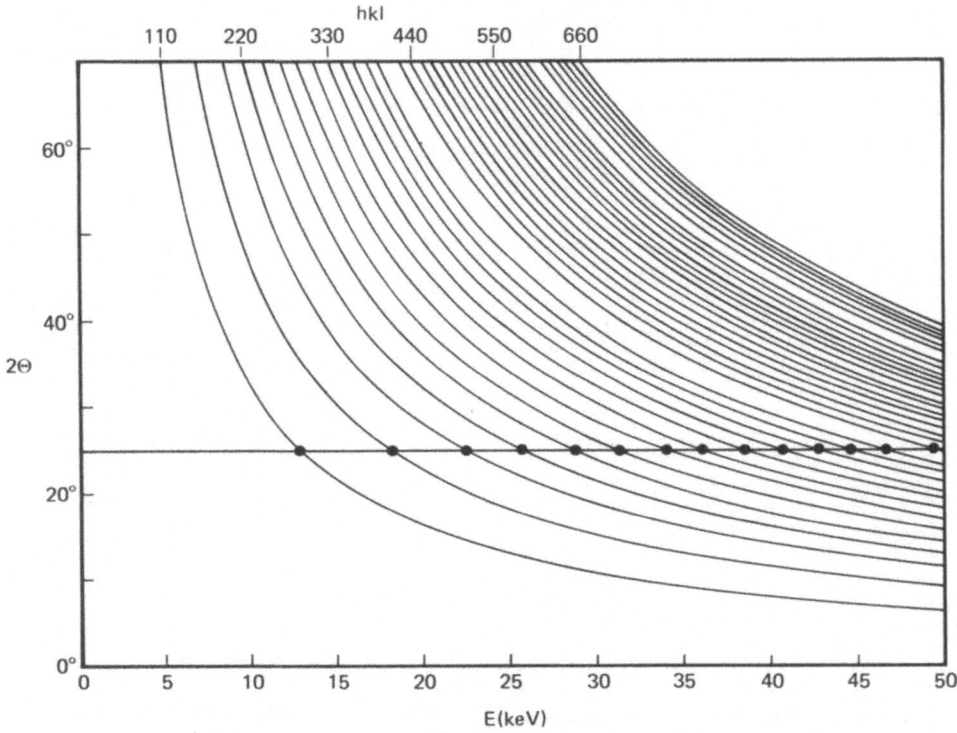

Figure 2. Dependence of energies of tungsten powder peaks on 2θ. The small circles
along the horizontal line at $2\theta = 25°$ correspond to the energies and indices of the reflections in Figure 1(b) to 50 keV.

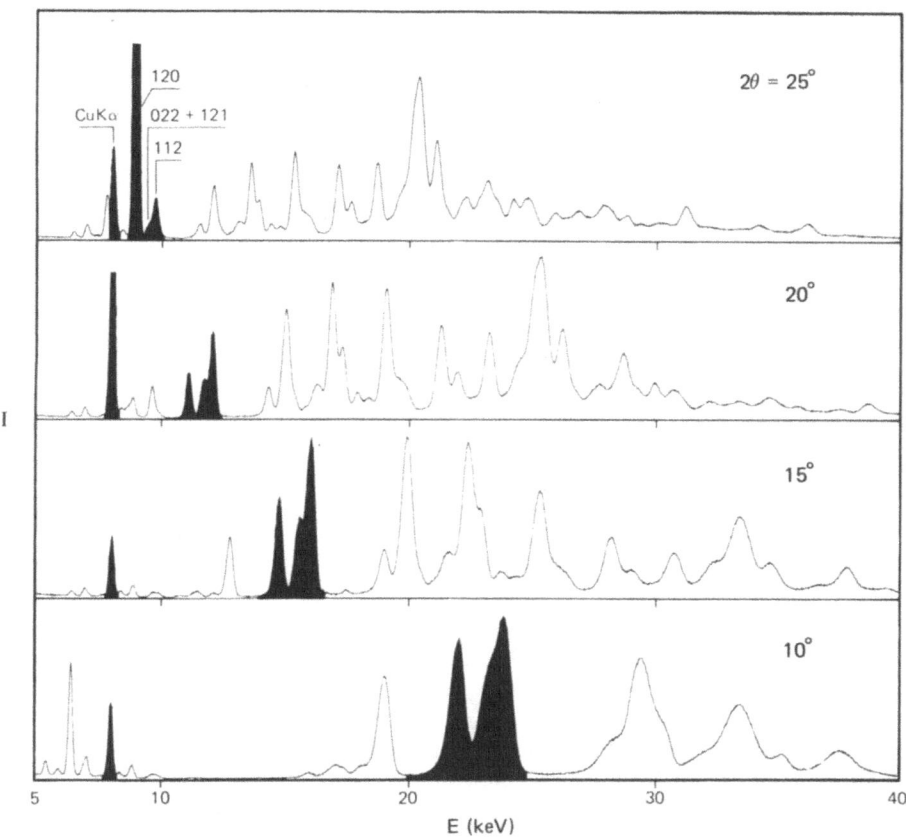

Figure 3. Energy dispersion pattern of topaz at various 2θ angles.

further decrease to 10°. This effect arises from two principal factors
which limit the gain in separation: (1) the detector resolution (and
quantum counting efficiency) decreases with increasing X-ray energy,
and (2) the specimen transparency and geometrical aberrations increase
and asymetrically broaden the peaks. The optimum separation in this
case appears to occur between $2\theta=15°$ and 20°.

Another important factor is the marked changes in relative and
absolute intensities which occur as the peaks are generated from
different X-ray energies as 2θ is changed. The effect becomes extreme
when an emission line from the X-ray tube happens to cause a
reflection. This occurs in the (120) peak generated by CuKβ radiation
in the $2\theta=25°$ pattern of Figure 3, and becomes the most intense peak
in the entire pattern.

It would be desirable to obtain relative intensities comparable
to those from angular dispersive patterns so that structural and
reference data such as the ASTM powder file could be used directly.

The term "intensity" is used here in a manner analogous to the common
usage in conventional X-ray diffraction as the number of detected
counts or photons with energies in a given energy interval ΔE.
Unfortunately, this is not as simple to determine as in the
conventional case where a characteristic X-ray line with a narrow
energy spread is used. In the energy dispersive case it is necessary
to find equivalent values of ΔE, to determine the shape of the X-ray
tube spectrum and to normalize the peak intensities to constant
generating intensities. This is difficult to do accurately because
specimen transparency and beam divergence result in a band of energies
entering the detector, and the band width varies from one peak to
another depending on the selected 2θ and the d's of the specimen. The
values of ΔE as a function of E which correspond to the wavelength
spread of the characteristic line $\Delta\lambda$ must be determined for all peaks
to obtain a direct comparison. The corrections for the scattered
background such as that in Figure 1 can be made relatively easily.

SPECIMEN TRANSPARENCY MODEL

The shape of a recorded energy diffraction peak can be described
by

$$P(E,\theta) = [X(E) \cdot T(E,\theta)] * S(E) * D(E) \qquad (3)$$

P(E) = Shape of peak, photons vs. energies,
X(E) = X-ray tube spectral distribution, photons vs. energies,
$T(E,\theta)$ = Geometrical aberrations including specimen transparency,
S(E) = Contributions from the specimen,
D(E) = Detector resolution function.

The shape of X(E) mainly effects the relative peak heights and
causes a slight asymmetric enhancement of the low energy side. X(E)
was determined in the vicinity of the peak by comparing the relative
peak heights of simple experimental patterns with those calculated
from theoretical models. The principal of the method was described
by Buras et al (7) who used Ge and Si single crystals; we used Si single
crystals cut parallel to (111) and (110) and a tungsten powder
specimen.

The function D(E) was determined by the analysis of the shapes of
emission lines at various energies as recorded on the multichannel
analyzer. Symmetrical Gaussian functions are a good approximation of
the experimental shapes. This is illustrated in Figure 4 which shows
the best fit of two Gaussians (appropriate widths vs. energies with
free variation of the positions and heights) to the $SnK\alpha_{1,2}$ doublet;
the peak positions agree with Bearden's values (8) within 2 eV. The
experimentally determined values of the resolution as a function of
energy were smoothed using the relationship

$$FWHM = (a + b\cdot E)^{1/2} \qquad (4)$$

thus determining a and b.

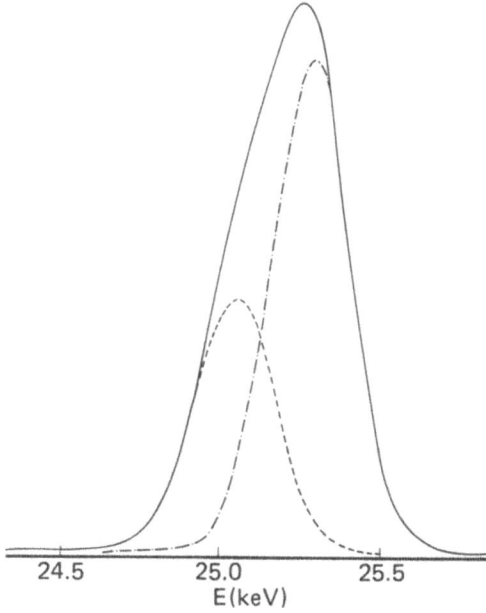

Figure 4. Profile fitting of $SnK\alpha_{1,2}$ doublet by
two Gaussians.

The main source of the asymmetric broadening is the function $T(E,\theta)$
which includes specimen transparency, beam divergence and parallel
specimen position displacement. The basic geometry is shown in
Figure 5. The slit system defines an "active volume" within the
specimen which is irradiated by the primary beam and can be seen by
the detector and thus is the portion of the specimen which contributes
to the diffracted beam. Each point within this volume, shown in more
detail in Figure 6, defines Bragg angles $2\theta_{min} \leq 2\theta \leq 2\theta_{max}$. Thus when the
diffractometer is set at $2\theta_{min}$ the line profiles consist of photons
with an energy $E_{max} = hc/2d \sin \theta_{min}$ as well as all energies $E_{min} \leq E \leq E_{max}$
which contribute mainly to the low energy tail of the profile. The
contribution of the transparency aberration can be minimized by
reducing the specimen thickness and angular apertures of the incident
and diffracted beams but this will cause a loss of intensity. The
precision of the model requires data for specimen thickness, density,
absorption coefficients and geometrical conditions and these are
usually known to a few per-cent.

The convolution of the functions $T(E,D)$ and $D(E)$ (neglecting any
influence of the shape of the primary radiation) results in an
asymmetric profile whose peak and centroid are shifted to lower
energies. The shift is large and may be of the order of -160 eV for
a low absorbing specimen. The calculated functions $T(E)$ and $T(E) *$
$D(E)$ using the absorption of a silicon specimen are shown in Figure 7.
The contribution of the specimen factors including small crystallite

M. Mantler and W. Parrish

sizes, strain, preferred orientation, etc., were negligible for the
ideal specimens used in this study.

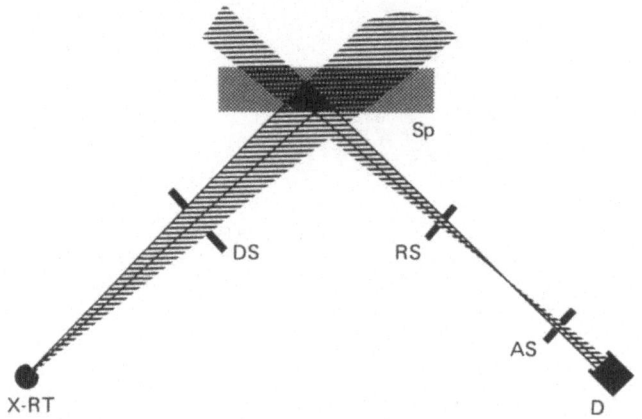

Figure 5. Basic geometry of the specimen transparency aberration. X-RT X-ray tube, DS divergence slit, Sp specimen, RS
receiving slit, AS anti-scatter slit, D detector.

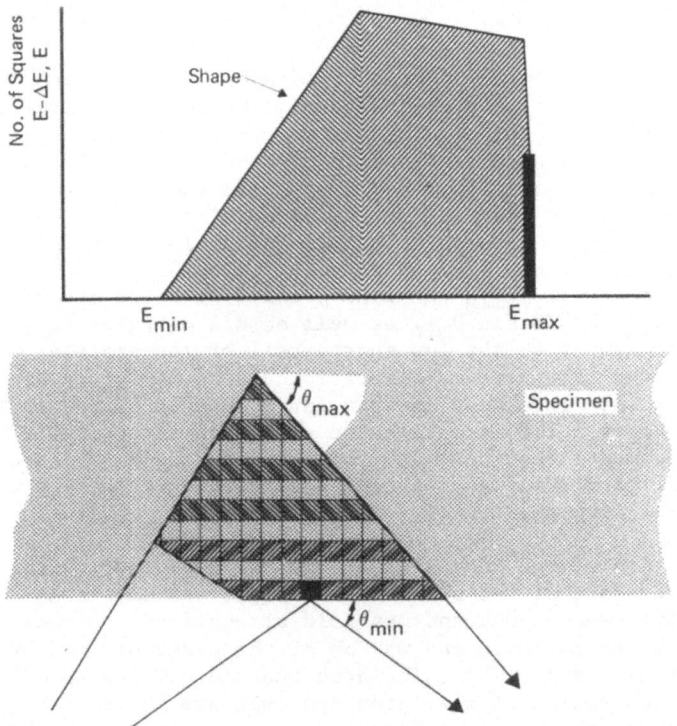

Figure 6. "Active volume" of specimen contributing to diffracted beam.
The profile shape (above) can be obtained by counting the numbers of
squares of each energy interval reflected.

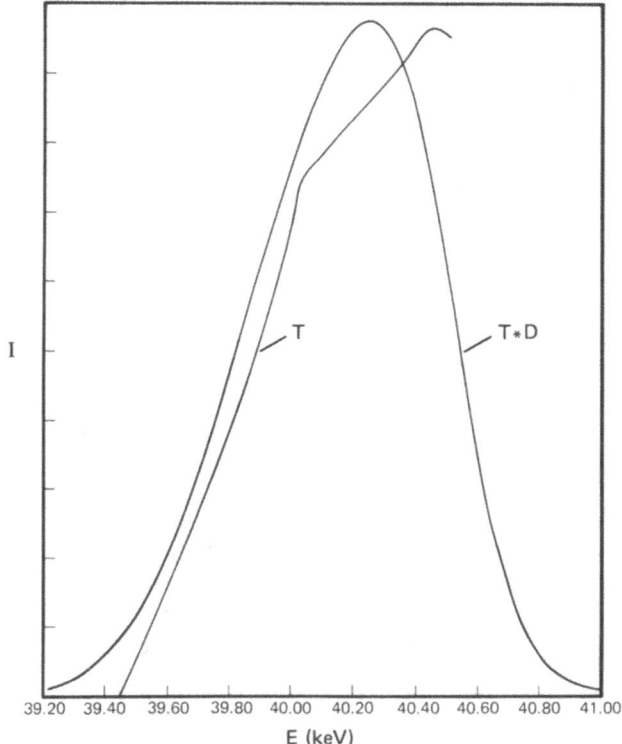

Figure 7. Theoretical profiles T due to specimen transparency and beam divergence, and T∗D the convolution with detector resolution.

PROFILE FITTING OF EXPERIMENTAL PEAKS

The basic principles of profile fitting applied to conventional scanning diffractometry have been recently described by Parrish et al (9) who use the experimental peaks of a standard specimen to determine the profile shapes. In the present case the shape of the peak is predicted by the model described above. A complex profile is then synthesized by varying the number, positions and amplitudes of the single peak profiles until the best match with the experimental profile is achieved. Powell and Zangwill's (10) technique was used to minimize the sums-of-error-squares function which defines the quality of the match.

The flow scheme in Figure 8 illustrates the steps of the computing process. Asymmetric Gaussians defined by

$$\phi(x) = \begin{cases} A \exp[(x-x_o)^2/s_1^2] & \text{for } x \leq 0 \\ A \exp[(x-x_o)^2/s_2^2] & \text{for } x \geq 0 \end{cases} \qquad (5)$$

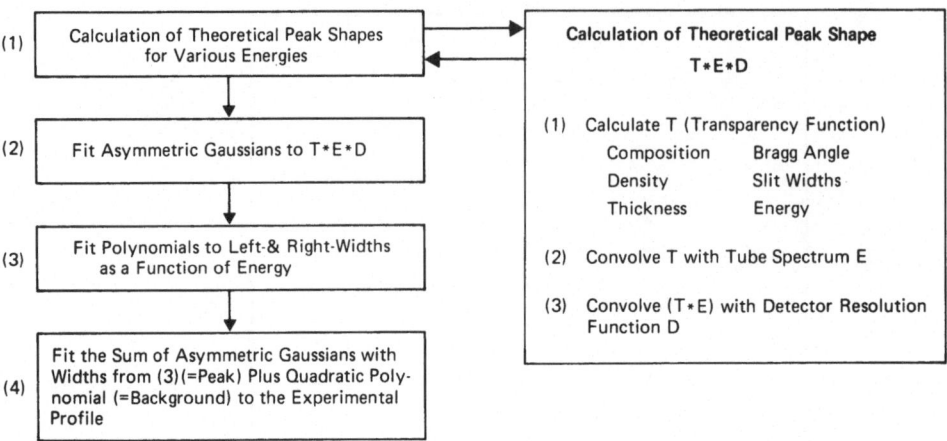

Figure 8. Flow scheme of the computer programs for profile fitting.

are used as a convenient description for the asymmetric (E·T) * D
function and fit well in most cases. The parameters of these
asymmetric Gaussians are calculated for several values of energies
within the region of interest (or parameter variation) and approximated
as a function of energy by best fit polynomials through these points.
This allows the interpolation of the line shape at any point of
interest, as required by the Powell-Zangwill iteration. In addition,
the shift of the maximum of the peak from E_{max} is also approximated
by a polynomial versus energy which allows the calculation of the
corrected peak positions.

The local background is matched by a quadratic polynomial, the
parameters of which are adjusted together with the parameter of the
peaks. This also compensates for embedded tails of adjacent peaks.

The method described here is relatively easily applied to simple
patterns in which there is little overlapping of the peaks. The
difficulties increase as the overlapping increases and the peak
separation should be at least the FWHM of the peaks for successful
data reduction. The ability to resolve overlapped peaks cannot be
stated simply because it depends on the degree of asymmetry of the
peaks, the number and relative intensities of the peaks, and the
counting statistical accuracy of the data, all factors which are
determined by the experimental conditions and the specimen itself.

The energy dispersive pattern of alpha quartz, Figure 9, illustrates
some of the results found in analyzing a pattern containing a wide
range of energies and intensities. The low energy peaks in the range
8 to 21 keV can be handled readily by the profile fitting method and
their peaks fixed to a few eV because they have much better counting
statistics, there are no close overlaps and the peaks are narrower

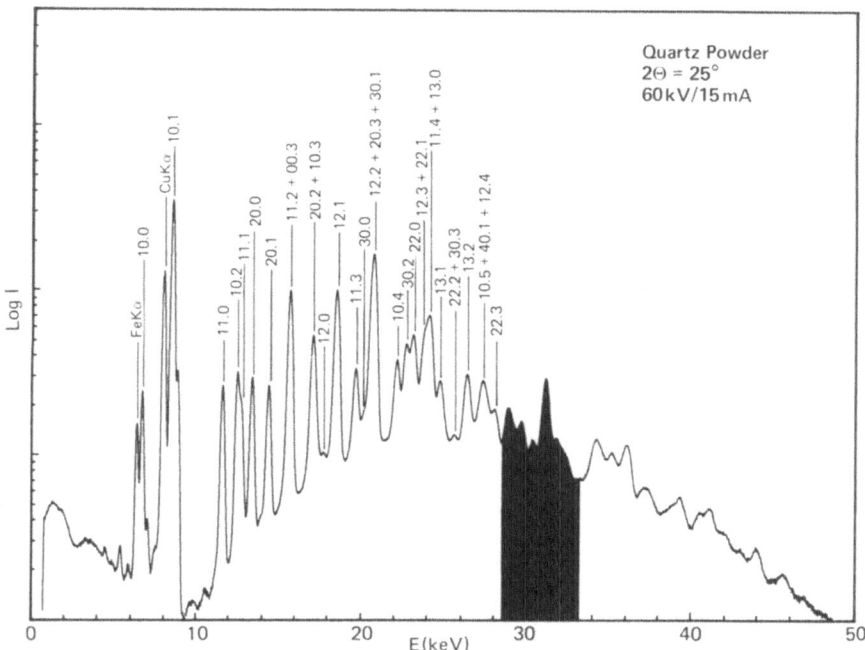

Figure 9. Energy dispersive pattern of quartz powder. The black portion is enlarged in Figure 10(b).

because of the smaller transparency aberration and better detector energy resolution than the higher energy peaks. The region 28 to 32 keV (black area) is shown in the corresponding scanning diffractometer pattern, Figure 10(a), and contains 13 reflections, each a CuKα$_{1,2}$ doublet in the angular range 102° to 120° (2θ). Figure 10(b) is an enlargement of the energy dispersive pattern for this region and shows low P/B ratios and low intensities and even though the recording was made in an overnight run. Figure 10(c) shows the results of the profile fitting in which six well defined quartz peaks and two extraneous aluminum peaks were found. The (23·1) peak is resolved but the remaining peaks contain two or three reflections and the weak (00·6) peak was not determined. Adding peaks to the synthesized profiles did not reduce the error function. The calculated positions of the (23·0), (23·1), (23·2) and (12·5) peaks match the expected values to 4, 1, 15 and 11 eV, respectively, corresponding to $\Delta E/E = 3 \cdot 10^{-5}$ to $5 \cdot 10^{-4}$; it should be noted that these values include large corrections of the order of −145 eV for the specimen transparency shift. The two aluminum reflections (331) and (420) come from behind the 0.5 mm thick quartz specimen which is mounted in an aluminum specimen holder. These required corrections analogous to parallel specimen surface displacement which together with transparency were nearly −500 eV.

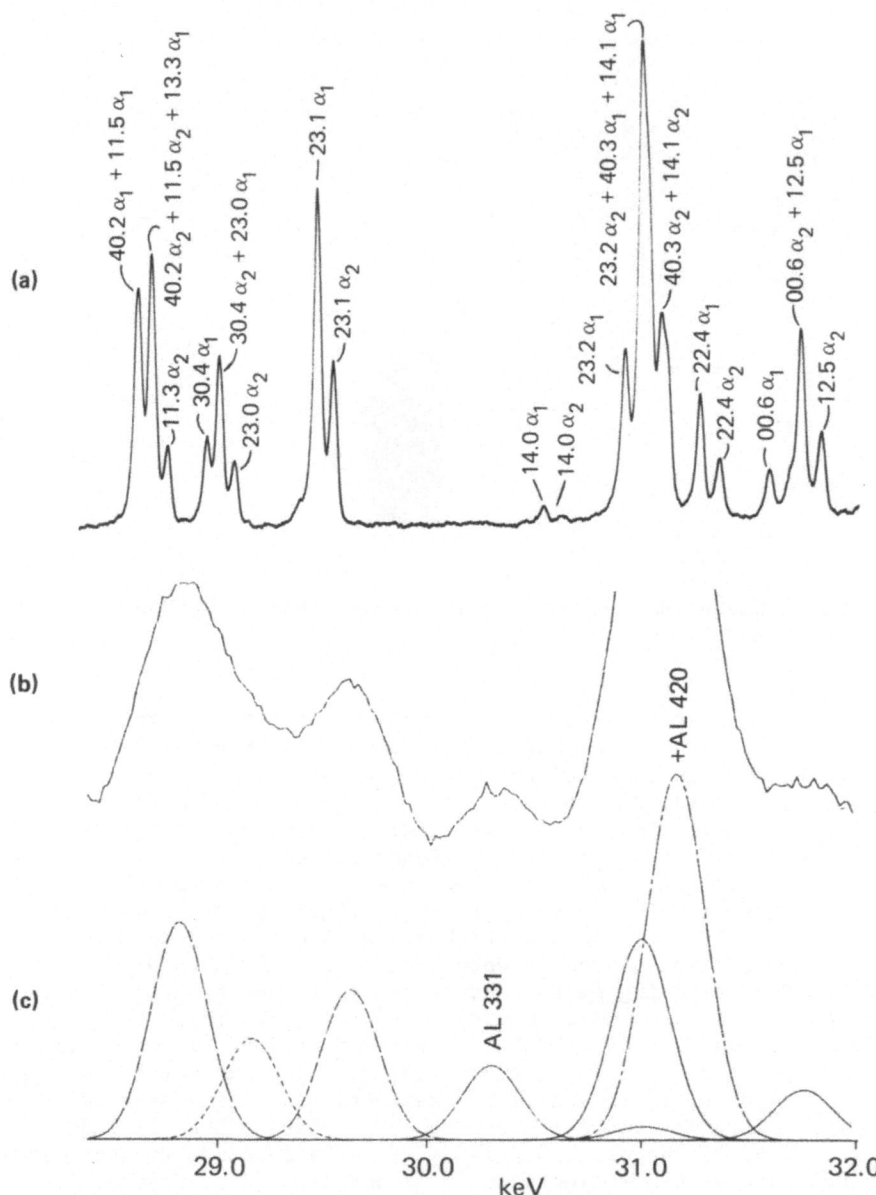

Figure 10. (a) Conventional diffractometer recording of quartz powder, 2θ = 102° to 120°, CuK$\alpha_{1,2}$ radiation. (b) Experimental energy dispersive pattern of same region (zero intensity at base of c), (c) Profile fitting of (b). All intensities on linear scale.

INSTRUMENTATION

A Rigaku horizontal diffractometer (SG-9D) was modified to support a solid state detector and allow interchangeability with a scintillation counter, Figure 11. This required the construction and counterbalancing of a special platform on the diffractometer on which the detector was mounted. A diffracted beam monochromator can be used optionally with both detectors. The specimen chamber and the slit system are designed to evacuate virtually the entire X-ray path. An important advantage of this instrumentation is that it permits a direct comparison of both methods using identical experimental conditions and specimens.

The specimen holders are solid aluminum cylinders, one inch diameter, undercut in the top plane to embed the powder. It should be noted that the higher energy X-rays may penetrate the powder and aluminum peaks may be detected; the holder should be designed with the proper powder depth or without backing to avoid this problem. The specimen holders are rotated at about 70 rpm to obtain better specimen statistical data.

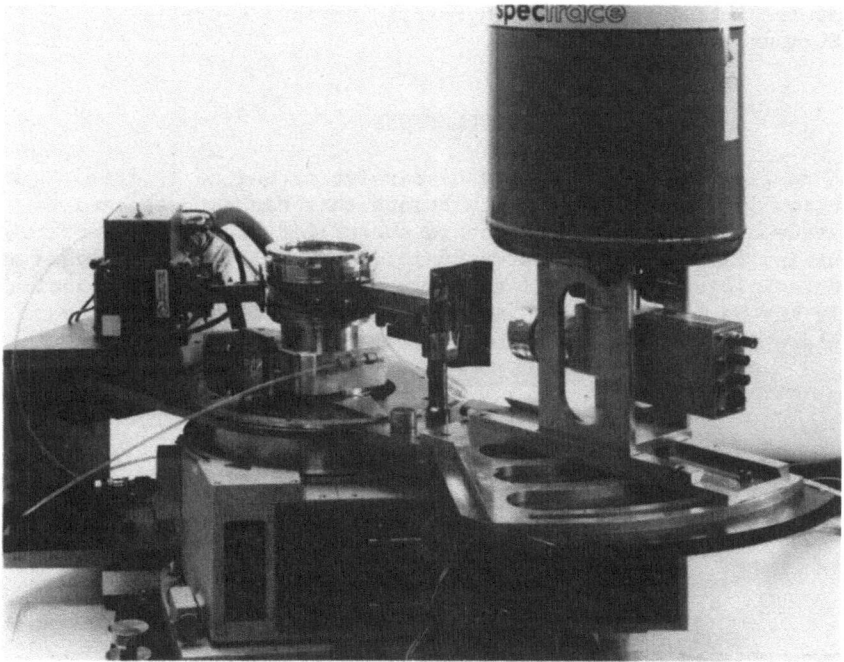

Figure 11. Rigaku diffractometer modified for use of solid state detector. The computer controlled pulser motors are coupled directly to the θ and 2θ gear shafts in front of the diffractometer.

The X-ray generator used was a Rigaku 3 kW DC constant potential unit operated at 60 kV, 15 to 25 mA. The line focus of a Norelco long fine focus Cu target X-ray tube was viewed at 12°. The angular aperture of the incident beam was 0.5° in the horizontal plane and the receiving aperture 0.3°. The incident beam parallel slit assembly had a 2δ vertical plane aperture of 4.1° and the diffracted beam assembly 4.4°. A Nuclear Semiconductor Si(Li) solid state detector, 80 mm^2 by 5 mm thick, resolution 172 eV for MnKα radiation at 1000 cps, and a Tracor Northern model 1700 8K multichannel analyzer were used.

An IBM System 7 computer, Figure 12, controls the θ to 2θ step motors on the diffractometer reads the data of the multichannel analyzer and operates the scaling circuit of the scintillation counter by means of an interface system described elsewhere (9). (The diffractometer can also be run with its internal synchronous motors for normal recording.) After completing the experiment the data are transferred to an IBM System 360/195 computer for storage and evaluation. All programs can be called (or generated) from a local keyboard terminal with CRT screen that is used for graphic displays of the spectra obtained and for comparison with stored data.

The experiments described in this paper were made at constant total counting rates of 5000 cps where no peak shifts, changes of line shape or resolution were observed for the various spectral distributions.

CONCLUSIONS

The application of the energy dispersive method to X-ray diffraction is a relatively new technique that has not yet been extensively developed. The method is useful for obtaining diffraction patterns and fluorescence spectra rapidly for qualitative analyses of known materials and in special cases where the conventional scanning geometry may be impractical as in high pressure, high temperature, rapid phase transitions and other special situations.

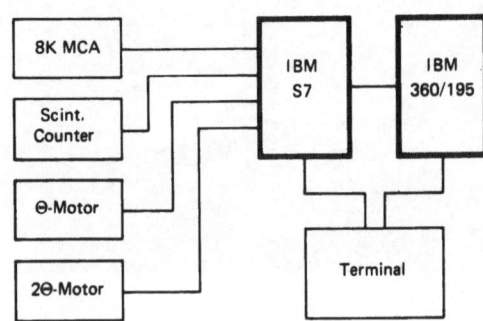

Figure 12. Block diagram of computer system.

The reflections in patterns which do not contain severe overlapping can be determined to a precision $\Delta E/E = \Delta d/d \approx 10^{-4}$. To obtain intensities equivalent to those in conventional diffractometry, corrections must be made for specimen transparency and other factors contributing to broadening and then a correction factor must be applied for comparing the intensity formulae for the Laue and Bragg methods (7,11-13). It should be noted that advanced computer methods are required to derive these data.

The possibility of extending the method to more complicated patterns depends on the ability to separate overlapping reflections which become a severe problem as the unit cell dimensions increase and symmetry decreases. The degree to which this is possible is determined by how well the model described above (which uses asymmetric Gaussians to fit E * T * D) describes the true peak shape, the statistical precision of the experimental data and the computer programs. In our present stage of development the first two are approximately equal and are the dominating factors for long overnight runs. The major instrumental limitations arise from the limited maximum counting rate (presently 5000 to 10000 cps) set by the electronics and intensity of the X-ray source. The counts in each peak accumulate at a very low rate and very long counting times are required to obtain good statistics. If the electronics were improved a synchroton source would be ideal.

A more detailed treatment of the transparency model and profile fitting method is being prepared. It is evident that profile fitting applied to conventional scanning diffractometry (9) provides for better data than EDXRD in its present state of development.

ACKNOWLEDGEMENTS

We are indebted to G. L. Ayers for aid in computer automation methods, G. C. Erickson for mechanical design, and T. C. Huang for discussions on profile fitting.

REFERENCES

1. B. C. Giessen and G. E. Gordon, "X-ray Diffraction: New High-Speed Technique Based on X-ray Spectrography," Science 159, 973-975 (1968).

2. J. P. Lauriat and P. Pério, "Adaptive d'un Ensemble de Détection Si(Li) á un Diffractometre X," J. Appl. Cryst. 5, 177-183 (1972).

3. C. J. Sparks and D. A. Gedcke, "Rapid Recording of Powder Diffraction Patterns with Si(Li) X-ray Energy Analysis System: W and Cu Targets and Error Analysis," Adv. in X-ray Anal. 15, 240-253 (1972).

4. T. Fukamachi, S. Hosoya and O. Terasaki, "The Precision of Interplanar Distances Measured by an Energy-Dispersive Diffractometer," J. Appl. Cryst. 6, 117-122 (1973).

5. E. Laine, I. Lähteenmäki, and M. Hämäläinen, "Si(Li) Semiconductor
    Detector in Angle and Energy Dispersive X-ray Diffractometry," J.
    of Phys. E: Sci. Instrum. 7, 951–954 (1974).

6. A. J. C. Wilson, "Note on the Aberrations of a Fixed-Angle Energy-
    Dispersive Powder Diffractometer," J. Appl. Cryst. 6, 230–237 (1973).

7. B. Buras, J. S. Olsen, L. Gerward, B. Selsmark and A. L. Andersen,
    "Energy-Dispersive Spectroscopic Methods Applied to X-ray Diffrac-
    tion in Single Crystals," Acta Cryst. A31, 327–333 (1975).

8. J. A. Bearden, X-ray Wavelengths, NYO–10586, U. S. Atomic Energy
    Comm. (1964).

9. W. Parrish, T. C. Huang and G. L. Ayers, "Profile Fitting: A Power-
    ful Method of Computer X-ray Instrumentation and Analysis," Trans.
    Am. Cryst. Assoc. 12, (1976).

10. IBM System/360 and System/370, IBM 1130 and IBM 1800 Subroutine
    Library – Mathematics, Manual SH 12-5300-1.

11. W. H. Zachariasen, Theory of X-ray Diffraction in Crystals, J.
    Wiley, New York 1945.

12. M. von Laue, Röntgenstrahl-Interferenzen, Akad. Verlag., Frankfurt
    (1960).

13. H. Cole, "Bragg's Law and Energy Sensitive Detectors," J. Appl.
    Cryst. 3, 405–406 (1970).

A NEW METHOD FOR THE DETERMINATION OF THE TEXTURE OF MATERIALS OF

CUBIC STRUCTURE FROM INCOMPLETE REFLECTION POLE FIGURES

Daniel RUER and Raymond BARO

Laboratoire de Métallurgie Structurale

Faculté des Sciences - Université de Metz

Ile du Saulcy - 57000 METZ (France)

ABSTRACT

The main characteristics of a new method of texture analysis are presented as well as a comparative study of the results obtained with three different methods.

INTRODUCTION

The crystallographic texture of polycrystalline materials of cubic structure can be determined by the use of the quantitative analysis methods which have been mainly developed by H.J. Bunge (1), R.J. Roe (2)(3), and R.O. Williams (4).

These methods enable the orientation distribution of the crystals to be calculated from the information which is contained in a minimum of two complete pole figures. Although the methods of Bunge and Roe are very versatile, the method of Williams can only lead to a result if both (100) and (111) complete pole figures are used.

J. Pospiech and J. Jura (5) and P.R. Morris (6) have implemented improvements of Bunge's and Roe's methods which permit the texture to be analysed from at least three incomplete pole figures.

All these methods are based on the existence of the orthorhombic symmetry normally found in rolled sheet samples. Recently however, A. Clément and G. Durand (7) have applied Bunge's method to the texture analysis of cubic materials which do not possess such symmetry properties.

The method which we present here is for the time being the only one which permits the orientation distribution of cubic polycrystalline materials to be calculated from one or more incomplete pole figures without restrictions concerning the macroscopic symmetry of the sample.

This new method is called the "vector method of texture analysis". It is based upon the geometric relationships which exist between the inverse pole figure of the normal to the sample plane and an arbitrary direct pole figure. The vector method is therefore of the same general type as the biaxial method due to Williams. As its detailed description will be published elsewhere (8), we will here give only its main characteristics as well as a comparative study on the same sample of the results obtained by three different methods.

THE BASIC ELEMENTS OF THE VECTOR METHOD

Each group of twenty four physically equivalent orientations of the cubic system is represented by a single set of three independent angular parameters (ψ, ω, ζ), as shown in figure 1-a). The parameters ψ and ω locate the normal $\overrightarrow{ON}$ to the sample plane in either spherical triangle T_1 or T_2. Each of these triangles has been divided in thirty six domains t_μ of equal area (μ = 1..., 36 for the domains of T_1 and μ = 1..., -36 for the symmetric domains of T_2). The domains t_μ of T_1 are figurated in figure 1-b. The parameter ζ which indicates the amplitude of the rotation about the normal $\overrightarrow{ON}$ varies between 0 and 2π if $\overrightarrow{ON}$ is located in T_1 and between 0 and -2π if $\overrightarrow{ON}$ is located in T_2.

The set G of these groups (ψ, ω, ζ) is then divided in N = 2592 equivalence classes $G_{\mu\nu}$ which are defined by a set of three parameter intervals ($\Delta\psi$, $\Delta\omega$, $\Delta\zeta$) (figure 1-c). The intervals $\Delta\psi$ and $\Delta\omega$ appear in figure 1-b. The intervals $\Delta\zeta$ are all equal to 10°. All the classes $G_{\mu\nu}$ contain the same number of orientations. It is moreover supposed, that two orientations belonging to the same class $G_{\mu\nu}$ cannot be distinguished from each other. The number N of the classes $G_{\mu\nu}$ consequently defines the resolution of the method.

For the computation it is necessary to replace the two index μ and ν by a single index n defined by :

$$n = (|\mu|-1).72 + k$$

where k = ν if ν > 0 and k = 36 - ν if ν < 0.

In figure 1-c, the class $G_{29,18}$ corresponds to domain 29 of T_1 (figure 1-b), the index ν = 18 th interval $\Delta\zeta$. The class G_{29}^* includes all the classes $G_{29,\nu}$ for ν = 1,....,36.

Each class G_n is associated with the partial volume $\Delta V = y_n.\Delta V°$ of the grains whose orientations belong to this class. The numbers y_n obey the two following relationships :

$$y_n \geqslant 0 \quad \text{and} \quad \sum_n y_n = N = 2592$$

In case of an isotropic sample y_n = 1 for all values of n. The volume unity $\Delta V°$ is equal to 1/N of the sample volume V.

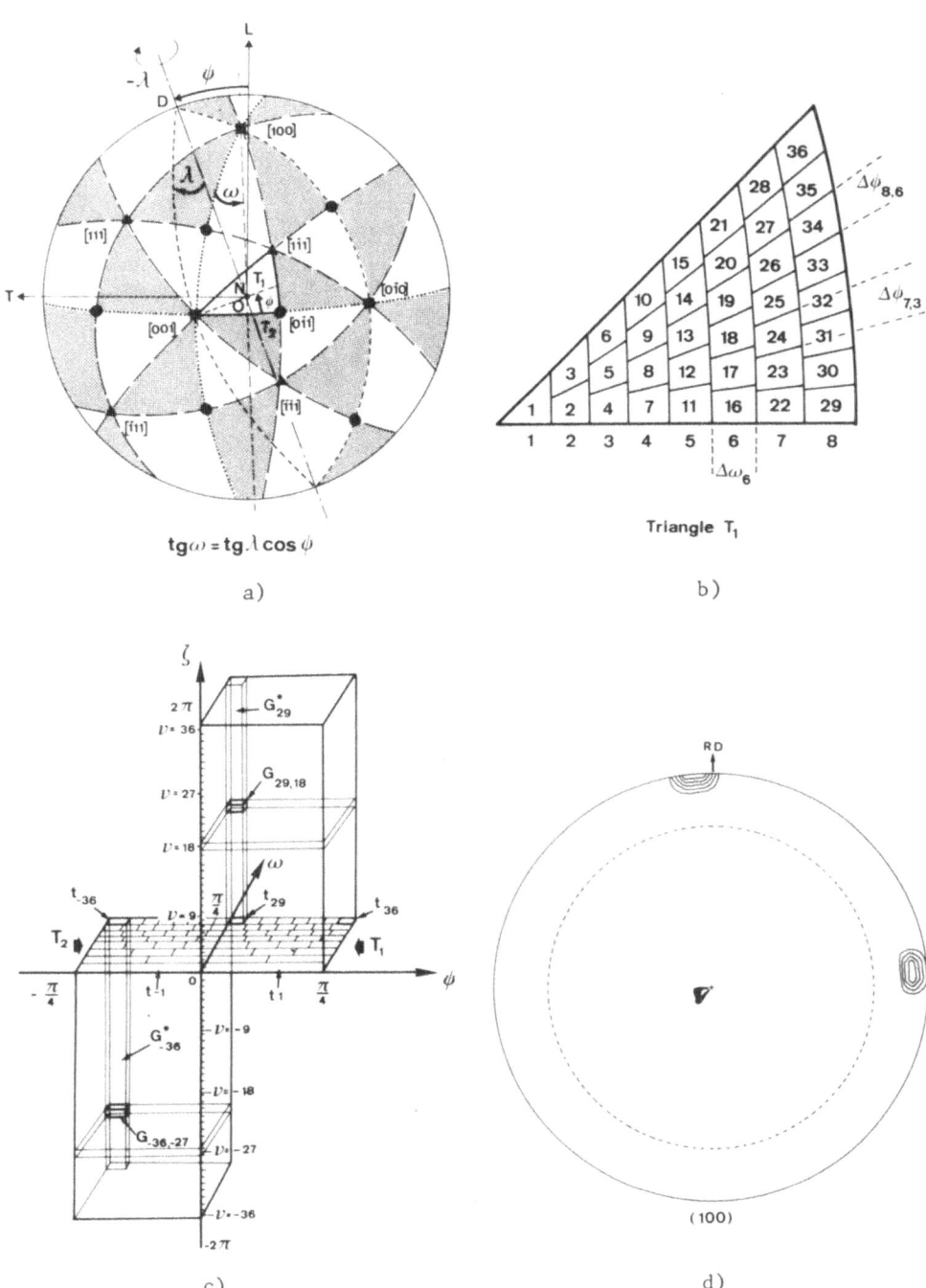

tgω = tg.λ cos ψ

a)

Triangle T_1

b)

c)

d)

Figure 1
a)- definition of the parameters and representation of an orientation
    (ψ,ω,O)
b)- stereographic projection of the domains t_μ of T_1
c)- symbolic representation of the 2592 classes $G_{\mu\nu}$
d)- the elementary (100) pole figure representing the class $G_{1,1}$

Each class G_n is equally associated with, for each plane (hkl), a pole figure which is called the "elementary pole figure". For example, figure 1-d shows the elementary pole figure representing the (100) pole densities of the class G_1. These elementary pole figures should not be confused with those defined by A. Clément and P. Coulomb (10). The superimposition of all elementary pole figures for a given plane (hkl) leads to a pole figure having an overall pole density equal to unity.

Each pole figure being divided into P = 3240 domains C_p, for each domain C_p, the density σ_{pn} of the poles (hkl) of the grains whose orientations belong to each class G_n are calculated. The texture of the sample can then be represented by the vector $\vec{Y}$ of the vector R^N defined by the matrix equation :

$$\vec{I}(hkl) = \left| \sigma_{pn}(hkl) \right| . \vec{Y}$$

where $\vec{I}(hkl)$ is a column vector representing the P measured intensities and $\left[\sigma_{pn}(hkl) \right]$ a non-negative P x N matrix whose N columns correspond to the N elementary pole figures and whose P arrows correspond to the P domains C_p of the (hkl) pole figure. R is discussed later.

To obtain the solution to this linear system of equations, the method of "the systematic reduction of the residue vector" is used (9). This method can be adapted in such a manner that it always converges towards the solution and so that the presence of negative components for the texture vector $\vec{Y}$ is not possible. For application, the fact that the dimension of R^P is greater than that of R^N allows us to compute the vector $\vec{Y}$ from a single incomplete pole figure obtained uniquely with the Schulz reflection technique.

EXAMPLE OF APPLICATION

The intensity of the diffracted X ray beam by the (220) planes of an aluminum sheet sample was measured by the Schulz reflection method. The sheet sample had previously been cold rolled to a height reduction of 90 %. This sample was arbitrarily chosen with the aim of testing the vector method.

The incomplete pole figures obtained after subtraction of the background intensity and correction of the weakening due to the defocalisation phenomena were then normalized. It should be noted that with the vector method the normalization of the pole figures is not necessary. Indeed, when the pole figure is not normalized, the computation leads to the solution of matrix equation :

$$K. \vec{I}(hkl) = \sigma_{pn}(hkl) . K.\vec{Y}$$

where K is an unknown positive number generally different from unity. The solution obtained is then not the vector Y but a vector K.Y. The coefficient K can however be easily determined since :

$$K = 1/N \sum_n K.y_n$$

The circle which limits the incomplete pole figure always has a
radius of 74° so that the experimental reflection technique is adequate
in any case.

The computation of the texture vector requires the recalculation of
the pole figure at each iterative step.Figures 2-a and 2-b show the
(220) experimental and calculated pole figures after thirty-nine
iterations at which point the numerical computation was stopped.

From the resulting texture vector, the (111) and (200) pole figures
were computed (figure 2-c and 2-d). These computed pole figures can
then be compared with those measured experimentally (figures 2-e and
2-f) permitting the accuracy of the present method to be evaluated. It
can be seen that a shifting of a few degrees of the rolling direction
has certainly occured in the measured pole figures.

It should be noted that the agreement is satisfactory considering
that the result was obtained with a single incomplete pole figure. For
greater accuracy, a second incomplete pole figure can be used. We will
see, in this event, that the vector method leads to results at least
as good as those obtained with other usual methods.

COMPARISON OF THE RESULTS OBTAINED WITH THE OTHER METHODS

The comparison with the other usual methods was carried out on
experimental data which have been kindly communicated to us by Professor
Verdeja of Oviedo (Spain). These data are composed of the (200) and
(222) complete pole figures of a rolled and recrystallised sample of
Fe containing 17 % Cr.

The comparison criterion used is the length R of the residue vector.
This is expressed in percent of the length of the vector $\vec{I}^m$ representing the experimental pole figure. The value of R is given by :

$$R = \left[\frac{\sum\limits_i (I_i^m - I_i^c)^2}{\sum\limits_i (I_i^m)^2} \right]^{1/2} \times 100$$

where I_i^c are the calculated intensities.

The results obtained with the methods of Bunge and Williams and the
vector method are summarized below in Table I which shows that the
smallest value of the arithmetical average $\bar{R}$ is the one corresponding
to the vector method.

The calculated pole figures (figure 3 and figure 4) illustrate the
results obtained from the different methods. In these figures, level
one represents one tenth of maximum intensity.

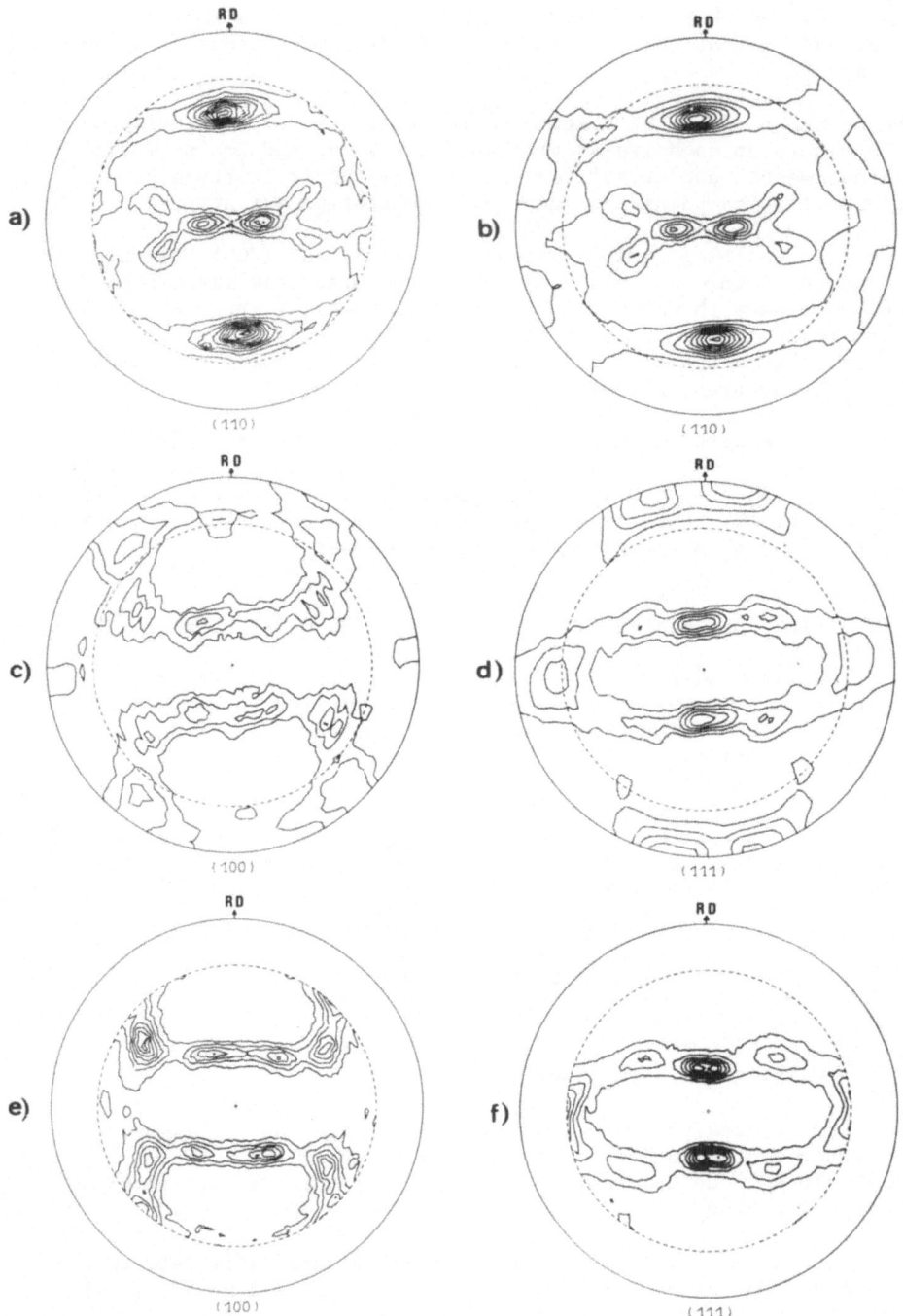

Figure 2
a)-experimental data from which the pole figures b), c) and d) have been
   calculated
e) and f)- experimental data permitting the accuracy evaluation of the
   calculus (Level 1 is equal to unity for all figures)

TABLE   I

METHOD	DATA	R(100) % of I(100)	R(111) % of I(111)	$\overline{R}$
BUNGE	(100) and (111) (complete)	22,83 %	10,48 %	16,65 %
WILLIAMS	(100) and (111) (complete)	21,42 %	13,57 %	17,49 %
VECTOR METHOD	(111) (incomplete)	28,23 %	6,92 %	17,57 %
	(111) (complete)	27,81 %	5,68 %	16,74 %
	(100) and (111) (incomplete)	12,78 %	12,72 %	12,75 %
	(100) and (111) (complete)	11,30 %	12,93 %	12,12 %

Compared accuracy of different methods

(The series development corresponding to Bunge's method was limited to the rank L = 22).

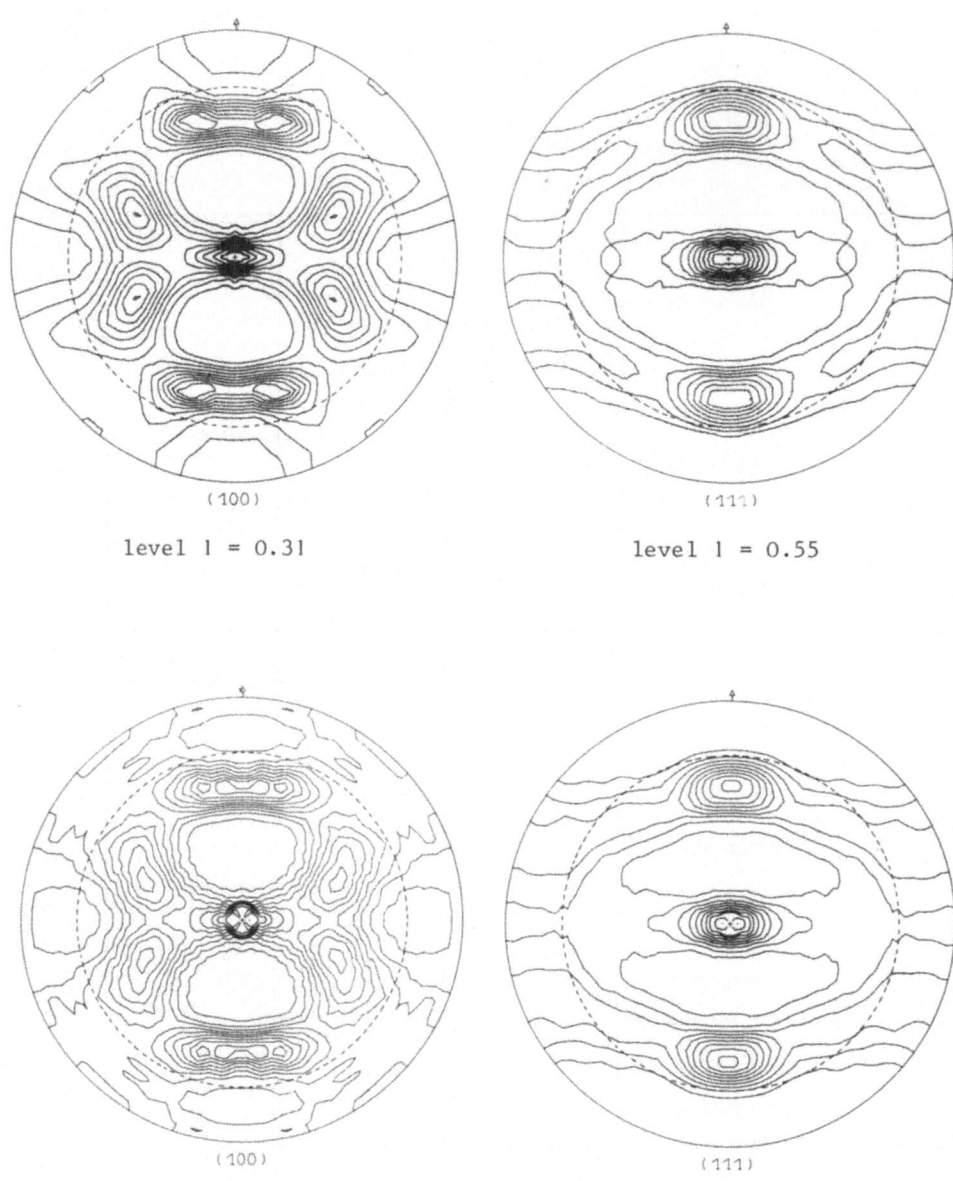

Figure 3
above : experimental data
below : pole figures recalculated with the vector method from the data
        appearing inside the dashed circles
        (Level 1 is equal to one tenth of the maximum intensity of each
        pole figure)

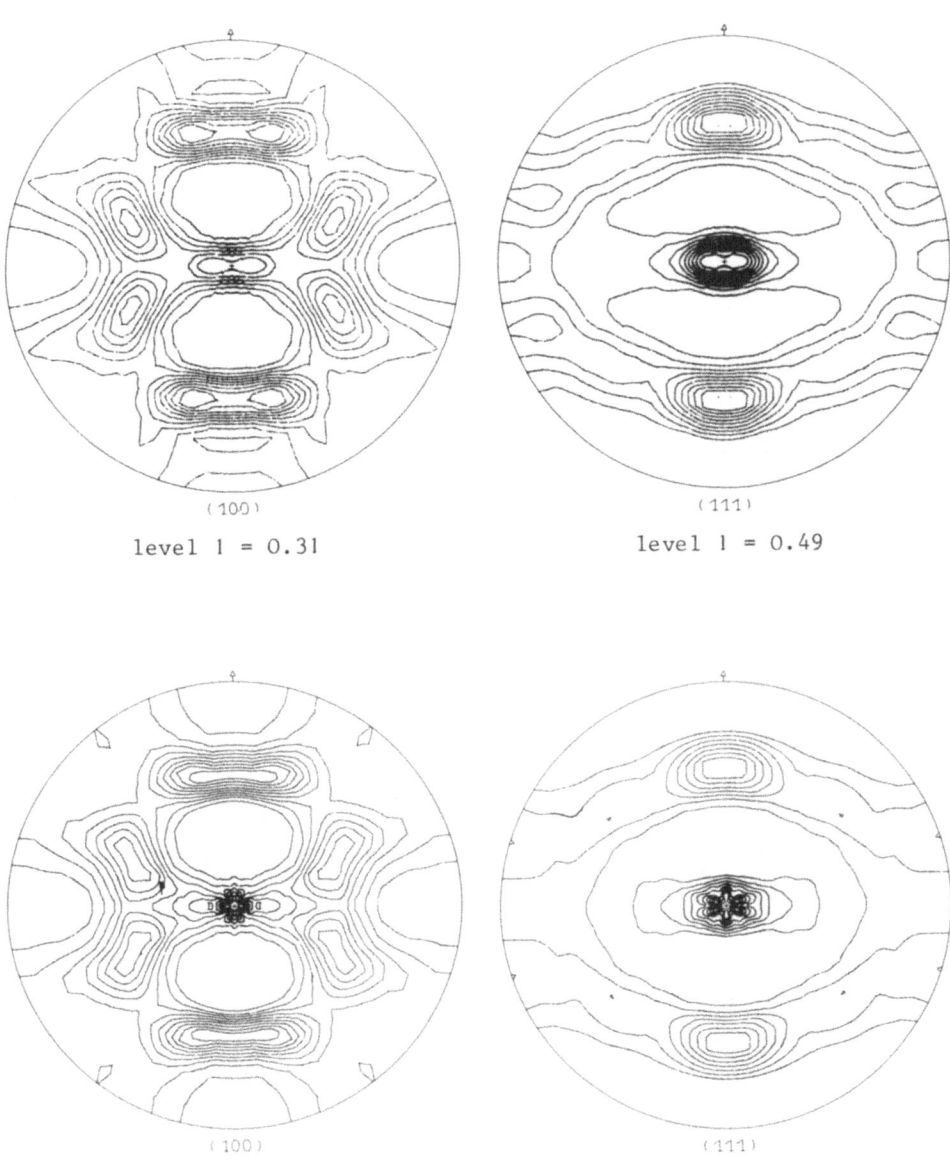

(100)

level 1 = 0.31

(111)

level 1 = 0.49

(100)

level 1 = 0.33

(111)

level 1 = 0.63

Figure 4
above : pole figures recalculated with Bunge's method
below : pole figures recalculated with William's method
(The computation with both methods have been carried out utilizing the
complete experimental pole figures of figure 3. Level 1 is defined as
in figure 3)

REPRESENTATION OF THE ORIENTATION DISTRIBUTION

Figure 5-a gives a symbolic representation of the texture vector $\vec{Y}_1$ computed from the single (222) incomplete pole figure and figure 5-b represents the texture vector $\vec{Y}_2$ computed from the two (200) and (222) incomplete pole figures. For the interpretation of this kind of diagram, it should be noted that the quadrants A and C correspond to the classes $G_{\mu\nu}$ with positive parameter values (right part of the figure 1-c). The quadrants B and D on the other hand correspond to the other half of the set G for which the parameter values are all negative. Moreover, the domains t_μ (or t-μ) of the triangle T_1 (or T_2) correspond to imaginary circles of wich only the intersections with four perpendicular radii are represented. The smallest circle corresponds to $|\mu| = 1$ and the greatest to $|\mu| = 36$. It should be noted that for lack of place the values of μ do not appear in the diagram. At the periphery, the values of ν correspond to 72 imaginary radii so that an intersection of a circle μ with a radius ν is associated with the class $G_{\mu\nu}$. Thus, each class $G_{\mu\nu}$ is associated with an arc of a circle whose thickness is proportional to the sample volume filled by the crystals whose orientations belong to the same class. In figures 5-a and 5-b, the thinnest lines correspond to the volume $\Delta V°$ which would be occupied in the isotropic case (for example $G_{29.2}$). The thickest lines correspond to a ten times larger volume.

It is obvious, as is here the case, that if the sample possesses orthorhombic symmetry, the quadrant A is sufficient for the description of the texture. However, to facilitate the understanding of the results, quadrant A can be represented by a (100) pole figure. Furthermore, if required, this quarter of the total distribution can be decomposed into smaller components, as shown in figures 5-c and 5-d.

In this example, the texture can be described as formed by the two main ideal orientations $(3\ \bar{4}\ 6)\ [\bar{2}\ \bar{3}\ \bar{1}]$ (figure 5-c) and approximately $(3\ \bar{3}\ 2)\ [\bar{1}\ 1\ 3]$ (figure 5-d). Taking into account that the sample possesses orthorhombic symmetry, the components A_3 and A_4 represent more thant 85 % of the total distribution.

This kind of representation also permits the orientations to be easily regrouped in four limited fibre textures. These fibre textures as well as the fibre axis have been numbered. It can be remarked that all the fibre axes are approximately located in macroscopic planes which are the most significant in regard to the rolling deformation. This remark and the fact that the fibre axes are parallel to the main crystallographic directions belonging to the iron glide systems confirm to us that the concept of limited fibre texture is not only a convenient means for describing the texture but that it must possess a real physical significance.

NEW POSSIBILITIES OFFERED BY THE VECTOR METHOD

One of the main advantages of the vector method is that with this technique it is possible to evaluate with facility and relative accuracy the texture modifications caused by mechanical or thermal treatments.

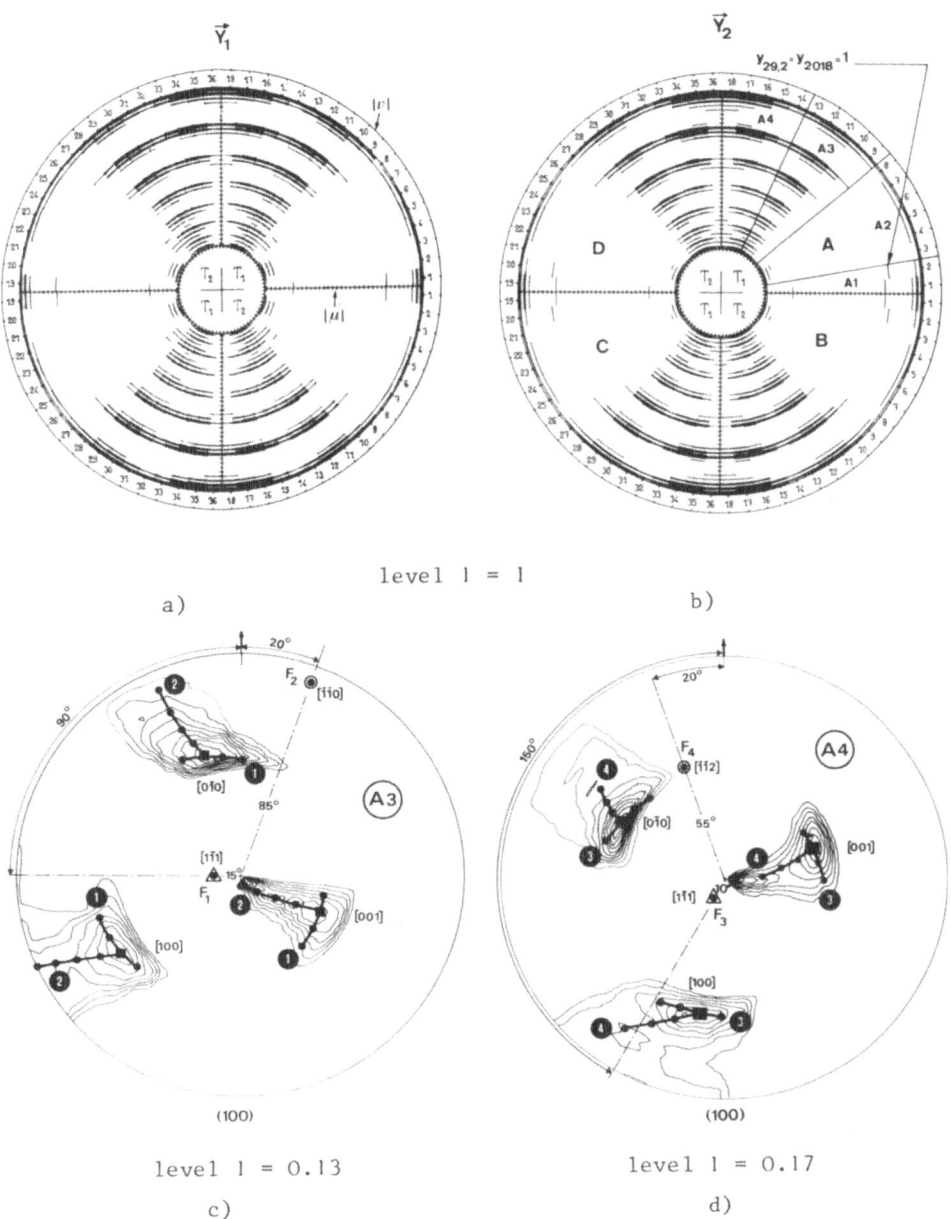

Figure 5

a) and b)- orientations distributions $\vec{Y}_1$ and $\vec{Y}_2$ calculated respectively from one and two incomplete pole figures

c) and d)- representation of partial orientation distribution of $\vec{Y}_2$

component A3 : 9,14 % of $\vec{Y}_2$

component A4 : 12,19 % of $\vec{Y}_2$

An example of the use of the vector method in this way can be presented here by considering that the orientation distribution $\vec{Y}_1$ and $\vec{Y}_2$ obtained after computation successively with one and with two incomplete pole figures may correspond to a texture modification. The vector method expresses the texture difference as the difference vector $\vec{\Delta Y} = \vec{Y}_2 - \vec{Y}_1$. This vector $\vec{\Delta Y}$ is in fact the sum of two vectors and we can write :

$$\vec{\Delta Y} = \vec{\Delta Y}^- + \vec{\Delta Y}^+$$

where $\vec{\Delta Y}^-$ represents the part of the orientation distribution $\vec{Y}_1$ that has disappeared and $\vec{\Delta Y}^+$ the set of the new orientations which have replaced the orientations represented by $\vec{\Delta Y}^-$ (figures 6-a and 6-b). The magnitude of the texture modification corresponds to 14,4 % of the original texture.

These last figures also show that the main modification observed consists of a diminution of the (111) fibre texture component and of a strengthening of the part of the orientation distribution including the orientations for which the (100) planes are nearly parallel to the rolling plane. It is furthermore naturally possible to decompose more finely the difference vector $\vec{\Delta Y}^+$ and $\vec{\Delta Y}^-$ (as already seen for the component A of $\vec{Y}_2$) when a more accurate description is required.

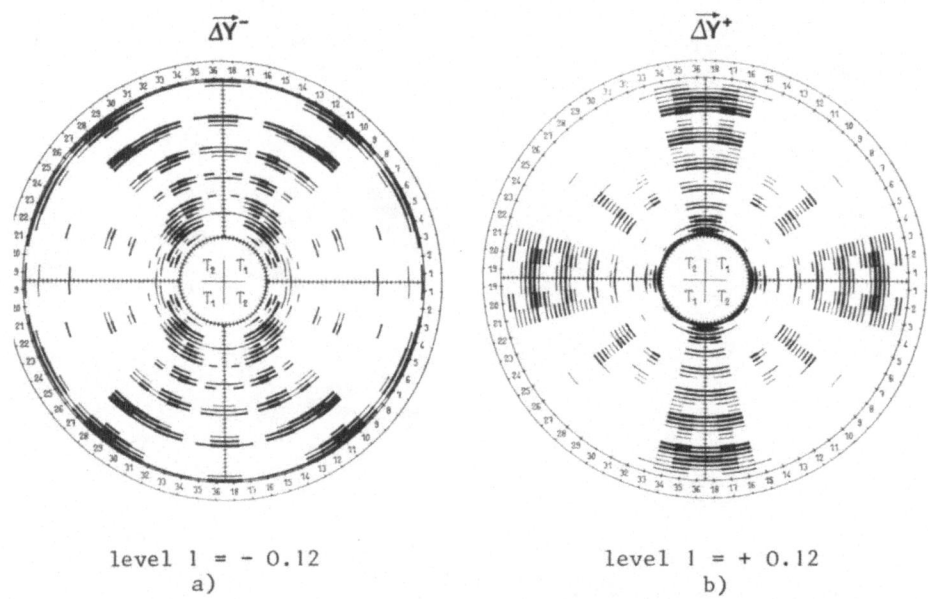

level 1 = - 0.12   level 1 = + 0.12
        a)   b)

Figure 6 - symbolic representation of the difference between the two
        textures corresponding to $\vec{Y}_1$ and $\vec{Y}_2$ in figure 5
a)- disappeared orientations = 14,4 % of $\vec{Y}_1$
b)- new orientations   = 14,4 % of $\vec{Y}_2$

CONCLUSION

The vector method of texture analysis which we have developed possesses several advantages in regard to the other usual methods. Firstly it is the only method which permits the orientation distribution to be calculated from one or several complete or incomplete pole figures without considerations concerning the macroscopic symmetry of the sample. Although this method has for the time being only been adapted to the analysis of the texture of materials of cubic structure, it can of course be generalized for the analysis of the texture of materials belonging to any crystallographic system.

Secondly, as we pointed out, the vector method leads to results of a good accuracy (comparatively with other methods). Furthermore, it gives the possibility to decompose the calculated orientation distribution into components whose identification is greatly facilitated by means of (100) pole figures. Moreover, this method seems to be very adequate for the description of texture modifications caused by physical treatments. Nevertheless, in regard to the numerical computation, the vector method is for the time being more time consuming than the other methods. We hope however, in a near future, to be able to perform the computation in approximately the same time as the other methods.

ACKNOWLEDGEMENTS

We express our thanks to Dr. P. Emery and Dr. P. Vayssière of the Institut de Recherches de la Sidérurgie (France) for the computing facilities and Prof. J.I. Verdeja of the Escuela de Minas of Oviedo (Spain) for the communication of experimental data as well as the corresponding pole figures calculated with the biaxial program.

REFERENCES

1. Hans-Joachim Bunge, "Mathematische Methoden der Textur Analyse", Akademie Verlag, Berlin, 1969.

2. Ryong-Joon Roe, "Description of Crystallite Orientation in Polycrystalline Materials. III. General Solution to Pole Figure Inversion", J. Appl. Phys., 36 : 2024, 1965.

3. Ryong-Joon Roe, "Inversion of Pole Figures for Materials Having Cubic Crystal Symmetry", J. Appl. Phys., 37 : 2069, 1966

4. R-O. Williams, "Analytical Methods for Representing Complex Textures by Biaxial Pole Figures", J. Appl. Phys., 39 : 4329, 1968.

5. Jan Pospiech and Jerzy Jura, "Determination of the Orientation Distribution Function from Incomplete Pole Figures", Zeitschrift für Metallkunde, 65 : 324, 1974.

6. Peter R. Morris, "Crystallite Orientation Analysis from Incomplete Pole Figures", Adv. in X-Ray Analysis, 18 : 514, 1974.

7. Alain Clément and Gérard Durand, "Programme FORTRAN pour l'Analyse Tridimensionnelle des Textures sans Hypothèse Restrictive sur les Symétries de l'Echantillon", J. Appl. Cryst., 8 : 589, 1975.

8. Daniel Ruer, Thèse d'Etat, University of Metz (France), to be published.

9. E. Durand, "Solutions Numériques des Equations Algébriques", Tome 2, Masson, Paris, 1961, 120.

10. Alain Clément and Pierre Coulomb, "Représentation de la Fonction de Répartition de la Texture par une Série de Projections Stéréographiques ou par une Série de Coefficients", Les Mémoires Scientifiques de la Revue de Métallurgie, 1 : 63, 1976.

CRYSTAL SUBGRAIN MISORIENTATIONS VIA X-RAY DIFFRACTION MICROSCOPY

R. W. Armstrong and C. Cm. Wu, University of Maryland,

College Park, Maryland, 20742; and, E. N. Farabaugh,

National Bureau of Standards, Washington, D. C., 20234

(C. Cm. Wu is currently on leave at the Naval Research

Laboratory, Washington, D. C., 20375.)

ABSTRACT

The "misorientation contrast" which occurs at boundaries marking the relative displacement of adjacent subgrain reflections in x-ray diffraction images is shown by a stereographic projection method of analysis to be useful for deciphering x-ray images obtained by the Berg-Barrett and Lang techniques. Experimental results are given for subgrain structures observed in Zn and Al_2O_3 crystals.

THE MISORIENTATION OF CRYSTAL SUBGRAINS

The mosaic pattern of subgrains within the microstructure of individual crystals is observed as clearly in Berg-Barrett (1;2) x-ray images as in images obtained with white radiation by means of the Schulz technique (3). In both cases, the principal reason for the subgrain units being revealed is due to their "misorientation contrast". This descriptive term applies for the observation whereby slightly differently directed x-ray beams are diffracted for a single lattice reflection from adjacent subgrains to give separated or overlapping x-ray intensities within a composite diffraction image.

Figure 1 shows this effect for a single subgrain within a Zn single crystal which had been carefully seeded and solidified along [0001]. The crystal surface orientation is (0001). The boundary is white where the reflected intensity from adjacent subgrains is separated. The varying misorientation contrast of this boundary has been analyzed by means of the stereographic projection method described previously by Wu and Armstrong (4,5). The rotation axis for the misorientation is [0001]. The subgrain boundary follows these differing crystallographic directions in accordance with the nature of its constituent dislocations and of the constraints at the boundary end points. The segmented boundary appears to have swept a region of the subgrain volume near the crystal surface, at least, free of individual dislocations in achieving the current position.

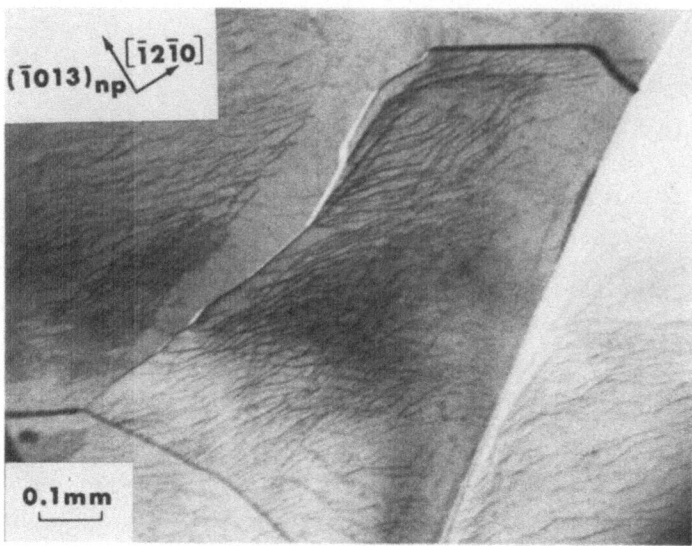

Figure 1.   Crystallographic direction dependence of dislocation
            subgrain boundary for Zn crystal solidified along [0001].

SUBGRAINS WITHIN AN Al_2O_3 CRYSTAL

Lang and Berg-Barrett x-ray micrographs are viewed along an approximate [0001] in Figures 2a., b., respectively, for an Al_2O_3 single
crystal (6). Limited projection Lang topographs have been employed to
characterize the nature of the individual dislocations which are shown
in great profusion through the crystal depth of Figure 2a. (7). The
white spot in Figure 2a. is at the center of an included group of
oddly-shaped subgrains. These subgrains are more clearly revealed along
with a number of others in Figure 2b.

The misorientation contrast of the enclosing boundaries of the
single subgrain volume which is encircled in Figure 2b. has been analyzed by employing all six of the available {11$\bar{2}$6} Berg-Barrett reflections. The results are indicated in Figure 3. For each reflection,
three separate boundary segments of the included subgrain bordering on
the matrix crystal volume were plotted according to the stereographic
projection method so as to determine the relative orientation of the
diffracted beams and, thence, the orientation of the line through the
reflecting plane normals. On the basis that the rotation axis for the
subgrain misorientation must lie along a great circle perpendicular to
this line, for each reflection, the various arcs were determined to give
an estimated rotation axis (RA) orthogonal to (0$\bar{1}$14). Hypothetical
diffracting plane normals for a 5° rotation about the RA are superposed
on the stereographic projection of Figure 3. With RA normal to (0$\bar{1}$14),
a value of b = (a/3)[2$\bar{1}\bar{1}$0] is suitable for the dislocations comprising
the misorientation. The elongated shape of the subgrain along [01$\bar{1}$0] is
a reasonable observation, both for this RA and corresponding b value for
the dislocations, because the principal boundary lengths would then

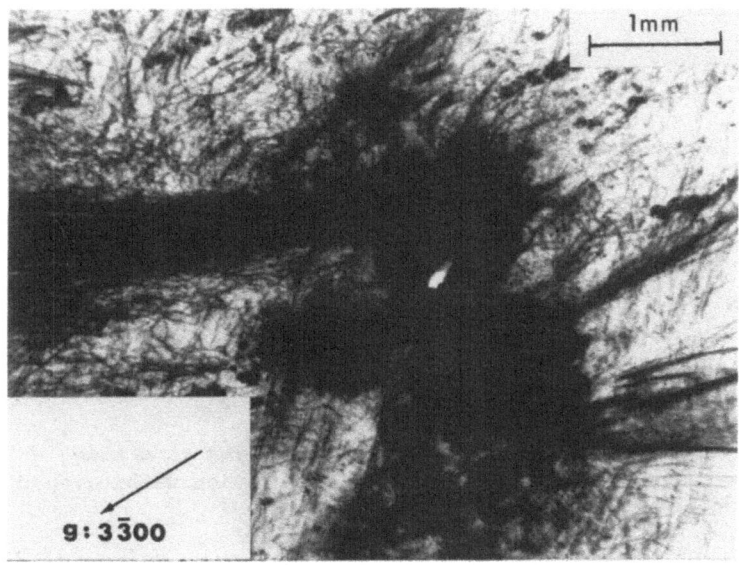

Figure 2 a.   Lang ($3\bar{3}00$) topograph with Ag Kα_1 radiation from (0001) Al$_2$O$_3$ crystal plate.

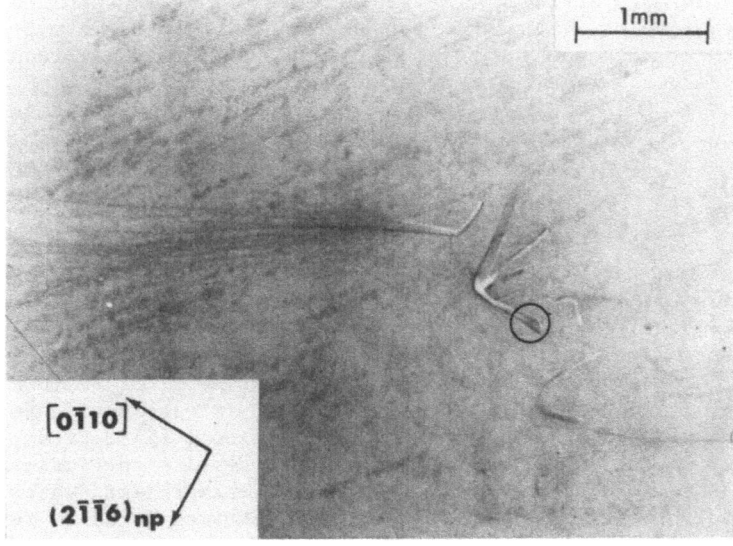

Figure 2 b.   ($2\bar{1}\bar{1}6$) Berg-Barrett reflection from crystal surface.

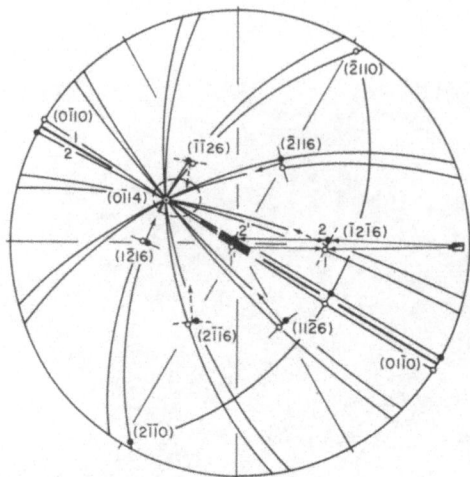

Figure 3. Stereographic projection analysis giving rotation axis
normal to (0$\bar{1}$14) for subgrain misorientation within vapor
grown Aℓ_2O$_3$ crystal.

constitute, essentially, walls of edge dislocations in a tilt orientation.
Of special significance, also, is the previous observation that the
growth of this crystal along [0001] occurs with surface facets at the
crystal-vapor interface being composed of either {0001} or {0$\bar{1}$14} ledges
(8).

MISORIENTATION CONTRAST WITHIN A LANG TOPOGRAPH

Because of the more stringent diffraction geometry which applies
for the Lang technique of obtaining x-ray topographs, very small misorientations between adjacent subgrains, say 0.005°, are normally required
for the simultaneous observation of adjacent subgrain reflections (9).
This general condition stems, by way of comparison, from the incident
x-ray beam divergence being, say, 0.3° for the Berg-Barrett technique
versus 0.01° for the Lang technique and, also, from the required presence
of a very narrow slit system behind the crystal in the Lang method so as
to selectively record the diffracted x-ray beam intensity (10).

A special circumstance in which misorientation contrast might be
fruitfully studies with the Lang method is indicated in Figure 4. It
applies for a silicon-iron alloy crystal plate having surfaces parallel
to the ($\bar{1}\bar{1}$0) whose normal is chosen as the center of the stereographic
projection. In this case, the x-ray source is shown as a dot on the
equatorial plane of incidence near the center of the projection. Two
(002) diffracting plane normals, for a 5° rotation about [$\bar{1}\bar{1}$0], are
shown to give the 1' and 2' diffracted beam images despite the small
divergence of the x-ray beam and the presence of the exit slit system.
For this technique, the crystal and film may be traversed so as to reveal
the boundary direction along a substantial length of the crystal. Such
boundary misorientations are possibly observed (9) for this specially
chosen misorientation of subgrains because all of the (a/3) <111> b

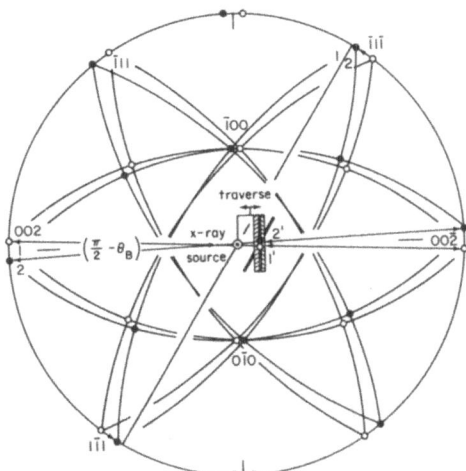

Figure 4.   Stereographic projection analysis of misorientation contrast
            within a Lang topograph for a special rotation between
            subgrains in an iron-silicon alloy crystal.

values for this crystal orientation are capable of being arranged, also,
in the form of tilt walls of edge dislocations perpendicular to the
crystal plate surfaces.

ACKNOWLEDGEMENTS

    Support for this research has been provided by the U.S. National
Science Foundation, Grant DMR73-07619-A01. The experimental results
were obtained within the X-ray Central Facility of the Center of Materials Research, University of Maryland, and at the Inorganic Materials
Division of the National Bureau of Standards, Washington, D.C.

REFERENCES

1. W. F. Berg, "Über eine rontgenographische Methode zur Untersuchung
    von Gitterstörungen an Kristallen," Naturwissenschaften 19, 391-396
    (1931).

2. C. S. Barrett, "A New Microscopy and Its Potentialities," Trans.
    AIME 161, 15-64 (1945).

3. L. G. Schulz, "Method of Using a Fine Focus X-ray Tube for Examining
    the Surface of Single Crystals," J. Met. 1082-1083 (1954).

4. C. Cm. Wu and R. W. Armstrong, "Misorientation Contrast of Crystal
    Subgrain Boundaries in Berg-Barrett X-ray Micrographs," J. Appl.
    Cryst. 8, 29-36 (1975).

5. C. Cm. Wu and R. W. Armstrong, "Quantitative Analysis of Subgrain
    Misorientations Observed in Berg-Barrett X-ray Micrographs," Phys.
    Stat. Sol. (a)29, 259-263 (1975).

6. E. N. Farabaugh, Ph.D. Thesis Research, University of Maryland (1976).

7. E. N. Farabaugh and C. Cm. Wu, "Characterization of Growth Defects
    within a Vapor Grown Sapphire Crystal via the Lang X-ray Technique"
    Amer. Conf. on Crystal Growth, ACCG III, p. 116-117, Stanford,
    California (1975).

8. E. N. Farabaugh, H. S. Parker, and R. W. Armstrong, "Skew-Reflect-
    ion X-ray Microscopy of the Vapor Growth Surface of an $A\ell_2O_3$ Single
    Crystal, "J. Appl. Cryst. $\underline{6}$, 482-486 (1973).

9. A. R. Lang and M. Polcarova, "X-Ray Topographic Studies of Disloca-
    tions in Iron-Silicon Alloy Single Crystals," Proc. Roy. Soc. London
    $\underline{A\ 285}$, 297-311 (1965).

10. R. W. Armstrong and C. Cm. Wu, "X-Ray Diffraction Microscopy," in
    J. L. McCall and W. M. Mueller, Editors, "Microstructural Analysis:
    Tools and Techniques," p. 169-219, Plenum Press (1973).

SOME TOPOGRAPHIC OBSERVATIONS OF THE EFFECTS
OF DYNAMICAL DIFFRACTION IN IMPERFECT METAL CRYSTALS

William J. Boettinger, Harold E. Burdette,

Edward N. Farabaugh and Masao Kuriyama

Institute for Materials Research

National Bureau of Standards

Washington, D.C.   20234

ABSTRACT

X-ray diffraction topographs obtained in the anomalous transmission geometry contain images which are quite different from those obtained from thin crystals ($\mu L \leq 1$). In this paper various topographic images which are unique to thick crystals ($\mu L > 10$) will be presented and discussed in terms of dynamical diffraction in imperfect crystals. First, it is observed in topographs that images of crystal imperfections caused by disruption of the anomalous transmission effect are slightly broader or more diffuse in the H-diffracted (Bragg-diffracted) beam than in the O-diffracted (transmitted beam). These observations have been made in both copper and nickel crystals. Such broadening can be explained by the presence of unique extinction terms in the expression for the intensity of the dynamically diffracted beam in an imperfect crystal. Second, it is observed that magnetic domain walls in thick nickel crystals are always imaged as deficiencies of the anomalous transmission background. This is to be contrasted with images of such walls in thin crystals which are either white or black and are caused primarily by mosaic type image formation from the adjacent domains. Third, unusual images will be shown which are deficient in x-rays in the O-diffracted beam and enhanced in the H-diffracted beam or vice versa. It will be shown experimentally in copper crystals that such images are caused by defects near the exit surface of the crystal. These images are thought to be caused by kinematical scattering of the already existing dynamically diffracted beam near the x-ray exit surface. Such surface defects are often artifacts of sample preparation and hence this information provides a useful tool for their identification. All of these topographic images have been obtained using asymmetric crystal topography. This paper will also include a brief comparison of this technique with more traditional topographic techniques.

INTRODUCTION

In the past twenty years, x-ray diffraction topography has become a valuable tool for the study of defects in crystalline materials. In the transmission geometry, the majority of these studies have been applied to thin crystals ($\mu L \leq 1$) where the kinematical image of defects lends itself to easy analysis. Spurred on by the desire to assess the perfection of metal crystals in a non-destructive way and by the availability of better quality metal crystals, x-ray topography of thick crystals using the Borrmann effect has today become more routinely possible and even desirable. In this paper, using nickel and copper, various topographic images unique to thick crystals ($\mu L > 10$) will be presented and discussed in terms of dynamical diffraction in imperfect crystals. The topographs used for this presentation were obtained using asymmetric crystal topography (ACT). This technique provides a simple yet strain sensitive alternative to scanning topographic techniques. A brief comparison of topographs taken with ACT will be made with topographs taken with a Lang camera.

ASYMMETRIC   CRYSTAL   TOPOGRAPHY

Asymmetric crystal topography (ACT) uses (111) asymmetric diffraction from the surface of silicon crystal to obtain a wide, monochromatic and highly parallel beam which is then diffracted from the sample crystal. A schematic of an ACT camera is shown in Figure 1. The wide beam permits the formation of topographic images of sample crystals without scanning. Further details of the technique are given elsewhere (1).

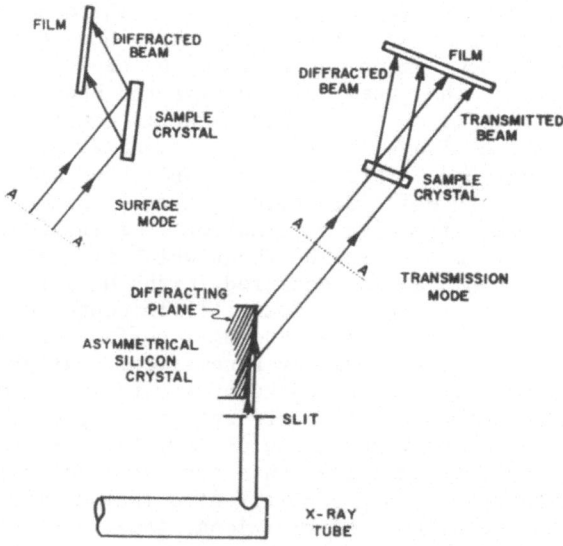

Figure 1.   Schematic of Asymmetric Crystal Topographic (ACT) Camera showing alignment for the sample crystal in surface reflection and transmission geometries.

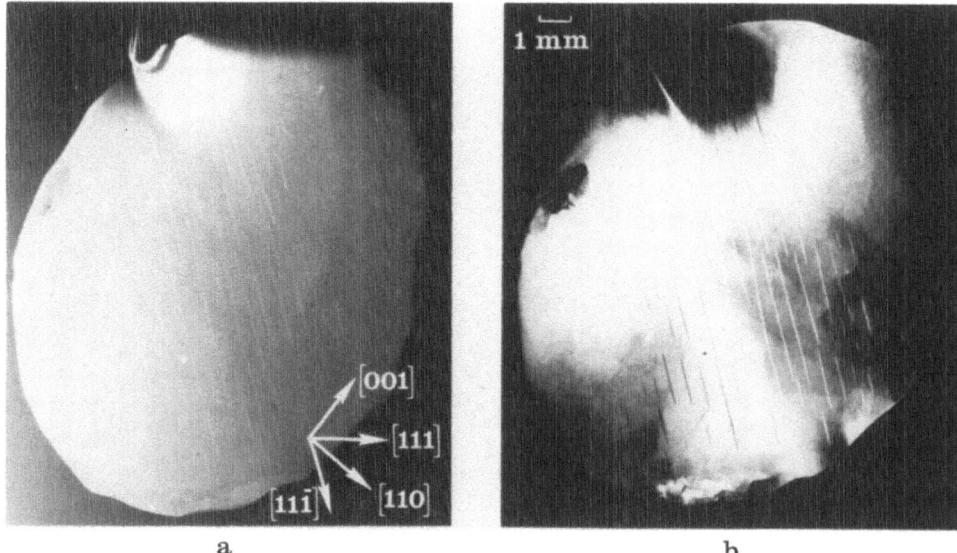

a   b

Figure 2.   ($\bar{2}$20) surface reflection topographs of a nickel crystal under
             the same diffraction conditions using a) a Lang camera and b)
             an ACT camera.

     For the purpose of demonstration we will compare ACT topographs to
topographs taken with a Lang camera for the same crystal and diffracting
conditions using both surface reflection and transmission geometries.
Nuclear plates are always set parallel to the crystal surface and symmetric diffraction is always used. Figure 2 shows ($\bar{2}$20) surface reflection topographs of a nickel crystal with a) a Lang camera and b) an ACT
camera. In these topographs, black represents fewer photons. Both
were taken with CuKα radiation and identical nuclear plates. The collimating slit used in the Lang camera was 100 μm and the microfocus tube
was run at 40 kv,2.5mA. The exposure time was 16 h. The ACT uses a
small focus tube and was run at 40kv,20mA. The exposure time was 3 h.
First, one should note in a) the de-emphasis of the overall strain field
and sub-grain boundaries which are clearly visible in b). Second, the
images parallel to [11$\bar{1}$] possess much more detail in b) than in a).
Specifically, there is more black-white contrast in these images in b).
These straight-line images are caused by the intersections of oblique
magnetic domain walls with the crystal surface and will be described
in more detail in a later paper (2) in this volume.

     Figure 3 shows (11$\bar{1}$) transmission topographs of a nickel crystal
taken with a) a Lang camera and b), c) an ACT camera. This crystal is
quite thick with μL = 22. Exposure times were 90 h and 65 h, respectively. Because these topographs are taken using the anomalous transmission effect, both the 0-diffracted (transmitted) and the H-diffracted
(Bragg-diffracted) beams contain topographic images. These two beams
are shown in b) and c) respectively and are recorded simultaneously on
the same plate. On the other hand, the 0-beam in the Lang topograph
was not useable because the topographic images were masked by those

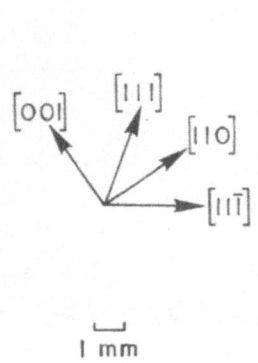

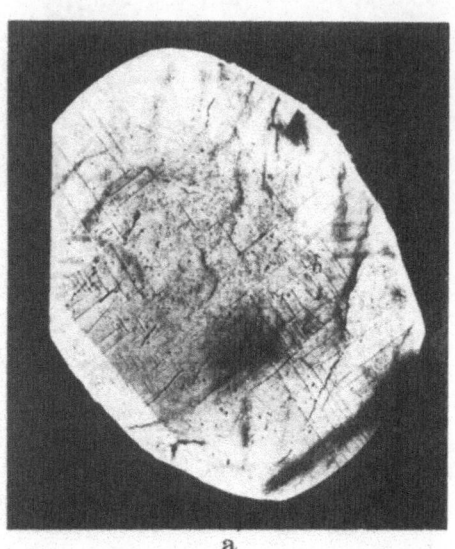

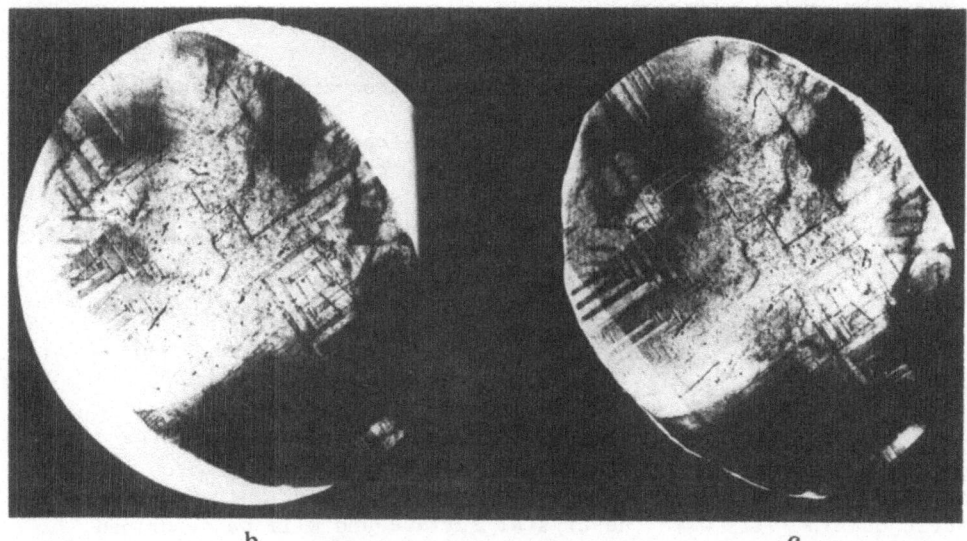

Figure 3.   (11$\bar{1}$) transmission topographs of a nickel crystal; a) the H-
diffracted (Bragg-diffracted) beam using a Lang camera and
b), c) the O-diffracted (transmitted) beam and H-diffracted
beam using an ACT camera.   μL = 22.   Black represents fewer
photons.

radiations for which μL was small. Figure 3a is the H-diffracted beam of the Lang topograph. A simple advantage of ACT is that both beams can routinely be obtained. As described later, the availability of the O-topograph assists uniquely in identifying the nature of crystal imperfections. In comparing the Lang and ACT topographs we again note that the visibility of the strain contour is lessened in the Lang topograph. Note the contrast between adjacent magnetic domains in a) and b), c), where the straight-line images parallel to [001] and [110] are the magnetic domain walls separating one domain from another. This contrast is much reduced in the Lang topograph. On the other hand, the dislocation images, especially those lying on (111) at the top, are more crisp in the Lang topograph. This crispness and the general uniformity of the intensity give the Lang topograph good visual appeal and coverage of the entire crystal. In general, however, the ACT picture appears to give more information on the details of the crystal imperfections. The use of a finer collimating slit for the Lang camera will, of course, enhance its strain sensitivity also. However, such fine slits introduce difficulties and increase exposure times for the Lang camera. ACT achieves such fine optics without scanning and without such fine slits.

DISRUPTION IMAGES

Effect of Crystal Thickness

For diffraction from crystals in the transmission or Laue geometry, three regimes of crystal thickness have been identified (3): $\mu L < 1$, $1 < \mu L < 10$, and $\mu L > 10$. The regime, $\mu L < 1$, is familiar to most and is dominated by kinematical scattering of the beam incident on the crystal. In the intermediate regime, dynamical effects become important; but topographic images are very complex because of a mixture of kinematical and dynamical effects. In the regime, $\mu L > 10$, topographic images become more simple and, except for images from imperfections located near the crystal exit surface, are due to disruption or shadowing of the anomalously transmitted background.

In the two regimes where $\mu L > 1$, dynamical imaging of crystal imperfections can be interpreted by considering two terms, the first term $|S^{(1)}|^2$ and the second term $|S^{(2)} + S^{(3)}|^2$ given in (3) and shown schematically in Figure 4. The first term, shown in Figure 4a, consists of the dynamical background and its local disruption by the imperfection V. The second term, shown in Figure 4b, consists of the effects of kinematical scattering of the already existing dynamically diffracted beam by the imperfection V. Figure 4c is the superposition of a) and b) to give a rather complex defect image with both white and black components imposed on the anomalous transmission background intensity. This is the type of image which appears in the range $1 < \mu L < 10$. For $\mu L > 10$, the second term is small and images of defects are caused primarily by disruption of the anomalous transmission effect. An exception to this is when the defect is located near the crystal exit surface.

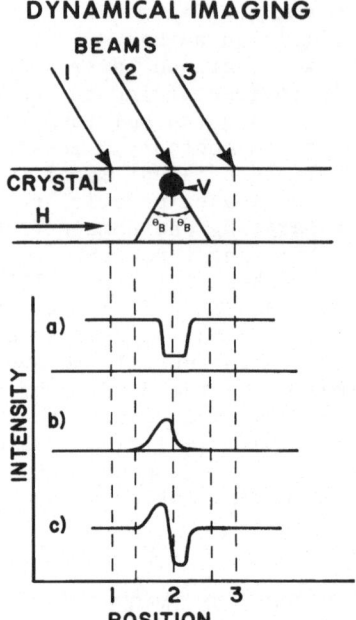

Figure 4.   Illustration of dynamical imaging of imperfection V; a) the
            disruption of the dynamical diffraction effect by the imper-
            fection and b) the kinematical scattering of the dynamical
            beam by the imperfection and c) the superposition to form a
            complex black-white image superimposed on the anomalous trans-
            mission background. When μL>10, the effect of b) is small
            and the image of imperfections is caused primarily by disrup-
            tion of the dynamical diffraction effect as shown in a).

        Examples of disruption images are shown in Figure 5a and b which
are enlargements from an ACT (11$\bar{1}$) transmission topograph where a) is
the O-diffracted (transmitted) beam and b) is the H-diffracted (Bragg-
diffracted) beam. In this and following topographs white represents
fewer photons. The random white lines are dislocation images and the
straight white lines parallel to [001] are the images of magnetic
domain walls perpendicular to the crystal surface. It has been demonstrated (4) that these images are in fact caused by domain walls. Note
that the images are essentially identical in the two beams. This is
characteristic of disruption images.

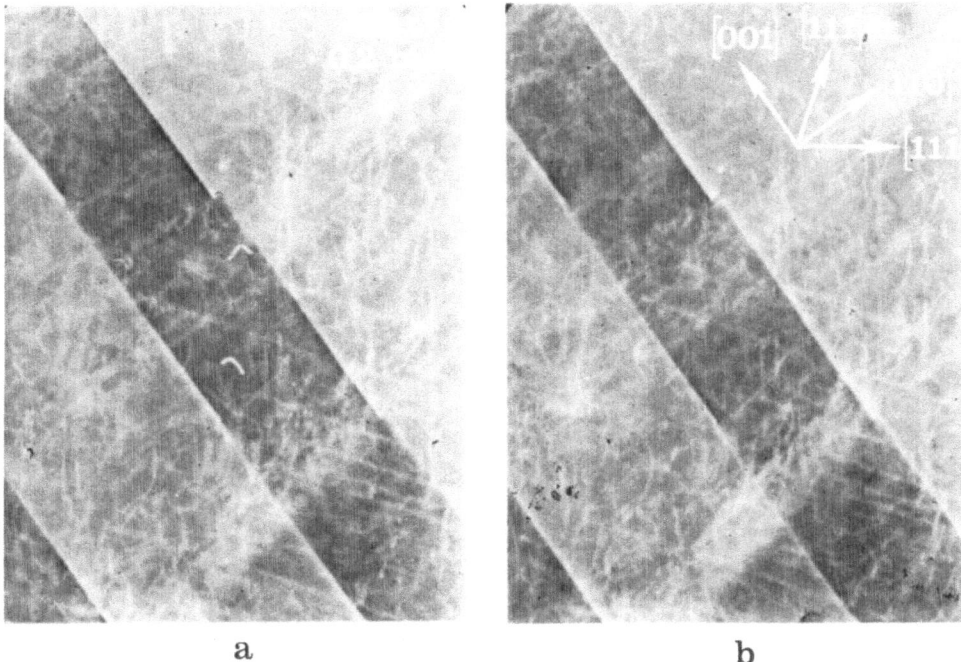

a   b

Figure 5. ACT (11$\bar{1}$) transmission topograph of a) the O-beam and b) the
H-beam of a nickel crystal showing disruption of the anoma-
lous transmission by dislocations (the random white lines)
and by magnetic domain walls (white images) parallel to
[001]. μL = 20.3. White represents fewer photons.

Broadening in H Beam

Even though the images in O and H topographs are essentially identical, we have observed a very slight broadening or diffuseness of
topographic images in the H beam as compared to the O beam. This often
lends to a choice of the O beam for better topographic quality when
only one of the beams is used for presentation. This broadening is
demonstrated in Figure 6a) and b) which are enlargements using high
contrast film from the O and H beams, respectively, from an ACT (11$\bar{1}$)
transmission topograph of a nickel crystal with μL = 32. The straight-
line images perpendicular to [111] are glissile dislocations lying on
the (111) plane which is perpendicular to the crystal surface (no magnetic domain walls are shown here).

This broadening is thought to be caused by a slight contribution
of the kinematical term to the primarily disruption image. Although
calculations have been performed to confirm this slight broadening in
the H beam, the cause of this broadening can be seen by examining the
form of an approximate expression for the amplitude, S_O and S_H of the
O and H beams given (2) by

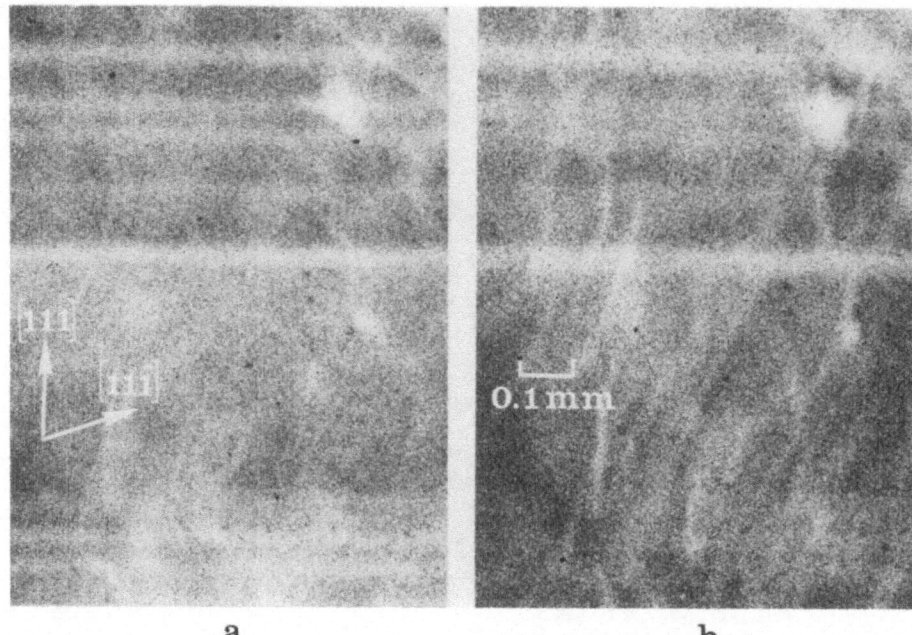

Figure 6.   Enlargements from ACT (11$\bar{1}$) transmission topograph of a) the
O-diffracted beam and b) the H-diffracted beam of a nickel
crystal showing glissile dislocation images lying on (111).
Note that the width of the dislocation images is slightly
broader in b) than in a). $\mu L = 32$.

$$S_0 = D_0^{(1)} \exp\{-\mu_e^{(1)}L\} + (LV_R)\sin(H{\cdot}u) \; v(-H) \; D_H^{(2)} \exp\{-\mu_0^{(2)}L\} \qquad (1)$$

$$S_H = D_H^{(1)} \exp\{-\mu_e^{(1)}L\} + (LV_R)\sin(H{\cdot}u) \; v(H) \; D_0^{(1)} \exp\{-\mu_H^{(1)}L\}, \qquad (2)$$

where

$D_K^{(i)}$ = Dynamical Field Functions equivalent to the
dynamical solution in a perfect crystal for
mode i for the K diffracted beam

$\mu_e^{(i)}$ = $\mu\{1+(-1)^i (1+\varepsilon^2)^{-\frac{1}{2}} e^{-\gamma}\}$ = effective absorption
coefficient for the i mode

μ = the ordinary linear absorption coefficient

ε = normalized deviation from the Bragg angle for the
incident beam

$e^{-\gamma}$ = $1 + V_R(\cos(H{\cdot}u)-1)$

V_R = ratio of imperfection volume to crystal volume

H = reciprocal lattice vector for the diffraction

u = displacement of atoms from perfect crystal lattice sites

L = crystal thickness

v(K) = Fourier transform of atomic polarizability, and $v(K) = v(-K)$ for the present crystals

$\mu_0^{(2)} = \mu_H^{(1)} = \mu\{1+(1-2p)\ (1+\eta^2)^{-\frac{1}{2}}\}$

pL = distance of imperfection in crystal from entrance surface

η = "scattering angle"--corresponds to positions on the exit surface measured from the imperfection as angles deviating from the diffracting plane.

The first terms in Equations (1) and (2) are not functions of η; i.e., scattering angle and hence no broadening due to the imperfection can be expected from them. These are the $S^{(1)}$ terms and describe the anomalous transmission and disruption of it by the imperfection. On the other hand, the second terms in Equations (1) and (2) are functions of η. These are the contributions in this approximation of the $S^{(2)}$ and $S^{(3)}$ terms. The dependence on scattering angle comes from the terms

$$D_H^{(2)}\ \exp\{-\mu_0^{(2)}L\}\quad\text{and}\tag{3}$$

$$D_0^{(1)}\ \exp\{-\mu_H^{(1)}L\}.\tag{4}$$

In these terms, the dynamical field functions are weighted for different scattering angles η by the exponential functions which are equal. This weighting function is shown schematically in Figure 7 and has finite value at $\eta = 0$ and vanishes for large $|\eta|$. The dynamical field functions have quite different form from one another and give rise to the broadening in the H beam. D_H for mode 2 has finite value at $\eta = 0$ and vanishes for large $|\eta|$, while D_0 for mode 1 vanishes for $\eta<<0$ and has non-zero (positive) value for $\eta>>0$. These functions are also shown schematically in Figure 7. Because of the non-zero nature of D_0 for mode 1 for $\eta>0$, Expression (4) is slightly broader in η than is (3). This broadening should be more noticeable for defects located toward the exit surface of the crystal. The dependence of the image on η for the O and H directions can be observed in topographs as a spatial broadening on the topographic plate in the O and H beams.

Magnetic Domain Walls

The images of magnetic domains as shown in Figure 5 are to be contrasted with the observation of magnetic domain walls by Polcarová and co-workers (5). They have imaged magnetic domains primarily in crystals where $\mu L<1$ and have shown images which are either black or white. They have explained these images based on the change in magnetostriction across a domain wall from one domain to another. This then

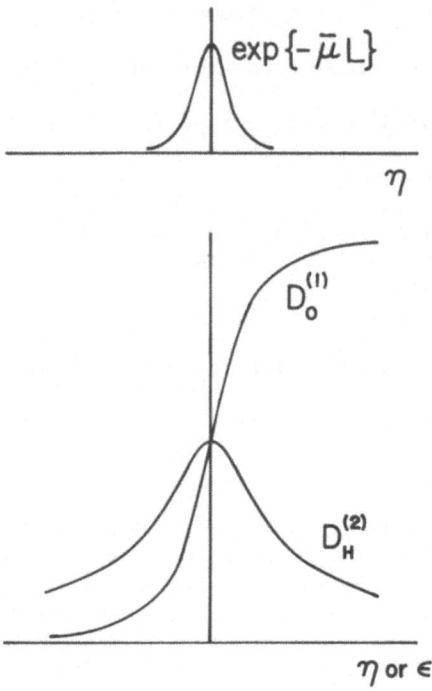

Figure 7. Schematic of the exponential weight-
ing function where $\bar{\mu} = \mu_O^{(2)} = \mu_H^{(1)}$ and
the dynamical field functions D_O for
mode 1 and D_H for mode 2. The non-zero
nature of D_O for mode 1 when $\eta \gg 0$ is
the source of the slight broadening
of disruption images in the H-beam.

produces "underlap" or "overlap" of the diffracted beam using ordinary
contrast theory. In thick crystals with $\mu L > 10$, magnetic domain walls
are always white or disruption images. Such images cannot accurately
be modeled using the change in magnetostriction from one domain to the
other but must be modeled by the displacements in and near the domain
wall itself. Such local distortion leads quite naturally to disruption
images.

SCATTERING IMAGES

As pointed out in the previous section, the effects of kinematical
scattering are usually confined to thin crystals. However when a defect is very close to the exit surface of a thick crystal, interesting
images can appear. For these images the kinematically scattered beam
does not travel a sufficient distance in the crystal to be absorbed and
thus exits the crystal and forms topographic images.

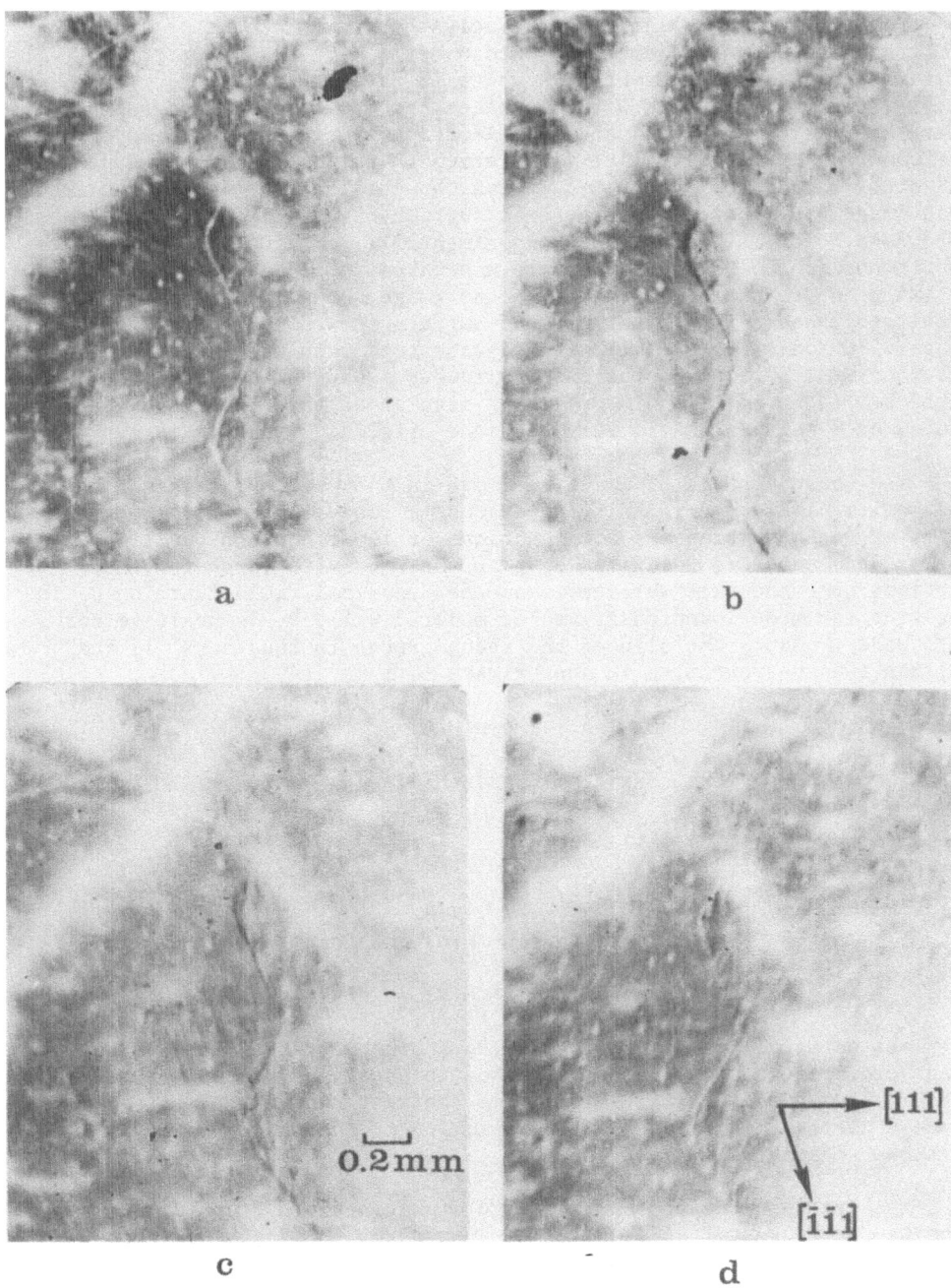

Figure 8.   ACT (111) transmission topograph of a) the O-beam and b) the
            H-beam and ACT ($\overline{1}11$) transmission topograph of c) the O-beam
            and d) the H-beam of a copper crystal. Note the wavy line
            image which is white in a) and d) and black in b) and c)
            while all other images are white. $\mu L = 35$.

Figure 8 shows transmission topographs of a copper crystal with
$\mu L = 35$ where a) and b) are respectively the 0 and H beams of a (111)
topograph and c) and d) are the 0 and H beams of a ($\overline{1}11$) topograph.
Note the wavy line image which is white in a) and black in b). This
image is to be contrasted with all other images in these topographs
which are white in both a) and b). Additionally in the ($\overline{1}11$) topograph
the contrast of the wavy line is reversed in the 0 and H beams and again
all other images remain white. It was thought that the defect causing
this image was located near the exit surface of the crystal for reasons
discussed below. To prove this experimentally, a ($\overline{2}20$) surface reflection topograph of this exit surface was taken. This topograph reveals
an image corresponding to the wavy line image shown in Figure 8. Surface topographs of the other crystal surface showed no such wavy line
image. In addition, a (111) transmission topograph was taken with the
other crystal surface as the exit surface. Again no wavy line image
appears. These additional topographs give proof that the image was
caused by a defect located very near the original crystal exit surface.

The cause of this black-white image in the 0 and H beams can be
seen quite nicely by examining Equations (1) and (2) again. The first
terms of each equation are not functions of imperfections position, p.
The second terms are functions of p and become quite large for imperfections near the exit surface. Now the dynamical field function D_H is
negative for mode 1 and positive for mode 2, while D_0 is positive for
both modes. Hence the sign of the second terms in Equations (1) and
(2) can thus be rewritten in approximate form

$$S_0 \sim A + B \sin(H \cdot u) \tag{5}$$

$$S_H \sim -A + B \sin(H \cdot u), \tag{6}$$

where A and B are positive;

and

$$|S_0|^2 \sim A^2 + B^2 \sin^2(H \cdot u) + 2AB\sin(H \cdot u) \tag{7}$$

$$|S_H|^2 \sim A^2 + B^2 \sin^2(H \cdot u) - 2AB\sin(H \cdot u) \tag{8}$$

The last terms of Equations (7) and (8) account for the black-white
image of a defect near the exit surface in the 0 and H beams. Because
$\sin(H \cdot u)$ is an odd function of H, this contrast changes to white-black
when -H diffraction is used. The formation of such black-white images
is shown schematically in Figure 9.

We have demonstrated an advantage of having both 0 and H topographs
from the same crystal. If a ($\overline{1}11$) transmission topograph had been
taken and only the H beam recorded (Figure 8d), one might have miscon-
strued the wave line image as being an internal defect. Since such
surface defects are often artifacts of sample preparation, quick assessment of a defect as such is quite useful.

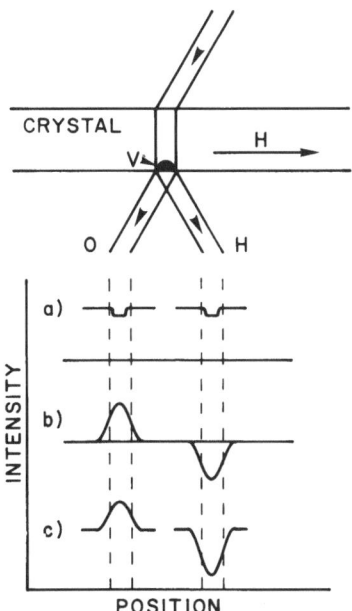

Figure 9.   Illustration of dynamical imaging in the O and H beams of a
            defect V located very near the crystal exit surface; a) the
            disruption of the dynamical diffraction effect by the imper-
            fection which is essentially identical in the O and H beams;
            b) the rather large in magnitude kinematical scattering of
            the dynamical beam by the imperfection, which is dependent
            on the sign of $\sin(\underline{H}\cdot\underline{u})$; and c) the superposition to form
            images which are white (or black) in the O-beam and black
            (or white) in the H-beam.

REFERENCES

1. W. J. Boettinger, H. E. Burdette, M. Kuriyama and R. E. Green, Jr.,
    "Asymmetric Topographic Camera," Rev. Sci. Instrum. <u>47</u>, 906-911 (1976).

2. M. Kuriyama, W. J. Boettinger and H. E. Burdette, "Crystal Imper-
    fections and Magnetic Domain Walls in Thick Czochralski-Grown Nickel
    Single Crystals," in <u>Advances in X-Ray Analysis</u>, Vol. 20 (1976).

3. M. Kuriyama, "X-Ray Diffraction from a Crystal Containing Isolated
    Imperfections," Acta Cryst. <u>A25</u>, 682-693 (1969).

4. M. Kuriyama, W. J. Boettinger and H. E. Burdette, "X-Ray Topographic
    Observation of Magnetic Domains in Czochralski-Grown Nickel Single
    Crystals in Anomalous Transmission Geometry," J. Appl. Phys. <u>47</u> (11)
    (1976).

5. M. Polcarová and J. Gemperlová, "Distortion of an Fe-Si Single
    Crystal and X-Ray Topographic Contrast due to a 90° Ferromagnetic
    Domain Wall," phys. stat. sol. <u>32</u>, 769-778 (1969).

DIRECT DISPLAY OF X-RAY TOPOGRAPHIC IMAGES

Robert E. Green, Jr.

Department of Mechanics and Materials Science

The Johns Hopkins University

Baltimore, Maryland 21218

ABSTRACT

The purpose of the present paper is to give a comprehensive state-of-the-art review of all electro-optical systems used to date for direct viewing of X-ray topographic images. Consideration is given to both direct conversion X-ray sensitive vidicon systems and to indirect conversion systems which use fluorescent screens to convert the X-ray image into a visible one. Included in this review is a discussion of the relative advantages and disadvantages of the various electro-optical systems, including cost, versatility, portability, simplicity of operation, sensitivity, and resolution capability.

INTRODUCTION

In 1970 the present author presented a review paper (1) concerning electro-optical systems for dynamic display of X-ray diffraction images. At that time only a few experiments had been performed in which such systems were successfully used for X-ray topographic applications (2-7). Since that time a number of additional papers have appeared in which electro-optical systems have been used to view X-ray topographic images. All of the systems reported to date may be grouped into two broad categories depending on the general principle used to permit rapid viewing and recording of X-ray topographic images, namely the direct method and the indirect method.

DIRECT METHOD

In the direct method an X-ray sensitive vidicon television camera directly converts the X-ray topographic image into an electronic charge pattern on a lead oxide or silicon diode array target; this charge pattern is read out by a scanning electron beam and displayed as a visible image on a television monitor. This method yields a topographic image of approximately 15 μm resolution, which is limited by either the bandwidth of the television system, the thickness of the X-ray sensitive target, or the size of the electron beam at the target. Due to the low

221

gain of the vidicon camera tubes, it must be used in conjunction with a
high intensity rotating anode X-ray generator capable of emitting more
than 5 x 10⁵ X-ray photons per sec per mm². Thus the direct method com-
bines a high intensity X-ray source with a low gain electro-optical
detector.

The first investigators to employ the direct method for viewing and
recording X-ray topographic images were Chikawa and Fujimoto (2) in
1968. Since that time Chikawa and associates (8-11) have continued to
use and improve their direct system. Chikawa et al. used a rotating
anode X-ray generator operating at 60 kV and 500 mA with a molybdenum
target for Lang transmission topography and a copper-chromium (1%)
target for Berg-Barrett and double-crystal methods. Their detector was
a lead-oxide photoconductive target vidicon camera tube with a beryllium
input faceplate.

Figure 1 shows a schematic drawing of the direct X-ray topography
video display system used by Chikawa. The television system is basically a standard closed-circuit vidicon television system incorporating
a television camera with the special beryllium faceplate lead-oxide
target vidicon camera tube, a video amplifier, camera control unit, and
picture monitor. A standard 525-line, interlaced, 30 frame per sec

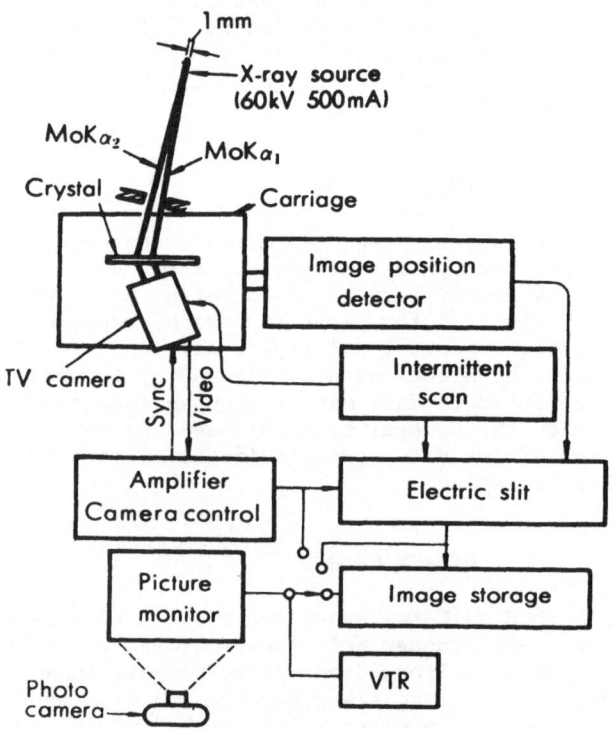

Figure 1.   Schematic of direct X-ray topography video
            display system (after Chikawa (10))

scanning system is employed. For weak images, an intermittent scanning unit is used which permits a charge build-up on the photo-conductive target before read-out by the scanning electron beam. The amount of charge build-up and hence the signal current is higher the longer the intervals between the intermittent scans. For transmission topography, the television camera and crystal specimen are placed on the carriage of the topographic camera. By orienting the crystal so as to satisfy the Bragg condition for the slightly divergent incident X-ray beam, images 1 mm wide from both the $K\alpha_1$ and $K\alpha_2$ diffracted beams are received by the vidicon target and displayed simultaneously on the television monitor. Chikawa refers to these images as "direct-view images." In order to image Lang topographs, the video signals due to the $K\alpha_1$ image is selected by the electric slit and stored in the image storage unit, while the carriage is moved for 3 to 10 sec. The position of the electric slit is moved in synchronization with the carriage motion, and the initial position, slit width, and traversing speed are adjusted while viewing the topographic images on the television monitor. At the end of each carriage traverse a Lang topograph, of dimensions 9 by 13 mm at the vidicon target, is displayed on the monitor. Chikawa refers to this image as a "synthesized image." Note that the "direct images" are due to both $K\alpha_1$ and $K\alpha_2$ wavelengths, while the "synthesized images" are made using the $K\alpha_1$ wavelength only.

Figure 2 serves to illustrate several features of Chikawa's system. Figure 2(a) shows the direct video image of dislocations in a silicon crystal wafer as obtained in the Lang transmission topographic geometry, but without scanning. The portions of the topographic image due to the $K\alpha_1$ and $K\alpha_2$ wavelengths are clearly delineated. The horizontal band across the top of the figure is due to the electric slit. By viewing this direct image, the silicon crystal was reoriented so that the topographic images were parallel with the image of the electric slit, as shown in Figure 2(b), where the electric slit has also been moved down vertically to overlap the $K\alpha_1$ portion of the topographic image. Figure 2(c) shows the synthesized video image obtained from the $K\alpha_1$ portion with about a 3 sec carriage traverse and a normal television scanning rate of 30 frames per sec. Figure 2(d) shows the Lang topograph of the same region of the silicon crystal as recorded on an Ilford L-4 nuclear emulsion plate. Figure 2(e) shows the video waveform corresponding to the television scan along the dashed line shown in Figure 2(c). The peak-to-peak noise height as indicated yielded a signal-to-noise ratio of about 2 along this typical television scan line. Chikawa was able to increase the output signal considerably by using both the intermittent television scan mode and the image storage unit and thus obtain a synthesized video topograph with an improved signal-to-noise ratio in about 10 sec.

Recently Chikawa (12) has installed a more powerful rotating anode X-ray generator capable of operating continuously at 60 kV and 1000 mA, and Imura (13) is installing the most powerful rotating anode X-ray generator yet designed which is capable of operating continuously at 60 kV and 1500 mA. Needless to say, the transformers associated with these high power generators are huge compared with conventional low power generators. The present author has seen the new system installed by Chikawa and the X-ray generator takes up most of the space in a fairly large laboratory. These systems are available commercially from

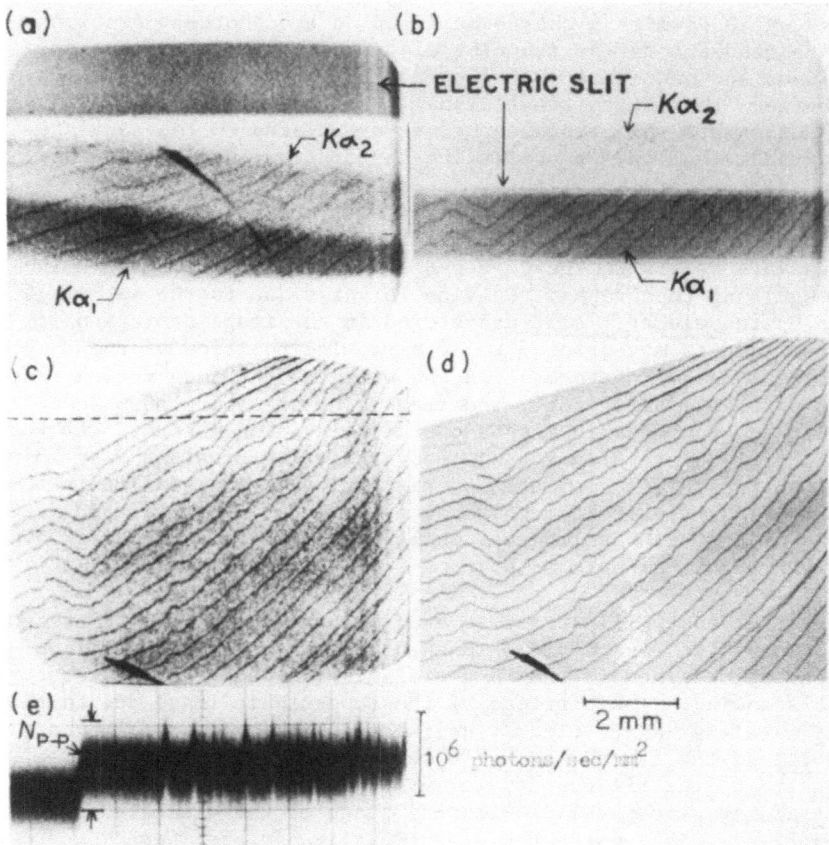

Figure 2.   Topographic images of static dislocations in a silicon
             crystal (after Chikawa (9)).

             (a)   Direct video image (non-scanning)
             (b)   Direct video image with crystal in correct
                   orientation and electric slit superimposed
                   on $K\alpha_1$ portion of image (non-scanning)
             (c)   Synthesized video image obtained in about
                   3 sec (standard scan)
             (d)   Scanning topograph recorded on nuclear plate
             (e)   Video waveform along dashed line in (c).

Rigaku Denki (14) in Tokyo. The 500 mA generator sells for over
$120,000 and the 1000 mA generator for over $150,000. The price of the
new 1500 mA generator is unknown by the present author. Replacement of
the standard copper anode with a molybdenum anode increases the price
by $5,000. The Lang goniometer system, including necessary accessories,
costs over $45,000, while the price for the complete television system
incorporating the image synthesizer is about $45,000. Imura is having
installed a four circle goniometer with computer control and Lang-type
topogoniometer with remote control including a television-video-tape
system at a cost of $700,000.

INDIRECT METHOD

In the indirect method the X-ray topographic image is converted into a visible light image by a fluorescent screen. This visible light image is then optically coupled either by a lens or a fiber-optic faceplate to the input photocathode of a low-light-level electro-optical device. Depending on the type of electro-optical device used, the output image may either be viewed directly or displayed on a television monitor. This method has been reported to yield a topographic image of approximately 10 μm resolution which is limited by the grain size and thickness of the fluorescent screen. This high resolution requires a fine grained thin fluorescent screen which has low conversion efficiency and hence necessitates use of a high intensity rotating anode X-ray generator. By increasing the grain size or thickness of the fluorescent screen the conversion efficiency can be increased to the extent that conventional low intensity X-ray generators may be used. However, the thicker the fluorescent screen the lower the resolution obtainable. A resolution of about 40 μm has been reported with only 10 X-ray photons per sec per mm^2. Thus the indirect method operates in one of two modes: (1) For high sensitivity applications a conventional low intensity X-ray source is combined with a high conversion efficiency, low resolution, fluorescent screen and a high gain electro-optical detector. (2) For high resolution applications a high intensity X-ray source is combined with a high resolution, low conversion efficiency, fluorescent screen and a high gain electro-optical detector.

There are two general categories of high gain electro-optical detectors used with the indirect X-ray topographic imaging method. In the first category fall those systems which use a low-light-level television camera tube as a detector, display the topographic image on a television monitor, and then use a video-tape recorder or a film camera to make a permanent record of the topograph. In the second category are those systems which use a low-light-level image intensifier tube as a detector, then either use a film camera to photograph the output image from the image intensifier tube directly or use an auxiliary conventional closed-circuit vidicon television camera to display the image on a television monitor and to record it on a video-tape recorder.

The first investigators to employ the indirect method using a low-light-level television camera tube as a detector were Meieran, Landre, and O'Hara (5) in 1969. Since that time Meieran has published one paper (15) in which he includes a section on direct imaging of X-ray topographs, and a review paper (16) devoted completely to this subject. Meieran used a conventional G.E. XRD-5 X-ray generator with either a copper or molybdenum anode X-ray tube operating at 40 or 45 kV and 15 or 16 mA. His detector was a SEC (secondary electron conduction) vidicon camera tube fiber optically coupled to a single stage image intensifier which in turn was coupled by a high-aperture lens to a fine-grained X-ray sensitive phosphor screen. A description of the operation of the SEC camera tube was presented earlier (1).

Kohra and associates (17-20) have utilized the low-light-level television camera tube indirect method since 1971 to image X-ray topographs. They used a conventional X-ray generator with a copper target tube operating at 40 or 45 kV and 2 or 13 mA to obtain double-crystal topographs and a Rigaku Denki Microflex X-ray generator with a tungsten target tube operating at 50 kV and 230 μA to obtain divergent Laue topographs. Their detector was an image orthicon camera tube with an X-ray sensitive phosphor sprayed directly on the fiber-optic input face-plate. The television system used a 525 line scan, 2 to 1 interlace, with 30 frames/sec and had a bandwidth of 7 MHz. The system possessed a sensitivity such that 3×10^4 X-ray photons per sec per mm^2 were required to give a video signal with a 10 to 1 signal-to-noise ratio. Intermittent scanning of the magnesium oxide target permitted electronic integration of the image and improved the signal-to-noise ratio by a factor of 10. The limiting resolution of the system was about 150 μm due primarily to the resolution of the phosphor screen.

Figure 3 serves as an example of the type of topographic images capable of being displayed with an indirect method incorporating a low-light-level television camera. This figure, taken from the work of Kozaki et al. (19), shows dislocation arrays observed in the $2\overline{2}0$ topo-graph of a silicon web-crystal 98 μm in thickness. Figure 3(a) is the video display, while Figure 3(b) shows a topograph of the same specimen area, indicated by the black circle, as obtained with the conventional Lang method using $MoK\alpha_1$ and a Sakura Nuclear Plate with an El emulsion 50 μm thick. Note that dislocations as close as 25 μm are seen resolved in the television display.

The most recent investigators to use the low-light-level television camera tube indirect method are Hartmann, Markewitz, Rettenmaier, and Queisser (21). Their 60 kV 500 mA rotating anode X-ray source had a focal spot size of 0.5 mm by 10 mm and a double slit limited the horizontal and vertical divergences of the primary beam. The specimen was set for Bragg reflection and a diaphragm stopped the primary beam. The source-sample and sample-screen spacings were closer than used in regular topographic arrangements and, therefore, the separation between $K\alpha_1$ and $K\alpha_2$ images was unresolvable. The diffracted X-ray topographic image was converted into a visible image by means of a special fluo-rescent screen which had a grain size of about 5 μm. Since the resolution of the low-light-level EIC (electron induced conductivity) television camera tube used was about 25 μm, a lens from an optical microscope with an aperture of 0.65 was used to magnify the visible image about 40 times. In this manner a topographic image from a crystal region about 0.35 mm by 1.5 mm was displayed on the television monitor within the frame time of 1/25 sec at a resolution of about 10 μm. It should be emphasized here that Hartmann et al. achieved this high resolution by looking at a very small region of the specimen and using a high-intensity X-ray source, a high-resolution low conversion efficiency fluorescent screen, and a high-gain low-light-level television camera detector.

Recently Vreeland (22) has installed an indirect method topographic system consisting of a single crystal cesium iodide fluorescent screen directly attached to the input fiber optic faceplate of a single stage image intensifier tube which in turn is directly coupled, fiber optic output faceplate to fiber optic input faceplate, to a SIT (silicon intensifier tube) low-light-level television camera tube. A system to digitize the television picture is under construction which will provide for integration on the silicon target.

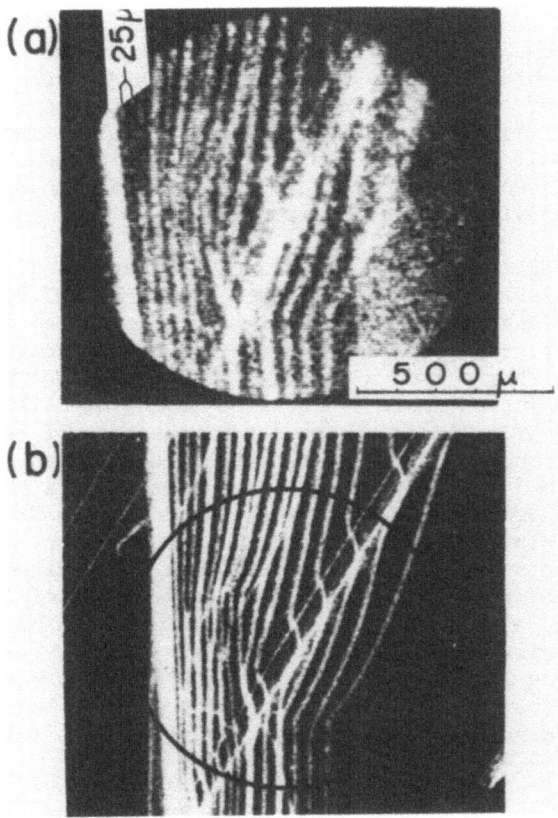

Figure 3.   Dislocation arrays observed in a topographic image of
            a silicon web-crystal (after Kozaki et al. (19))

       (a)   Video display (Note dislocations as close as
             25 μm are resolved)

       (b)   Topograph of same specimen area, indicated by
             black circle, as obtained with Lang method on
             nuclear plate.

    The primary advantage of any topographic video display system is
the capability of integrating the video signal on the camera tube
target. However, different targets are capable of different integration
times and, as with all integrating systems, the longer the integration
time the less direct view the systems actually become. One disadvantage
is that in order to obtain a hard copy of the video image either a fairly
expensive device has to be used or else the image on the television
monitor has to be photographed with a special technique. The prices for
commercial low-light-level television camera systems range from about
$10,000 for relatively inferior systems to over $60,000 for the better
ones.

Finally, let us consider those indirect X-ray topographic imaging
systems which use a low-light-level image intensifier tube as a detector.
One advantage that an image-intensifier tube has over a television camera
system is that it is much simpler to operate and the entire image is
instantaneously available for viewing, since no electron beam scanning
system is used. All of the electrons liberated from the photocathode
are simultaneously converted into an intensified visible light image by
the output phosphor. If a remote television display is desired, the
simplest, cheapest vidicon camera closed circuit television system and
monitor may be used to view this output image.

The first investigators to employ this method were Lang and
Reifsnider (6,7) in 1969. These authors used a 4-stage EMI 9694
magnetically focused image intensifier tube lens coupled to a fluorescent screen. In order to obtain the brightest output image they
used a ZnS(Ag) screen formed from a polycrystalline phosphor powder,
while the highest resolution image was obtained using a CsI(Tℓ) single
crystal as the screen. The image on the output phosphor of the image
intensifier was lens coupled to an EMI 9A vidicon camera and displayed
on a television monitor. Using this detection system and a Phillips
fine focus X-ray tube operated at 800 W they obtained Lang projection
topographs showing dislocations in a natural diamond crystal and twins
and other growth defects in a natural quartz plate. The video topographic images were obtained with specimen traverse times of 12 sec as
compared with topographs recorded on Ilford L4 50 µm thick nuclear
emulsion plates requiring 11 hr. and 9 hr. exposure times, respectively.
Using this system, Lang and Reifsnider reported a measured resolution
of 42 µm. Moreover, they reported that X-ray topographic images were
clearly seen on the image intensifier output phosphor with incident
X-ray fluxes of only 10 X-ray photons per sec per mm^2.

In 1971 Reifsnider (23) reported use of an RCA 8606 3-stage
electrostatically focused image intensifier tube with a fluorescent
screen attached directly to the input fiber optic faceplate to record
a reflection topograph from a silicon crystal. A Siemens fine-focus
X-ray tube with a copper target operating at 40 kV and 25 mA was used
as the X-ray source. Reifsnider reported that a distinct topographic
image could be instantly seen on the image tube output phosphor when
the X-ray generator was operated as low as 20 kV 20 mA.

That same year, Lang (24) used a single-stage electrostatically
focused image intensifier tube (Mullard Type XX1050) with the fluorescent screen in direct contact with the fiber optic input faceplate
as a preamplifier for the earlier system used by Lang and Reifsnider
(6,7). With this system he was able to view directly X-ray moire
fringes produced by a pair of simultaneously Bragg reflecting silicon
crystals. Even though this system required an increased X-ray flux of
35 X-ray photons per sec per mm^2 the resolution was improved to 35 µm.

Also in 1971, Parpai and Tanner (25) reported use of a channel
plate in image intensifier tubes operating in both the magnetic and
proximity focusing modes for instantaneous display of X-ray topographs.
Using the magnetic focus image tube and a Philips fine focus generator
with a copper target tube operating at 1200 W, they obtained a transmission topographic image of a growth boundary in a synthetic quartz
specimen with an exposure time of 12 sec. With the proximity focus

tube and an Elliott GX6 X-ray generator with a copper target tube
operated at 50 kV and 10 mA, they obtained a wide beam reflection
topograph of a boron diffused silicon wafer with an exposure time of
30 sec.

In 1975 Vreeland (26) reported use of an RCA 4549 3-stage electrostatically focused image intensifier tube to view Berg-Barrett topographs
of zinc and silicon crystals. The X-ray sensitive phosphor was deposited
on a well-polished fiber optic plate which was coupled to the input fiber
optic faceplate of the image intensifier tube with either silicone grease
or optical cement. He was able to resolve 40 μm gold dots flashed onto
a zinc crystal with basal surface orientation from an X-ray topographic
image taken from the output faceplate of the image intensifier tube.

The same year, Green, Farabaugh, and Crissman (27) reported use of
an RCA 8606 3-stage electrostatically focused image intensifier tube to
view Lang transmission section topographs of paraffin single crystals.
A high resolution X-ray sensitive fluorescent screen was pressed
directly against the fiber optic input faceplate of the image intensifier tube. Using a silver target microfocus X-ray generator operated
at 40 kV and 2 mA, the topographic image could be viewed directly on
the output phosphor of the image tube.

The most recent publication in which an image intensifier was used
to directly view X-ray topographic images is that of Boettinger,
Burdette, Kuriyama, and Green (28). These authors used the same
electro-optical detection system as used by Green et al. (27) earlier
to obtain instantaneous topographic images of copper, zinc, and nickel
crystals. These images were obtained with an asymmetric crystal topographic camera in both the surface reflection and transmission
geometries. A G.E. XRD-5 X-ray generator with a point-focus copper
target tube operated at 40 kV and 20 mA was used in all cases. Figure 4
shows a schematic drawing of the asymmetric crystal topographic (ACT)
system as used by Boettinger et al. (28) showing the positions of the
first and second crystals and the image intensifier detector in both
the surface reflection and transmission geometries.

Because of the relatively low intensity of X-rays which are diffracted from sample crystals using topographic cameras with conventional
X-ray generators, an X-ray image intensifier system is an extremely
convenient tool for the rapid assessment of long range strains and for
rapid optimum alignment of sample crystals for long exposures on
standard high-resolution nuclear emulsion plates. Although, as has been
pointed out in the present paper, many sophisticated electro-optical
systems can be employed for this task, only those systems which operate
on the indirect method of conversion of the X-ray topographic image into
a visible one by using a high-conversion efficiency fluorescent screen
have sufficient gain to be useful with conventional X-ray generators.
Since the resolution of these systems is limited by the fluorescent
screen, it is impractical to employ low-light-level television systems
which are relatively expensive such as SEC, image orthicon, EIC, or SIT
cameras. High gain multiple stage image intensifiers are readily
available for less than $10,000 and hence serve as the least expensive
devices available for use in direct view X-ray topographic systems. The
three stage 40 mm electrostatic focus image intensifier used by the

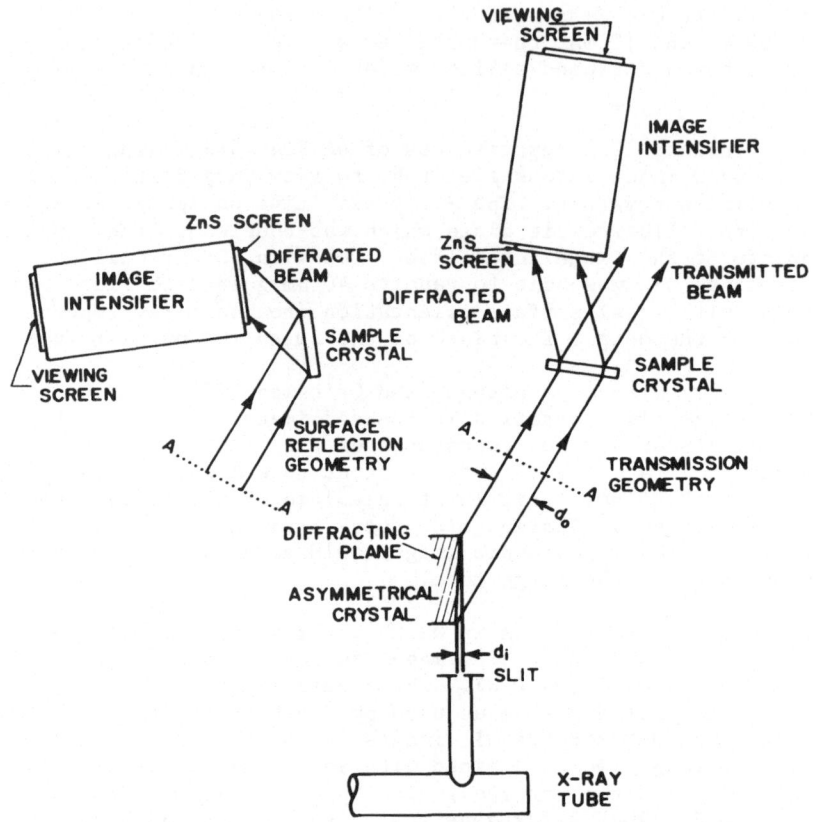

Figure 4.   Schematic diagram of asymmetric crystal topographic
            (ACT) camera combined with an indirect X-ray
            topography display system incorporating a high gain
            image intensifier tube.   (after Boettinger et al.(28)).

present author and others (1, 23, 26, 27, 28) with an overall luminance
gain of over 2×10^5 and a center resolution of 20 μm is currently
available for less than $5,000. The tube is extremely portable and
operates at maximum gain with a 6.75 V dc power input. It should also
be pointed out that although it is impossible to electronically integrate the X-ray topographic image with an image intensifier tube, it is
a simple matter to photographically integrate the image displayed on
the output phosphor.

     In order to demonstrate the convenience of the indirect X-ray
topographic image intensifier display system used by the author and
colleagues, topographs of a copper crystal with several sub-grains
will be shown. Figure 5 is a 2$\overline{2}$0 surface reflection topograph
obtained of this crystal using the ACT camera as recorded on a high
resolution nuclear emulsion plate. Figure 6 shows a series of photographs taken from the output phosphor of the image intensifier tube

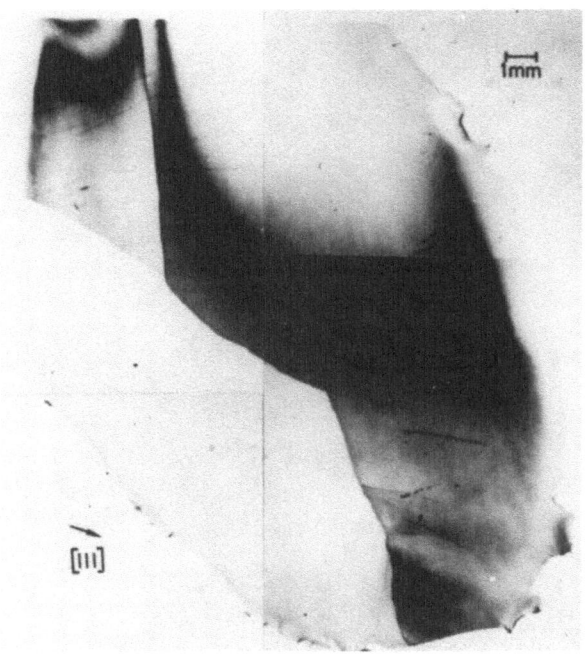

Figure 5. Surface reflection topograph of copper crystal
with sub-grains obtained with the ACT camera on
high resolution nuclear plate. (after Boettinger
et al. (28)).

using a 35 mm camera. These photographs, which show the same $2\bar{2}0$
surface reflection topographs taken at different glancing angles,
yield quantitative information about the degree of lattice strain and
misorientation in the sample crystal. Figure 6(c) indicates the best
glancing angle to use for a topograph if good coverage of the largest
area of the copper crystal is desired. If highest intensity from any
particular sub-grain is desired either one of Figures 6(a), (b) or (d)
is most suitable depending on the sub-grain of interest. Additionally,
observation of the topograph on the image intensifier tube while rocking the crystal yields a series of topographs (a topographical analog
to a rocking curve) each with good sensitivity to strain. This yields
instantaneous information about the distribution of strain over the
entire area of the sample crystal. The image intensifier system is
equally useful for viewing topographs in the transmission geometry.

ELECTRO-OPTICAL SYSTEMS

In his 1970 review paper (1) the present author described in more
detail a number of the electro-optical systems referred to in the
present article. Additional information on these systems as well as
other electro-optical systems not yet used for X-ray topographic imaging
applications may be found in the literature (29,30).

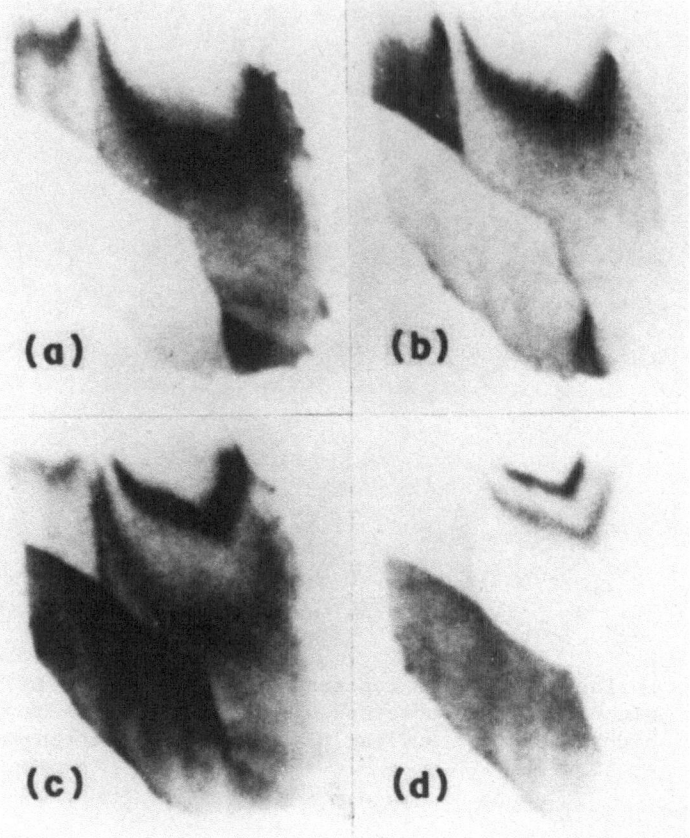

Figure 6.   The same surface reflection topographs as Figure 5
photographed from the output phosphor of the image
intensifier tube at different glancing angles.
(after Boettinger et al. (28)).

(a)   initial angle
(b)   84 sec rotation
(c)   120 sec additional rotation
(d)   234 sec additional rotation

CONCLUSIONS

1.   The direct method of displaying X-ray topographic images uses a
low gain X-ray sensitive vidicon as a detector and a high intensity
rotating anode X-ray generator as a source. Optimum performance
also requires use of electronic integration on the vidicon target
and frame storage with a scan-converter tube. The overall requirements make this system bulky, complex to operate, and very
expensive.

2. One indirect method of displaying X-ray topographic images uses a high conversion efficiency fluorescent screen optically coupled to a low-light-level television camera system. This system may be used with conventional X-ray generators, is moderately portable, relatively complex to operate, and fairly expensive.

3. A second indirect method of displaying X-ray topographic images uses a high conversion efficiency fluorescent screen optically coupled to a high gain image intensifier tube. This system may be used with conventional X-ray generators, is extremely portable, simple to operate, and the least expensive.

ACKNOWLEDGEMENTS

The author would like to thank the various investigators whose systems he has discussed for supplying him with technical information and copies of original figures. He is particularly grateful to his colleagues at the National Bureau of Standards, M. Kuriyama, W. J. Boettinger, H. E. Burdette, and E. N. Farabaugh, who have advised him about various topographic systems. Finally, he would like to thank Corinne Harness for her skill and patience in typing and preparing the manuscript.

REFERENCES

1. R. E. Green, Jr., "Electro-Optical Systems for Dynamic Display of X-ray Diffraction Images," in C. S. Barrett, J. B. Newkirk, and C. O. Ruud, Editors, Advances in X-ray Analysis, Vol. 14, p. 311-337, Plenum Press (1971).

2. J. Chikawa and I. Fujimoto, "X-ray Diffraction Topography with a Vidicon Television Image System," Appl. Phys. Lett. 13, 387-389 (1968).

3. A. N. Chester and F. B. Koch, "Instantaneous Display of X-ray Diffraction Using a Diode Array Camera Tube," in C. S. Barrett, J. B. Newkirk, and G. R. Mallett, Editors, Advances in X-ray Analysis, Vol. 12, p. 165-173, Plenum Press (1969).

4. G. A. Rozgonyi, S. E. Haszlo, and J. L. Statile, "Instantaneous Video Display of X-ray Topographic Images with Resolving Capabilities Better than 15μ," Appl. Phys. Lett. 16, 443-446 (1970).

5. E. S. Meieran, J. K. Landre, and S. O'Hara, "Direct Video Imaging of X-ray Topographs," Appl. Phys. Lett. 14, 368-371 (1969).

6. A. R. Lang and K. Reifsnider, "Rapid X-ray Diffraction Topography Using a High-Gain Image Intensifier," Appl. Phys. Lett. 15, 258-260 (1969).

7. K. Reifsnider, "Time-Resolved X-ray Diffraction Microscopy: Development of a New Technique," Department of Engineering Mechanics, Virginia Polytechnic Institute, DEMVIP Rept. No. 4-2 (1969).

8. J. Chikawa, I. Fujimoto, and T. Abe, "X-ray Topographic Observa-
    tion of Moving Dislocations in Silicon Crystals," Appl. Phys. Lett.
    21, 295-298 (1972).

9. J. Chikawa, I. Fujimoto, S. Endo, and K. Mase, "X-ray Television
    Topography for Quick Inspection of Si Crystals," in H. R. Huff
    and R. R. Burgess, Editors, Semiconductor Silicon 1973, p. 448,
    Electrochemical Society, Princeton (1973).

10. J. Chikawa and I. Fujimoto, "Video Display Technique for X-ray
    Diffraction Topography," Nippon Hoso Kyokai (Japan Broadcasting
    Corporation) Technical Research Laboratories, Technical Monograph
    No. 33, March (1974).

11. J. Chikawa, "Live Topography Using a Television System," Inter-
    national Summer School on X-ray Dynamical Theory and Topography,
    Limoges, France, August (1975).

12. J. Chikawa, NHK Technical Research Laboratories, 1-10-11, Kinuta,
    Setagaya-ku, Tokyo 157, Japan (private communication).

13. T. Imura, Department of Metallurgy, Faculty of Engineering,
    Nagoya University, Furo-cho, Chikusa-Ku, Nagoya 464, Japan (private
    communication).

14. Rigaku/USA
    Lakeside Office Park, Door 16
    607 North Avenue
    Wakefield, Massachusetts 01880

15. E. S. Meieran, "The Application of X-ray Topographical Techniques
    to the Study of Semi-conductor Crystals and Devices," Siemens
    Review, Fourth Special Issue, Vol. 37 (1970).

16. E. S. Meieran, "Video Display of X-ray Images 1. X-ray Topo-
    graphs," J. Electrochem. Soc. 118, 619-631 (1971).

17. H. Hashizume, K. Kohra, T. Yamaguchi, and K. Kinoshita, "Applica-
    tion of an Image Orthicon Camera Tube to X-ray Diffraction
    Topography Utilizing the Double-Crystal Arrangement," Appl. Phys.
    Lett. 18, 213-214 (1971).

18. H. Hashizume, H. Ishida, and K. Kohra, "A Study on Equal-Thickness
    Fringes in a Silicon Crystal by Means of an X-ray Video Imaging
    Device," Japan. J. Appl. Phys. 10, 514-515 (1971).

19. S. Kozaki, H. Hashizume, and K. Kohra, "High Resolution Video
    Display of X-ray Topographs with the Divergent Laue Method,"
    Japan. J. Appl. Phys. 11, 1514-1521 (1972).

20. H. Hashizume, K. Kohra, T. Yamaguchi, and K. Kinoshita, "A Few
    Applications of an X-ray Television System to Diffraction Studies,"
    in G. Shinoda, K. Kohra, and T. Ichinokawa, Editors, Proceedings
    of the Sixth International Conference on X-ray Optics and
    Microanalysis, p. 695-701, University of Tokyo Press (1972).

21. W. Hartmann, G. Markiewitz, U. Rettenmaier, and H. J. Queisser, "High-resolution direct-display X-ray topography," Appl. Phys. Lett. 27, 308-309 (1975).

22. T. Vreeland, Jr., W. M. Keck Laboratory of Engineering Materials, California Institute of Technology, Pasadena, California 91125 (private communication).

23. K. Reifsnider, "Recent Developments in the Use of High-Gain Optical Image Intensifiers for High-Speed Photography of Dynamic X-ray Diffraction Patterns," J. Soc. Motion Picture and Television Engineers 80, 18-22 (1971).

24. A. R. Lang, "A Versatile Electronic Image Intensification System for X-ray Topography," J. Phys. E: Sci. Instrum. 4, 921-922 (1971).

25. D. Y. Parpia and B. K. Tanner, "Display of X-ray Topographic Images Using a Channel Plate," phys. stat. sol. (a) 6, 689-692 (1971).

26. T. Vreeland, Jr., and S. S. Lau, "A Simple Image Intensifying System for Berg-Barrett Topography," Rev. Sci. Instrum. 46, 41-43 (1975).

27. R. E. Green, Jr., E. N. Farabaugh, and J. M. Crissman, "X-ray Topographic Examination of Large Paraffin Single Crystals," J. Appl. Phys. 46, 4173-4180 (1975).

28. W. J. Boettinger, H. E. Burdette, M. Kuriyama, and R. E. Green, Jr., "Asymmetric Crystal Topographic Camera," Rev. Sci. Instrum. 47 906-911 (1976).

29. L. M. Biberman and S. Nudelman, Editors, Photoelectronic Imaging Devices, Plenum Press (1971).

30. L. Marton, Editor, Advances in Electronics and Electron Physics, Vol. 12 (1960), Vol. 16 (1962), Vol. 22 (1966), Vol. 28 (1969), Vol. 33 (1972), Academic Press.

CHARACTERIZATION OF STRAIN DISTRIBUTION AND ANNEALING RESPONSE IN

DEFORMED SILICON CRYSTALS

S. Weissmann and T. Saka

College of Engineering, Rutgers University

New Brunswick, New Jersey 08903

ABSTRACT

The strain distribution associated with incipient microplasticity and its response to subsequent annealing were studied in smooth and notched silicon crystals. The characterization of plastic and elastic residual strains was carried out by X-ray topography based on Pendel-lösung Fringe (PF) patterns and was supplemented by X-ray double-crystal diffractometry and specimen scanning with automatic Bragg angle control (ABAC). It was shown that the sites at or near the surface were preferred sites where, upon mechanical deformation, dislocation sources became activated. Microplastic zones thus formed constrained long-range, residual, elastic strains. Deposition of a sputtered tungsten layer of about 1500 A at 200°C introduced lattice bending of the outer layers of the silicon crystal to a radius of curvature of about 48 m and generated, besides elastic strains, dislocations at the interface between the deposit and the host lattice. For both the mechanically deformed and the sputtered crystals, strain accommodation in response to annealing occurred by emission of dislocations from the constraining microplastic zones.

INTRODUCTION

Previous studies of the deformation behavior of silicon have shown that the surface has a special propensity for generating microplastic zones (1-3). These microplastic zones, when formed at temperatures at which work-hardening could take place, played an important role in constraining long-range, residual, elastic strains. The present investigation is particularly concerned with the response of the surface and bulk properties to the deformation behavior of silicon and to the subsequent effects of annealing.

EXPERIMENTAL PROCEDURE

Wafers of 1.5-mm thickness with [1$\bar{1}$2] surface orientation were cut from dislocation-free crystals by means of a diamond saw. The resistivity of the crystals was 0.0032 ohm-cm. Smooth and notched flat tensile specimens with gage dimensions of 5 X 20 mm and tensile axis

parallel to [110] were shaped by means of a Servomet spark erosion
machine. Wedge-shaped, notched tensile specimens were also prepared
using the shaping and polishing procedures described previously (1-3).
The specimens were used for the characterization and mapping of the
microplastic zones and residual elastic strains over large specimen
areas by employing Lang traverse X-ray topography. The topographs were
obtained with AgK_α and MoK_α radiation. Using the double-crystal diffractometer in the n,-n arrangement, rocking curves were also taken
with $CuK_{\alpha 1}$ radiation to supplement the topographs.

A modification of a scanning X-ray topographic camera with a simple comparator circuit and associate equipment providing automatic
Bragg angle control (ABAC) (4) was used to determine lattice curvature.
The unit operated as an electromechanical feedback system which maintained the orientation of the crystal such that the diffracted X-ray
intensity remained within preset limits (ABAC window) while a finely
collimated beam traversed the crystal (5). The sensitivity of the
control of the ABAC window, set to the value corresponding to 75 percent of the peak height, was such that changes in lattice curvature of
± 1.5 seconds of arc could be detected. For the ABAC curvature and
topography studies, the Rigaku-Denki 100-mA rotating anode X-ray
machine was employed. The tensile tests were carried out in a modified
Instron tensile tester (2) with argon as protective atmosphere. Controlled lattice bending was induced by sputtered film deposits using a
tungsten target at 200°C and a vacuum of ~5 X 10^{-7} torr.

RESULTS

Distribution of Microplastic Zones and Residual Elastic
Strains in Relation to Surface

To characterize the overall distribution of the elastic strains in
relationship to the microplastic zones, traverse topographs were taken
from wedge-shaped crystals because in flat specimens the PF would overlap on traversing the specimen, resulting merely in a diffuse background
of equal intensity. Figure 1 exhibits the traverse topographs of a
notched, wedge-shaped tensile specimen which was deformed at 700°C with
an applied stress of 2 kg mm^{-2}. Figure 1a shows the distribution of
microplastic zones and constrained, residual elastic strains prior to
the removal of the surface layers, while Figures 1b and 1c exhibit it
after removal of 82 and 210 μm of surface layers, respectively. The
images of the microplastic zones are characterized by the total destruc-
tion of the PF, and those of the elastic strains by the bending and distortions of the PF. Comparison of the topographs of Figure 1 shows that
the density of the microplastic zones has decreased progressively with
removal of the surface layers. The long-range, elastic strains became
more conspicuous, and the bending of the PF's toward the surface, together with the diminution of interfringe spacing as the microplastic
zones are approached, discloses the constraining influence of the zones
located at or near the surface. The lattice misalignment associated
with the microplastic zones was studied by taking rocking curves in
reflection at discrete specimen areas. Figure 2 shows a plot of half-
width values as the specimen was traversed in the plane of the notch
along the [111] direction, that is, perpendicular to the tensile

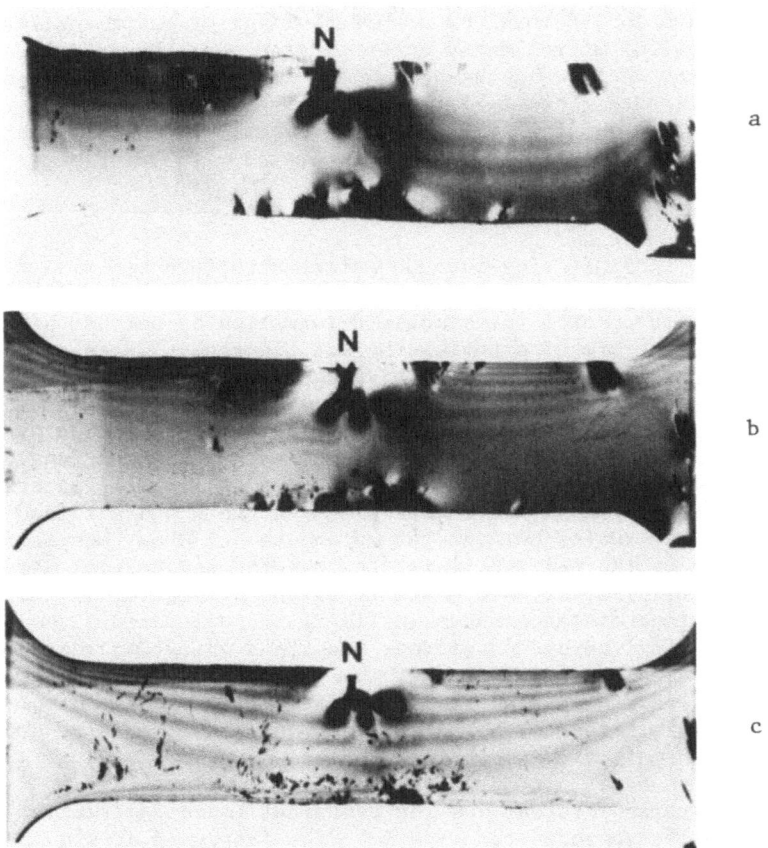

Figure 1. Traverse topographs of wedge-shaped silicon deformed in
tension at 700°C. (a) Without surface removal. (b) 82 μm removed.
(c) 210 μm removed.

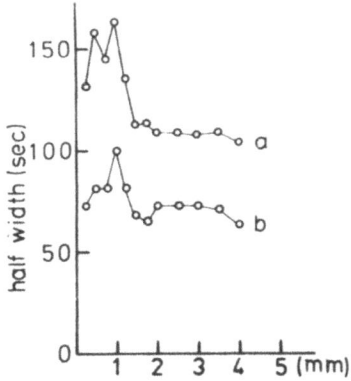

Figure 2. Plot of X-ray rocking curve half-widths in direction perpendicular to tensile axis. CuK$_\alpha$, (422). (a) Without surface removal.
(b) 82 μm removed.

axis. The profile of this plot reflects the configuration of the zones
shown in Figure 1b. Removal of a surface layer of 82 µm resulted in a
partial removal of microplastic zones located near the surface, and in
a corresponding decline of the half-width values. The profile of the
half-width-distance plot of Figure 2b virtually replicates, on a diminished scale, that obtained prior to the removal of this surface layer,
retaining the characteristic maxima associated with the plastic zone in
front of the notch tip.

Sputtering-Induced Defect Structure

In order to control the bending deformation of notched silicon
crystals with the aim of studying the incipient development of a microplastic zone at the notch tip, a tungsten layer of about 1500 A was
deposited on the crystal by sputtering. The radius of curvature of the
uniformly bent crystal induced by the deposit was determined by ABAC
scanning. Figure 3a gives the plot of the change in Bragg angle $\Delta\theta$, in
seconds of arc, as a function of the distance ΔX, in millimeters, traversed on the specimen. The radius of curvature R was obtained from
the relation $R = 206(\Delta X/\Delta\theta)$ and turned out to be 48 m. Despite the
uniform slope of the ABAC analysis, the bending deformation was not
purely elastic in nature, since the topograph of Figure 4a, traversed
over the specimen distance shown in Figure 3a, disclosed a few lattice
defects characterized by conspicuous black and white contrast effects.
With the aid of section topographs, the induced lattice defects could
be identified as dislocations. The experimental arrangement of the
incident beam, relative to the diffracting net planes and the deposited
layer, is schematically shown in Figures 4a and 4b.

Of particular interest was the dynamical image contrast at the
notch tip, which appeared to arise from the increased strain which the

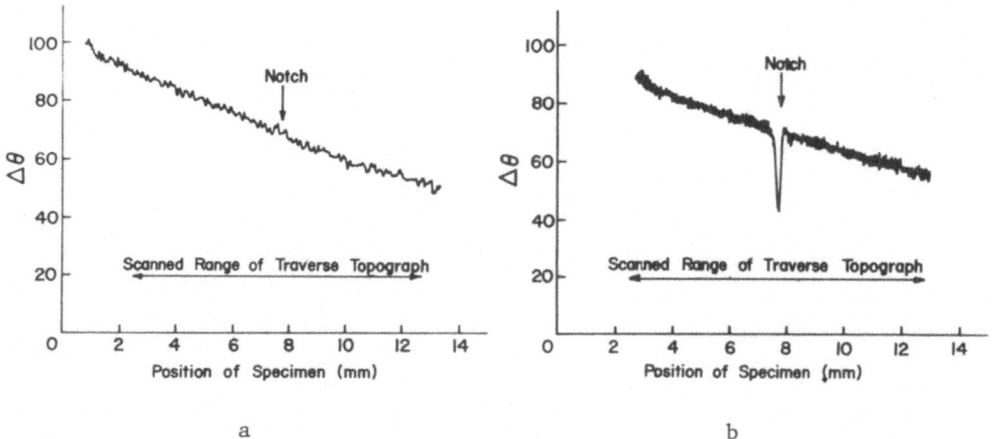

a   b

Figure 3. ABAC determination of curvature of Si induced by sputtering.
(a) g_{220}. (b) $g_{\bar{2}\bar{2}0}$.

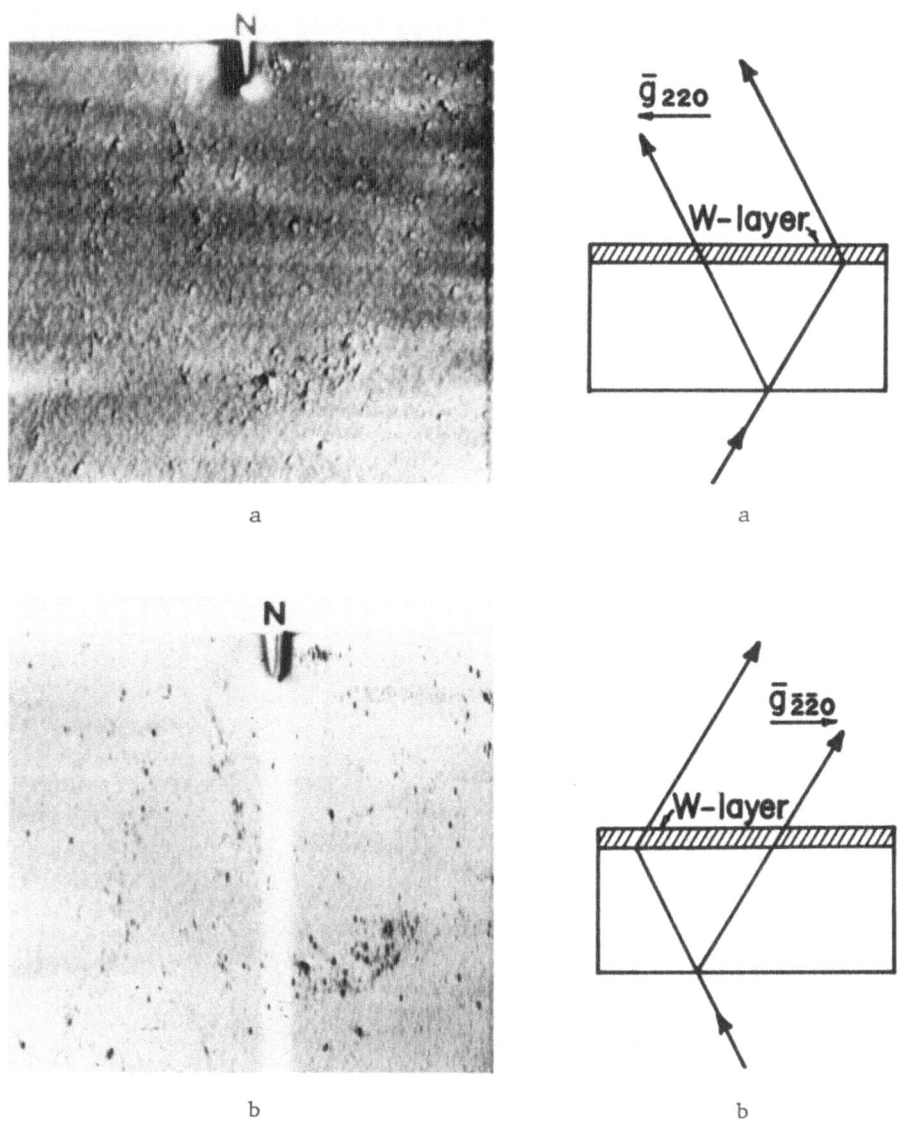

Figure 4. Defect structure induced in notched Si by sputtering.
Traverse ABAC topograph. Deposit on exit surface. (a) g_{220} (standard).
(b) $g_{\bar{2}\bar{2}0}$ (standard reversed).

sputtered layer induced at this site. To elucidate this strain effect
at the notch tip, a standard reverse ABAC traverse topograph was taken,
as shown in Figure 4b. It will be seen that a white line contrast
appeared in this topograph which corresponded to the beam position at
the notch. This line contrast was manifested in the corresponding ABAC

scan profile of Figure 3b by a conspicuous dip. This phenomenon is interpreted to mean that with g$_{\bar{2}20}$ operative, the lattice strain at the notch tip induced a convex curvature to the net planes so as to cause a focusing effect of the diffracted beam. The resultant intensity increase at the notch-tip region became large enough to overstep the intensity requirements of the preset ABAC window, thus giving rise to the sudden dip in Figure 3b. Since the rest of the crystal at this beam position did not contribute to the sudden intensity increase, a white line contrast appeared in the traverse topograph of Figure 4b.

Annealing Response and Strain Accommodation

Since it was shown that the sputtered tungsten deposit introduced a lattice curvature and generated some dislocations in the silicon crystal, the question of the origin of these defects and of the extent of the induced strain field immediately became important. To elucidate these points, annealing experiments were performed. These were motivated by the idea that annealing will give rise to strain accommodation, and that as a result of these accommodations, pertinent information regarding the location and extent of the strain field might emerge.

The traverse topograph of Figure 5 shows the effect of annealing at 700°C for 1 hour upon the silicon crystal with the sputtered deposit on the exit surface (standard topograph). It will be seen that a great many dislocations with strong black-white contrasts have emerged. ABAC scanning of the annealed crystal showed that the slope $\Delta\theta/\Delta X$ was zero

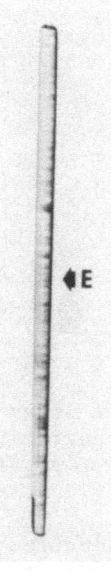

Figure 5. Effect of annealing on sputtered Si traverse topograph (standard).

Figure 6. Section topograph of annealed Si, $\bar{g}$ (reciprocal). Deposit on entrance surface.

and that, therefore, the lattice curvature had disappeared. A section topograph was taken with the deposited tungsten layer at the entrance surface (reciprocal $\bar{g}_{220}$). This experimental arrangement was selected so that any defects which might have formed at the interface between the deposit and the host crystal, as a result of the strain accommodation, will give rise to scattering, which has to traverse the entire Borrmann triangle with concomitant contrast effects. Figure 6 shows such a section topograph. It will be seen that diffuse, banded streaks emanated from the interface between the tungsten deposit (on the entrance surface) and the silicon substrate. These derived from the scattering of the dislocations shown in Figure 4. It is quite apparent that these dislocations originated from the interface in response to the strain accommodations induced by annealing.

DISCUSSION

The study has shown that sites at or near the surface are preferred locations where, upon tensile deformation, dislocation sources become activated. The ease of activation of these sources is presumably due to the special role which image forces exert on the free surface (6). Furthermore, the surface, viewed on an atomic scale, contains steps which may function as effective stress-raisers activating dislocation half-loops. It has been previously shown (2) that when microplastic zones were formed by multiple-slip activity, they became work-hardened, and in this state were able to constrain long-range, residual, elastic strains. Upon annealing, strain accommodation occurred. This was accomplished by relaxation of the boundary condition between the microplastic zone and the adjacent, elastically strained lattice domain. This relaxation took place by emission of dislocation loops from the microplastic region into the strained domain or via polygonization of the boundary region (2).

When the specimen was deformed by bending using a sputtered tungsten deposit as a means of inducing a lattice curvature of about 48 m, it was shown from the annealing response that only a small outer layer of the crystal became strained (Figure 6). The ABAC and correlated topographic studies have shown that the outer layers of the silicon crystal became expanded, undoubtedly by the incorporation of the sputtered tungsten atoms, giving rise to an enhanced focusing effect of the X-ray beam diffracted by $\vec{g}_{2\bar{2}\bar{0}}$ (Figures 3b and 4b). These strains induced by sputtering could not be accommodated solely by elastic strains, and microplastic zones in the form of dislocations were also introduced (Figures 4a and 4b). If the stresses in the silicon substrate were purely elastic, one could apply the equation developed by Stoney (7) to determine, approximately, the stress in a beam curved by a film on one surface: $\sigma_{max} = 2/3 \cdot E \cdot 1/R$, where R is the radius of curvature and E is the elastic modulus. Using $E = 24 \times 10^6$ psi (8), one would obtain for the minimum stress in the silicon substrate $\sigma_s = 2.9 \times 10^7$ dyn/cm^2. This stress value may be considered as the limit of elastic stress in the outer layer of the substrate which had to be exceeded before dislocations could be generated to form the microplastic zones.

One may now view the defect structure induced by sputtering analogous to that obtained by tensile deformation of the crystal, namely as
consisting of microplastic zones constraining regions of residual,
elastic strains. Based on this concept, the strain accommodations resulting from high-temperature annealing are predictable: dislocations
will emanate from the microplastic zones to relieve the residual, elastic strains. Figure 5, showing the profuse generation of dislocations
after annealing, and Figure 6, locating these dislocations at the interface between the tungsten deposit and the host lattice, seem to corrob-
orate this conclusion. Thus, the principle emerged that annealing can
be used as a diagnostic procedure to locate microplastic regions and
the surrounding elastic strain fields.

ACKNOWLEDGMENT

The authors wish to thank Dr. W. J. Slusark for his aid in the
sputtering experiments. The support of this work by the Metallurgy and
Materials Division of NSF is gratefully acknowledged.

REFERENCES

1. S. Weissmann, Y. Tsunekawa, and V. C. Kannan, "Fracture Studies in
    Silicon Crystals by X-Ray Pendellösung Fringes and Double Crystal
    Diffractometry," Met. Trans. 4, 376-377 (1973).

2. Y. Tsunekawa and S. Weissmann, "Importance of Microplasticity in
    the Fracture of Silicon," Met. Trans. 5, 1585-1593 (1974).

3. Y. Tsunekawa and S. Weissmann, "Dislocation Generation Associated
    with Crack Growth of Silicon Crystals Containing Precipitate Defect
    Structure," Mat. Sci. Eng. 17, 51-56 (1975).

4. A. R. Storm, "Automatic Bragg Angle Control with Simple, Inexpensive
    Electronics," Rev. Sci. Instrum. 46, 883-885 (1975).

5. G. A. Rozgonyi and T. J. Ciesielka, "X-Ray Determination of Stresses
    in Thin Films and Substrates by Automatic Bragg Angle Control," Rev.
    Sci. Instrum. 44, 1053-1057 (1973).

6. J. P. Hirth and J. Lothe, Theory of Dislocations, p. 129, McGraw-
    Hill (1968).

7. G. G. Stoney, "The Tension of Metallic Films Deposited by
    Electrolysis," Proc. Roy. Soc. (London) A82, 172-175 (1909).

8. A. S. Tetelman and A. J. McEvily, Jr., Fracture of Structural
    Materials, p. 47, John Wiley (1967).

CRYSTAL IMPERFECTIONS AND MAGNETIC DOMAIN WALLS

IN THICK CZOCHRALSKI-GROWN NICKEL SINGLE CRYSTALS

Masao Kuriyama, William J. Boettinger and Harold E. Burdette

Institute for Materials Research

National Bureau of Standards

Washington, D.C. 20234

ABSTRACT

To study the relationship between crystal growth conditions and resultant crystalline perfection, large nickel single crystals more than 12 cm long and 2 to 3 cm in diameter have been grown from the melt by the Czochralski method. Unlike semiconducting materials, one cannot easily thin metal crystals, without straining them, for the purpose of applying ordinary Lang topography with $\mu L \sim 1$ where kinematical scattering in imperfect crystals is a good approximation. This situation with metal crystals necessitates the use of dynamical diffraction effects in imperfect crystals to permit sample crystals to be thick enough to demonstrate their imperfections as in the bulk.

The crystal perfection in as-grown nickel single crystals has, therefore, been assessed by x-ray dynamical diffraction topography with an asymmetrical (double) crystal topographic (ACT) camera. Transmission topographs were obtained from crystals of thickness ranging from 0.37 mm ($\mu L=18$) to 1.03 mm ($\mu L=52$) using {111}, (002) and (220) diffraction. The crystals grown under favorable conditions have shown strong anomalous transmission. The 0-diffracted (transmitted) and the H-diffracted (Bragg-diffracted) beams display almost identical disruption images of crystal imperfections in the interior of the crystals. The types of imperfection and the degree of crystal perfection will be sorted by a set of crystal growth parameters, such as seed orientation and rotation rate.

In addition to the images of crystal imperfections, there are disruption images of 71°, 109° and 180° magnetic domain walls in the topographs. These images in transmission topographs will be compared with those in surface topographs obtained with the surface reflection geometry. Contrast conditions and image formation mechanism will be discussed by a general extinction theory.

INTRODUCTION

X-ray double crystal diffraction topography has recently been employed to establish the fact that crystal growth conditions play a significant role in producing large copper single crystals of high perfection from the melt by the Czochralski technique (1,2,3,4). In particular, several interesting crystallographic effects, such as seed orientation and dislocation type, have been found to influence the resultant crystalline perfection. However, these results alone are not sufficient to draw the conclusion that the observed effects take place in most metal crystals. Nickel has a face-centered cubic crystal structure like copper, while the material properties such as dislocation mobility, stacking fault energy and dislocation velocities are considerably different from copper. We have thus extended our work to large nickel single crystals grown from the melt by the Czochralski technique.

Nickel is ferromagnetic. Production of highly perfect nickel single crystals certainly provides the opportunity to study magnetic domain configurations in the interior of bulk nickel crystals by x-ray dynamical diffraction topography. In the past most of the magnetic domain observations in the interior of metal crystals were restricted to iron crystals or iron-silicon crystals; thus this new opportunity with nickel will be quite significant in understanding not only the behavior of ferromagnetic domains and their walls, but also the imaging mechanism of x-ray dynamical diffraction topography in imperfect crystals.

When compared with iron, nickel has smaller saturation magnetization, larger magnetostriction and a substantially smaller magnetic anisotropy energy. Nickel crystals thus deviate more significantly from cubic symmetry than iron or iron-silicon crystals. The magnetic domain walls would be broader since the wall thickness is inversely proportional to the square root of the magnetic anisotropy energy (5). These different magnetic properties may display diffraction phenomena in such a favorable way that one can approximate a real domain wall by a simple mathematical model of crystal imperfections, to which a general dynamical diffraction theory (6,7,8) can be applied rather rigorously.

CRYSTAL IMPERFECTIONS VS. CRYSTAL GROWTH

The procedures of nickel single crystal growth and sample preparation for x-ray diffraction topography have been described in detail previously (9). Sample discs have $(1\bar{1}0)$ planes as their parallel faces in this work. X-ray diffraction topographs were taken in both the transmission and the surface reflection geometry with an asymmetric (double) crystal topographic (ACT) camera (2,10). In the transmission geometry both the O-diffracted (transmitted) beam and the H-diffracted (Bragg-diffracted) beam were recorded simultaneously. CuKα_1 radiation was used throughout the present work. For the crystals used in this work the product of the ordinary linear absorption coefficient, μ = 49.6 mm^{-1}, and the thickness, L, ranges from 18 to 52. Topographs were taken using symmetric $(\bar{2}20)$ diffracting planes in the surface reflection geometry and (111) and $(11\bar{1})$ planes in the transmission geometry and occasionally, (220) or $(00\bar{2})$.

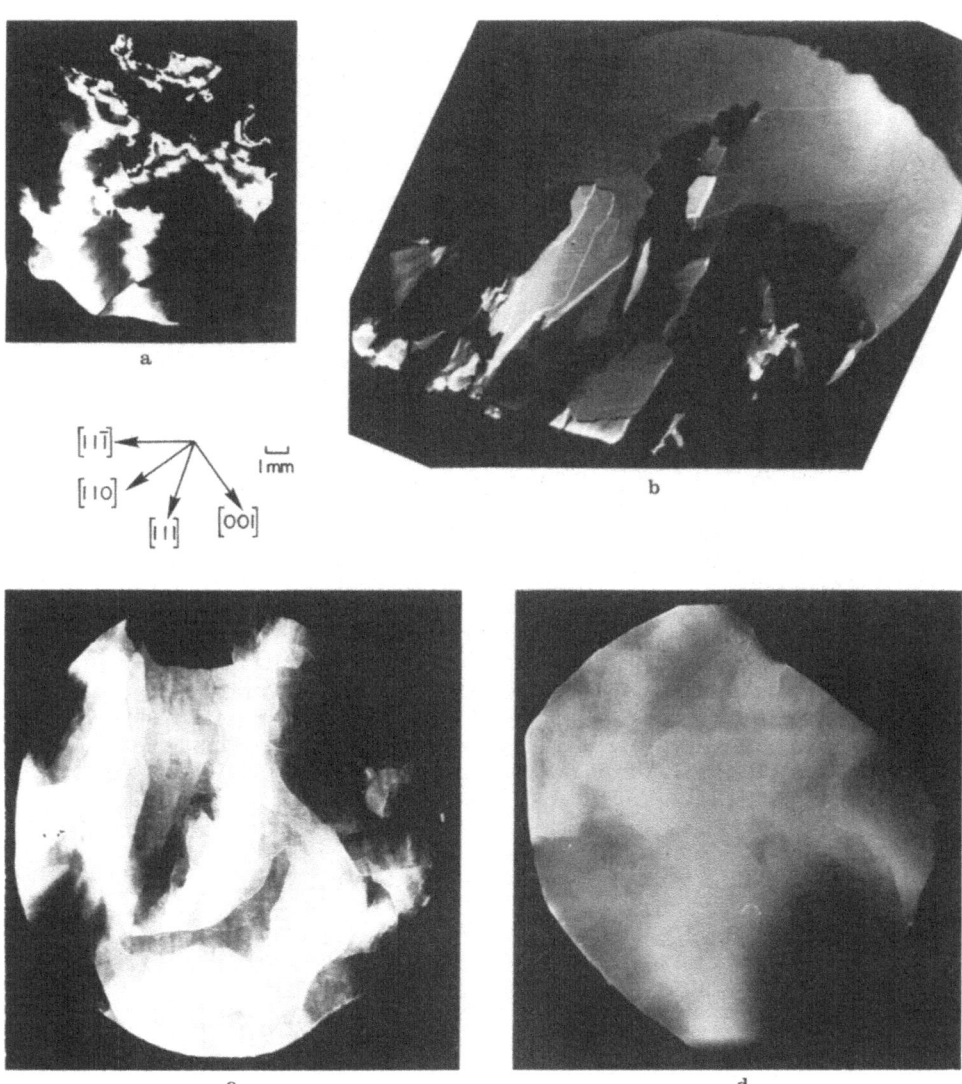

Figure 1. ACT ($\bar{2}$20) surface reflection topographs of nickel crystals of increasing perfection; a) grown with a [110] seed; b) grown with a [100] seed; c) grown with a [111] seed under poor diameter control; d) grown with a [111] seed and good diameter control.

Figure 1 shows drastic changes in the gross aspects of crystal perfection as the crystal growth conditions vary. Figure la shows that the crystal consists of many grains when a [110] oriented seed was used. Figure 1b indicates that the crystal consists of several subgrains whose misorientations are within 100 seconds of arc; this crystal was grown from a [100] oriented seed. Figure 1c represents a typical surface topograph from crystals grown from a polycrystalline seed or a [111] oriented seed under poor diameter control; small angle grain boundaries are visible. Figure 1d is an example of surface reflection topographs obtained from crystals grown under satisfactory diameter control from either a polycrystalline or a [111] oriented seed. For all of these crystals the growth conditions characterized by the rotational speeds and the pulling speed are almost identical. These crystals have a successful bottleneck.

When the gross aspects of crystal perfection shown in surface topographs surpass the level represented by Figure 1b, crystals can display anomalous transmission. Although the documentation of resultant crystal perfection with varying growth conditions has not been completed at this stage, it can be concluded from the present results that in order to obtain highly perfect nickel crystals, I) bottlenecking at the initial stage of crystal growth is necessary, II) satisfactory diameter control is desirable during the entire period of growth, III) a polycrystalline or [111] oriented seed is preferrable, in contrast with the previous results of copper single crystal growth (1), and IV) there is a general trend towards a reproducible relationship between crystal perfection and growth conditions characterized by the practical process parameters. It is also worth mentioning that, when crystals are grown from a polycrystalline seed, the resultant growth direction usually turns out to be very close to [111]. This may explain why the results obtained from a polycrystalline seed are similar to those from a [111] seed. More complete documentation of crystal perfection vs. growth conditions is being continued.

MAGNETIC DOMAIN WALLS

In the past, two attempts (11,12) were made to observe magnetic domains in nickel crystals with x-ray diffraction topography, while many extensive topographic studies were carried out with Fe-Si and iron crystals. The primary difficulty of x-ray topographic observation of magnetic domains and domain walls in nickel crystals lies in the required degree of crystal perfection. As demonstrated in the previous section, it is now possible to achieve a sufficiently high level of crystal perfection in large nickel crystals grown by the Czochralski technique. There exist, in fact, extremely straight-line images along with rather random arrangements of crystal imperfection images in the transmission topographs. These straight-line images usually form rectangular arrangements, and they disappear upon application of a magnetic field in one of the easy magnetization directions ⟨111⟩ as shown previously (9,13).

It may be helpful for the identification of images to correlate crystallographic directions with the easy magnetization directions. The magnetization vectors in the sample crystals probably are the directions of easy magnetization ([111] and [11Ī]) in the (1Ī0) plane which is parallel to the surfaces of all our crystal discs. In such crystals, 71° and 109° domain walls intersect the crystal surfaces in lines parallel to [001] and [110], respectively, while 180° walls intersect the crystal in lines parallel to [111] and [11Ī] as shown in Figure 2.

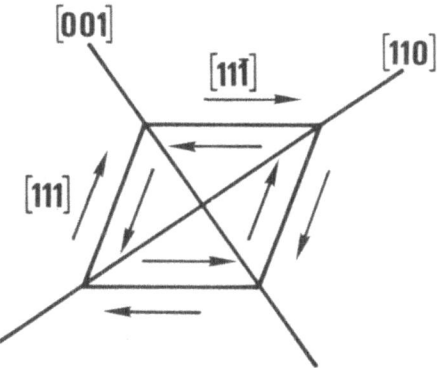

Figure 2.   Schematic of easy magnetization directions [111] and [11Ī] contained in (1Ī0) crystal surface showing possible directions of intersections with the crystal surface of 71°, 109° and 180° magnetic domain walls.

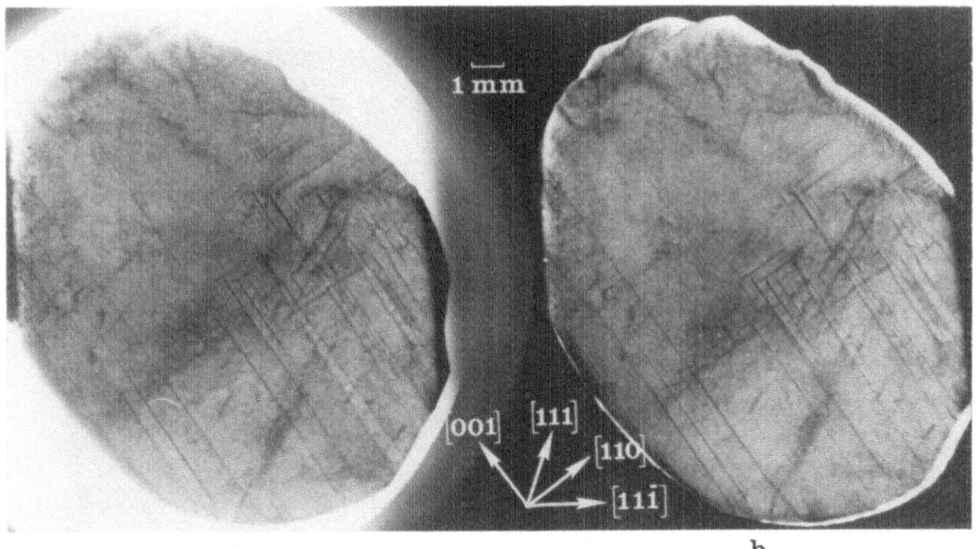

a   b

Figure 3.   ACT (11Ī) transmission topograph of a) the O-beam and b) the H-beam from a very good quality nickel crystal showing isolated dislocations and a few small angle grain boundaries. Also shown are the images of 71° and 109° magnetic domain walls parallel to [110] and [001] respectively. μL = 36.0.

Figure 3 shows a typical example of ACT 11$\bar{1}$ transmission topographs. A topograph of the O-diffracted (transmitted) beam from a sample is shown in Figure 3a and that of the H-diffracted (Bragg-diffracted) beam is in Figure 3b. A majority of the images corresponds to the 71° and/or 109° walls. Some of the short-line segment images possibly have a width almost as wide as a magnetic domain itself. Their image contrast, however, is created by the image contrast of narrowly spaced domain walls. All the images of magnetic domain walls are primarily black except for the details; i.e., photons have been disrupted by the magnetic domain walls, thus forming images with contrast similar to crystal imperfections, such as dislocations and very small angle grain boundaries.

The visibility of the magnetic domain walls changes as different diffraction conditions are applied. One example is shown in Figure 4, which was obtained with the (00$\bar{2}$) diffracting plane in transmission. All the 71° and 109° walls are not visible in both the O-diffracted (Figure 4a) and the H-diffracted (Figure 4b) topographs. Only visible are the short-segment line images parallel to [11$\bar{1}$], perhaps corresponding to 180° walls. Almost all the possible diffracting conditions were applied to crystals. Here we summarize the results briefly.

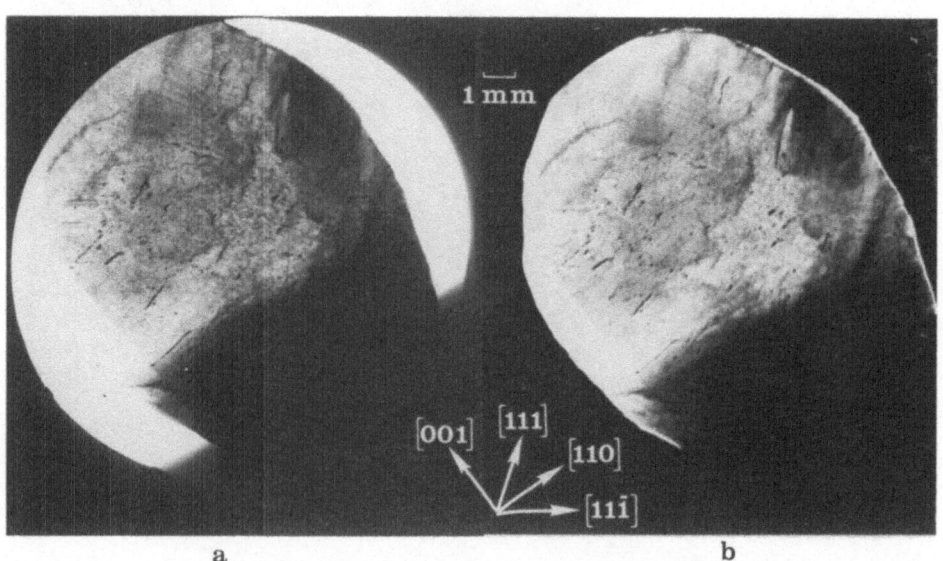

Figure 4.   ACT (00$\bar{2}$) transmission topograph of a) the O-beam and b) the H-beam, showing the effect of diffracting plane on image visibility.

1.   When 71° and 109° walls intersect the crystal surfaces at a
     right angle, their images are primarily caused by disruption.
     When the diffraction vector H is perpendicular to the walls,
     no images are visible for this diffraction condition, and the
     images are most visible when the walls contain the H vector.

2.   As far as their imaging condition is concerned, magnetic
     domain walls can be approximated by a single factor, possibly
     an average atomic displacement vector u_m. This vector likely
     lies inside the wall (parallel to the wall).

3.   When 180° walls intersect the crystal surfaces at a right
     angle, they cannot be visible, regardless of diffraction
     conditions (14). This is consistent with observations made
     in iron and Fe-Si crystals. The displacement vector u_m in
     180° walls in this geometry must be extremely small.

4.   In contrast with the above results, 180° walls become visible
     when their walls intersect the crystal surfaces at an oblique
     angle. The visibility of such 180° walls in surface reflec-
     tion topographs have been reported previously (14). When this
     situation arises, even the topographs taken in the transmission
     geometry show the 180° wall frequently as a black (disrupted)
     band, as shown in Figure 5. Figure 5a and b show the O-
     diffracted and the H-diffracted topographs taken from a
     sample in the (11$\bar{1}$) transmission diffraction geometry. The
     bands running in the [11$\bar{1}$] direction correspond to those
     oblique 180° walls, while 71° and 109° walls are also visible

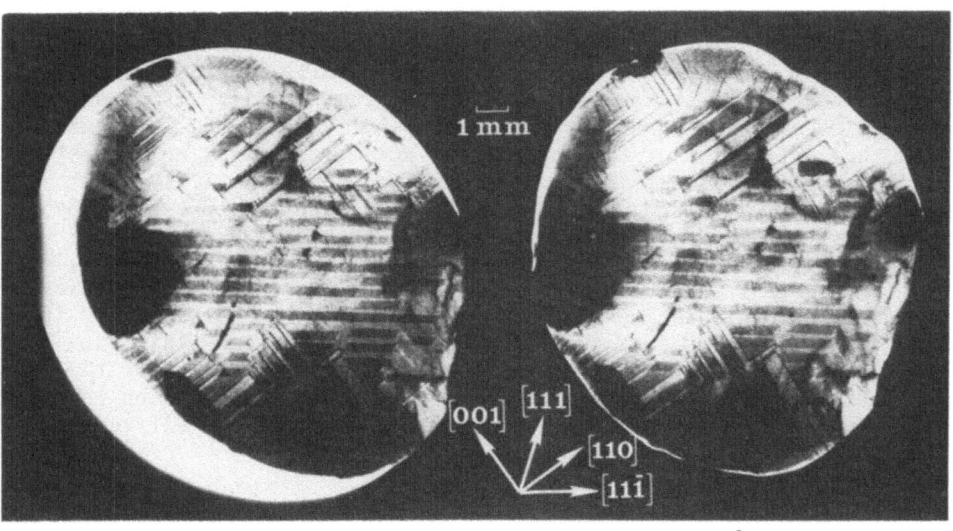

Figure 5.   ACT (11$\bar{1}$) transmission topographs of a) the O-beam and b)
            the H-beam of a nickel crystal showing bands parallel to
            [11$\bar{1}$]. These bands are 180° domain walls which are oblique
            to the crystal surface and extend through the entire thick-
            ness of the crystal. μL = 18.5.

in their typical rectangular formation along the [001] and [110] directions. To the present authors' knowledge, Figure 5 is, perhaps, the first transmission topograph of 180° magnetic domain walls. It should be noted that the band seems to have uniform contrast, in spite of the inclination of the walls through the crystal thickness. The band images become most visible when the diffraction vector H is contained in the oblique domain walls.

5. It is worth mentioning that the 180° walls intersecting the crystal at a right angle cannot be made visible, even when the crystal is tilted so that <u>the plane of diffraction</u> (defined by the incident beam and the <u>H</u>-vector) does not coincide with the walls. This observation strongly suggests that the inclination of 180° walls against the crystal surfaces plays a significant role in imaging the 180° walls. The 71° and 109° walls also appear as a band image with a black or white component whenever they are inclined. The present observation is consistent with the previous observation made in the surface reflection topographs (14).

IMAGE FORMATION MECHANISM

Almost ten years ago, one of the present authors (15,16) formulated a general dynamical diffraction theory for imperfect crystals (a general extinction theory). This theory was applied to diffraction image problems associated with several types of crystal imperfections, where the theory did not lose its rigor (6,7,8,17). These applications explained the images of crystal imperfections in terms of the dynamical diffraction effect with disruption (disruption images) and the scattering (or divergence) from imperfection domains (black and white contrast images). In the present problem it appears from the experimental results obtained so far that magnetic domain walls provide another theoretical rigorous case for the application of the general extinction theory. The theoretical results will be almost identical to the previous ones (7,17), but with less complexity, since magnetic domain walls can be represented by a simpler crystal imperfection model. In the theory one of the important factors is the local imperfection factor, $\alpha_K = \{\exp[+iK \cdot u(r)]-1\}$, (16,17). As shown in Figure 6, a magnetic domain wall is characterized by a unique atomic displacement u_m within the wall. In a given diffracting domain of atoms (7), the atomic displacement $u(r)$ within the wall of finite thickness is given by $u(r_p) \equiv u_m$ at the depth pL of the center of the wall thickness. The thickness of the wall is taken into account when one sums up the diffracted amplitude within the diffracting domain of atoms.

The scattering amplitude is given by

$$<k_K, R'; \text{out}|k, R; \text{in}> = N_K [\underline{\underline{F}}(R',R)]_{K,0} , \qquad (1)$$

where

$$\underline{\underline{F}}(R',R) = \underline{\underline{W}}^{-1}(R')\underline{\underline{A}}^* \ \underline{\underline{W}}(R') \ \underline{\underline{S}} \ \underline{\underline{W}}^{-1}(R) \ \underline{\underline{A}} \ \underline{\underline{W}}(R) . \qquad (2)$$

Here we use the matrix notation introduced previously (18); $\underline{W}$'s stand
for phase difference factors associated with the entrance and observa-
tion positions, $\underline{S}$ is the dynamic scattering amplitude and $\underline{A}$'s are the
energy-momentum distributions (response functions) in the incident and
detector systems. k_K indicates the outgoing beam in the K-diffracted
beam at the observation point R´, and k, the incoming beam at the
entrance point R.

In this paper we consider plane wave incidence and ideal detector
resolution; A and A* are given by a product of spherical Bessel functions of zero-order. Then, the matrix $\underline{F}$ is given by

$$[\underline{F}]_{K,O} = \sum_i D^{(i)}_{KO}(\epsilon) \exp[-i\beta_i(\epsilon)L]$$

$$+ (-iL/2k\cos\theta_B) \sum_{i,j} \sum_{I,J} D^{(i)}_{KI}(\eta) D^{(j)}_{JO}(\epsilon) \sum_\ell{}^´ [\underline{V}(\ell,j)]_{IJ}$$

$$\times \exp[-i\beta_i(\eta)(L-\ell_z)], \tag{3}$$

where D's are the dynamical field functions, β_i's represent the dynamical propagation for the mode i, and their variables ϵ and η correspond
to the deviation from the ideal Bragg angle. In particular, η corresponds to the position of the observation point measured from the imperfection center in angles. The matrix $\underline{V}$ is the dynamical imperfection
potential; and this can be given in the first order approximation by

$$[\underline{V}^{(1)}(\ell,j)]_{I,J} = -(1/N)v(I-J)\alpha_{I-J}\exp[-i\beta_j(\epsilon)\ell_z], \tag{4}$$

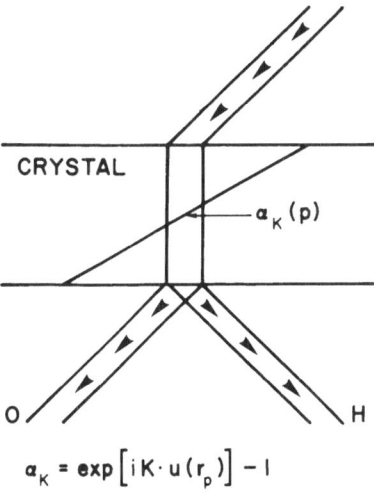

Figure 6. Schematic of oblique magnetic domain wall extending through
            the thickness of the crystal. Portion of the domain wall
            contained in a given diffracting domain of atoms is repre-
            sented by an imperfection factor $\alpha_K(p)$.

where N is the number of atoms in the crystal, and v is the Fourier
transform of atomic polarizability. The prime sign on the summation
indicates that the sum should be taken over the diffracting domain of
atoms. Equation (3) is still exact for any crystal imperfections.

Now we introduce a magnetic domain wall characterized by the α_K in
Figure 6. If we drop a common phase factor in Equation (3) and denote
such new scattering amplitudes by S_K, then we can write in this approximation for a thick crystal

$$S_0 = D_0^{(1)} \exp\{-\mu_e^{(1)}L\} + (LV_R) \text{ Im}\{\alpha_H\}v(-H)D_H^{(2)} \exp\{-\mu_0^{(2)}L\}$$

$$\tag{5}$$

$$S_H = D_H^{(1)} \exp\{-\mu_e^{(1)}L\} + (LV_R) \text{ Im}\{\alpha_H\}v(H)D_0^{(1)} \exp\{-\mu_H^{(1)}L\},$$

where we have dropped the second 0 subscript on the dynamical field
functions, D_{K0}. The first term depends mainly on ϵ and the second on η
in this approximation. Here the $\mu_e^{(i)}$ is the effective absorption coefficient for the mode i and is given by

$$\mu_e^{(i)} = (1 + 2D_H^{(i)} e^{-\gamma}) \tag{6}$$

and the $\mu_K^{(i)}$ is the absorption coefficient associated with the position,
pL, of the domain wall (or crystal imperfection) and is given by

$$\mu_0^{(i)} = \{1 + 2(1-2p)D_H^{(i)}\}\mu \tag{7}$$

with the properties,

$$\mu_0^{(1)} = \mu_H^{(2)} \text{ and } \mu_0^{(2)} = \mu_H^{(1)}. \tag{8}$$

The mode i = 1 corresponds to the anomalous transmission mode. The
ordinary linear absorption coefficient is given by μ, and V_R is the
volume ratio of the domain wall to the relevant diffracting domain of
atoms. In Equation (5), L actually stands for $L/\cos\theta_B$. It should be
noted that S_0 and S_H are all real; and

$$\text{Im}\{\alpha_H\} = \sin H \cdot u_m. \tag{9}$$

The quantity γ in Equation (6) is the distortion correction factor (6),
and is related to α_H by

$$\gamma = - \ln[1 + \delta_{\epsilon\eta} V_R \text{ Re}\{\alpha_H\}]. \tag{10}$$

It is now obvious that we have reproduced the previous results
(7,17) in a simpler form through Equation (5). The first term in S_K
describes the dynamical diffraction effect with disruption caused by
magnetic domain walls and/or crystal imperfections, and the second term
describes the additional scattering due to the defects. As illustrated
in Figure 7, the first term produces the disruption image (curve a) of
a localized crystal imperfection, when the incident beam is at position
2; i.e., the diffracting domain of atoms for this beam includes the
crystal imperfection. Curve b is the contribution from the second term
and creates black and white contrast (curve c) in the imperfection
image. A more detailed description of these imaging mechanisms has

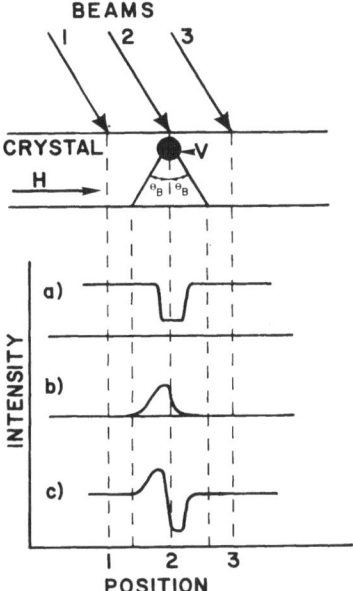

Figure 7.   Illustration of dynamical imaging of imperfection V; a) the
             disruption of the dynamical diffraction effect by the im-
             perfection and b) the kinematical scattering of the dyna-
             mical beam by the imperfection and c) the superposition to
             form a complex black-white image superimposed on the anoma-
             lous transmission background. When $\mu L > 10$, the effect of b)
             is small and the image of imperfections is caused primarily
             by disruption of the dynamical diffraction effect as shown
             in a).

been given previously (7). In this paper we summarize the results of
calculation pertinent to the present experiment.

1. The first term corresponding to a disruption image is inde-
    pendent of the position, p, of the local imperfection (magnetic
    domain wall). This is consistent with the observation made on
    inclined magnetic domain walls; the band image appears to have
    uniform contrast across the band width, even though the band
    extends from the entrance surface to the exit surface.

2. The second term in S_H usually causes very subtle broadening in
    images (or more precisely, divergence in the H-diffracted beam),
    because of the functional behavior of D_O compared to D_H. This
    effect becomes more prominent when p approaches 1; imperfections
    (or position of the domain wall) located near the exit surface
    tend to show more broadened images in the H-diffracted beam
    than in the O-diffracted beam. This phenomenon is observed.

3. The second term in both S_O and S_H becomes dominant when p
    approaches 1; the black and white contrast appears at the edges

of the disruption image of the domain wall, as seen in the topographs.

4.   When the second term effect becomes dominant, the image contrast in the O and H-diffracted beams is usually reversed; the black and white contrast of the detailed image in the edge portion of the band reverses their contrast from the O-image to the H-image. This effect will be discussed in detail elsewhere. The -H image obtained from the -H Bragg diffraction beam is very similar to the O-diffracted beam under the H-Bragg diffraction condition. Although topographs are not shown in this paper, this phenomenon has also been confirmed. This effect becomes particularly noticeable when the plane of diffraction does not coincide with the walls, as expected from this calculation.

5.   Finally, it should be noted that Equation (5) is no longer a good approximation if a magnetic domain wall intersects the crystal surfaces at a right angle (18). The first term in S_K, however, retains its correct form (7,8,17), although the distortion correction factor, γ, may not be given by a simple form like Equation (10). For such a domain wall, the first term becomes prominent; and the image is formed by the disruption effect. For 71° and 109° walls intersecting the surfaces at a right angle, all the images are distinctly black as the previous topographs show. However, there is no image from 180° walls intersecting the surface at a right angle. This indicates that $\text{Im}\{\alpha_H\}$ is negligibly small, and γ is zero; and thus the atomic displacement in the wall is negligible. On the other hand, the inclined walls definitely form the uniform images due to the disruption effect along with the black and white side band presumably created by the position of the wall near the exit crystal surface. This indicates that the inclined domain walls have a non-vanishing α_H.

REFERENCES

1.   M. Kuriyama, W. J. Boettinger, H. E. Burdette and R. M. Eaton, "Crystal Perfection in Czochralski Growth," NBSIR 74-611, 3-21 (1974).

2.   M. Kuriyama, J. G. Early and H. E. Burdette, "Fluid Flow Effects on Crystalline Perfection," Proc. of AIAA 12th Aerospace Sciences Meeting, Paper No. 74-204 (1974).

3.   M. Kuriyama, J. G. Early and H. E. Burdette, "An Immobile Dislocation Arrangement in As-Grown Copper Single Crystals Observed by X-Ray Topography," J. Appl. Cryst. 7, 535-540 (1974).

4.   W. J. Boettinger, H. E. Burdette and M. Kuriyama, "Application of Contrast Conditions to Dynamical Images of Immobile Dislocations," Phil. Mag. 34, 1-9 (1976).

5. C. Kittel, "Physical Theory of Ferromagnetic Domains," Rev. Mod.
    Phys. 21, 541-583 (1949).

6. M. Kuriyama, "Distortion Correction in Anomalous Absorption
    Coefficients," phys. stat. sol. 24, 743-748 (1967).

7. M. Kuriyama, "X-Ray Diffraction from a Crystal Containing Isolated
    Imperfections," Acta Cryst. A25, S204 and 682-693 (1969).

8. M. Kuriyama and T. Miyakawa, "Primary and Secondary Extinctions in
    the Dynamical Theory for an Imperfect Crystal," Acta Cryst. A26,
    667-673 (1970).

9. M. Kuriyama, W. J. Boettinger and H. E. Burdette, "Crystal Perfec-
    tion in Czochralski Growth," NBSIR 76-980, 3-21 (1976).

10. W. J. Boettinger, H. E. Burdette, M. Kuriyama and R. E. Green, Jr.,
    "Asymmetric Crystal Topographic Camera," Rev. Sci. Instrum. 47,
    906-911 (1976).

11. Y. Chikaura, H. Fukumori and S. Nagakura, "Growth and Structure of
    Nickel Plate Crystals from Nickel Bromide by Reduction," Japan J.
    Phys. 11, 1582-1583 (1972).

12. V. Alex, L. V. Tikhonov and O. Brümmer, "Growth, Preparation and
    X-Ray Topography of Nickel Crystals," Kristall u. Technik 9, 643-
    645 (1974).

13. M. Kuriyama, W. J. Boettinger and H. E. Burdette, "X-Ray Topo-
    graphic Observation of Magnetic Domains in Czochralski-Grown
    Nickel Single Crystals in Anomalous Transmission Geometry," J.
    Appl. Phys. 47 (11) (1976).

14. M. Kuriyama, W. J. Boettinger and H. E. Burdette, "X-Ray Surface
    Reflection and Transmission Topography of Magnetic Domain Walls in
    Czochralski-Grown Nickel Single Crystals," J. Mater. Sci. 12
    (1977).

15. M. Ashkin and M. Kuriyama, "Quantum Theory of X-Ray Diffraction by
    a Crystal," J. Phys. Soc. Japan 21, 1549-1558 (1966).

16. M. Kuriyama, "Theory of X-Ray Diffraction by a Distorted Crystal,"
    J. Phys. Soc. Japan 23, 1369-1379 (1967).

17. M. Kuriyama and T. Miyakawa, "Theory of X-Ray Diffraction by a
    Vibrating Crystal," J. Appl. Phys. 40, 1697-1702 (1969).

18. M. Kuriyama and J. G. Early, "The Dynamical Scattering Amplitude
    of an Imperfect Crystal, III. A Dynamical Diffraction Equation
    for Topography in the Spatial Coordinate Representation," Acta
    Cryst. A30, 525-535 (1974).

SOME PROBLEMS IN X-RAY STRESS MEASUREMENTS

B. D. Cullity

University of Notre Dame

Notre Dame, Indiana 46556

ABSTRACT

Some kinds of plastic deformation leave a residual pseudo-macro-stress in the material. The existence of this stress in certain speci-mens introduces errors into the standard x-ray measurement of residual macrostress, namely, in specimens that have been stretched, compressed, bent, rolled, or drawn.

No error is involved in x-ray measurements on peened, ground, or machined surfaces, or in specimens where the source of the stress is remote from the surface examined.

The x-ray and magnetic evidence for pseudo-macrostress is reviewed.

INTRODUCTION

The x-ray method of measuring residual stress, in essentially its present form, is now forty years old. In 1936 Gisen, Glocker, and Osswald (1) first measured stress without reference to a stress-free standard; they measured lattice strains by x-rays in two directions in the residually stressed object and computed the stress from these two strains.

This method has proved very valuable and is universally used. It is the only non-destructive method available that is applicable to all crystalline materials.

However, the x-ray method yields erroneous results for certain kinds of specimens, namely, those which have been subjected to a par-ticular kind of plastic deformation in the region examined by x-rays. The phenomenon causing this error, which is of both practical and fundamental interest, is examined below.

DEFINITIONS

There are only two practical methods of measuring residual stress:
dissection (mechanical relaxation) and x-ray diffraction.

Two kinds of residual stress may be distinguished on the basis of
scale:
(1) Macrostress is reasonably constant (in magnitude, sign, and direction) over a fairly large distance (many grain diameters). It is measurable by dissection, because removal of part of the specimen causes
strain in the remainder. It produces an x-ray line shift when the
specimen is rotated to a new orientation with respect to the incident
x-ray beam.
(2) Microstress varies over a small distance (one grain diameter or
less). It is not revealed by dissection. It causes x-ray line broadening.

When engineers use the term "residual stress," they generally
mean macrostress, and this is the kind of stress that the x-ray method
is designed to measure. Macrostress is an inevitable result of non-
uniform plastic deformation, and such deformation is its chief cause.

In the 1930s, however, x-ray workers observed a peculiar effect:
when a metal bar was stretched plastically a few percent and then unloaded, x-ray measurements showed a line shift that indicated residual
compressive macrostress in the direction of prestrain. This effect was
first studied systematically by Smith and Wood, who reported their
results in a series of papers beginning in 1940 (2). They also found
that the effect was symmetrical: after plastic compression, x-rays
indicated residual tensile stress (3).

This effect, which has been described in various reviews (4-7), is
indeed strange. Uniaxial plastic deformation would not be expected to
produce macrostress, and dissection experiments do not reveal any.
The stress indicated by x-rays has variously been called fictitious
macrostress (6), anomalous stress (8), and pseudo-macrostress (9).
The last term seems most descriptive ("pseudo" because it is not a
true macrostress causing strain on dissection, and "macrostress" because it causes an x-ray line shift), and it will be used in what follows.

PSEUDO-MACROSTRESS

Pseudo-macrostress (PMS) has been observed in a number of metals
and alloys after uniaxial deformation, both by means of x-ray diffraction and by magnetic measurements.

X-ray Evidence

Figure 1 shows the genesis and nature of PMS. X-rays were reflected from (310) planes nearly parallel to the surface of flat specimens of mild steel (0.1 percent C). Let y be the direction of stretching and z the surface normal, as in Figure 2 (top). Then Figure 1(a)

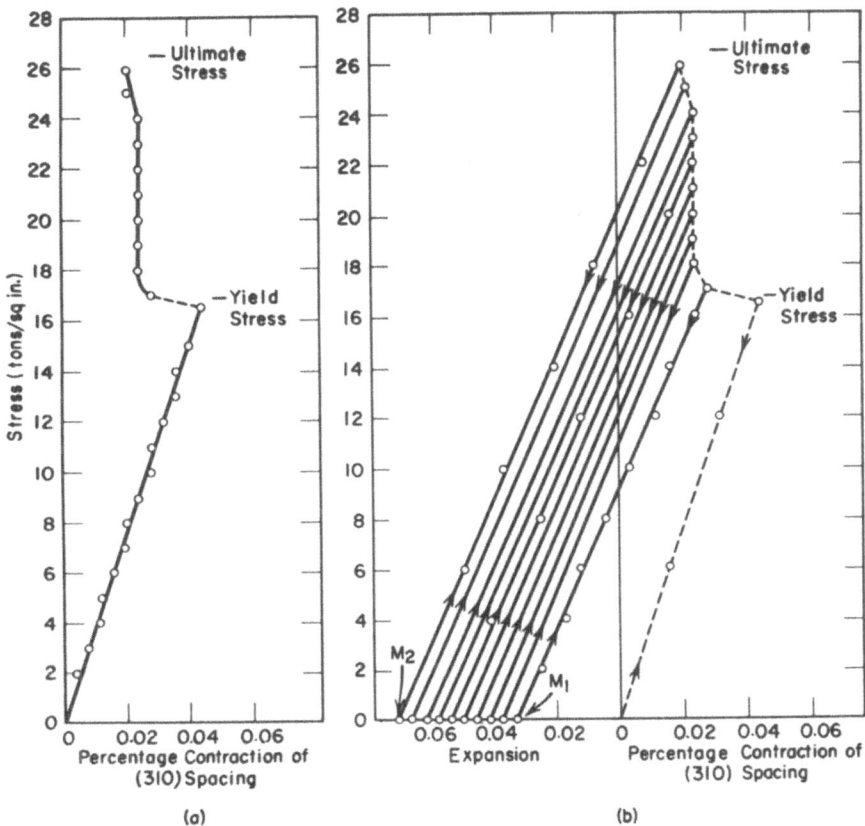

Figure 1.   Lattice strain, in a direction almost normal to the applied
            stress, as a function of applied stress in mild steel.
            1 ton/sq. in. = 2.24 ksi. Smith and Wood (10).

shows the strain ε_z during loading. The strain varies linearly with
stress up to the elastic limit, decreases abruptly as the specimen
suddenly yields, and then remains roughly constant in the plastic region. (In materials that do not show a yield point, this abrupt change
in lattice strain does not occur; the strain above the elastic limit
merely changes more slowly with applied stress than it did below the
elastic limit). The unloading curves in (b) show how PMS develops:
positive values of ε_z remain at zero applied stress and increase in
magnitude with the amount of plastic prestrain. The residual stress
σ_y is therefore compressive.

   Actually, the residual stress system is not simply uniaxial but
biaxial at the surface because a compressive transverse stress σ_x also
develops on unloading. (In the interior of the specimen the residual
stress is triaxial). From data in a later paper by Smith and Wood (11)
on the same mild steel, both stresses may be calculated; they are
$\sigma_y = -28$ ksi and $\sigma_x = -15$ ksi after unloading from 45 ksi (9 ksi over
the yield point); the ratio σ_x/σ_y is 0.5. Other values for this ratio

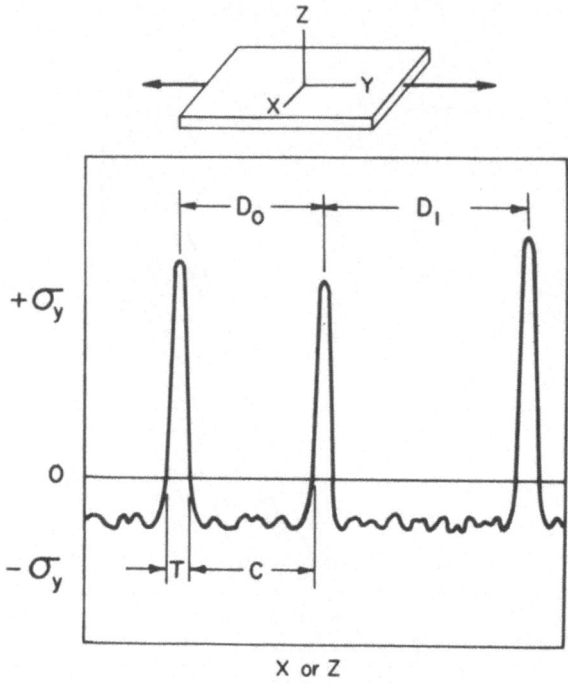

Figure 2. Assumed variation of residual stress σ_y in
directions $\underline{x}$ or $\underline{z}$ at right angles to previous
stretching.

are 0.5 for iron (12),0.2 for annealed 1045 steel (9), 0.3 for quenched
and tempered 1045 steel (8), and 0.6 for nickel (13).

PMS after uniaxial deformation has been found, not only in BCC
iron and ferritic steels, but also in FCC metals and alloys: copper
(14), nickel (13,14), aluminum (14,15), and the aluminum alloy 2024 in
the aged condition (16).

The observation that σ_y is compressive at the surface of a specimen after tensile prestrain in the $\underline{y}$ direction might mean that a real
macrostress exists, compressive at and near the surface and tensile in
the interior. This possibility has been excluded by the following kinds
of measurements.

(1) Dissection. Removal of a layer from one side of a specimen does
not cause bending of the remainder, as would occur if macrostress were
present. This has been done for iron (16), 1045 steel (8), and 2024
aluminum alloy (16), among others.
(2) X-raying the interior. Stress measurements made on interior surfaces exposed by cutting or etching have shown that the surface stress
persists in the interior, unchanged or only moderately decreased, in
iron (9,16,17,18) mild steel (11), 1045 steel (8,9), and 2024 aluminum

alloy (16). These measurements show that PMS exists throughout the volume of the specimen.

On the other hand, Macherauch (14) has concluded, from x-ray measurements on specimens progressively thinned by etching, that stretched aluminum, copper, and nickel contain macrostress, compressive at the surface and tensile in the interior. Dissection measurements were not made to support this conclusion, and it conflicts with the investigations referenced above and with the magnetic measurements to be described later. (Such a macrostress could be generated if the surface of a specimen behaves differently from the interior during plastic straining. This difference undoubtedly exists, but the weight of the evidence shows that any macrostress that might result from this effect is very much smaller than PMS).

The existence of compressive PMS after plastic tension requires that tensile stresses be distributed throughout the material in such a way as to form a balanced-force system and be effectively invisible to x-rays. A stress distribution that meets these requirements is shown in Figure 2. The prestrained material is assumed to consist of two kinds of regions within a single grain: C regions, stressed more or less uniformly in compression and comprising most of the volume (90 percent or more), and T regions highly stressed in tension but of small volume and cross-section. The relative uniformity of stress in the C regions produces a line shift, and the minor stress variations within these regions cause line broadening. The T regions have volumes so small, and stresses and stress gradients so large, that they scatter x-rays into the tails of the diffraction line or into the general background and do not affect the line position. The stress distribution after plastic compression is assumed to be like that of Figure 2, only inverted so that most of the volume is in tension. Either distribution accounts for the Bauschinger effect. [This effect is the deviation from elastic straining beginning at a lower stress when a preceding straining in the opposite direction has exceeded the elastic range.--Ed.]

PMS is therefore a special kind of microstress. PMS is "macro" in that it is constant through a large volume. But the stress is not constant through a large continuous volume, and for that reason is not a true macrostress. PMS cannot be revealed by dissection because we do not have a cutting instrument fine enough to separate the C and T regions.

The stress distribution of Figure 2 is a description and not an explanation of the experimental results, and it says nothing about the physical identity of the C and T regions. In 1944 Smith and Wood (11) believed that the C regions were "crystalline" portions bounded by "highly distorted" layers (T regions). The subsequent advent of high-magnification electron microscopy allowed this view to be sharpened, and it was later suggested (9,19,20) that the C regions were the subgrains into which a metal breaks up during plastic flow and that the T regions were the subgrain boundaries, where the dislocation density is very high. In this view, the distances D_0, D_1, ... between the tensile peaks of Figure 2 are equal to the subgrain diameters. The stressed state of the C regions arises from back stresses due to dislocations piled up in the subgrain walls.

It is tacitly assumed above that the material is single-phase, either a pure metal or a homogeneous solid solution. When a two-phase alloy is plastically stretched, the residual stress state after unloading is presumably the sum of two contributions:
(1) PMS in the matrix due to subgrain formation, as described above. (The matrix is assumed to be more voluminous than the precipitate, and weaker).
(2) Interphase microstress, caused by the strong precipitate forcing the weak matrix into compression on unloading, as occurs for example, for cementite particles in ferrite (21). Perhaps the stress distribution of Figure 2 is adequate to describe this effect also, with C representing the matrix and T the precipitate. If so, the interphase stress system is also a PMS, insofar as x-ray stress measurements on the matrix are concerned.

Magnetic Evidence

Many magnetic properties of ferromagnetic materials are stress sensitive and can serve as indicators of the stress state of a specimen. If the specimen has been plastically deformed, the stress indication is usually qualitative. The following properties have been measured on specimens prestrained uniaxially:
(1) Magnetic loss. Alternating magnetization at 60 Hz of silicon steel (Fe + 3 percent Si in solid solution) causes domain wall motion that produces heat. This heat represents a power loss called core loss. In silicon steel tensile stress applied parallel to the direction of magnetization has little effect on core loss, but compressive stress greatly increases it. If this material is loaded in tension to just beyond the elastic limit, only 0.1 percent plastic flow, the loss remains constant, but it abruptly increases during unloading because a compressive PMS has developed (20).
(2) Magnetic anisotropy. Stress is only one of many possible causes of magnetic anisotropy, which can be measured by determining the torque required to rotate a disc-shaped specimen in a strong magnetic field. Annealed nickel has a low-amplitude torque curve (Figure 3); the anisotropy is small and due to preferred orientation; the angle θ is the angle between the magnetic field and the long axis of the strip from which the disc was cut (the y axis of Figure 2). Plastic flow profoundly modifies the torque curve, because the development of PMS overwhelms preferred orientation as a source of anisotropy. The symmetry of PMS is nicely demonstrated, and the positioning of the two curves relative to the θ axis shows that σ_y is compressive after plastic tension and tensile after plastic compression.
(3) Magnetostriction in nickel. Magnetostriction is a strain, symbolized by λ, caused by an applied field. If the material is mechanically stressed before the field is applied, the resulting magnetostriction may be quite different from the normal value. In nickel, λ is negative and applied compressive prestress makes it less negative. Plastic prestrain in tension has the same effect, showing that a compressive PMS has developed (23).
(4) Magnetostriction in iron. The variation of λ with field H is complex in iron, as shown by the central curve of Figure 4. Here the full curves refer to an annealed specimen subjected to applied stresses within the elastic limit; tension is seen to shift λ in the negative direction

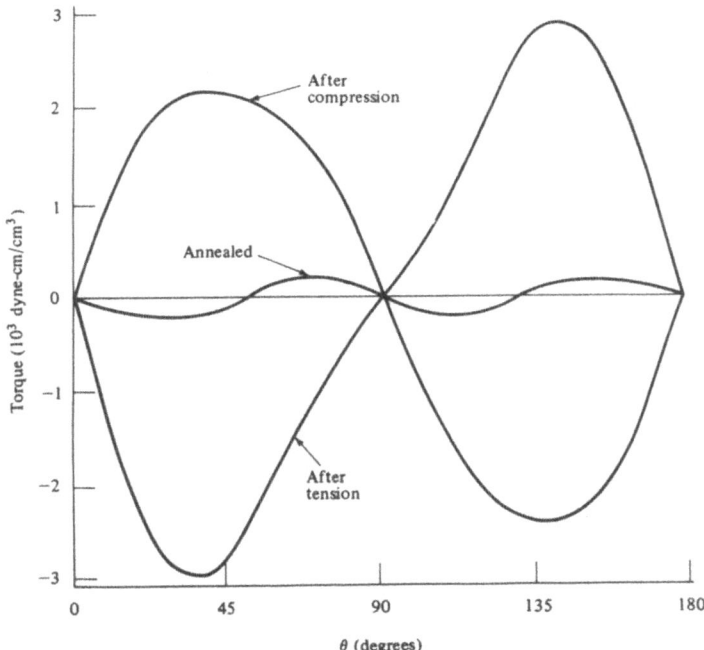

Figure 3.   Torque curves for nickel before and after
            plastic tension and compression of 2.5 percent.
            Krause et al. (13,22).

and compression in the positive direction, as in nickel. The dashed
curve is for a specimen prestrained in tension by 3.2 percent; it was
then unloaded and λ measured at zero applied stress. The placement of
the dashed curve on this graph, next to the curve for an annealed specimen under applied compression, demonstrates that prestrain in tension
produces a residual compressive PMS.
(5) Magnetoresistance in nickel. The magnetoresistance effect is a
change in electrical resistance caused by a magnetic field. As an experimental tool magnetoresistance is exactly analogous to magnetostriction. Measurements of magnetoresistance on prestretched nickel showed
that a residual compressive PMS existed (26).

     All of these magnetic measurements demonstrate that plastic pre-
strain in tension leaves most of the volume of the material in a state
of axial compression. (The balancing small-volume, high-stress tensile
regions are magnetically invisible, just as they are invisible to x-
rays, and for similar physical reasons). This conclusion permits only
two possible stress distributions:
(1) A macrostress distribution with a thin surface layer in a state of
high tension, which has never been observed by x-rays, balanced by a
thick interior in compression, or
(2) the PMS distribution of Figure 2.

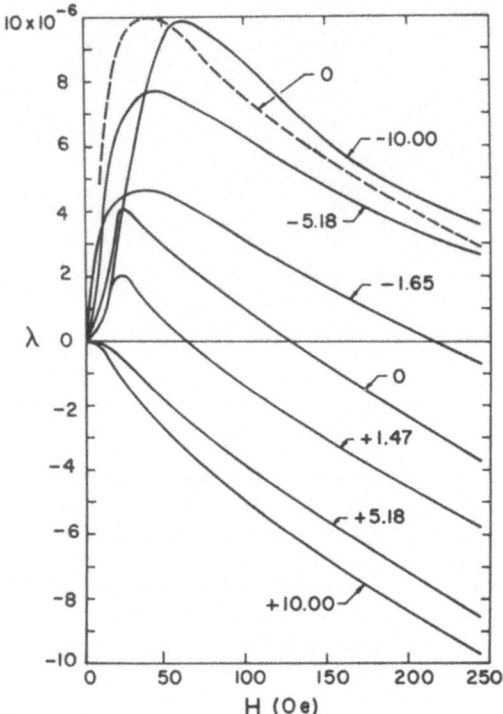

Figure 4. Magnetostriction λ as a function of
           field H for iron. The number on each
           curve is the applied stress in ksi.
           Kuruzar et al. (24, 25).

It may be concluded that the stress distribution of Figure 2 is consistent with both the x-ray and magnetic evidence.

Complications

Because PMS, whether due to the subgrain or interphase effect, mimics a true macrostress with respect to x-ray measurements, it follows that PMS should be characterized by a linear variation of lattice $\underline{d}$ spacings with $\sin^2 \psi$ according to:

$$d_i = \sigma_\phi \, d_n \, K_e \sin^2 \psi + d_n \qquad (1)$$

where d_i and d_n are the spacings of planes whose normals are inclined to the specimen-surface normal at angles of ψ and 0, respectively, σ_ϕ is the stress, and $K_e = (1+\nu)/E$, where ν is Poisson's ratio and E is Young's modulus. The stress is found from the slope of the plot of d_i vs $\sin^2 \psi$.

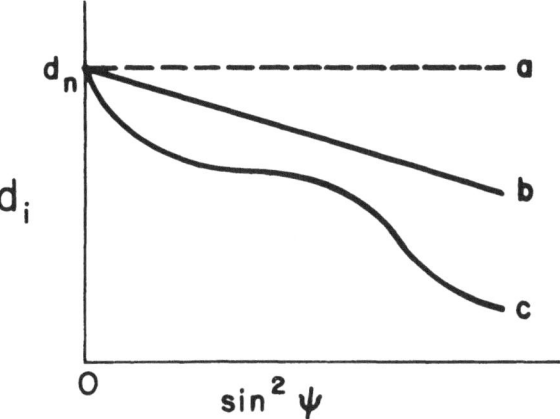

Figure 5. Variation of plane spacing with
plane orientation

Schematic plots of this kind, for specimens prestrained in tension, are shown in Figure 5. Curve (a) is what would be expected (zero stress), given the gross uniformity of the deformation process. Curve (b) represents compressive PMS, and is often observed. But sometimes curve (c) is observed; here the d_i values oscillate about a mean straight line drawn through the experimental points.

When oscillations in d_i occur, they are often found to correlate with the texture of the specimen, in the following way. If I is the pole density of the (hkl) planes being observed, then the spacing d_i of these planes is above the mean straight line through the experimental points at those values of ψ where I is large and below it where I is small (27,28). The larger-than-expected values of d_i are supposed to result from stress relaxation by dynamic recovery. Why this postulated process should be more effective at ψ angles where I is large rather than, for example, at a ψ angle having some fixed relation to the axis of prestrain is not entirely clear.

The circumstances that favor the existence of oscillations after tensile prestrain are obscure. They are observed in iron (16,28) and not observed in iron (12) or in mild steel (11,29). In 1045 steel small oscillations were observed (9,28) and not observed (29). They were not observed in nickel (14) but were very large in Cu_3Au, ordered before straining (28). It is sometimes difficult to distinguish true oscillations from normal scatter, for example, in the reported data for aluminum and copper (14).

Oscillations in d_i values represent a microstress distribution more complex than a $\overline{PMS}$, because PMS, by definition, causes a linear variation of d_i vs $\sin^2\psi$. When such oscillations exist, they will not be detected by the two exposure x-ray method because it involves measurements at only two ψ angles (0 and 45°). When d_i values oscillate, it is not clear what physical significance, if any, should be attached to the mean straight line through those values.

DEFORMATION BY BENDING

In contrast to plastic tension or compression, deformation by bending produces a true macrostress, as shown long ago by dissection measurements. If the bar in Fig. 2 is bent about the x axis, the residual stress σ_y will be compressive at the surface that was in tension during loading and tensile in the opposite surface.

In plastic bending the predominant deformation mode is a simple tension or compression of elements parallel to the y axis. The unloaded specimen should therefore contain PMS superimposed on a real macrostress, and the x-ray method will give an erroneous value of the true macrostress. The sign, but not the size, of the error is predictable because the pseudo and real macrostresses are of the same sign at each surface. The x-ray result will be numerically larger at either surface than the result obtained by dissection.

Few comparative data are available on bending stresses. On a bent rail steel x-rays indicated -26 ksi, compared to a value of -7.5 ksi computed from strains mechanically measured during loading and unloading (30). In a bent 1045 steel x-ray stresses on opposite sides were about +80 and -75 ksi, compared to +50 and -75 ksi by dissection (8).

Stresses resulting from plastic bending are of practical interest because of "fit up" problems. In assembling, some structures members that do not fit as they are supposed to are plastically bent to make them fit.

DEFORMATION BY ROLLING AND DRAWING

Rolled sheet and strip and drawn wire and rod contain real macrostress. But deformation by rolling or drawing is more complex than bending deformation, because the forces on the material at the roll or die surface are no longer axial, but inclined to the long axis of the material. Rolling and drawing add such a complex system of microstress to the real macrostress that x-ray measurement of the latter is inaccurate.

Plots of d_i vs $\sin^2\psi$ for rolled material often show large and unambiguous oscillations. These oscillations correlated with texture, in the sense described earlier, for iron (28) but not for mild steel (29), although both were given about the same reduction. The theories developed to account for the oscillations observed in these two investigations were therefore quite different.

A method for determining macrostress in cold-drawn steel rods from x-ray measurements has been developed, involving large corrections for subgrain and interphase PMS (31). The corrected x-ray stresses agreed with dissection measurements, although the comparison did not extend over the entire cross-section.

DEFORMATION BY PEENING, GRINDING, AND MACHINING

These processes produce large residual macrostresses in a thin surface layer, and many measurements have been made of the stress variation within the layer. Stresses measured by x-rays and by dissection are in excellent agreement both for shot peening (32) and grinding (33), showing the absence of PMS.

Deformation by these processes appears to be multiaxial, rather than uniaxial (tension, compression, bending) or partly uniaxial (rolling, drawing). As a result, the microstresses created by peening, grinding, and machining are non-oriented and therefore incapable of forming a PMS system. (It may be that the most useful classification of microstresses is one based on their direction. At one extreme, after tension, these stresses are all parallel; at the other, after peening, they are randomly oriented).

In a study of the stress produced by surface deformation of steel and aluminum alloys (34), plots of d_i vs $\sin^2\psi$ were found to be linear for a peened surface. The same was true after grinding and machining (turning), but some curvature developed for material near the bottom of the deformed layer.

CONCLUSIONS

The x-ray method of measuring macrostress is inaccurate when applied to specimens that have been plastically deformed by tension, compression, bending, rolling, or drawing. It may be possible to develop methods, adequate for all such specimens, of correcting the stresses determined by x-rays.

On the other hand, the accuracy of the x-ray method is unimpaired by plastic deformation caused by peening, grinding, and machining.

It therefore follows that the history of a plastically deformed specimen must be known before the applicability of the x-ray method can be decided.

The discussion in this paper of certain problems besetting the x-ray method should not be allowed to obscure the great value of this method in dealing with many problems. It gives an accurate and non-destructive measure of macrostress, not only at some surfaces which have been plastically deformed by multiaxial forces, as in peening, but also at surfaces remote from the plastic or elastic deformation that is the source of stress. Included in the latter are stresses due to casting, quenching, welding, and phase transformations.

REFERENCES

1. F. Gisen, R. Glocker, and E. Osswald, Z. Tech. Physik, 17, 145 (1936).

2. S. L. Smith and W. A. Wood, Proc. Roy. Soc. A, 176, 398 (1940).

3. S. L. Smith and W. A. Wood, Proc. Roy. Soc. A, 181, 72 (1942).

4. G. B. Greenough, Progress in Metal Physics, Vol. 3, Pergamon Press (1952).

5. D. M. Vasil'ev and B. I. Smirnov, Soviet Physics Uspekhi, 4, 226 (1961). In English.

6. B. D. Cullity, J. Appl. Phys., 35, 1915 (1964).

7. Eckard Macherauch, Exper. Mechanics, 6, 140 (1966).

8. R. E. Ricklefs and W. P. Evans, Adv. in X-Ray Analysis, 10, 273 (1967).

9. R. I. Garrod and G. A. Hawkes, Brit. J. Appl. Phys., 14, 422 (1963).

10. S. L. Smith and W. A. Wood, Proc. Roy. Soc. A, 179, 450 (1942).

11. S. L. Smith and W. A. Wood, Proc. Roy. Soc. A, 182, 404 (1944).

12. W. A. Wood, Proc. Roy. Soc. A, 192, 218 (1948).

13. R. F. Krause and B. D. Cullity, J. Appl. Phys. 39, 5532 (1968).

14. Eckard Macherauch, Adv. in X-Ray Analysis, 9, 103 (1966).

15. W. A. Wood and N. Dewsnap, J. Inst. Metals, 77, 65 (1950).

16. Matthew J. Donachie, Jr. and John T. Norton, Trans. AIME, 221, 962 (1961).

17. C. J. Newton and H. C. Vacher, Trans. AIME, 203, 1193 (1955).

18. S. L. Smith and W. A. Wood, Proc. Roy. Soc. A, 178, 93 (1941).

19. C. J. Newton, J. Research, Nat. Bur. Std., 67C, 101 (1963).

20. B. D. Cullity, Trans. AIME, 227, 356 (1963).

21. D. V. Wilson and Y. A. Konnan, Acta Met., 12, 617 (1964).

22. B. D. Cullity, Introduction to Magnetic Materials, Addison-Wesley (1972).

23. B. D. Cullity and O. P. Puri, Trans. AIME, 227, 359 (1963).

24. Michael E. Kuruzar and B. D. Cullity, Inter. J. of Magnetism, 1, 323 (1971).

25. B. D. Cullity, J. of Metals, AIME, 23, 35 (1971).

26. D.K. Bagchi and B. D. Cullity, J. Appl. Phys. 38, 999 (1967).

27. F. Bollenrath, V. Hauk, and W. Weidemann, Archiv. fur das Eisen-
     huttenwes., 38, 793 (1967).

28. R. H. Marion and J. B. Cohen, Adv. in X-Ray Analysis, 18, 466
     (1975).

29. Toshio Shiraiwa and Koshiyasu Sakamoto, p. 15 of Proc. of Seminar
     on X-Ray Study on Strength and Deformation of Metals, Soc. of
     Materials Science, Japan (1971).

30. W. S. Hyler and L. R. Jackson in Residual Stresses in Metals and
     Metal Construction, W. R. Osgood, ed., p. 297. Reinhold (1954).

31. Shuji Taira, Kozaburo Hayashi, and Shin Ozawa, p. 287 of Proc. of
     1973 Symposium on Mechanical Behavior of Materials, Soc. of
     Materials Science, Japan (1974).

32. M. Miller, E. Mantel, and W. Coleman, Proc. Soc. Exper. Stress
     Analysis, 15, 101 (1957).

33. D. P. Koistinen and R. E. Marburger, Trans. ASM, 51, 537 (1959).

34. Paul S. Prevey, Adv. in X-Ray Analysis, 19, 709 (1976).

STRESS MEASUREMENTS IN THIN FILMS

DEPOSITED ON SINGLE CRYSTAL SUBSTRATES

THROUGH X-RAY TOPOGRAPHY TECHNIQUES*

E. W. Hearn

IBM East Fishkill Laboratories

Hopewell Junction, New York   12533

ABSTRACT

The application of x-ray topographic techniques to the measurement of stress in thin films is discussed. Quantitative measurements of stresses in thin films deposited on semiconductor substrates, such as silicon, are also discussed. Double crystal and single crystal techniques are used for such measurements. Both techniques are applied to the measurements of stress in silicon oxide, silicon nitride and poly-crystalline silicon films on silicon. The double crystal technique is useful for measurements of stresses as low as 10^5 dynes/cm^2 in films only 1000A thick. The single crystal technique is less sensitive by one order of magnitude. The advantage of the single crystal technique is its simplicity and speed. It is useful for large scale measurements as encountered in the manufacture of silicon integrated circuit.

INTRODUCTION

In semiconductor technology, thin films are of great technological importance. Such films can vary in thickness from a few thousand angstroms to several microns. The films may be of crystalline nature or they may be amorphous and they are used either for insulation or for current conduction. The films are produced by different techniques, such as diffusion, epitaxy, vapor deposition, sputtering, and ion implantation. The films are deposited on semiconductor substrates, such as silicon, germanium, gallium arsenide, and others. The substrate crystals have a high degree of perfection and, therefore, x-ray topography techniques can be employed to obtain information about film properties, such as film adhesion, breakdown in film adhesion, tensile properties of films and how these properties change during semiconductor processing. Specific examples demonstrating the application of x-ray topography to a variety of thin film problems are discussed in Ref. 1. In addition to the examples discussed in Ref. 1 this paper presents new information related to the the application of x-ray topography techniques to obtain quantitative data about stress in films deposited on semiconductor substrates.

* Sponsored in part by AFCRL Contract Number F19628-75-C-0174.

EXPERIMENTAL

Quantitative measurements of stress in thin films deposited on semiconductor substrates rely on a "radius of curvature" measurement of the substrate through x-ray topography techniques. The "curvature" of the substrate is a direct result of the film deposition and is related to the film stress. The relationship can be expressed as shown (2) in equation (1):

$$\delta = \pm \, (1/6) \, (t_s^2/t_f R) \, (E_s/1-V)_{s \, T} \qquad (1)$$

+	=	Tensile, − = Compressive
t_s	=	Substrate thickness; t_f = film thickness;
R^s	=	Radius of curvature
E_s	=	Young's modules of substrate (3)
V^s	=	Poisson's ratio of substrate
T	=	Room temperature

The E and V values for substrate and film are taken to be identical according to Ref. 2 (thick substrate and thin film). The expression $E_s/1-V_{s(T)}$ in equation (1) is calculated for silicon of [100] orientation as 1.85×10^{12} dynes/cm^2.

Radius of curvature measurements of substrates are made through (a) double crystal and (b) through single crystal measurements. The principle of both techniques is well known (4,5) and is shown schematically in Figs. 1a and 1b. The convention used to describe "tensile" or "compressive" stress is given in Fig. 2.

Figure 1a shows a schematic of the experimental arrangement for the double-crystal diffractometer technique in reflection (4). FeK$_\alpha$ radiation impinges on a [100] silicon monochromator and is Bragg diffracted from the (440) planes. The diffracted beam is directed on to the [100] sample and undergoes another reflection. This twice diffracted beam, in the case of two perfect crystals, has a divergence of only a few secs of arc and will produce a large area topograph of the second crystal.

If the second crystal is replaced by a crystal having a curvature due to a thin film the resulting topograph of the twice diffracted beam will be a narrow vertical line. This results from the small divergence of the beam being diffracted from that narrow portion of the bent crystal which satisfies the Bragg condition.

If the sample is curved enough and is rotated, by a small angle, another section of the crystal will diffract and produce a parallel line, displaced a distance ℓ from the first line, in the topograph. In practice, the + and − positions of the half width of the rocking curve are used for this angle. (Fig. 1a) The amount of sample rotation ($\Delta\theta$) is a measure of the magnitude of the curvature of the crystal according to $R = \ell/\Delta\theta$.

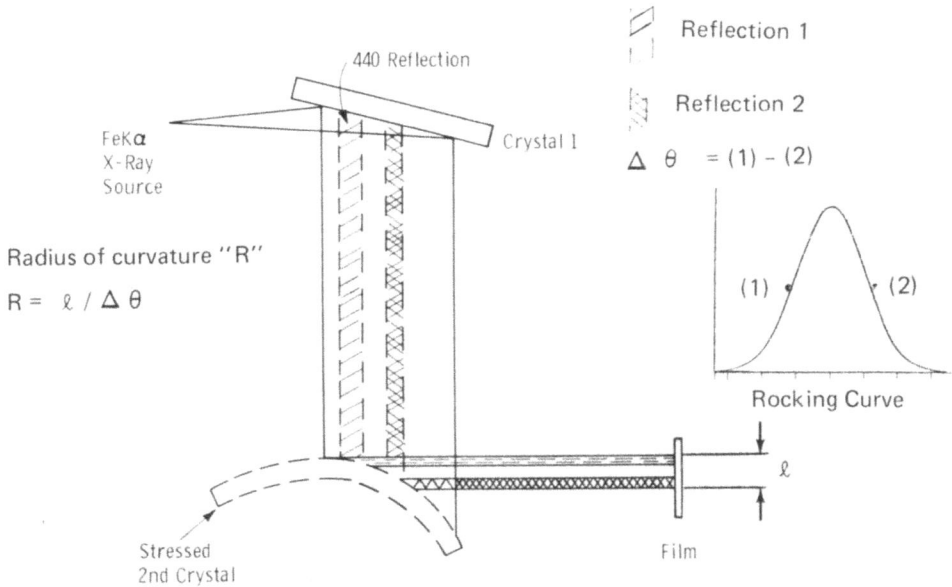

Figure 1a. Schematic of double-crystal (D.C.) reflection
stress measurement technique.

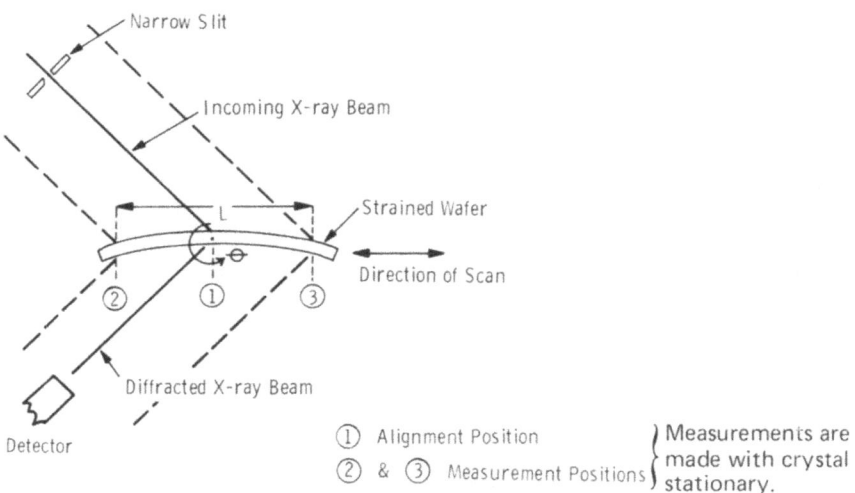

Figure 1b. Schematic of single-crystal transmission
stress measurement technique.

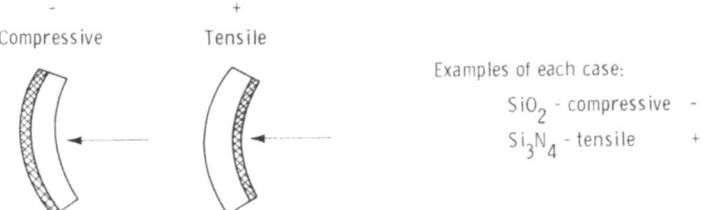

Figure 2. Convention used to describe stress of film.

If the sample curvature is too small to yield a double stripe pattern a pair of pinhole slits is introduced between the first and second crystal. This allows two beams of known distance separation to diffract from the sample crystal. The crystal rotation $\Delta\theta$ required to maximize the intensity from the two different diffracting positions is the measure of the radius of curvature. In this case the film is replaced by an electronic detector. The sample under investigation is measured before and after film deposition to correct for substrate effects.

The radius of curvature (R) is determined by $\ell/\Delta\theta$ where ℓ is the distance between pinholes or the distance between the lines in the topograph. $\Delta\theta$ is the measured sample rotation angle in both cases.

Using either one of these two approaches or a combination of both, sample rotation angles from a few seconds of arc up to a few degrees of arc can be measured quickly and reliably through double-crystal topography.

The method of measuring curvature of crystals as described above is very sensitive. The high sensitivity of the double-crystal technique can be a disadvantage because the monochromater crystal must be of the same orientation as the sample crystal. Therefore, the technique is limited to higher Bragg angles and requires exacting alignment procedures. The single-crystal case allows a broader range of samples to be used with greater ease and time savings. The sensitivity, while less than the double-crystal case, is well within the range of practical interests.

Figure 1b shows the experimental arrangement of the single-crystal techniques. The incident x-ray beam is directed to the middle of the sample through very narrow slits. The crystal is aligned as in standard topographic work. The sample is then positioned such that each edge of the crystal is exposed to the main beam and the θ dial is realigned for each position to obtain maximum readings. This procedure is repeated five times to obtain statistical data. The θ readings are made on the standard goniometer of the topographic camera. These readings are calibrated using the $\Delta\theta$ resulting from the measured distance of K_{α_1} - K_{α_2} separation of an unstrained crystal and have on our camera a resolution of 0.8 seconds of arc per division. The radius of curvature, and therefore stress, is calculated according to equation (1).

Both techniques are useful for exact stress measurements of films deposited on single-crystal substrates. The D.C. technique is more sensitive. The single-crystal technique is fast and useful for the evaluation of many samples. A comparison of the results obtainable with these two techniques is given in Table I.

The sensitivity of both techniques is compared in Fig. 3. The plot shown in Fig. 3 relates the lower limits of stress still measurable to a corresponding film thickness for both techniques. Accordingly, the D.C. technique can measure stresses lower by approximately one order of magnitude compared to the single crystal technique.

Table I.   Comparison of Stress Measurements Using Double- and
           Single- Crystal X-Ray Techniques

Sample #	Double Crystal $\sigma \times 10^9$ dynes/cm^2	Topography Camera $\sigma \times 10^9$ dynes/cm^2	% Diff. $= \dfrac{Dc - Topography}{Dc} \times 100$
1	8. 8	8. 5	+3. 4
2	8. 12	8. 44	+3. 9
3	9. 75	10. 2	-4. 6
4	11. 01	11. 01	0
			+3. 47 Avg.

Sample thickness in all cases = 0. 025 cm. Si_3N_4 Films $\simeq$ 600 – 800$\mathring{A}$

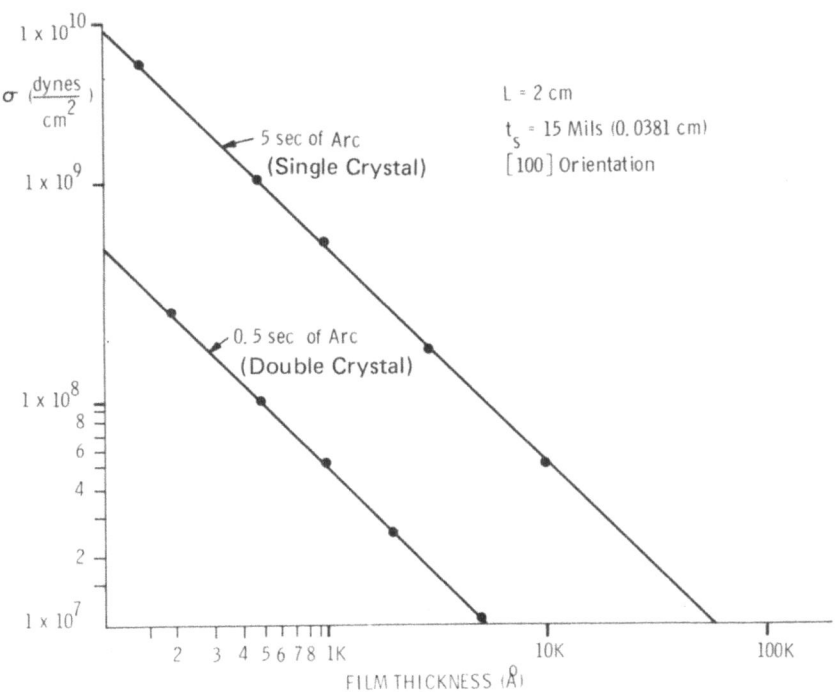

Figure 3. Relationship of stress vs. film thickness for both x-ray
           techniques. The curves are an indication of the
           minimum sensitivity for each system.

MEASUREMENTS

Pertinent examples of x-ray topographic stress measurements that relate to problems encountered in semiconductor manufacturing are discussed in the following. Silicon-nitride film deposition is an important manufacturing step in modern device fabrication. Manufacturing problems that relate to film composition and to film stress are known to occur (1,6). In this context quantitative measurements of film stress can lead to the control of film stress. Such a control can be achieved either through monitoring the Si_3N_4 composition or through the use of multiple films, such as silicon-nitride and silicon-oxide combinations. The influence of changing composition of a $Si_xO_yN_z$ system on the stress in a film deposited on silicon substrates is given in Fig. 4 which shows a plot of stress versus "equivalent" Si_3N_4 mole fraction (as controlled by gas flow volumes). An increase of stress is seen as the composition is varied between pure SiO_2 to pure Si_3N_4. The plot includes two deposition temperatures. Note that the stress decreases when the deposition temperature increases.

The influence of annealing on silicon-nitride film stress is shown in Fig. 5. Interesting is the finding that annealing does not reduce but increases the film stress. This is in agreement with the findings that film cracking due to film densification is aggravated through film annealing (6).

Stress measurement on single films (thermal SiO_2, Si_3N_4, and pyro-SiO_2) and composites of such films are summarized in Table II. The results indicate that lower stresses can be obtained in semiconductor films through use of film composites. This result is of great practical interest in the application of SiO_2-Si_3N_4 films for devices. SiO_2-Si_3N_4 combinations are very effective in reducing film stress.

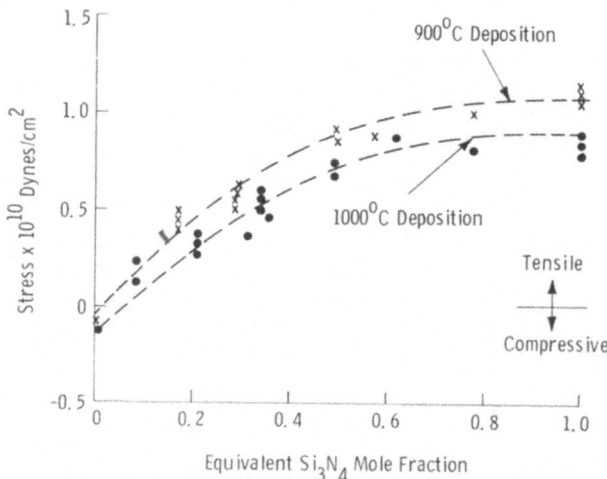

Figure 4. Relationship of stress vs. Si_3N_4 mole fraction.

The last column in Table II correlates the quantitative x-ray data with stress data obtained by stress x-ray topography as described in Ref. 6. The "sign" of the quantitative data is in good agreement with the tensile or compressive character of the films obtained from x-ray topographs. The contrast produced by the x+y+z film combination in the x-ray topographs is too low to allow an analysis. This is also in good agreement with the low value of $+0.52 \times 10^9$ dynes/cm^2 measured for this film combination.

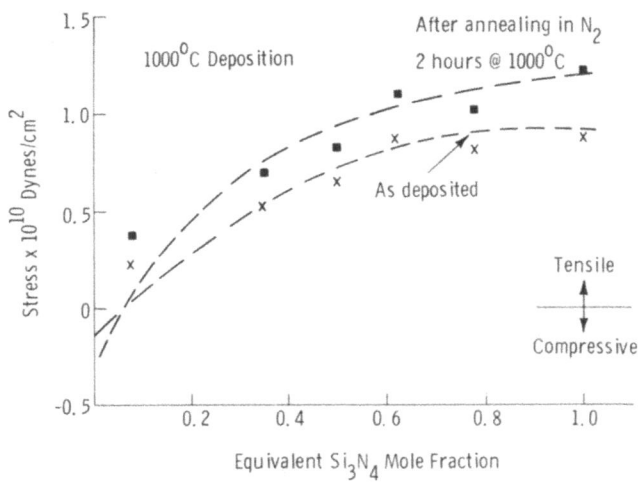

Figure 5. Stress vs. Si$_3$N$_4$ mole fraction after annealing at 1000°C for 1 hr stress increases due to densification.

Table II. Comparison of Pure and Composite Films

X = Pyro SiO$_2$ 1200 Å
Y = Si$_3$N$_4$ 850 Å
Z = Thermal SiO$_2$ 1200 Å

Combination	Stress by X-Ray Methods x 10^9 dynes/cm^2	Stress Topography (Ref. 5)
Z	- 1.43	Tensile
Y	+12.5	Compressive
X	- 4.4	Tensile
X + Y	+ 4.4	Compressive
X + Y + Z	+ 0.52	--

A final example relates to stresses in polycrystalline silicon films deposited on silicon substrates. Polycrystalline silicon films are of importance in advanced MOS devices. For the measurements a blanket layer of boron-doped polycrystalline silicon was deposited on a silicon wafer, which was subsequently oxidized. The curvature of the substrate was measured before and after oxidation. After oxidation, first the oxidized layer and then the polycrystalline silicon layer was stripped. X-ray topographs were recorded and stress measurements were made after every processing step.

Figure 6 shows the results. The radius of curvature is plotted as a function of process step with the accompanying topographs. The as-deposited sample shows a large amount of warpage: Substrate and film are defect free. After the oxidation step, the radius of curvature increases, indicating a relief of stress. The topograph obtained after this step shows line defects in [110] directions. Finally, after all polycrystalline silicon and oxide is removed the wafer curvature is negligible. The line defects are still visible in the topograph.

This example indicates that stresses in thin films deposited on silicon substrates can introduce dislocations into the silicon (Ref. 1). Such dislocations are undesirable and can be harmful to device operations (Ref. 1). In this context stress measurements as reported in this paper are very useful and allow process monitoring of stresses in film silicon combination.

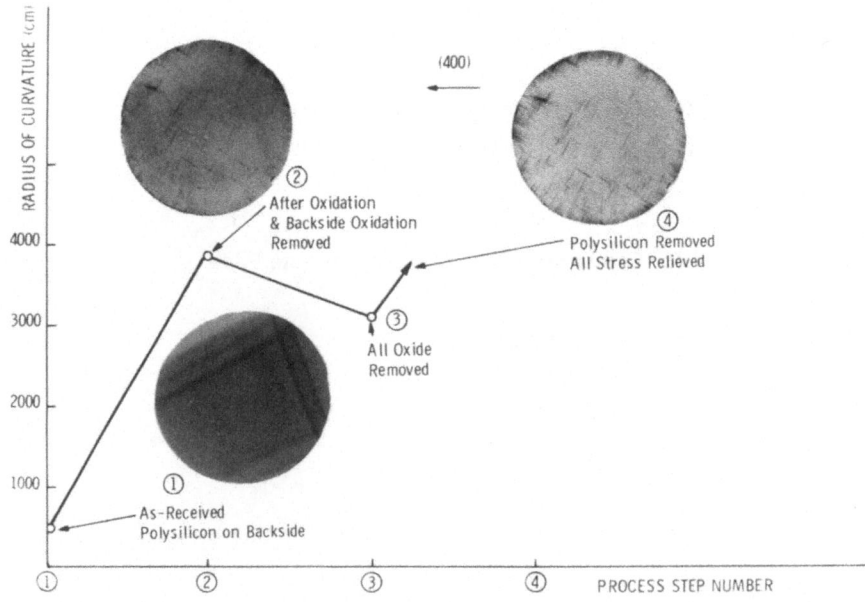

Figure 6. Plot of radius of curvature as a result of processing steps indicated. Topographs show quality of substrate.

SUMMARY

X-ray topography techniques useful for obtaining quantitative data of stresses in thin films deposited on single crystal substrates are described. The stress measurements are based on measurements of "curvature" induced in the substrate through film deposition. Radius of curvature measurements of silicon substrates are made through double crystal and single crystal x-ray topographic techniques. Experimental arrangements are described for both techniques. Double crystal and single crystal stress measurements are made on the same samples. Stress data obtained by the two techniques agree within $\pm 3.5\%$. The sensitivity of the double crystal technique is such that it can measure stresses lower than 10^9 dynes/cm^2 for a film thickness of 1000 A. This is one order of magnitude lower than obtainable by the single crystal technique. Application of both techniques to manufacturing problems that occur in the fabrication of silicon integrated circuits is discussed. Stresses in silicon nitride films and their dependency on film composition are determined. It is shown that stress in silicon nitride can be reduced through the deposition of "composite" films. It is also shown that stresses in films can generate defects in the silicon substrates.

ACKNOWLEDGMENTS

I would like to thank A. Gaind for providing the oxynitride samples and for much valuable discussion; also Dr. K. Yang for providing the polysilicon films and C. Hoogendoorn for some of the double-crystal measurements.

REFERENCES

1.   G. H. Schwuttke, Epitaxial Growth part A, Chapter 3.2, Edited by J. W. Matthews, Academic Press, New York, (1975).

2.   R. J. Jaccodine, W. A. Schlegel, "Measurements of Strain at Si-SiO$_2$ Interface," J. Appl. Phys. 37, 2429 (1966).

3.   W. A. Brantley, "Calculated Elastic Constants for Stress Problems Associated with Semiconductor Devices," J. Appl. Phys. 44, 534 (1973).

4.   R. Renninger, "Lattice Plane Interferometry" Phys. Letters 1, 104, 6, (1962).

5.   G. H. Schwuttke, J. K. Howard, "X-ray Stress Topography of Thin Films on Germanium and Silicon," J. Appl. Phys. 39, 1581 (1968).

6.   W. A. Westdorp, G. H. Schwuttke, "Stress Effects of Silicon Nitride Films on Silicon," Thin Film Dielectrics, p. 546, Edited by F. Vratny, Electrochemical Society, New York (1969).

LOCATION OF DIFFRACTOMETER PROFILES IN X-RAY STRESS ANALYSIS

D. Kirk and P. B. Caulfield

Lanchester Polytechnic

Coventry, England   CV1 5FB

ABSTRACT

The location of x-ray diffractometer line profiles is of fundamental importance in x-ray stress analysis. Objective estimates of the line profile position may be derived from diffractometer measurements at a number of equally-spaced points in the region of maximum intensity. These estimates are generally based upon the corresponding turning-point of a polynomial fitted to step-scanning data. Techniques are described which simplify the computation involved. A least-squares fitting of a quadratic is satisfactory for most routine applications provided that a reasonable number of points is involved. The three-point parabola method is less satisfactory. Expressions are presented, derived using the maximum-likelihood approach, for peak location and counting variance using both quadratic and cubic equations. It is shown that if a cubic equation is used to allow for peak asymmetry then the counting variance of peak location approaches 6.25 times that for a quadratic equation fitted to the same data.

INTRODUCTION

There are many possible ways of estimating the position of an intensity distribution such as a diffraction line profile. The most commonly used are the position of the peak of the profile, the median and the mean. At large Bragg angles only a part of the distribution may be available. This is often a feature of the reflections used in residual macrostress analysis. An added problem is the exaggerated breadth of profiles such as the 211 given by the martensite phase in hardened steels which may be as much as 10° 2Θ at the half-height using Cr K$_\alpha$radiation. In these cases an estimate of the position of the peak of the profile is the most practical measure of location.

An effective compromise between accuracy and simplicity of computation is the 'three-point parabola' method proposed by Koistinen and Marburger (1). Three points at equal angular intervals, δ, are chosen in the region of maximum intensity such that a parabola is assumed to be a good fit. The turning point, $2\Theta_{K(or\ L)}$, is then given simply by:

$$2\theta_{K(\text{or } L)} = 2\theta_1 + \delta(3y_1 + y_3 - 4y_2)/(2y_1 + 2y_3 - 4y_2) \tag{1}$$

where K and L denote trough or peak corresponding to the fixed-count timing and fixed-time counting procedures respectively, $2\theta_1$ is the angular position of the lowest angle point and y_1, y_2 and y_3 are the measures of intensity (counts or times) at the angular positions $2\theta_1$, $2\theta_1 + \delta$ and $2\theta_1 + 2\delta$ respectively.

For broad diffraction profiles several geometrical correction factors vary significantly within the range, 2δ, so that each data point must be individually corrected. A disadvantage of the three-point method is that the line profile must first have been scanned in order to determine both the approximate peak position and also the angular range, 2δ, which will give a good fit. This preliminary data is then not used for the actual turning-point estimate.

The turning-point of a polynomial, fitted by least-squares to a number of measurements in the neighbourhood of the maximum intensity, has the disadvantage of additional complexity of computation. This disadvantage can, however, be offset by the more efficient use of the x-ray equipment. The complexity of computation can be reduced using the techniques described in this paper.

For most applications some measure of accuracy is required. In the case of a particular line profile the method of peak location can affect both the precision and the bias of the corresponding experimental result. The precision is affected by the counting variance of peak location and the bias by the goodness-of-fit of the polynomial fitted to the experimental data. Several other factors affect the accuracy of peak location but these are beyond the scope of this paper.

An important advance in variance analysis was made by Wilson (2) who derived an expression for the counting variance of a parabola fitted by least-squares to a range of fixed-time counts made in the neighbourhood of the maximum intensity. This work has been extended in the present paper to include the counting variance of both quadratic and cubic equations fitted to either fixed-count timing or fixed-time counting data. Asymmetry, which is a feature of many diffraction peaks, introduces a bias into the estimate of the peak position if a quadratic is used. A cubic is an appropriate polynomial which helps to remove the bias provided that the data used is restricted to the region of maximum intensity.

The difference between the indicated angle and the true angle using commercial diffractometers is a smooth and repeatable function of the angle and rarely exceeds $0.005°2\theta$. Commercial stepping devices can give rise to steps of varying lengths. The standard deviation of the step length has been estimated to be about $0.0005°2\theta$ for a mechanical device and should be even smaller if a stepping motor is used. It can therefore be assumed that for most practical applications the variance of the angular measurements does not limit the accuracy of the peak location.

Estimates of the peak location and the counting variance are derived in the following account for both quadratic and cubic equations fitted to data obtained in the region of maximum intensity. The expressions are of fairly general application but the particular situation considered here is that of an x-ray diffractometer using either fixed-time counting of fixed-count timing procedures. The maximum likelihood approach has been used as this leads directly to the required expressions and it is easily generalised for the case when the variance of the data is not constant.

TURNING POINTS OF QUADRATIC AND CUBIC EQUATIONS

If the intensity is measured at an odd number of equally-spaced points considerable simplifications in calculation are effected. Suppose then that the intensity is measured at $2n + 1$ points distributed approximately equally about the turning point and are all within about 20% of the maximum intensity. The step length, δ, is the angular interval between adjacent points and its magnitude need only be introduced into the final working equations. The notation x_i, is standardised and has integral values, denoted by the subscript, which covers the range $i=n$ to $i=-n$. Since summation is always over the range $i=n$ to $i=-n$ we shall let Σ denote summation between the limits $i=n$ and $i=-n$. The relevant properties of x_i are given in Table II.

The observed variable is y_i and is either the number of counts or the time depending upon the step-scanning mode which is employed and in both cases can be assumed normal. An estimate of the variance of y_i has to be obtained from the observed data and approximations are necessary at this stage.

For fixed-time counting y_i has a variance equal to its mean. As y_i is not allowed to vary by more than about 20% it is reasonable to assume, for the purpose of calculation, that a good estimate of the variance of y_i is given by

$$\hat{\sigma}^2 = \bar{y}_i$$

For fixed-count timing on the other hand, the variance of y_i is rather more complicated. The estimated variance, $\hat{\sigma}^2$, can be derived using the fact that $2y_i.I_i$ has a χ_{2N} distribution where I_i is the intensity and N is the fixed number of counts. For the purpose of estimating the variance we take I_i as being constant. It follows then that a good estimate of the variance of y_i is given by

$$\hat{\sigma}^2 = \bar{y}_i^2/N$$

If, however, the background intensity is significantly high then the corresponding correction should be applied to $\hat{\sigma}$.

A) Turning Point of Quadratic

If we assume that a quadratic will be a good fit to the observed data then the relationship between y_i and x_i can be expressed in the form

$$y_i = a + bx_i + c(x_i^2 - \bar{x}^2) \tag{2}$$

The turning point, K, occurs when

$$0 = b + 2cK \quad \text{so that}$$

$$K = -b/2c$$

Equation (2) can therefore be rewritten as

$$y_i = a - 2Kcx_i + c(x_i^2 - \bar{x}^2) \tag{3}$$

Equation (3) gives the quadratic in terms of the turning-point without loss of generality.

The likelihood, L, of a random sample is given by

$$L = \Pi f(x_i, a, K, c)$$

where f is the probability density function in terms of the variable x_i and the unknown parameters a, K and c. Π is the product of all functions for values of i from -n to n. With the assumptions made previously we can write the likelihood as

$$L = \left| (2\Pi)^{(2n+1)/2}.\delta^{2n+1} \right|^{-1}.\exp \left| -(2\delta^2)^{-1}\Sigma\{y_i - a + 2Kcx_i - c(x_i^2 - \bar{x}^2)\}^2 \right.$$

and hence it follows that

$$L^* = A - (2\delta^2)^{-1}\Sigma\{y_i - a + 2Kcx_i - c(x_i^2 - \bar{x}^2)\}^2 \tag{4}$$

where L* is the natural logarithm of the likelihood function and A is a constant. The maximum likelihood estimators, $\hat{a}$, $\hat{K}$ and $\hat{c}$ of the parameters in equation (4) are obtained by equating the corresponding partial derivatives of L* to 0 and solving. The solutions are given in Table I.

The estimated variance of $\hat{K}$ can be obtained* using the second partial derivatives of L*. This estimate is given in Table I. For most practical cases the second term can be ignored so that a good approximation for the variance of K is that

$$v|\hat{K}| = \delta^2/\{4\hat{c}^2\Sigma(x_i^2)\} \tag{5}$$

B) Turning Point of Cubic

If we now assume that a cubic will be a good fit then the equation can be expressed in the form

$$y_i = \bar{y} + bx_i + c(x_i^2 - \bar{x}^2) + dx_i^3 \tag{6}$$

*Details of the mathematical stages involved are available from the authors together with worked examples of curve-fitting.

Table I Working Expressions for Peak Location and Counting Variance using Quadratic and Cubic Equations

Parameter	Equation	Working Expression
$\hat{K}$	Quadratic	$\dfrac{-\delta(2n+3)(2n-1)}{30} \cdot \dfrac{\Sigma(x_i y_i)}{\Sigma\{(x_i^2-\bar{x}^2)y_i\}}$
$V\lvert\hat{K}\rvert$	Quadratic	$\dfrac{\delta^2\sigma^2}{4\hat{c}^2\Sigma(x_i^2)}\left\lvert 1 + \dfrac{60\hat{K}^2}{(2n+3)(2n-1)}\right\rvert$
$V\lvert\hat{K}\rvert$ (approx.)	Quadratic	$\dfrac{\delta^2\sigma^2}{4\hat{c}^2\Sigma(x_i^2)}$
$\hat{c}$	Quadratic	$\dfrac{45\Sigma\{(x_i^2-\bar{x}^2)y_i\}}{n(n+1)(2n+1)(2n+3)(2n-1)}$
$\hat{K}$	Cubic	$\dfrac{\delta\{\Sigma(x_i^4).\Sigma(x_i^3 y_i)-\Sigma(x_i^6).\Sigma(x_i y_i)\}.\Sigma\{(x_i^2-\bar{x}^2)^2\}}{2\Sigma\{(x_i^2-\bar{x}^2)y_i\}\lvert\Sigma(x_i^2).\Sigma(x_i^6)-\{\Sigma(x_i^4)\}^2\rvert}$
$V\lvert\hat{K}\rvert$	Cubic	$\delta^2\sigma^2\left\lvert\dfrac{\Sigma(x_i^6)}{4\hat{c}^2\lvert\Sigma(x_i^2).\Sigma(x_i^6)-\{\Sigma(x_i^4)\}^2\rvert} + \dfrac{\hat{K}^2}{\hat{c}^2\Sigma\{(x_i^2-\bar{x}^2)^2\}}\right\rvert$
$V\lvert\hat{K}\rvert$ (approx.)	Cubic	$\dfrac{\delta^2\sigma^2\Sigma(x_i^6)}{4\hat{c}^2\lvert\Sigma(x_i^2).\Sigma(x_i^6)-\{\Sigma(x_i^4)\}^2\rvert}$
$\hat{c}$	Cubic	$\dfrac{45\Sigma\{(x_i^2-\bar{x}^2)y_i\}}{n(n+1)(2n+1)(2n+3)(2n-1)}$

Table II Relevant Properties of x_i

$$\Sigma(x_i) = 0$$

$$\Sigma(x_i^2) = n(n+1)(2n+1)/3$$

$$\Sigma(x_i^4) = n(n+1)(2n+1)(3n^2 + 3n-1)/15$$

$$\Sigma(x_i^6) = n(n+1)(2n+1)(3n^4 + 6n^3 - 3n+1)/21$$

$$\bar{x}^2 = n(n+1)/3$$

and $\Sigma\{(x_i^2 - \bar{x}^2)\}^2 = n(n+1)(2n+1)(2n+3)(2n-1)/45$

Expressions for the turning-point and the estimated variance can be derived by again using the maximum-likelihood approach. These are given in Table I together with the good approximation to $V|\hat{K}|$ which follows from omitting the second term.

COMPARISON OF VARIANCE USING QUADRATIC AND CUBIC EQUATIONS

A direct comparison can be made of the variance of peak location for quadratic and cubic equations fitted to the same data. Using the approximate equations given in Table I then

$$V|\hat{K}|_{cubic}/V|\hat{K}|_{quadratic} = \Sigma(x_i^2)\Sigma(x_i^6)/|\Sigma(x_i^2)\Sigma(x_i^6)-\{\Sigma(x_i^4)\}^2|$$

The magnitude of this ratio is a simple algebraic function of n which has values of, for example, 7.35, 6.45 and 6.25 when n is equal to 3, 7 and ∞ respectively. These results reflect the relatively unstable character of a cubic equation as compared to a quadratic equation.

A comparison of the ratio of variances obtained using the more exact expressions can only be made for individual cases. The ratio obtained is significantly affected by the value of $\hat{K}$. In an extreme case with $\hat{K}$ approaching unity a ratio of 5.72 was obtained as compared to 6.45 using the approximate expressions.

A series of 26 replicate measurements at 29 points were obtained for the 211 martensite peak of an air-hardened steel sample using Cr K_α radiation. Estimates were made of the corresponding turning point for both quadratic and cubic equations fitted to each of the sets of data. The calculated value of $V|\hat{K}|$ was found to be 2.76 times greater for cubic curve-fitting than for quadratic curve-fitting. This ratio is lower than would be predicted using equation (5) because both variances include other random errors such as those due to the mechanical stepping mechanism employed. This experimental result is, however, confirmation of the prediction of a substantial decrease in the precision of peak location which results if a cubic equation is fitted rather than a quadratic.

DISCUSSION

The equations for peak location shown in Table I give the angular difference between the peak position and the position of the working origin (when $x_i = 0$). These equations represent a considerable reduction in the complexity of computation normally involved in curve-fitting. This is a result of using an odd number of data points and of employing the relatively unfamiliar forms of equation given as (2) and (6). The constant term is not a dependant variable so that the number of normal equations which have to be solved is reduced by one for both the quadratic and cubic solutions. The equations for peak location are exactly the same as those which would result from the use of a least-squares approach.

It cannot be assumed that a quadratic is necessarily a good fit to a given set of experimental data points. The goodness-of-fit can be examined using 'residual analysis' techniques. In almost all practical cases the quadratic is an adequate approximation provided that a reasonable number of data points is involved. If a three-point parabola is fitted to data which deviates substantially from a quadratic then it has been shown (3) that the estimated peak position is dependant upon the position of the centre point relative to the 'true' peak.

The solution of the approximate equations for the variance of the peak location is only slightly more complicated than that for peak location. Almost all of the terms are common so that a reasonable estimate of the variance can be readily obtained. A cubic equation can be used for the location of an asymmetric profile. Additional counts, will, however, be needed in order to achieve the precision of location which would be given by quadratic curve-fitting. Higher order poly-nomials, for example a quartic, could be used but they would require much more computational effort.

REFERENCES

1. D. P. Koistinen and R. E. Marburger, "A Simplified Procedure for
    calculating Peak Position in X-Ray Residual Stress Measurements on
    Hardened Steel", Trans. ASM, 51, 537-555, (1959).

2. A. J. C. Wilson, "The Location of Peaks", Brit. J. App. Phys.,
    16, 665-674, (1965).

3. D. Kirk, "Experimental Features of Residual Stress Measurement by
    X-Ray Diffractometry", Strain, 7, 7-14, 1971.

STUDY OF THE PRECISION OF X-RAY STRESS ANALYSIS

M. R. James[o] and J. B. Cohen[*]

The Technological Institute, Northwestern University

Evanston, Illinois 60201

ABSTRACT

Software is described for complete computer control of residual stress measurements. One program (that incorporates either the two tilt method, the $\sin^2\psi$ procedure, or the Cohen-Marion technique) has been developed for use with either a normal detector or a position sensitive detector. The operator inputs the desired error in stress and various instrumental parameters that determine systematic errors. The counting strategy to obtain the total error is then determined by the software.

Employing this automated system, an investigation of a parabolic fit to the top of a diffraction profile indicates that a three point fit is satisfactory only for sharp profiles. Surprisingly, with a standard detector for fixed total time of data accumulation, the $\sin^2\psi$ procedure gives better precision than the two tilt method.

A position sensitive detector system exhibited excellent precision in replicate residual stress measurements. Errors of ± 6000 psi were obtained on quenched and tempered steel samples having broad diffraction profiles in only 30 seconds.

It was also found that sample displacement is less important for stationary slit geometry than with the parafocusing technique. Parallel beam geometry shows minimal effects due to sample displacement (as is well known) but the precision of the residual stress measurement decreases because this procedure broadens the diffraction profile.

Design parameters for a portable unit based on these results are discussed.

This research was supported by the U.S. Office of Naval Research.

[o]Research Assistant; [*]Frank C. Engelhart Professor of Materials Science

INTRODUCTION

Factors influencing the accuracy of X-ray measurements of residual stress have been extensively treated. Consideration has been given to systematic aberrations from both sample and instrumental sources (1,2). Corrections for 2θ dependent factors arising from the geometry of the diffraction process have been developed (3,4). Some comparisons between the different techniques of parafocusing, stationary-slit or "non-focusing", and parallel beam geometry (2,5) have been made. However, the precision or closeness of agreement among repeated measurements has never been reported. This lends itself to study with an automated diffractometer, and such automation is clearly desirable to industry.

A completely computer controlled stress analysis system giving the user maximum flexibility of operation has therefore been developed. The time of data collection is optimized for an error selected by the operator. The computer program provides a more complete and versatile package than previously reported (6,7) in that analysis can be carried out using the 'two-tilt' or '$\sin^2\psi$' technique. Also, if oscillations in d vs. $\sin^2\psi$ are present the method of Cohen and Marion (8) is implemented to obtain the true macrostress. The system can be employed using a normal detector with or without movement of the receiving slit or with a one-dimensional position sensitive detector. Results with this system on the precision of stress measurements with X-rays are reported. A portable residual stress analyzer based on these results is discussed.

THE AUTOMATED RESIDUAL STRESS PROGRAM

The hardware employed in this study includes a Picker diffractometer equipped with SLO-SYN stepping motors for the 2θ and $\omega(\psi)$ axes of the goniometer and one for positioning the receiving slit when parafocusing optics (9) are employed. A 16 K PDP8/E minicomputer (Digital Equipment Corporation) with dual tape drives was interfaced to the scintillation detector and the motors. For the (Tennelec, Inc.) position sensitive detector (PSD), the electronics were connected to a multichannel analyzer which was also interfaced to the computer.

A version of OMSI-FOCAL, a Digital Equipment Corporation high level language modified initially by the Oregon Museum of Science and Industry (10) was then further modified by the X-ray group at Northwestern University to provide an all-purpose software package to enable rapid development of all kinds of X-ray measurements. This language allows for extended variable storage, motor positioning and counting-timing functions and several operations such as turning on and off a teletype, receiving data from a multichannel analyzer and monitoring a temperature controller.

The residual stress program incorporates the following features which are chosen by the user by means of an initial dialog:

1) The experimental parameters are input (approximate peak location, divergent slit, sample displacement and ψ-axis missetting) to calculate geometric aberrations (1,11).
2) Either the 'two tilt' or '$\sin^2\psi$' methods of residual stress analysis may be applied, with a normal detector or the PSD.

3) The Cohen-Marion technique is implemented automatically if a least-squares fit to d vs. $\sin^2 \psi$ indicates oscillations.

4) The operator specifies the counting accuracy in terms of degrees 2θ for each peak or in stress (PSI) and the counting strategy is determined taking into account both statistical and geometric errors.

5) The peak is fit to a parabola with an operator specified number of points and background subtraction. A parabolic fit was found to yield an appreciably smaller error in replicate tests than the center of gravity or position of half height.[24]

6) Parafocusing or stationary slit geometry can be chosen.

7) An estimate of the overall time of analysis is given after a preliminary scan of the peak. The operator can then change the error if the time is too long.

The formulae (based on Reference 12, 13) for the least-squares parabolic curve fitting and the associated statistical counting errors for the two-tilt and $\sin^2 \psi$ procedures are given in Appendix A.

Five successive step scans are involved in the procedure for determining a peak position, as described below. The data is collected in fixed count mode and corrected for deadtime and angular dependent intensity factors.

1) A step scan is made in large increments ($\sim .2° 2\theta$) set by the operator, accumulating 1000 counts at each position from the initial 2θ, until a count rate of less than 90 pct of the maximum is obtained.

2) Two step scans follow in smaller increments (set initially by the user and typically $0.05°$) down each side of the peak to locate two angles, $2\theta_1$ and $2\theta_3$ at 85 pct of the maximum intensity. (The background measured at an angular position specified by the user, may be subtracted from the data in determining the two end points. This procedure serves to define the region of parabolic fit more accurately.)

3) A third step scan at the two angles from Step 2 and the central angle between them is made with a preset count of 5000. A three point parabola is fit to the data. Then angles are calculated for the desired number of data points such that the central data point will be very close to the center of the parabolic fit. This minimizes odd order terms in the error equations. These preliminary scans typically take 60 seconds.

4) A step-scan at the final angular positions for a preset 1000 counts is used to back calculate the necessary counts from Equation A-1 to obtain the desired precision. Steps 3 and 4 constitute a multiple pass procedure (12,14) which serves to improve the reliability of peak location and minimize the time required for a given precision.

5) The final data is acquired for the calculated preset counts at each angular position.

This procedure of peak location is similar to that presented by Kelly and Eichen (6) except that the preset count mode of data collection

enables the program to work equally well over a wide range of peak
breadths, intensities and peak to background ratios.

MATERIALS AND PROCEDURES

Five flat steel samples covering a wide range of stress and peak
breadth were studied. The residual stress and some characteristics of
the diffraction peak from each specimen are included in Table 1. The
notation for each specimen refers to the SAE designation with the last
number used to differentiate samples of the same composition. A 'stress
free' sample designated TBA G-5 was supplied by Timken Co., Canton
Ohio. The remaining samples, supplied by W. P. Evans, Caterpillar Tractor
Co., Peoria, Illinois, were austenized in a neutral atmosphere, quenched
and tempered. The 1045-2 and 1045-3 samples were shot peened to various
degrees. The 1090-1 sample was stress relieved while the 1045-1 sample
was deformed in tension to a true strain of 13% by R. Marion (11).

The 211 peak was examined with CrK_{α} radiation (40 kV - 14 mA) in all
cases. A scintillation detector with a pulse height analyzer set to 90
pct acceptance and a K_{β} filter in the diffracted beam was used in conjunction with the Picker diffractometer, or a one-dimensional position
sensitive detector. The characteristics, experimental procedures and
calibration techniques for this detector have been described previously
(13).

Angles of 0° and 45° were used for the two tilt method. For the
$\sin^2 \psi$ procedure four ψ angles of 0°, 26.57°, 39.23° and 50.77° were
employed. The values of Young's modulus and Poisson's ratio were taken
to be 3×10^7 psi and .29 respectively (17) giving a stress constant of
approximately 86000 psi/°2θ for the two tilt technique.

After careful diffractometer alignment each sample was positioned
for minimum displacement from the center of the diffractometer by examining the slope of the Nelson-Riley factor vs. lattice parameter (9). No
instrumental changes were made during the replicate testing. The
stationary slit method was utilized in all precision studies so as not
to include random errors due to repositioning of the receiving slit.
However, it was subsequently found that the loss in precision due to
this repositioning was only ± 600 psi.

PRECISION OF THE X-RAY STRESS MEASUREMENT

The statistical counting error (see Appendix A) is often stated as
the precision of a measurement (9,15), but this is not usually justified.
Replicate measurements on one sample will always indicate a somewhat
larger error than that predicted by counting statistics alone because additional errors such as X-ray source stability may be present. Significant
errors will also arise if the diffraction profile is not well represented
by the curve fitting procedure.

Traditionally in stress measurements, the apex of a three point
parabola fit to the uppermost 15 pct of the diffraction profile is used
to define the Bragg angle (9). Jatczak and Boehm (1) concluded that
more elaborate procedures using 2nd, 3rd and 4th order curves to five

TABLE 1

RESULTS OF REPLICATE FIXED TIME TESTS
(TWO-TILTS)

Sample	Breadth at Half Maximum Intensity (°2θ)	PK/BKGD Ratio	PK. INT (CPS)	Number of Data Points	Total Time (SEC)	Stress (PSI)	Observed Error 10 Tests (PSI)	Statistical Error (PSI) EQ.A-9
1090-1	.45	6	1180	3	360	+4690	±203	±141
				7	360	+4580	±173	±171
				15	360	+4550	±223	±195
1045-1	.76	3.5	850	3	360	-26080	±1457	±500
				7	360	-24266	±1006	±583
				15	360	-23967	±888	±628
1045-3	3.4	2.0	2230	3	1000	-101470	±1987	±663
				7	1000	-101556	±1347	±973
				15	1000	-101432	±953	±1077
1045-2	5.1	1.6	1450	3	1000	-57396	±4160	±1325
				7	1000	-57958	±1925	±1894
				15	1000	-57107	±1857	±2049
TBA G-5	5.8	1.3	1180	3	1000	+2300	±4898	±1336
				7	1000	+1898	±2276	±1726
				15	1000	+1829	±1791	±1818

data points produce adequate agreement with the three point method and are
therefore not worth the extra time needed to accumulate the data. However, Marion (11) found a least-squares parabolic fit to 10 to 20 data
points gave better reproducibility than just the three point fit. This
apparent contradiction is important to resolve because the curve fitting
procedure will affect the precision. It is obvious that many data points
are preferable to only a few given unlimited time. For a fixed total
time, however, this is not necessarily true. Yap (16) concluded that if
the profile is a perfect parabola the optimum procedure for collecting
data is to spend most of the time at the two end points of the parabola
and only a short time at the peak in a three-point fit. Since random
errors dictate that the observed peak is not a perfect parabola the effect of
the number of observation points on the precision was experimentally
tested in this study. The total time for data accumulation was fixed
and tests were run using the 'two tilt' method.

The $\sin^2\psi$ method involves measurement of the peak position at more
than two inclinations which could minimize the random errors associated
with individual measurements of the peak. The same total time of data
accumulation was used as in the two tilt technique, enabling the precision of both procedures to be compared.

Table 1 summarizes the effect of the number of data points employed
to determine the peak maximum on the precision with a normal detector
and the two tilt procedure. The number of data points, column 5 and the
total time of data accumulation for the two tilts,[*] column 6, were set
in the initial dialog with the computer program. The residual stress
given in column 7 is the average value over 10 measurements. The
observed error, one standard deviation from the mean stress for the
10 replicate measurements is given in column 8 and the statistical
counting error for one standard deviation (Equations A-1 and A-9 in
Appendix A) is given in the last column.

The data for the 1090-1 sample which exhibits a sharp 211 diffraction
profile with excellent $K\alpha_1$-$K\alpha_2$ separation illustrates that for such a
profile, the number of data points has little effect on the precision.
The counting error is slightly less than the observed precision, as
expected.

As the breadth of the diffraction profile increases the peak
position becomes more sensitive to small fluctuations in the data and
the counting statistics for the three point fit do not predict the true
error. But as the number of data points in the least-squares fit
increase, the observed precision once again approaches the counting
error. These results show that counting statistics are not a good
measure of the precision when using only three points for a parabolic
fit on samples having broad profiles. Only where the number of data
points is increased is the observed precision represented by this counting error.

The reason for this result is not immediately obvious. If the
angular position of the data points is fixed and replicate measurements
are made using these fixed positions, the observed error is indeed very

---

[*]The total time is for the final step scan only and does not include the
time spent in the preliminary scans as this is independent of the number of data points.

close to that predicted by counting statistics alone. But this is not
the case when the angular position of the data points are rechosen by
automatic peak search procedures. It then becomes important to analyze
the goodness of fit of a parabola to the top of the diffraction profile.
Asymmetry of the data, even after correction for the angular dependent
intensity factors, can be treated by examining the odd powers of the
intensity in the statistical formula (see Equation A-1). The odd powers
have been found to be small in this study indicating negligible asymmetry. (If the odd terms are completely neglected, the formulae can be
simplified as suggested by Wilson (12). This was found to be quite
satisfactory when a multiple pass method of peak location is used as
was the case in this study).

The curve fitting is also dependent on the extent of the region
chosen. A detailed study has been made by Yap (16) on the range of
validity of a parabolic fit. Assuming a quasi-Lorentzian line profile,
Yap concludes that a parabola will have a satisfactory fit up to V=.32
where:

$$V = \frac{2(2\theta p - 2\theta min)}{W} \; . \tag{1}$$

Here W is the full width at half maximum intensity, $2\theta p$ the peak position
and $2\theta min$ the minimum setting lying on the parabola. The empirical
method of utilizing the top 15 pct of the peak common to X-ray residual
stress analysis is compared to Equation 1 in Table 2. The peak location
at $\psi=0°$ is given in column three. Column four records the minimum 2θ
setting given by Equation 1. The value for the minimum 2θ setting with
"the top 15 pct rule" is given in column five and the value of $2\theta min$
calculated by applying "the top 15 pct rule" to data corrected for background is given in column six.

All three procedures yield a similar region of parabolic fit for the
1090-1 sample which has a sharp 211 peak. As the breadth of the peak
(column 2) increases, the top 15 pct rule tends to predict a region
larger than that obtained from Equation 1. The region obtained by first

TABLE 2

THE REGION OF PARABOLIC FIT, 211 PEAK

	Breadth ($°2\theta$)	2θ Peak ($°2\theta$)	2θ min EQ.1 ($°2\theta$)	2θ min 15% Rule ($°2\theta$)	2θ min Background Subtraction ($°2\theta$)
1090-1	.45	156.14	156.07	156.05	156.05
1045-1	.76	155.89	155.65	155.61	155.65
1045-3	3.4	155.53	154.99	154.76	154.93
1045-2	5.1	155.50	154.68	153.20	154.40
TBA G-5	5.8	155.29	154.33	152.69	154.29

subtracting the background before determining the top 15 pct region
shows good correspondence to Equation 1. Not subtracting the background
on broad peaks or peaks exhibiting a poor peak to background ratio
results in data being collected outside the true region of parabolic
fit. The peak position then becomes more susceptible to the angular
position of the observation points and the precision is less than that
expected by statistical counting errors alone. The improvement in the
precision that is possible by subtracting the background when determining the region of fit was tested with replicate measurements. This
correction need not be highly accurate because it is only done to
determine the region of curve fitting. This was accomplished in the
second step scan as indicated earlier. The replicate results indicate
an improvement by a factor of almost two in the precision as compared
to Table 1 for the samples exhibiting broad profiles when using a three
point parabolic fit. However, the error for the 1045-2 and TBA G-5
samples was still almost twice that predicted by counting statistics
alone.

The precision of the $\sin^2 \psi$ method was determined in the same manner
as described for the two tilt method, the total time of data accumulation
being identical. Only a three point fit was employed, the replicate
measurements being tested on the 1045-1, 1045-2 and TBA G-5 samples. If
the results given in Table 3 are compared to those in Table 1, it is seen
that the $\sin^2 \psi$ technique has better precision over the ten measurements
than the two tilt technique even when the total time of data accumulation
is identical. The effect of errors in peak location are minimized with
multiple tilts. Increasing the number of ψ inclinations to more than
four was tested but improvement in the precision was only nominal. The
standard error in the mean stress (column 4, Table 3) is still 30 pct
greater than that predicted by counting statistics (column 5) for the
three point fit on the broader profiled samples.

The best results could be expected to be found by combining both
the $\sin^2 \psi$ method and background subtraction. The results from such
measurements are given in Table 4. The precision is indeed improved and
compares quite well with that expected from counting statistics alone.

TABLE 3

PRECISION USING $\sin^2 \psi$ METHOD
(THREE POINT PARABOLIC FIT TO PEAK)

	Total Time (SEC)	Stress (PSI)	Observed Error 10 Tests (PSI)	Statistical Error (PSI)
1045-1	360	-27988	±822	±866
1045-2	1000	-58430	±2310	±1581
TBA G-5	1000	+3300	±2630	±2100

TABLE 4

PRECISION USING $SIN^2 \psi$ METHOD AND BACKGROUND CORRECTION
(THREE POINT PARABOLIC FIT TO PEAK)

	Total Time (SEC)	Stress (PSI)	Observed Error 10 Tests (PSI)	Statistical Error (PSI)
1045-1	360	-28952	±898	±694
1045-2	1000	-61529	±1620	±1539
TBA G-5	1000	2134	±2340	±2415

The position sensitive detector collects data simultaneously across the entire diffraction profile. This eliminates the necessity for preliminary step scans in locating a peak and enables many data points to be used in the least-square parabolic fitting procedure. These factors combine to improve the speed of data accumulation by up to a factor of 20 (13). But the inherent precision of such a system has never been reported.

The results tabulated in Table 5 indicate that the accuracy of the PSD system is excellent. The observed error (column 5) in the mean stress is very close to that predicted by the counting statistics (column 6) alone. Thus the $sin^2 \psi$ method is not needed. Obtaining many data points (column 2) within the region of parabolic fit eliminates the.necessity for subtracting background. The best precision obtained with the normal detector comes from using the $sin^2 \psi$ method and subtracting background. The time for the final data collection for this procedure (not including the 60 seconds or so taken during the preliminary scans) is given in Table 4 and can be compared with the total time of analysis using the PSD in Table 5. This demonstrates the remarkable speed of the PSD, even when only a normal X-ray tube is used.

To judge the most rapid time possible with a PSD, ten replicate measurements, each for a total time of 30 sec, were run on the 1045-2 sample with the two tilt method. The average stress, -63,130 psi, had a standard error of ± 6156 psi while the counting statistics predicted an error of

TABLE 5

PRECISION OBTAINED WITH POSITION SENSITIVE DETECTOR
(TWO TILT METHOD)

	Number of Data Points	Time (SEC)	Stress (PSI)	Observed Error 10 Tests (PSI)	Statistical Error (PSI)
1045-1	31	100	-28795	±829	±631
1045-3	87	200	-104200	±1071	±996
1045-2	141	200	-59805	±1336	±1115

± 5868 psi. Because this sample has a very broad profile, it is clearly possible to obtain a precision of better than ± 6000 psi on many kinds of samples in well under 30 sec, and in much less time with higher powered tubes and distances shorter than those on a normal diffractometer.

SAMPLE DISPLACEMENT

Perhaps the largest single source of instrumental error associated with the X-ray measurement of residual stresses is sample displacement. In focusing geometry, if the effective diffracting volume is not located at the center of the diffraction circle, there is a relative shift between $\psi=0°$ and $\psi=\psi°$ given by (18,19):

$$\delta(2\theta)_{SD} = \frac{360}{\pi} \Delta X \cos\theta \left[\frac{1}{R_{GC}} - \frac{\sin\theta}{R_p' \sin(\theta+\psi)}\right] \qquad (2)$$

where ΔX represents the displacement of the sample*, R_{GC} the radius of the goniometer circle and R_p' the position of the receiving slit at $\psi=\psi°$. The position depends upon whether the parafocusing or the stationary slit (2) method is used and therefore the relative effect of sample displacement will be different for the two techniques.

A third beam optics technique known as the parallel beam method (20) is widely used in Japan. The beam is made parallel by placing fine slits perpendicular to the diffractometer plane. This technique was used in this study by placing two sets of Soller slits together and rotated 90° to the usual position. This yielded a divergence of .5° similar to the values used in Japan. Such slit systems were placed in the primary and diffracted beams.

The effect of sample displacement was tested for each technique. A dial gauge graduated in .0001 inch was mounted behind the sample holder to determine the displacement. Three replicate measurements were made at each sample position up to a total displacement of ± .08 inches in increments of .02 inches. The 1045-1 sample was utilized because it gave a reasonably sharp profile. The residual stress was measured using the two tilt procedure and a three point parabolic fit without background correction.

The error for both the parafocusing and stationary slit methods predicted by Equation 2 are drawn as solid lines in Figure 1. The data for each technique closely follows the predicted error. The stationary slit technique is seen to be less susceptible to sample displacement by almost a factor of five over the parafocusing technique.

The parallel beam geometry is seen to be insensitive to reasonable sample displacements in agreement with Aoyama, et al (21). [Of course, if the sample is displaced far enough, the diffracted beam will not fall completely in the receiving Soller slits introducing a large error.] But there is an effect on precision; compare Table 1 and Table 6. The parallel beam technique besides decreasing the intensities, severely

---

*A negative displacement is defined as being towards the back surface of the sample.

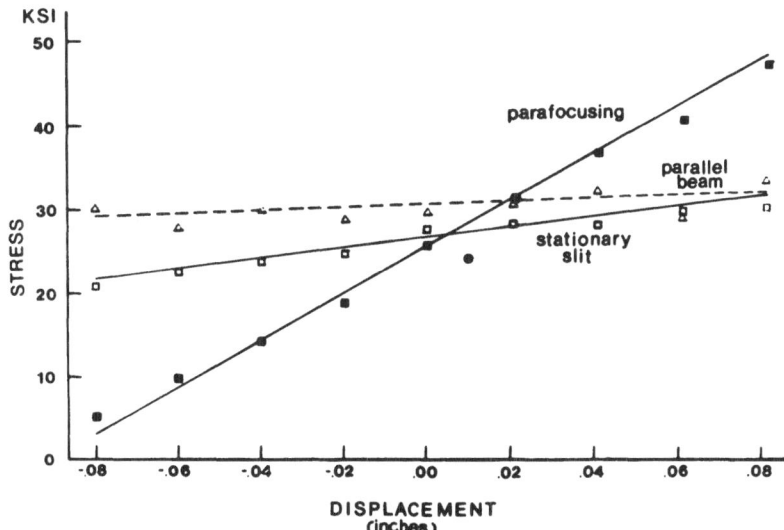

Figure 1. Dependence of stress on sample displacement for three geometric focusing techniques, 1045-1 sample. Solid lines represent the error as calculated from Equation 1. The dashed line represents the slope of the actual parallel beam data.

broadens the diffraction profile. This instrumental broadening is most important on samples exhibiting sharp profiles when focusing geometry is used. The data for Table 6 was acquired for a fixed time to give a statistical error close to that in Table 1 using the two tilt procedure and a three point fit (no subtraction of background). The precision for the 1045-1 sample is definately poorer but the error approaches that of the focusing techniques for samples having broad profiles.

DESIGN OF A PORTABLE RESIDUAL STRESS SYSTEM

The design of any mechanical system is based on those features one desires to optimize, in this case time and accuracy. Both the speed and precision of the position sensitive detector (PSD) indicate that such a device should form the basis of a portable stress analysis system.

TABLE 6

PRECISION USING PARALLEL BEAM GEOMETRY
(TWO TILT PROCEDURE - THREE POINT PARABOLIC FIT TO PEAK)

	Breadth (°2θ)	Time	Stress (PSI)	Observed Error (PSI)	Statistical Error (PSI)
1045-1	2.1	500	-23359	±2245	±583
1045-3	3.8	1300	-101922	±2401	±728
1045-2	5.3	1300	-60523	±4453	±1026
TBA G-5	6.0	1300	+1682	±3613	±2442

The two tilt procedure has been shown to be adequate with the PSD. In addition, the stationary slit geometry has only nominal susceptibility to sample displacement while yielding better precision than parallel beam geometry for sharp diffraction profiles. Using the two-tilt procedure with stationary-slit geometry the design parameters of a rapid residual stress device can be treated.*

The statistical precision, being a function of the total number of counts is dependent on the intensity of the diffracted beam and time of data collection (T). The systematic errors include resolution and calibration of the detector system and instrumental aberrations such as sample positioning. The relationship between the systematic errors, the intensity, and the sample to detector distance, R, has been considered in order to optimize the distance for a minimum time of analysis. The variance in stress, $S^2(\sigma_\varphi)$ is given by

$$S^2(\sigma_\varphi) = \left[K_1 (R/T \exp (-.07R))^{\frac{1}{2}} \right]^2$$
$$+ K^2 \left\langle \frac{\Delta 2\theta}{C(R)} (C(R) - C(R+\Delta X)) - \frac{360}{\pi} \cdot \frac{\Delta X}{R} \right.$$
$$\left. \cos\theta \left[1 - \frac{\sin\theta}{\sin(\theta+\psi)} \right] \right\rangle^2 \tag{3}$$

The first bracketed term relates the change in intensity and statistical precision to R. The value of K_1 is dependent on the sample characteristics (breadth, step increment, etc.) and the intensity through the statistical formulae (Equation A1). Its value may be found experimentally by considering the time necessary to obtain a given statistical error. The exponential term accounts for the change in intensity due to air absorption from sample to detector.

The term in carats accounts for the instrumental errors arising from a displacement, ΔX, of the sample from the true distance, R. The term K is the stress constant (taken as 86000 psi/$°2\theta$). The functions C(R) and C(R+ΔX) deal with the change in the calibration constant (converting position along the PSD into degrees 2θ) if the sample is displaced. The last term accounts for the peak shift due to the sample displacement. Derivations of this equation can be found in Reference 24.

From Equation 3, the optimum radius, R, to achieve a given error, $S(\sigma_\varphi)$, in the minimum time can be found. Results based on the TBA G-5 sample are given in Figure 2. The total error, including both the statistical precision and instrumental errors, was specified to be ± 8000 psi. The fixed parameter for each curve, ΔX, is considered for displacements up to .1 inch.

---

*Another technique, the single exposure method, exploits the fact that for exposure at oblique incidence, opposite sides of a diffraction ring correspond to different angles between the diffracting planes and the surface normal. Although being a very rapid technique, Norton (23) has shown that the stress constant for the one exposure method is more than twice as large as that for the two tilt method and therefore more dependent on the elastic constants which are not known precisely for residual stress determination. This fact, and the requirement of two position sensitive detectors for the one exposure method, make the two tilt technique more desirable.

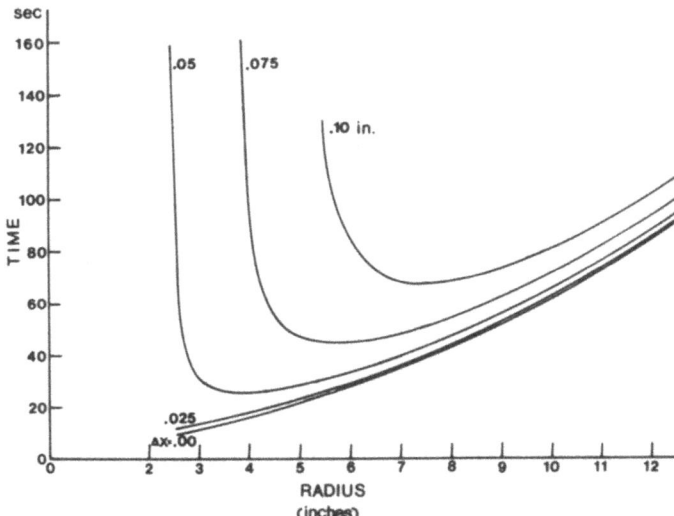

Figure 2. The relation between the sample to detector distance and time
           of analysis for a total error of ± 8000 psi. The error in
           sample displacement is given by ΔX. Results based on TBA G-5
           sample.

A reasonable degree of alignment for a portable residual stress
device is about ± .05 inches. Figure 2 shows the optimum radius of
four inches will enable the measurement to be made in 30 seconds. For
better alignment the time decreases to about 10 seconds at a radius of
2.5 inches, at which the angular resolution of the PSD becomes insufficient.

This analysis demonstrates that a very rapid device can be built
to determine the residual stress in under 30 seconds with only a normal
X-ray tube. More intense X-ray sources (affecting the term K_1 in
Equation 3) would reduce the speed even further. The new miniature
X-ray tubes (25) could be employed to increase the portability of the
system. Such a system is now being developed.

ACKNOWLEDGEMENTS

This research was supnsored by the Office of Naval Research under
contract no. N00014-75-C-0580. Mr. W. P. Evans, Caterpillar Tractor Co.,
and Mr. C. F. Jatczak, Timken Co., kindly provided some of the samples.
This study (and Reference 13) constitutes portions of a Ph.D. thesis
submitted (by M. James) in partial fulfillment of the requirements for
the Ph.D. degree at Northwestern University.

REFERENCES

1. C. Jatczak and H. Boehm, "The Effects of X-Ray Optics on Residual
    Stress Measurements in Steel", Advances in X-Ray Analysis, Vol. 17,
    p. 354, Plenum Press (1973).

2. C. Jatczak and H. Zantopulos, "Systematic Errors in Residual Stress
    Measurements Due to Specimen Geometry", Advances in X-Ray Analysis,

Vol. 14, p. 360, Plenum Press (1970).

3. D. P. Koistinen and R. E. Marburger, "Calculating Peak Positions in X-Ray R.S. Measurements on Hardened Steel", Trans. ASM 51, 537 (1959).

4. M. Short and C. Kelly, "Intensity Correction Factors for X-Ray Diffraction Measurements of Residual Stress", Advances in X-Ray Analysis, Vol. 16, p. 379 (1972).

5. J. Fukura and H. Fumiwara, "Evaluation of Various Techniques of X-Ray Stress Measruement With the Diffractometer", Journal Soc. Mat. Sci., Japan, 15, No. 159, 825 (1966).

6. C. Kelly and E. Eichen, "Computer Controlled X-Ray Diffraction Measurements of Residual Stress", Advances in X-Ray Analysis, Vol. 16, p. 344, Plenum Press (1972).

7. T. Hayama and S. Hashimoto, "Automation of X-Ray Stress Measurement by Small Digital Computer", Journal Soc. Mat. Sci., Japan, 24, 75 (1975).

8. R. H. Marion and J. B. Cohen, "Anomalies in Measurement of Residual Stress by X-Ray Diffraction", Advances in X-Ray Analysis, Vol. 18, p. 466, Plenum Press (1974).

9. H. S. 182, J784, "Residual Stress Measurement by X-Ray Diffraction", Soc. Autom. Engr., SAE, New York (1971).

10. D. Schneider and B. Smith, "PS/8 FOCAL 1971", STUDENT RESEARCH CENTER, OMSI (1971).

11. R. H. Marion, Ph.D. Thesis, Northwestern University, Evanston, Illinois (1972).

12. A. J. C. Wilson, "The Location of Peaks", Brit. J. Appl. Phys., 16, 665 (1975).

13. M. R. James and J. B. Cohen, "The Application of a Position Sensitive Detector to the Measurement of Residual Stresses", Advances in X-Ray Analysis, Vol. 19, p. 695 (1975).

14. D. Kirk, "Experimental Features of Residual Stress Measurement by X-Ray Diffractometry", Strain, January 1971, p. 7.

15. C. Kelly and M. Short, "Error in Residual Stress Measurement due to Random Counting Statistics", Advances in X-Ray Analysis, Vol. 14, p. 377 (1972).

16. F. Y. Yap, Dissertation, The John Hopkins University, Baltimore, Maryland (1967).

17. F. A. McClintock and A. S. Argon, "Mechanical Behavior of Metals", p. 86, Addison-Wesley (1966).

18. J. B. Cohen, unpublished results.

19. D. N. French, "X-Ray Stress Analysis of WC-Co Cermats: II, Temperature Stresses", J. Amer. Cer. Soc., <u>52</u>, 271 (1969).

20. "X-Ray Studies on Mechanical Behavior of Materials", Journal Soc. Mat. Sci., Japan, p. 89 (1974).

21. S. Aoyama, K. Satta and M. Tada, "The Effect of Setup Errors on the Accuracy of Stress Measured with the Parallel Beam Diffractometer", Journal Soc. Mat. Sci., Japan, <u>17</u>, No. 183, 1071 (1968).

22. E. Macherauch, "X-Ray Stress Analysis", Proc. Soc. Exp. Mechanics, <u>23</u>, 140 (1966).

23. J. T. Norton, "X-Ray Stress Measurement by the Single-Exposure Technique", <u>Advances in X-Ray Analysis</u>, Vol. 11, p. 401, Plenum Press (1967).

24. M. R. James, Ph.D. Thesis, Northwestern University, Evanston, Illinois (1977).

25. H. K. Herglotz, "Miniature X-Ray Equipment for Diffraction and Fluorescence Analysis", <u>Advances in X-Ray Analysis</u>, Vol. 16, p. 260, Plenum Press (1972).

APPENDIX A

FORMULAE FOR PARABOLIC CURVE FIT AND STATISTICAL COUNTING ERRORS

It is assumed that data is obtained at an odd number of observation points and is also taken in equal increments of 2θ. The data can then be said to be measured at $2n+1$ points, the center point being taken as a working origin $2\theta_0$. The data accumulated at each jth point is I_j, the power in counts/second. Call the 2θ step increment δ. The variance in the peak location due to random counting errors is (12,13):

$$S^2(2\theta p) = \frac{\delta^4 (n_2^2 - n_0 n_4)^2}{4 n_2^2 (n_2 \delta^2 M_0 - n_0 M_2)^4} \sum_{j=-n}^{n} (n_0 \delta^3 M_1 j^2 + (n_2 \delta^2 M_0 - n_0 M_2) \delta^2 j \quad \text{(A-1)}$$
$$- n_2 \delta^3 M_1)^2 S^2(I_j)$$

and

$$n_i = \sum_{j=-n}^{n} j^i$$

$$\text{(A-2)}$$

$$M_i = \delta^{i+1} \sum_{j=-n}^{n} j^i I_j \ .$$

The variance in the power, $S^2(I_j)$ used in the derivation of Equation A-1, depends on the method of data accumulation and is given by (12):

$$S_{FT}^2(I_j) = I_j/t \quad \text{for fixed time, t}$$

$$S_{FC}^2(I_j) = I_j^2/c \quad \text{for fixed counts, c.}$$

(A-3)

The error, $S(2\theta p)$, introduces an error in the calculated interplanar spacing given by:

$$S^2(d) = \left[\frac{\lambda \cos\theta}{2 \sin^2\theta}\right]^2 \frac{S^2(2\theta p)}{2} \left(\frac{\pi}{180}\right)^2$$

(A-4)

where λ represents the characteristic wavelength and $S(2\theta)$ is in degree 2θ.

Sin$^2\psi$ Method

The surface stress as determined by the $\sin^2\psi$ method is given by (9):

$$\sigma_\varphi = \frac{m^*}{d_o \cdot S_2/2} \quad \text{and} \quad m^* = \frac{\partial d_\psi}{\partial \sin^2\psi} .$$

(A-5)

The term m^* represents the slope of d_ψ vs. $\sin^2\psi$ and d_o and d_ψ are the interplanar spacings at $\psi=0°$ and $\psi=\psi°$ respectively. The quantity $S_2/2$ represents the X-ray elastic constant and may be experimentally (9) or theoretically (23) determined.

The variance in σ_φ as determined by the $\sin^2\psi$ method is:

$$S^2(\sigma_\varphi) = S^2(m^*)/[d_o \cdot S_2/2]^2 .$$

(A-6)

The variance of the slope, $S^2(m^*)$, is given by:

$$S^2(m^*) = \frac{\sum_\psi (\sin^2\psi - \overline{\sin^2\psi})^2 \cdot S^2(d_\psi)}{[\sum_\psi (\sin^2\psi - \overline{\sin^2\psi})^2]^2} .$$

(A-7)

The term $S^2(d_\psi)$ is given by Equation A-4 and the bar denotes an average.

Two Tilt Technique

The formula relating the surface stress to the strain is given by (9):

$$\sigma_\varphi = 1/S_2/2 \cdot 1/\sin^2\psi \cdot (d_\psi - d_o)/d_o .$$

(A-8)

The term $1/S_2/2 \cdot 1/\sin^2\psi$ is sometimes referred to as a stress constant K'. The statistical counting error is given by:

$$S^2(\sigma_\varphi) = \left(\frac{K'}{d_o}\right)^2 \left[S^2(d_\psi) + S^2(d_o)\right] . \qquad \text{(A-9)}$$

It is (unfortunately) common practice to approximate $(d_\psi-d_o)$ in Equation A-8 by $-\cot\theta \cdot (\theta_\psi-\theta_o)$ on the basis of Bragg's law to obtain a formula in terms of the peak position 2θ (9):

$$\sigma_\varphi = K(2\theta_\psi - 2\theta_o) \qquad \text{(A-10)}$$

where $K = \frac{1}{2} \frac{\pi}{180} \cdot 1/S_2/2 \cdot 1/\sin^2\psi \cdot \cot \frac{1}{2}(\theta_o+\theta_\psi)$. The variance in the surface stress is given by (13):

$$S^2(\sigma_\varphi) = K^2 \left[S^2(2\theta_o) + S^2(2\theta_\psi)\right] . \qquad \text{(A-11)}$$

Details on the derivations of these equations can be found in Reference 24.

THE EFFECT OF TEMPERATURE AND LOAD CYCLING ON THE RELAXATION OF RESIDUAL STRESSES

J. M. Potter

Air Force Flight Dynamics Laboratory

Wright-Patterson Air Force Base

R. A. Millard

Aeronautical Systems Division

Wright-Patterson Air Force Base

ABSTRACT

The Fastress automatic residual stress measurement system was used to measure residual stress in unnotched shot peened 7075-T6 aluminum alloy specimens for a study of the effect of load and thermal exposure on residual stresses. Cyclic loads at zero and high mean stress conditions at stress levels in excess of those used in aircraft structures gave no apparent change in residual stress. Thermal exposure above 200°F resulted in significant changes in residual stress; at 225°F, 30 to 50 hours were required to cause the residual stress to reduce to one-half the original value and less than 15 hours at 250°F was sufficient to cause a 50% reduction. The data indicate that the cyclic load history that may be expected in aircraft structures is not sufficient to cause residual stress relaxation in areas remote from fasteners and other stress concentrations but that thermal exposure can cause significant relaxation.

INTRODUCTION

Shot peening is a reasonably common practice for critical structural components in the aircraft industry. The shot peening process creates a beneficial compressive residual stress at the surface of the component that has the effect of improving the fatigue life and reducing stress corrosion (1,2). Periodically, there is some question as to whether the residual stress changes during the life of the structure and to what extent the working load and temperature environment play in the relaxation. The major mechanisms for changing the residual stress due to shot peening are metal removal, applied load and thermal exposure (3). The metal removal technique, whether accomplished by mechanical or chemical means, has been extensively studied and is a standard method used to determine the residual stress distribution through the thickness of a specimen (4) and will not be studied herein.

The mechanism of load induced reduction in residual stress can be broken into two sub-categories; those loads high enough to cause net section plasticity in the structure and the lower level cyclically applied working loads. Loads that are high enough to cause net section yielding occur infrequently, if at all, in durable structures although they may cause problems in a few individual structures affected. A limited study was accomplished in this program to determine the tensile and compressive stress level necessary to reduce stresses caused by shot peening.

The cyclically applied working loads of lesser magnitude may still be sufficient in number to cause the residual stress to relax during the normal lifetime of the structure. The literature concerning cyclic stress induced residual stress relaxation (5-7) models contains conflicting information as to the mechanism of the relaxation. Of these models, only that derived by Rotvel (6) is based on actual residual stress measurements and his were not based on stresses due to shot peening. The other models (5,7) were based on indirect observations of mechanical behavior of simpler specimens. The data from Rotvel indicate that the amplitude of loading is most important in residual stress relaxation. This data has its limitations. The measurements showed in general only a change in residual stress of 10 ksi from -20 ksi initial stress during cycling. This data was derived from steel coupons where Rotvel's estimated error in the process used to measure the residual stress is +3 ksi and the change in residual stress is approximately three times the error inherent in the procedure used.

The indirect data by Morrow, Ross and Sinclair (5) was derived using unnotched, strain controlled specimens that were not shot peened. The resultant stress history under strain controlled cycling was presumed by Morrow, et al. to be representative of residual stresses in shot peened structures. This analogy combined with their data indicated that the residual stresses relaxed mainly due to cyclic amplitude with very little effect of the applied mean stress.

Potter (7) developed a model of residual stress relaxation using fatigue initiation life data of notched specimens that experienced periodic overloads. This data indicated a strong dependence on applied mean stress and a smaller effect of the amplitude. Essentially, the Morrow, et al. model of relaxation gives the most relaxation under loading conditions where Potter's finds little or none and vice versa. Therefore, the cyclic loading phase of this program was initiated to determine the mechanism of cyclic load residual stress relaxation in shot peened specimens.

The third major mechanism, thermal exposure, is extensively utilized in many industries to relax detrimental residual stresses in structural components (3) and thus it may also be considered to have the ability of reducing the beneficial residual stresses caused by shot peening. Therefore, the thermal exposure portion of this study was initiated to determine the extent of the residual stress reduction that may be expected with the thermal environment of the aircraft wing skin.

ORIGINAL SPECIMEN

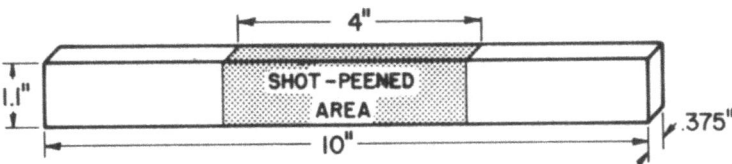

7075-T6 ALUMINUM

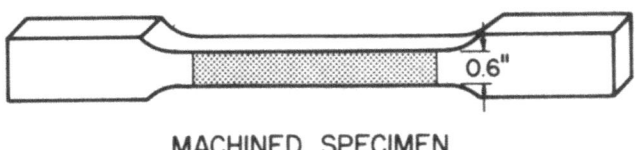

MACHINED SPECIMEN

Figure 1.   Test Specimen Configurations

There exists, in the literature, very little data on residual
stresses in structures and their relaxation. Part of the reason for the
lack of data is that the measurement of residual stresses is a very time
consuming and exacting process when done with conventional equipment.
Recently, the Fastress* automatic X-ray diffraction based equipment (8)
has been developed for residual stress measurement. By comparison to
conventional equipment, the Fastress requires 1 to 3 minutes to make a
residual stress measurement in aluminum alloy materials whereas one hour
is necessary with conventional equipment. Thus, with the Fastress equipment, much more quantitative information can be obtained in less time.

This program was initiated as a part of a feasibility program to
evaluate the capability of the Fastress equipment to solve aircraft
structural problems. The specific study involved the measurement of
residual stresses to determine the extent of the changes due to load and
temperature exposure in unnotched shot peened 7075-T6 aluminum alloy
specimens.

DESCRIPTION OF SPECIMENS AND INITIAL RESIDUAL STRESS

Twenty-two specimens of 7075-T6 aluminum alloy material were
prepared for the test program. The specimens were configured as rectangular bars nominally 3/8" thick, 1.1" wide and 10" long as shown in
Figure 1. An area of 4" long in the middle of the specimen was shot-
peened inducing a compressive residual stress of -20 to -40 ksi. Before
any testing began, residual stress measurements were taken in the shot
peened area to determine if the specimens met the required residual
stress specifications.

* Trademark of General Motors Corporation.

Five residual stress measurements were taken near the center of the shot
peened area of each specimen for a total of 110 measurements. The measured residual stresses ranged from -23 to -35 ksi. This data had an
arithmetic mean of -28.44 ksi and a standard deviation of 2.435 ksi.
The largest stress variation of the five measurements in a given specimen was 6 ksi, while the average variation was 2 to 3 ksi. Figure 2
shows the histogram of the residual stress measurements made on the as-
received specimens.

    After some initial testing it was decided that the rectangular
configuration was inappropriate for the cyclic load tests since the
specimens were breaking in the grip area. Therefore, the remaining
specimens that were to be used in the cyclic load tests were machined
into a dogbone configuration from 1.1 inch to 0.6 inch width using a
high speed metal router-type machine. This reduced the effective cross-
sectional area to almost one-half of the original, correspondingly
reducing the load requirements. Spot checks were made in the reduced
section specimens to determine if major changes in residual stress were
apparent due to the machining. Residual stress changes of less than
3 ksi were found in these checks and thus, the original residual stress
levels were accepted. The specimens that were to experience thermal
exposure were not reduced in cross section as there was to be no applied
loading.

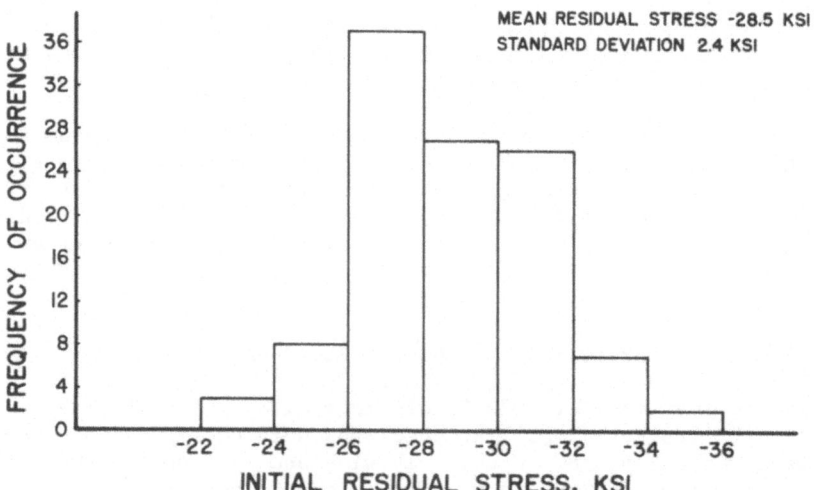

Figure 2.   Initial Residual Stress Level Histogram

TEST PROCEDURE

All cyclic load tests were accomplished in a closed loop hydrauli-
cally operated test frame with a load cell as the feedback element. The
specimens were held with hydraulic squeeze grips and the loads were
applied along the longitudinal axis of the specimen.

For all residual stress measurements, the specimens were removed
from the test frame. A simple specimen positioning device was developed
so that the same areas of the specimens could be repeatedly examined for
the measurements. Three measurements were taken from each specimen at
each condition; one measurement at each of three positions one-half inch
apart along the longitudinal axis of the specimen.

For the thermal effects studies, the procedure used was to insert
each specimen in a laboratory oven at the desired temperature for a
period of time. At that time, the specimen was removed, allowed to air
cool to room temperature, and then measurements of residual stresses
were taken using the sample positioning fixture and procedures noted
above.

All measurements were taken using the Fastress X-Ray Stress Analyzer
manufactured by the American Analytical Corp. The equipment used in
these tests contained two chromium target X-ray sources. The measurements were all made at input voltages of 35KVP and 10 ma current in each
source. Simple cylindrical X-ray beam collimators of 0.060 inch diameter
were used at each source. The Fastress measures stress by comparing the
output of two geiger tubes. The geiger tubes are physically driven
through the diffraction cone and when the output of each tube is balanced
with the other it is assumed that the diffraction peak is midway between
the centers of the geiger tubes. In the Fastress the angular distance
between the geiger tube centers is adjustable. For the measurements
made on these aluminum specimens the centers were set at 6° for the ψ_0
peak and 4° for the ψ_{45} peak. Chromium radiation makes a diffraction
peak with aluminum at a 2θ angle of 156.9° with the 222 crystallographic
plane. A vanadium β filter was used on each geiger tube. Pulse height
discrimination was not used. The bulk elastic properties for aluminum
alloy materials, 10^7 psi elastic modulus and 0.3 Poisson's Ratio, were
assumed for the stress calculations.

RESULTS

Preload Effects

One of the ways that residual stresses produced by shot peening can
be reduced is by causing net section plastic straining of the structure.
This disturbs the highly localized plasticity that originally causes the
residual stress. Two specimens were chosen for a study of the change in
residual stress due to preloading. Specimen 2 was subjected to a series
of monotonically increasing tensile preloads and specimen 10 experienced
a series of monotonically decreasing compressive preloads. The procedure
used was to load each specimen incrementally until failure occurred. The
specimens were removed from the test frame for stress measurement after
each load increment.

The data obtained in this phase of the program is shown in Figure 3.
The data indicate that almost no residual stress relaxation occurred
until both the tensile and compressive applied loads were greater than
60 ksi. At loads greater than $\pm$65 ksi, a sharp dropoff in residual
stress is apparent. The compressively loaded specimen buckled at a load
slightly greater than −70 ksi. No measurements are reported for the
compression specimen after failure because the buckling caused severe
localized plastic flow. Most measurements on this specimen were near
zero.

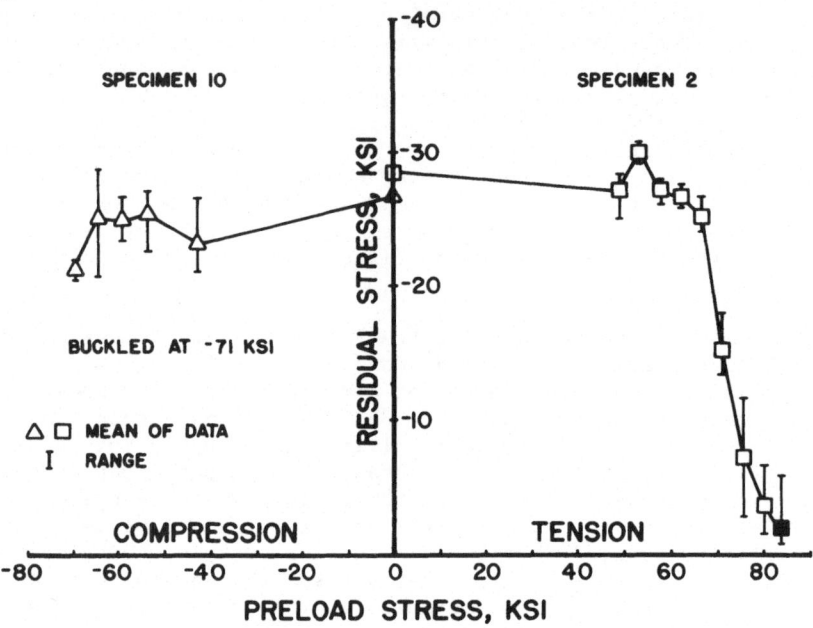

Figure 3.   Effect of Preload Stress on Residual Stress Level

Cyclic Load Effects

To test the competing models, two test conditions were utilized;
fully reversed cycling ($S_{min} = -S_{max}$) where the model of Morrow, et al.
would expect the most relaxation and high mean stress ($S_{min} = 0.5S_{max}$)
where the model of Potter would expect the most relaxation.

Test parameters for the high mean stress tests were set so that the
maximum stress level was slightly below the level where the residual
stress changed due to a tensile preload as in figure 3. The test parameters for the completely reversed conditions were set such that the
stress amplitude would be approximately twice that of the high mean
stress tests. For the high mean stress tests the stress amplitude ranged
from 13 to 16 ksi whereas for the fully reversed tests the stress amplitude ranged from 33 to 38 ksi. All specimens were cycled to failure.
Residual stresses were measured after failure on some, but not all, of the
specimens.

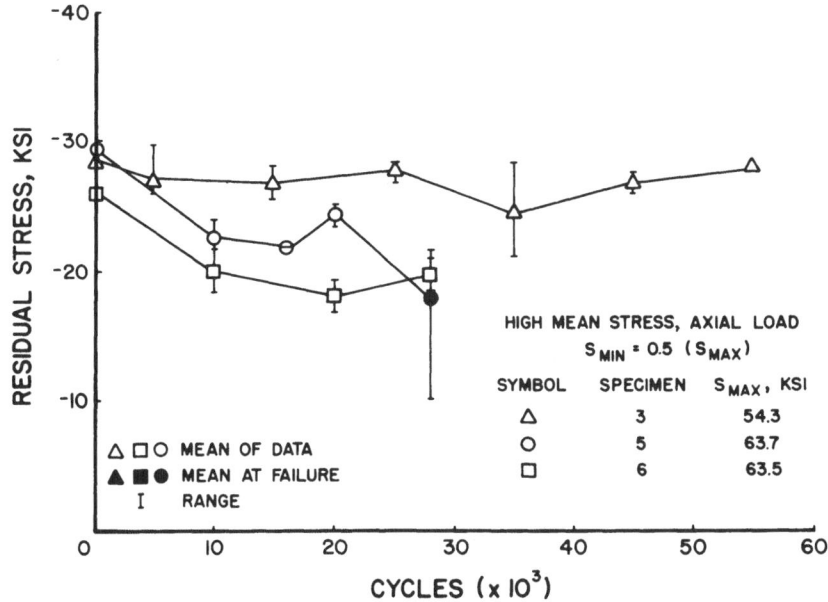

Figure 4. Effect of High Mean Stress Loading on Residual Stress

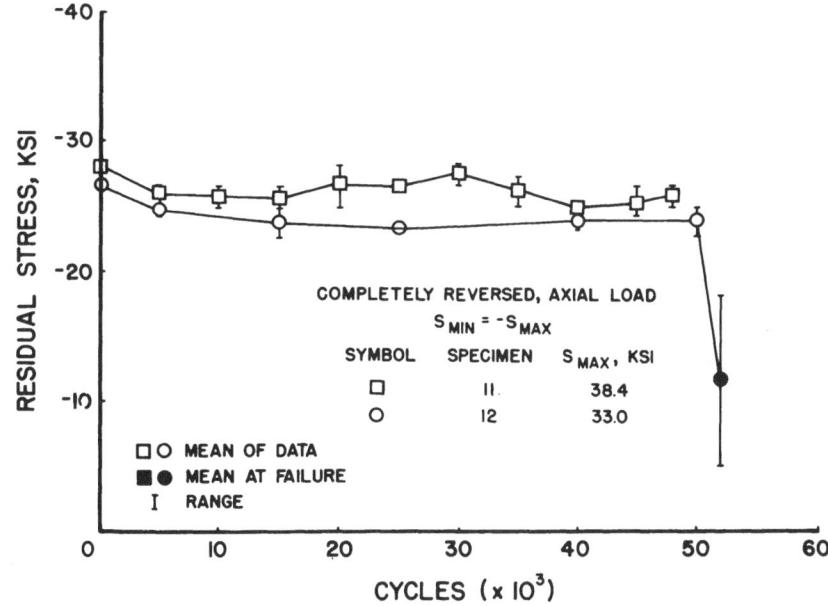

Figure 5. Effect of Completely Reversed Loading on Residual Stress

The results of the high mean stress tests are shown in Figure 4. The data at a maximum stress level of 54 ksi shows no appreciable change in residual stress over its lifetime. The two specimens that have a maximum applied stress level of greater than 63 ksi show a small amount of residual stress relaxation between the initial level and the measurements at 10,000 cycles and then no further appreciable relaxation. This apparent initial relaxation can be attributed to the closeness of the maximum stress level to that of the single preload level that would cause a similar relaxation. The fact that no further relaxation is apparent indicates that cyclic load relaxation does not occur for this condition.

The results of the fully reversed tests are shown in Figure 5. These results also indicate that there is no appreciable change in residual stress with applied cycles for these test conditions.

Thermal Effects

The aluminum alloy material used in this study is utilized in the shot peened condition in many aircraft applications where occasional and sometimes prolonged duration thermal exposure is experienced. The purpose of this phase of the program is to determine the thermal environment necessary to relax these residual stresses. Thermal exposure at 200°F, 225°F, and 250°F were chosen for test in this series. The 7075-T6 material is not normally used in applications that exceed 250°F so that is the maximum temperature used.

Specimen 9 was first exposed to 16 hours at 200°F. When no change in residual stress was apparent the oven temperature was increased to 250°F. The resultant data are shown in Figure 6. The residual stress was shown to drop to approximately one-half of the original value within 15 hours.

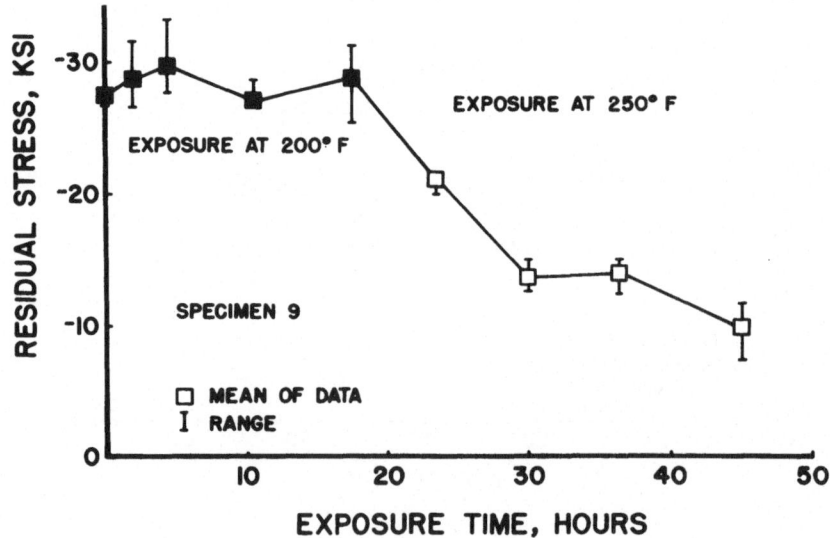

Figure 6.   Effect of Thermal Exposure on Residual Stress

Figure 7 shows the residual stress reduction with exposure time at
225°F for specimen 20. The residual stress is seen to decrease to 50%
of the original value within 30 to 50 hours at this temperature.

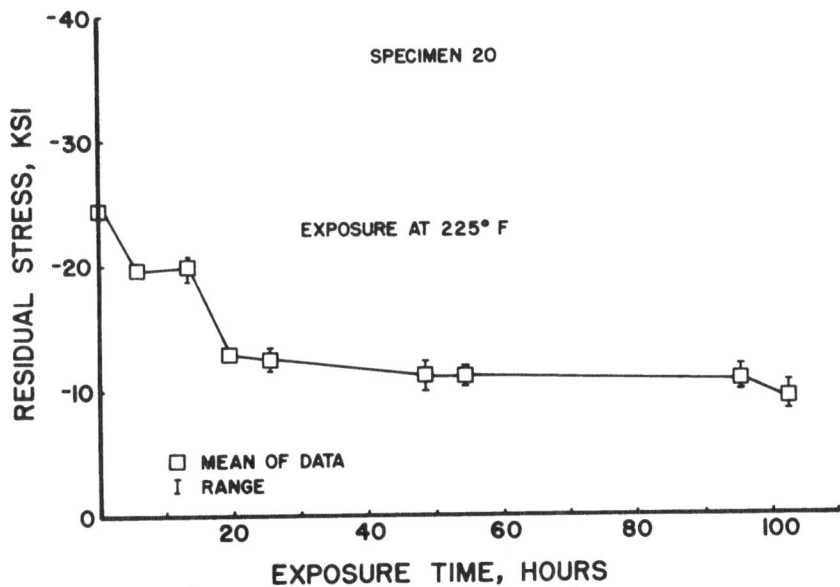

Figure 7.   Effect of Thermal Exposure at 225°F on
            Residual Stress

DISCUSSION

The accuracy of the measurements presented in this paper can be
affected both by variations in shot peening intensity and because of
basic errors in the process of X-ray diffraction (XRD). The errors in
the process of XRD are generally accepted as being on the order of 2 to
3 ksi in aluminum based alloys. That the standard deviation of the
measured residual stresses shown in Figure 2 is on the order of the
accepted error for XRD is indicative both of the uniformity of the shot
peening and of the repeatability of the Fastress. The shot peening was
done in the Blue Ash, Ohio plant of the Metal Improvement Company.

The residual stress reduction due to preload shown in Figure 3
indicates that applied stress levels of approximately 65 ksi magnitude
in both compression and tension were necessary to cause the residual
stress to relax. Because the yield strength of this material is approximately equal to 70 ksi in compression and tension, the data indicate
that net section plasticity is necessary to cause the relaxation. The
authors were somewhat surprised that the relaxation in the compression
loaded specimen did not occur at stress that were lower in magnitude than
-65 ksi. It was expected that the combination of applied compressive
stresses and the compressive residual stress would lead to yield behavior
within the peened material at relatively low applied stress. This would
have resulted in a relaxation of residual stress with the compressive

preload at a magnitude of applied stress significantly less than that of
the tensile loaded specimen.

The cyclic load data shown in Figures 4 and 5 indicated no change
in residual stress due to these two loading conditions. Both loading
conditions are simplifications of those seen in different sections of
many structures. The applied stresses used in these tests are much
higher than would prudently be utilized in structures that are expected
to be durable. The fact that the residual stresses did not relax under
these conditions means that they will probably not relax in less
demanding applications.

These results do not form the test of the Morrow , et al. (5),
Rotvel (6), or Potter (7) models of residual stress relaxation that was
desired at the initiation of the program. Only the Morrow, et al. model
purported to be applicable to relaxation of residual stresses due to
shot peening and, as stated before, their model was not based on measurements from shot peened specimens. In this model, relaxation was not
expected until the amplitude of cycling approached the yield stress
level. This condition was not directly investigated herein but is similar to conditions where a single cycle preload could remove the residual
stress. Thus, the applicability of the Morrow, et al. model to relaxation at these levels has not been tested in this study.

The application of the Potter model with the relaxation constants
derivable from that reference would have resulted in significant residual
stress relaxation in the high mean stress tests. This data indicates
that Potter's model of residual stress relaxation is not applicable to
the prediction of residual stress behavior in shot peened specimens.

The Rotvel model was developed from measurements on steel specimens
and he presents no relaxation constants for aluminum so no estimate of
relaxation can be made. Taken together, the data indicate that the
residual stresses created by shot peening are highly resistant to relaxation due to cyclic loading provided that net section yielding does not
occur.

The thermal exposure data shown in Figures 6 and 7 indicate a rapid
relaxation of residual stresses at these modest temperatures. The authors
are not sure if the relaxation seen at 225°F and 250°F is caused by the
shot peening losing its effectiveness or if it is an artifact of material
changes. The material 7075-T6 receives a final aging heat treatment at
a temperature of 250°F ±10°F (9). Whenever a specimen made from this
material approaches that temperature, the material will age and accelerated relaxation may be possible. Nevertheless, this material is commonly
used in aircraft in locations where these temperatures may be encountered.
Thus, it may be that thermal exposure could cause significant relaxation
of residual stresses in these structures. For a study of residual stress
relaxation without the complication of material change at 250°F and
higher, the authors recommend the use of the 7000 series aluminum alloys
in the overaged condition or the 2000 series where the aging temperature
is typically more than 100°F higher.

This paper does not address the relaxation in shot peen induced
residual stresses that could exist due to the structure containing

fastener holes and other stress concentrations. Localized yielding
could occur at these discontinuities that would cause the residual stress
to change. The results presented herein basically apply to residual
stresses in locations that are remote from fasteners holes and other
stress concentrations.

CONCLUSIONS

1. The Fastress automatic stress analyzer provides a fast, repeatable
and accurate means of determining residual stresses.

2. Cyclic stresses at levels well above those used in aircraft structures
result in no apparent relaxation in residual stresses.

3. Thermal exposure at temperatures above 200°F caused rapid relaxation
of residual stresses in this material.

REFERENCES

1. Fuchs, H.O., "Shotpeening Effects and Specifications", in American
    Society for Testing and Materials, Special Technical Publication
    (1962).

2. Alman, J.O. and Black, P.H., Residual Stresses and Fatigue in
    Metals, McGraw-Hill, New York (1963).

3. Richards, D.G., "Relief and Redistribution of Residual Stress in
    Metals", pp. 129-204 in Residual Stress Measurement, American
    Society for Metals, Cleveland (1952).

4. "Residual Stress Measurement by X-Ray Diffraction", SAE Information
    Report J784a, Society of Automotive Engineers, New York (1971).

5. Morrow, JoDean, Ross, A.S. and Sinclair, G.M., "Relaxation of
    Residual Stresses Due to Fatigue Loading", SAE Transactions, Society
    of Automotive Engineers, Vol. 68, pp. 40-48 (1960).

6. Rotvel, Find, "On Residual Stresses During Random Load Fatigue",
    Symposium on Random Load Fatigue, Advisory Group for Aerospace
    Research and Development Conference Proceedings, AGARD CP-118 (1972).

7. Potter, J.M., "The Effect of Load Interaction and Sequence on the
    Fatigue Behavior of Notched Coupons", Cyclic Stress-Strain Behavior;
    Analysis, Experimentation, and Failure Prediction, ASTM STP 519,
    American Society for Testing and Materials, pp. 109-132 (1972).

8. Weinman, E.W., Hunter, J.E. and McCormack, D.D., "Determining
    Residual Stresses Rapidly", Metal Progress, pp. 88-90, July 1969.

9. ALCOA Aluminum Handbook, Aluminum Company of America, Pittsburgh
    (1957).

STRESS MEASUREMENTS ON COLD-WORKED FASTENER HOLES

G. Dietrich and J. M. Potter

Air Force Flight Dynamics Laboratory

Wright-Patterson Air Force Base, Ohio 45433

ABSTRACT

Magnitude and distribution of residual stresses around cold-worked holes in 1045 steel were measured with the Fastress automatic stress X-ray analyzer. Permanent interferences between 0.8% and 6.2% at a hole with 0.5 in. diameter produced compressive hoop and radial stresses up to -60 ksi and -40 ksi respectively. The size of the zone with compressive stress is significantly wider than existing analysis predicts.

INTRODUCTION

Many construction parts in airplanes must be joined together with fasteners. But fasteners require holes and these holes reduce fatigue life. Therefore, many efforts have been undertaken to increase the life-time of structures with fasteners. Various methods, for example interference fits, bolt preload and cold-working of the hole, have shown remarkable improvements in fatigue lifetime of bolted and riveted joints [1, 2, 3]. This investigation will deal with the cold-working process.

It is generally agreed that the increase in fatigue life observed for cold-worked holes is due to residual compressive hoop stresses at the hole. Those residual hoop stresses reduce the high tensile stresses produced by stress concentration at holes under remote tension loads. Current methods of determining the increased life are empirical and require costly fatigue tests. Analytical methods of predicting life extension need a good knowledge of the magnitude and distribution of induced residual stresses. For a experimental solution of this problem residual stress measurements around cold-worked holes were taken nondestructively with the Fastress automatic stress X-ray analyzer. Then these data were correlated with analytical stress predictions.

SPECIMEN MATERIAL AND PREPARATION

The material chosen for the tests is a plain carbon steel with C = 0.45%. To enhance the reproducibility of the X-ray stress measurements, the material was heat treated to obtain a fine grain size (ASTM 10-11), a very fine pearlite microstructure and stress relief.

The specimens were cut to a size 3 in. by 3 in. from a 0.25 in.
thick plate (75 x 75 x 6.25mm). They were prepared for the stress
measurement by grinding to remove a 0.005 in. (125µm) thick decarburized
zone; subsequent etching was necessary to get rid of the residual
stresses generated with the grinding process. An 0.003 in. (75µm) layer
was etched away with a 1:1 mixture of concentrated hydrochloric and
nitric acids. After this preparation the fastener hole was drilled and
cold-worked in each specimen.

Prior to cold-working, measurements were made on one sample to
determine the stress level reproducibility of the unstressed material.
Statistical evaluation of stress measurements on 25 points within an
area of 0.2 by 0.2 in. (5 by 5mm) showed a mean value of +0.3 ksi (+2MPa)
and a standard deviation of 4.4 ksi (30MPa). The measured yield
strength of the material was 104 ksi (717MPa) and the hardness was 25
on the Rockwell C-Scale.

The hardware for the cold-working process is very simple. An
oversize tapered mandrel is pulled or driven through the hole. The
mandrel used for the cold-working had a diameter of 0.495 in. (12.57mm),
therefore, the initial hole diameter was chosen to have values from
0.458 in. (11.63mm) to 0.486 in. (12.34mm) to achieve various permanent
interferences with cold-working between approximately 1 and 6%. Cold-
working using a mandrel causes a permanent increase in hole diameter.
Permanent interference is defined as this measured increase in hole diameter divided by the initial hole diameter. The relationship between
the initial diameter and the measured permanent interference was an almost linear function within this range of hole deformation (Figure 1).

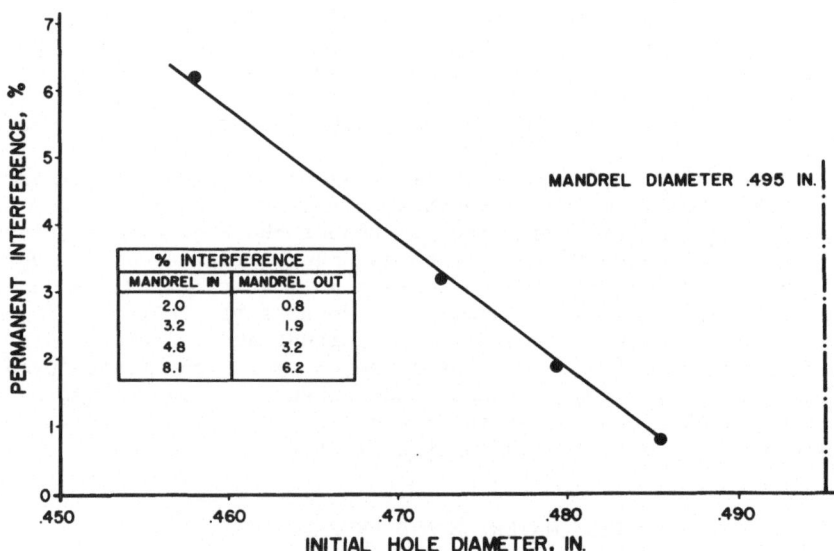

Figure 1.   Measured Permanent Interference vs. Initial Hole
            Diameter.

EXPERIMENTAL PROCEDURE

The stress measurements were taken with the Fastress unit which has
two preset ψ angles of 0° and 45°. Chromium radiation was used as the
X-ray source which makes a diffraction peak with ferritic steel at an
angle 2θ at 156.2° with the crystallographic plane (211). The voltage
for the X-ray tubes was set to 35KVP, the current was approximately 5mA.
Cylindrical beam collimators of 0.030 in. (0.75mm) diameter narrowed the
X-ray beam so that a good resolution of stress distribution could be
expected within the steep stress gradient. The Geiger tubes that were
used had no slits.

For the stress measurement the Fastress device was mounted on a
traveling stage so that it could be moved in increments of 0.001 in.
(0.025mm) in the X, Y, and Z direction to position the focal spot of
the X-ray unit. The distance between two measurement points was normally
0.02 to 0.05 in. (0.5 to 1.25mm) depending on the stress gradient.
The time to get one stress value was about 2 minutes, thus a complete
row with 30 single measurements to a distance of 1 in. (25mm) took about
1 hour. For the evaluation of the stress measurement the direct indication of stress units on the Fastress was taken assuming that the X-ray
elastic constants calculated with the bulk values for the elastic modulus
$(30 \times 10^6$ psi $= 207 \times 10^3$ MPa) and Poisson's ratio (0.3) are valid in
this case.

RESULTS

The results of a hoop stress measurement on a cold-worked hole
with 6.2% permanent interference are shown in Figure 2. For a more

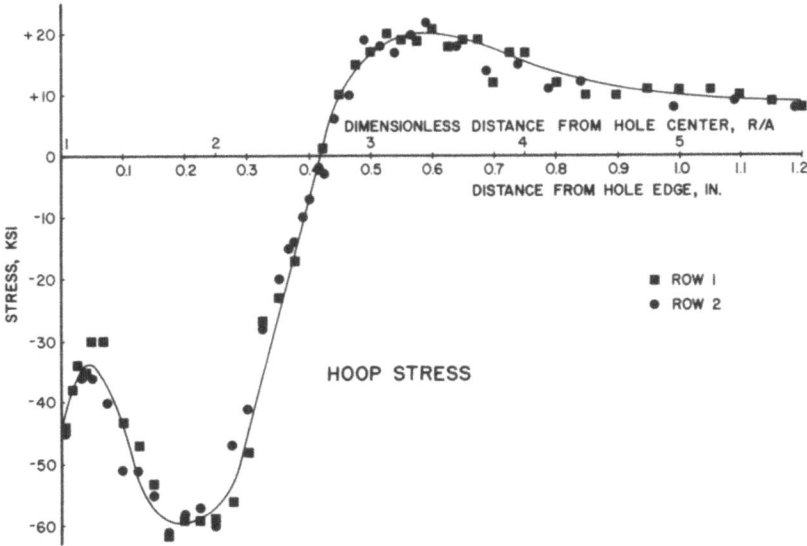

Figure 2.   Hoop Stress Distribution at a Cold-Worked Hole
            with 6.2% Permanent Interference.

common presentation of the data and an easier comparison with analytical
results, a dimensionless scale r/a for the position is introduced
(distance "r" of a measurement point from the center of the hole to "a",
the radius of the hole). The hoop stress reaches a maximum compressive
stress of about -60 ksi (-414MPa) near r/a =1.8 distance from the edge
of the hole, passes the zero stress after a relatively steep increase,
reaches a maximum value at +20 ksi (+138MPa), near r/a = 3.4, then tends
to go down. Because of the wide zone under compressive stress and the
necessary equilibrium of forces, zero stress is not reached after the
tension stress maximum. The two different symbols along the curve stand
for the two measurement rows starting from opposing sides of the hole.
The diagram shows good reproducibility even within the steeper part of
the curve.

The radial stress distribution for the same specimen is shown in
Figure 3. The stress curve starts at the edge of the hole (r/a = 1)
with a zero stress, as can be expected because of the boundary condition
(i.e., no external forces in the radial direction around the edge of the
hole). A maximum compressive stress of -42 ksi (-290MPa) is reached at
a distance of about the size of the hole radius (r/a = 2). Moving
radially away from this point, the radial stress approaches zero value
with a gradual slope, but is always in the compressive range.

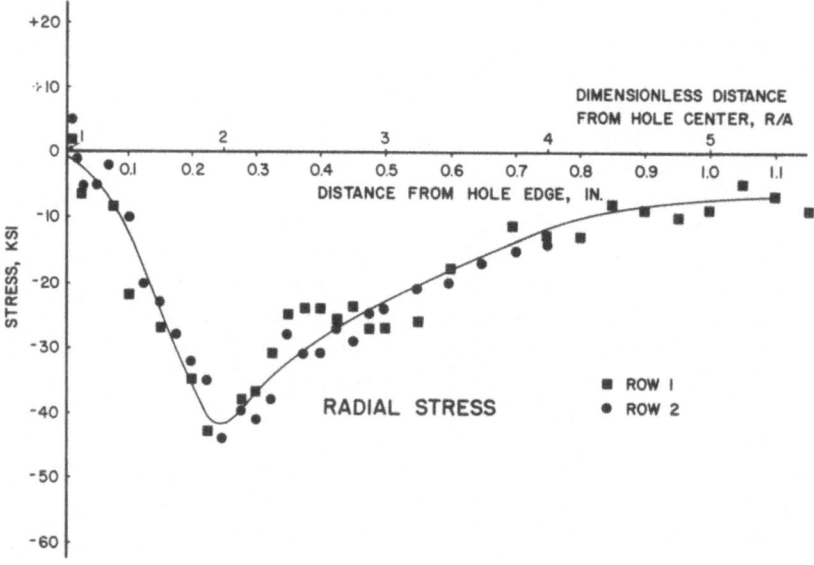

Figure 3. Radial Stress Distribution at a Cold-Worked Hole
           with 6.2% Permanent Interference.

Additional hoop and radial stress measurements were made on specimens with permanent interferences of 0.8%, 1,9% and 3.2%. They all
exhibited a hoop and radial stress distribution similar to that at the
6.2% level shown above.

An unexpected stress distribution was measured near the edge of the
hole in the hoop direction (Figure 4). There is an initial increase of
stress in the same order of magnitude for all interference levels that
cannot be explained at this time. The peak of this initial stress
increase rises with decreasing interference.

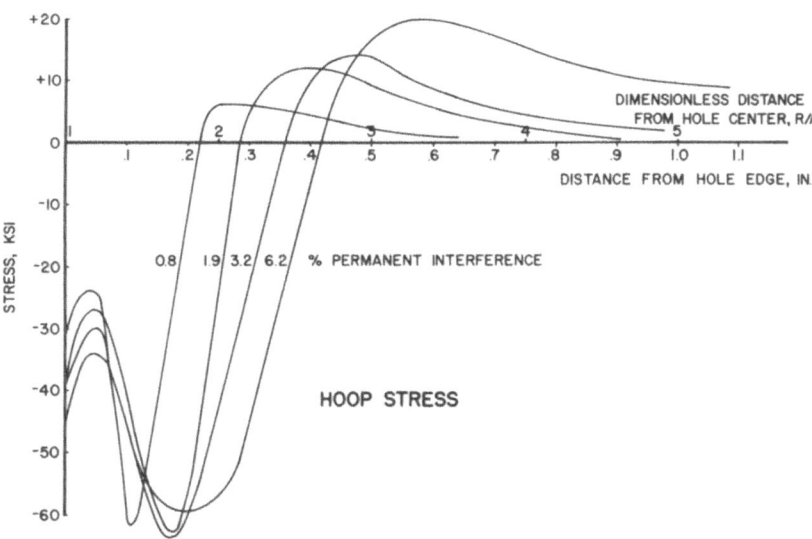

Figure 4. Hoop Stress Distribution at Cold-Worked Holes
           with Various Permanent Interferences.

The maximum compressive hoop stress is reached after a steep slope
at a value of approximately -60 ksi (-414MPa) which seems to be independent of the interference level within the chosen range. The end of the
plastically deformed area is definable by the position of the tension
stress peak. At a distance greater than this peak position, the material
appears to be only elastically deformed. The peak position is shifted
from over 1/4 in. to 1/2 in. (6mm to 12mm) distance from the hole edge
for the extreme values of investigated interferences 0.8% and 6.2%,
respectively. The magnitude of the peak tensile stress increases from
+6 ksi to +20 ksi (+41 MPa to +138MPa) with increasing interference.

For the radial stress (Figure 5) the zone with highly compressive
stress became wider with increasing interference, as it was expected.
There was little difference in the magnitude of the peak compressive
stress: for all interference levels a value near -40 ksi (-276MPa) was
reached. Near the edge of the hole a relatively small zone with tension
stresses over +10 ksi (+69MPa) was found at lower interferences. There
is no explanation for this behavior.

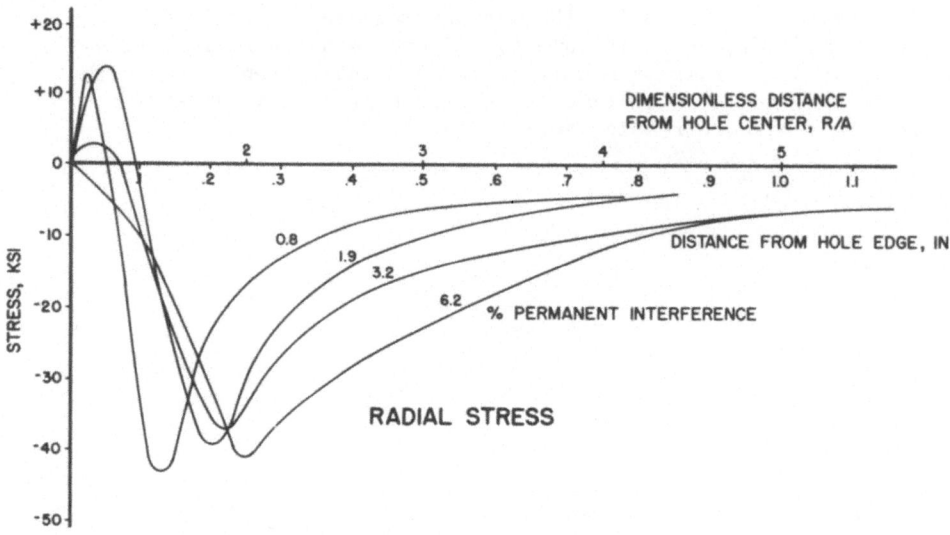

Figure 5.   Radial Stress Distribution at Cold-Worked Holes
with Various Permanent Interferences.

COMPARISON WITH ANALYTICAL RESULTS

There are few analytical solutions available which predict the
stress distribution around cold-worked holes. Different types of
material, different hole sizes and interference levels make a comparison
rather difficult and allow at best a qualitative comparative evaluation.

Potter and Grandt [4] have published a theoretical analysis of
residual stresses due to radial expansion of fastener holes. They came
up with a graphical dimensionless form for the presentation of their
results. Thus, a reasonable comparison with these test results is
possible.

The results of their analysis (Figure 6) show a shape of the radial
and hoop stress distribution similar to the experiment. One major
difference in the experiment results is the low compressive stress maximum for an interference near 1%. The experiment gave in this range
a relation for $(\sigma_{radial}/\sigma_{hoop})_{max}$ of 40 ksi/60 ksi, i.e., 2/3, whereas
the analysis predicts a relation of 1/5.

A second important difference is in the width of the plastically
deformed area which ends where the maximum tension stress in hoop
direction occurs. A graphical comparison between analytical and
experimental results shows a much smaller plastic zone predicted by the
analysis (Figure 7). The analysis curve ends with a maximum permanent
interference of 1.8%. There was no solution for higher interferences
with the elastic-perfectly plastic model used by Potter and Grandt.

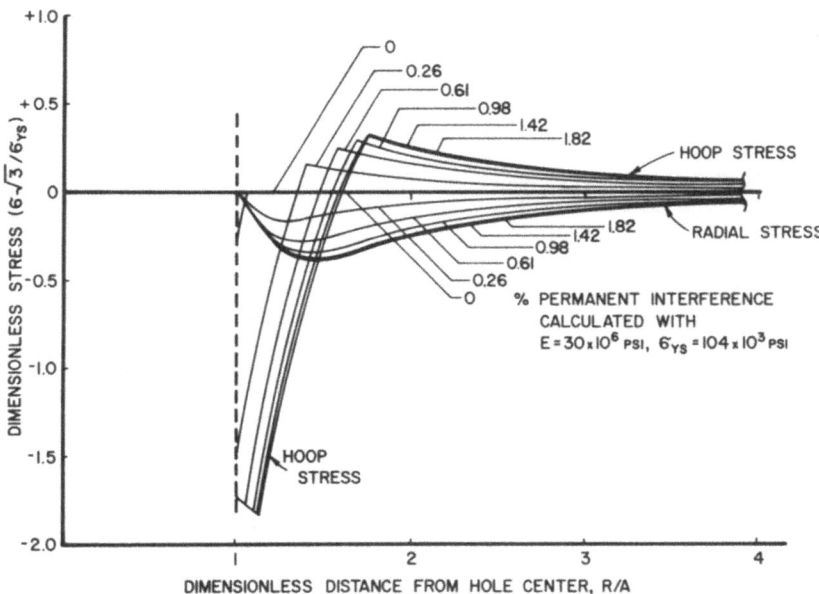

Figure 6.   Dimensionless Stress Distribution after Cold-
            Working [4].

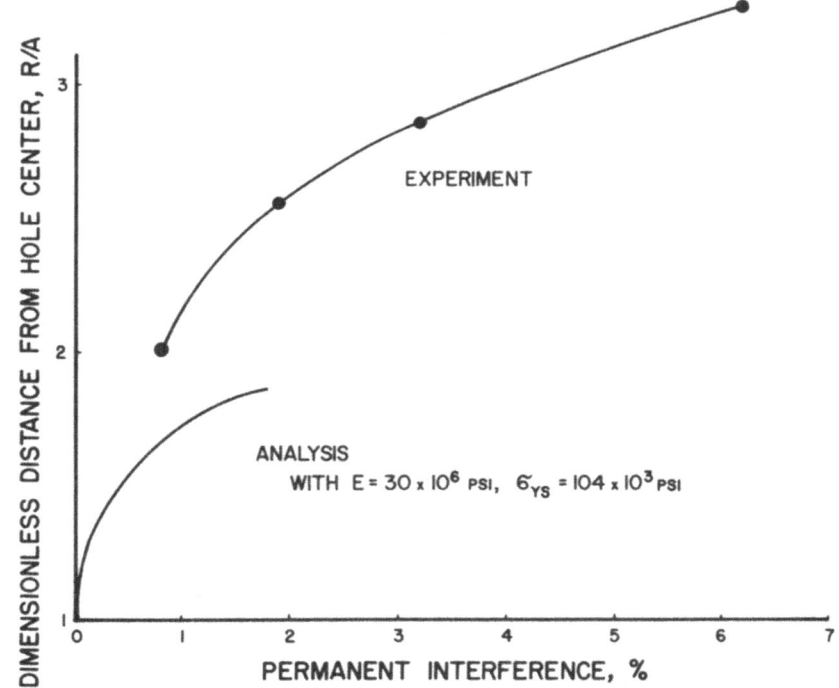

Figure 7.   Peak Position for Tension Hoop Stress vs.
            Permanent Interferences.

CONCLUSIONS

It was demonstrated that the Fastress automatic stress analyzer is a suitable instrument to measure the stress distribution on cold-worked fastener holes with sufficient accuracy, reproducibility and a reasonable expense of time.

With a permanent interference of over 1%, an area with a diameter twice the hole diameter is plastically deformed and mostly under compressive stress in the hoop direction.

The experimental results showed a shape of the stress profile similar to analysis results but the measured size of the zone with compressive stress is wider and the maximum radial stress is more compressive than analysis results predict. Thus, the model used for the analysis seems to be inadequate to completely describe the stress behavior within the plastically deformed area.

ACKNOWLEDGEMENTS

The authors would like to acknowledge the generous assistance of Mr. J.O. King of J.O. King, Inc. for supplying the large mandrels used in this investigation and of Mr. V. DePierre of the Air Force Materials Laboratory for his efforts in determining the proper heat treatment for this material.

REFERENCES

1. Regalbuto, J.A., and Wheeler, O.E., "Stress Distribution from Interference Fits and Uniaxial Tension", Experimental Mechanics, Vol. 10, No. 7, pp. 274-280, July 1970.

2. Phillips, J.L., "Sleeve Cold-Working Fastener Holes", Technical Report, AFML-TR-74-10, Wright-Patterson Air Force Base, OH 45433, February 1974.

3. Petrak, G.J. and Stewart, R.P., "Retardation of Cracks Emanating from Fastener Holes", Engineering Fracture Mechanics, Vol. 6, No. 2, pp. 275-282, September 1974.

4. Potter, R.M. and Grandt, A.F., Jr., "An Analysis of Residual Stresses and Displacements Due to Radial Expansion of Fastener Holes", Paper submitted for presentation at the ASME 1975 Failure Prevention and Reliability Conference, 17-18 Sept 1975, Washington, D.C.

DIFFRACTION TECHNIQUE FOR STRESS MEASUREMENT IN POLYMERIC MATERIALS

Charles S. Barrett

Denver Research Institute

University of Denver

Denver, Colorado

ABSTRACT

Stresses in polymeric materials can be measured by diffracting X-rays from embedded crystallites that yield suitable powder diffraction lines. An appropriate technique for this has been reported by C. S. Barrett and Paul Predecki elsewhere and is reviewed here with additional details. It is based on a diffractometer modified so as to operate at very high 2θ angles, with specimens prepared in various ways. Results show that adequate precision is reached by scanning $K\alpha$ doublet in a few minutes and simply noting any displacement of peaks compared with peaks from an unstressed specimen; higher precision can be reached with slower scans or more detailed analysis of the peaks. Tests with embedded metallic particles were successful when the particles were (A) distributed throughout a moulded polymethyl methacrylate (Lucite) sheet of the order of a millimeter thick, alone or sandwiched between layers of particle-free material, or (B) embedded in polyester resin applied to a surface of Plexiglas bars, or (C) embedded in graphite-fiber epoxy composites in a layer at a chosen position within a specimen. The method should prove to be useful in a wide range of practical problems. With calibration runs and suitably prepared specimens, the method can be used for measuring residual stresses after loading or after curing, measuring static or dynamic stresses externally applied, determining stress distributions, measuring stresses from temperature changes and temperature gradients in specimens, and determining the progress of any relaxation of stresses; the specimen matrix may be homogeneous or may contain other fillers or reinforcing fibers.

INTRODUCTION

This paper reviews briefly some research by Paul Predecki and me on the X-ray measurement of stresses in polymeric materials. A report of the work is being published elsewhere [1], but some added details of the work are given here.

The lack of sharp diffraction lines from polymers inhibits X-ray stress analysis in such materials, although some Japanese workers have found it possible to determine elastic moduli in highly crystalline, oriented polymers by applying uniaxial stresses and measuring low 2θ reflections from the crystallized regions by transmission through thin samples [2]. Their conclusions had to be based on the assumption that stresses were homogeneous throughout their samples. Some Russian workers have detected residual strains in crystalline fillers dispersed in amorphous aluminosilicate. After formation of a gel, thin slices were cut from it, dried, washed free of electrolyte, dried at 100°C, then reflections at low 2θ angles were recorded on film [3].

These two groups of investigators appear not to have realized the potentialities for stress measurement when suitable diffracting particles are embedded in homogeneous or reinforced polymers and high angle powder diffraction lines from them are measured, as in our work. The importance of using high 2θ reflections lies in the fact that not only are the changes in 2θ from elastic strains maximized, but also any difficulties from line widening due to X-ray penetration into the specimen are minimized. We have found that both residual and applied stresses can be measured if suitable particles are embedded in homogeneous polymers before curing, or when particles are inserted between the plies of graphite fiber/epoxy composites before curing.

APPARATUS AND TECHNIQUE

Our experiments were done on a Siemens diffractometer, modified so as to reach powder diffraction lines at higher 2θ angles than normally possible. As sketched in Fig. 1, an arm was added which carried a counter, B, at a radius somewhat larger than the radius normally used (see counter A, Fig. 1); the added arm could be quickly clamped on, and removed when normal diffractometer operation was resumed. This modification permitted scanning through the 511 + 333 CuKα reflection at $2\theta = 162.5°$ from powders of aluminum and from silver, $2\theta = 156.7°$, with specimens placed at the center of the diffractometer circle, 21 cm from the 0.070" slit at counter B; the slit at the X-ray tube was also 0.070" wide.

The specimens that were graphite fiber reinforced epoxy composite bars, 0.1" x 1.18" x 5.6" with embedded diffracting particles, could be stressed by bending in the jig shown at the center of the diffractometer. A composite bar was bent to a uniform curvature by bending moments applied near the ends. A pair of bolts and a pair of round steel rods between the composite bar and a heavy steel bar at the back (Fig. 1) served to provide the bending moments. Curvatures were determined by dial gauge measurement of deflection at the center of a 2.125" gauge length. Other specimens consisting of Plexiglas bars 0.25" x 1.125" x 4" with diffracting particles embedded in a polyester resin coating were stressed in the same jig.

Longitudinal strains in the bulk material containing the diffracting particles were computed as $e_\ell = \Delta R/(R-\Delta R)$ where R is the radius of curvature at the particles and ΔR is the distance from the neutral axis in the bar to the particles. If the particles were dispersed

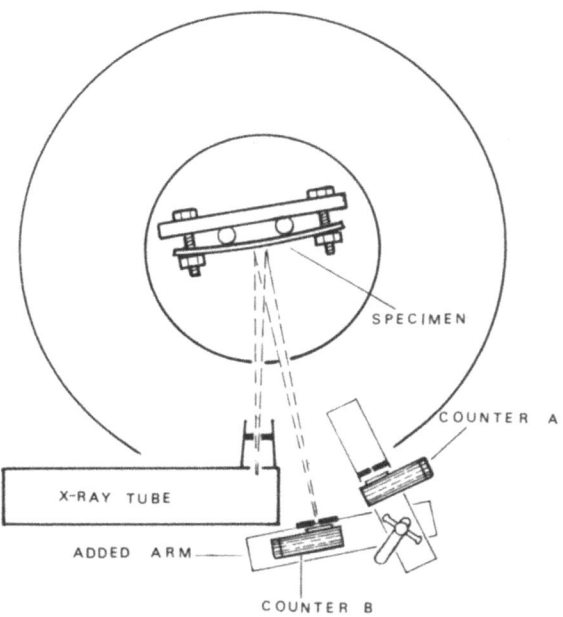

Fig. 1. Modification of a Siemens diffractometer for the present work.
         An added arm was clamped onto the arm that carries the usual
         counter (A), and carries a slit and counter (B) to higher
         angles, with negligible loss of diffracted beam focussing.

throughout a layer of non-negligible thickness, ΔR was the distance
from the neutral axis to an effective mean diffracting position. R
was obtained from the relation

$$R = (h^2 + \ell^2)^{1/2} \, / \, 2 \sin [\text{arc tan } (h/\ell)] \qquad (1)$$

where h is the deflection at the center of the gauge length and ℓ is
half the gauge length, i.e. 1.0625". All specimens were assumed to
have uniform modulus throughout, so that the neutral axis was along the
center of the bar. (In some specimens there could have been a slight
neutral axis displacement, which we considered was small enough to be
neglected, caused by thin layers of resin added to one surface of a
bar). The strains in the radial direction, e_r, were calculated from
e_ℓ obtained from deflection measurements by assuming that the bars were
isotropic elastic beams. Under this assumption the radial strain is
given by

$$e_r = -\nu \, e_\ell \qquad (2)$$

and the longitudinal stress by

$$\sigma_\ell = E e_\ell \qquad (3)$$

where ν and E are Poisson's ratio and Young's modulus in the bulk material containing the diffracting particles. Values thus computed were
given in Ref. (1).

The width to thickness ratio of the bars was rather large, 4.5/1 for the Plexiglas and 10/1 for the graphite epoxy; therefore, the bars might be considered to behave more like plates than beams, i.e. in plane strain. The radial strain in the bulk material for the isotropic elastic case would then be given by

$$e_r = \frac{-\nu e_\ell}{1 - \nu} \qquad (4)$$

and if ν for the bulk material is taken as 0.33 this gives a computed radial strain 1.5 times that for the isotropic elastic beam case. The corresponding longitudinal stress computed with the plane strain assumption would be

$$\sigma_\ell = E e_\ell / (1 - \nu^2) \qquad (5)$$

and is only 1.12 times that computed for the isotropic elastic beam case.

Longitudinal strain and stress in the particles were calculated from equations 2 and 3 for uniaxial tension (the beam case) and from equations 4 and 5 for plane strain (the plate case) using E and ν values for the particles.

The X-ray reflections were recorded on strip charts during scans through the $K\alpha_1$ and $K\alpha_2$ peaks, with the effects of stresses being revealed by the displacement that was required to bring the upper parts of the peaks from a stressed specimen into coincidence with the upper parts of the peaks from a stress-free specimen. Sufficient precision was obtained in scans that required only a few minutes to make and that were read in this simple way; far greater precision would of course be possible if more time were spent on scanning and reading peak positions, but only with increased danger of stress relaxation during the tests, since detectable relaxation occurred in a period of only three minutes in some experiments.

The strip chart records contained their own calibration for reading strains from peak shifts since the $K\alpha$ doublet spacing served this purpose. The principal strain in the radial direction of a bent bar, normal to the length of the bar, is given by

$$e_r = 0.00249 \ \delta/q \qquad (5)$$

where δ is the peak displacement due to stresses and q is the spacing between $CuK\alpha_1$ and $CuK\alpha_2$ peaks, since 0.00249 is the difference in wavelength of $CuK\alpha_1$ and $CuK\alpha_2$ divided by the wavelength of $CuK\alpha_1$.

RESULTS

Particles in Polyester Resin on Plexiglas Bars

Polyester resin was applied to a thickness of approximately 0.026" on Plexiglas (polymethyl methacrylate) bars. Filings of 7039 aluminum alloy 50 to 250 microns in size that had been held at 425°C for 40 minutes were dispersed in the resin at a concentration of 20 weight percent. The resin layer faced the X-ray tube in the jig of Fig. 1.

Bending experiments produced the results shown in Fig. 2, where the deflections at the center of the gauge length are plotted against the peak shifts of $K\alpha_1$ from the peak position for the bar before it was bent. Strains in the particles increased 2θ with increasing applied stress, up to a maximum 2θ displacement of $0.13°$ from the peak position for the bar before it was bent; then with more severe bending the peak shifts fell to the $0.10°$-$0.11°$ range. A similar apparent yield point such as is seen in Fig. 2 was also seen on curves from other specimens, and was found to occur at lower applied stresses with particles of silver or of aluminum in the resin than with the particles of alloyed aluminum, thus suggesting a correlation with the yield point in the metals. We use the term "apparent yield point" because possibly other factors than metal yielding were partially responsible.

When a bar was released after high strains and it returned nearly to its original straightness, the peaks shifted to the opposite side of their original position, showing that longitudinal compressive strains residual from the bending existed in the particles for many minutes. These results led us to conclude that provided peak displacements are calibrated by tests with known applied stresses, they can be used to measure both applied stresses and residual stresses when their magnitudes are less than the apparent yield point.

These conclusions were confirmed by similar tests on other bars. It was found, however, that when a bar was maintained at a constant bend radius above the apparent yield point for three minutes or more, the peak shift noticeably decreased indicating stress relaxation; therefore the drop in stress shown at the higher strains in Fig. 2 can be ascribed at least in part to such stress relaxation. In Fig. 2 the times at the beginning of each measurement are indicated at each data point, and the size of the circles indicates the estimated precision of 2θ readings. Constantly changing stresses were also seen after releasing the bar, the changes gradually slowing down.

The materials used had 30 minute relaxation modulus values as follows: 0.72×10^6 in the Plexiglas, 0.68×10^6 in the polyester resin and 0.79×10^6 in a resin sample containing aluminum alloy filings 8.46% by volume; Poisson's ratio was taken as 0.33. For the particles, E and ν were taken as 10×10^6 and 0.33 for aluminum, 10.4×10^6 and 0.33 for 7039 aluminum aluminum alloy, and 11×10^6 and 0.33 for silver.

Radial and longitudinal strains and longitudinal stress for the bulk composite and for the diffracting particles at a deflection corresponding to the apparent yield point of the particles are summarized in Table I. The calculated values are for both the isotropic elastic beam and plate cases.

By comparing the radial strain in the bulk composite (either beam of plate assumption) with that measured in the particles, it is evident that the composite radial strain is almost an order of magnitude greater, whereas the longitudinal stresses are somewhat closer together. It is clear that to use this method to measure stresses in bulk material

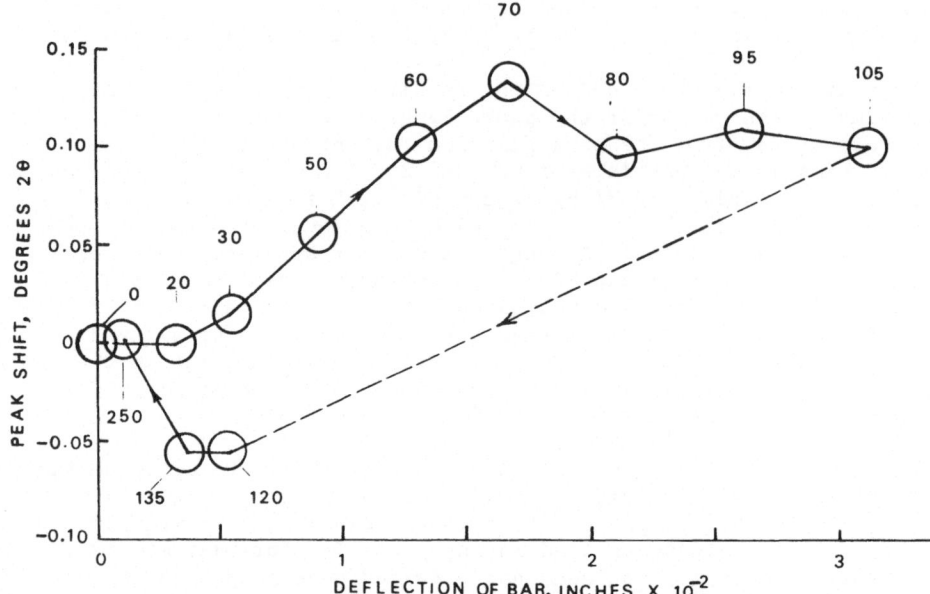

Fig. 2.   Bar bending experiment. Aluminum alloy filings in a 0.026"
          thick polyester resin layer applied to the surface of a
          Plexiglas bar 0.249" thick strained in longitudinal tension
          by bending; deflections of the bar at the center of a 2.125"
          gauge length vs. peak shift of the $CuK\alpha_1$ and $K\alpha_2$ 511+333 re-
          flections. Numbers at points are minutes after start of test.

Table I

Strain and Stress Values for Plexiglas Bars

	In Bulk Composite	In Diffracting Particles
e_ℓ (beam)	0.00443 (measured)	0.000776
e_r (beam)	−0.00146	−0.000256 (measured)
σ_ℓ (beam)	3190 psi	8070 psi
e_ℓ (plate)	0.00443 (measured)	0.00052
e_r (plate)	−0.00218	−0.000256 (measured)
σ_ℓ (plate)	3580 psi	6069 psi

it will be necessary to calibrate it with known stresses, since neither
stresses nor strains are matched on the two sides of a polymer-particle
interface.

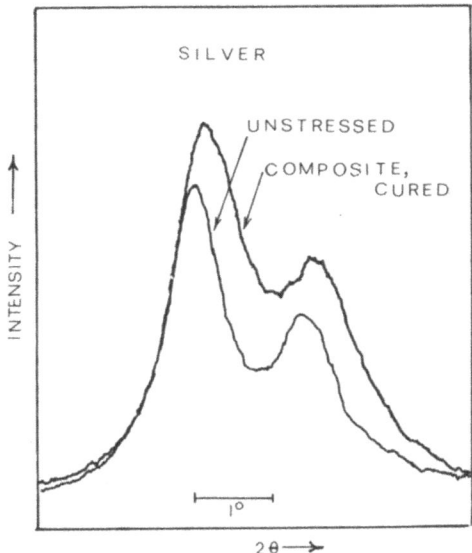

Fig. 3.   The 511+333 CuKα_1 and Kα_2 peaks from silver powder
          embedded in a graphite fiber epoxy composite and
          cured, compared with unstressed powder. Peak shifts
          indicate compressive strains normal to the specimen
          surface, after cooling to room temperature from curing
          at 350°F. Increased shifts followed when the bar
          was bent.

Particles in Graphite Fiber/Epoxy Composites

The method was applied to unidirectional fiber reinforced composite
samples. A graphite fiber/epoxy composite was prepared by stacking 20
layers of Hercules 3501 tape laid longitudinally, to make specimens
0.10" x 1.18" x 5.6", in which the fibers constituted 62% of the volume.
Silver powder of 6 micron average size was dusted on two sides of the
19th tape. After curing at 350°C for 1.7 hours under 100 psi pressure,
the Young's modulus in the longitudinal, x, direction determined mech-
anically was $E_x = 25 \times 10^6$ and in the transverse direction, y, was $E_y = 1 \times 10^6$, with $\nu_{xy} = 0.3$.

Diffraction peaks from the silver in the cured bar at room temperature and those from the silver particles laid stress free on cardboard
are shown in Fig. 3. The 511+333 CuKα peaks from the cured bar were
shifted 0.21° higher in 2θ than those from the unstressed particles, due
to compression from the curing and cooling process. An idea of the
order of magnitude of the stresses can be had by making an arbitrary
assumption of isotropic conditions with each particle subjected to
hydrostatic compression (i.e. principal stresses equal) arising from
differential thermal contraction. For this hypothetical condition the
hydrostatic pressure would be given by $\sigma = E\, e_r/(1-2\nu) = -12,520$ psi.

When the bar was bent, with the embedded silver on the outer side
of the bend, the compressive strains $-e_r$ were increased; for a deflection of 0.0173" at the center of the 2.125" gauge length the 2θ peak

shift increased to 0.25° and for a deflection of 0.0380" the shift increased to 0.27°. Upon release, the bar straightened and the shift
reduced to 0.10° but remained in the direction of higher 2θ than with
unstressed particles. A similar bar was prepared with aluminum alloy
filings, -250 mesh size, of 7039 alloy, which had been held 1/2 hr at
350°C after filing. The amount of filings that were dusted on the 19th
layer was not sufficient to yield good diffraction peaks, but a peak
shift amounting to about 0.75°2θ was judged to be present.

CONCLUSIONS

From the experiments cited here and in the detailed report [1] we
conclude that the method reveals both residual and applied strains in
the particles, of both tensile and compressive type. Neither strains
nor stresses are of equal magnitude on the two sides of a metal parti-
cle/polymer interface, but by the use of calibration experiments the
strains that are measured by diffraction in the particles can be used
to measure applied stresses and residual stresses in a homogeneous matrix around them such as polyester resin or polymethyl methacrylate, or
stresses in a reinforced polymer such as, for example, the graphite fiber reinforced epoxy of these experiments. The method can be used to
reveal stress relaxation, since the time required for a sufficiently
precise measurement of diffraction angle is small compared with most
relaxation times. We conclude that particles need not be added in
quantities so large that they cause serious modifications of the
strength properties of either a homogeneous or a reinforced matrix.
The method opens the possibility of measuring applied or residual
stresses, static or dynamic, not only in surface layers, but also at
some depth below the surface. The residual stresses may be those from
the curing process, differential thermal contraction, thermal gradients,
or external loads.

ACKNOWLEDGMENT

The help of J. Peng of the Martin Marietta Company, Denver, in preparing the graphite-epoxy composites and measuring their elastic properties is gratefully acknowledged. The discussion of stresses for the
case of plate behavior (not given in Ref. 1) was kindly contributed by
Paul Predecki.

REFERENCES

1. C. S. Barrett and Paul Predecki, "Stress Measurement in Polymeric
    Materials by X-Ray Diffraction," to appear in Polymer Science and
    Engineering.
2. I.Sakurada, Y. Nukushina and T. Ito, "Experimental Determination of
    the Elastic Modulus of Crystalline Regions in Oriented Polymers,"
    Jour. Polymer Science 57, 651-660 (1962).
3. S. I. Kontorovich, K. A. Lavrova, V. V. Davidov, G. M. Plavnik and
    E. D. Shukin, "X-ray Investigation of Internal Stresses in Filled
    Polymers" (in Russian), Strukt. Svoista Proverkh. Sloev. Polim,
    143-147 (1972); Chemical Abstracts 79, 79660c.

X-RAY DIFFRACTION STUDIES OF SHOCKED LUNAR ANALOGS

R.E. Hanss, B.R. Montague, and C.P. Galindo

St. Mary's University

San Antonio, Texas   78284

ABSTRACT

X-ray diffractometer studies of single-crystal quartz and ortho-clase reveal the peak shock pressure experienced by the samples. This procedure may facilitate rapid, quantitative interpretation of the peak shock pressures experienced by materials occurring in lunar or terrestrial impact structures. Shocked specimens were obtained from the NASA 20 mm flat-plate accelerator at Johnson Space Center, Houston. Orthoclase single crystals were shocked normal to the (001) plane at pressure intervals between 0 and 297 kb. The amplitude of the 27.6° two-theta (002) maximum decreases as a function of increasing shock pressure. Quartz single crystals were shocked normal to the (0001) plane at pressure intervals between 0 and 310 kb. Examination of the peak amplitude/half-width ratios for the 26.6° two-theta (101) and the 20.8° two-theta (100) reveals a general correlation of these ratios with shock pressure. This method seems appropriate for the calibration of shock pressures experienced by crystalline materials.

SHOCK PROCEDURE

Two important and common minerals found in terrestrial impact craters are quartz (SiO_2) and orthoclase ($KAlSi_3O_8$), which, although not very prevalent upon the lunar surface, were used to initiate this diffractometer study of shock effects. Profound shock effects ranging from fracturing and density changes to the formation of glasses are produced by meteoroid bombardment of the earth, moon and other terrestrial bodies. Quantitative measures of these effects, however, can only be obtained through controlled shock recovery experiments. Thus, samples of quartz and orthoclase utilized for this research were shocked in a 20 mm flat-plate accelerator located at the Johnson Space Center, Houston, which can generate shock pressures up to 1300 kb. The samples consisted of single-crystal discs, approximately 6 mm in diameter and 1 mm thick, mounted in stainless steel sample holders. This assembly serves as the target to be struck by a plastic projectile faced with a metal "flyer plate", accelerated via electrically-ignited conventional pistol powder. As the projectile nears the target it interrupts three He/Ne laser beams which enable its velocity to be determined to better

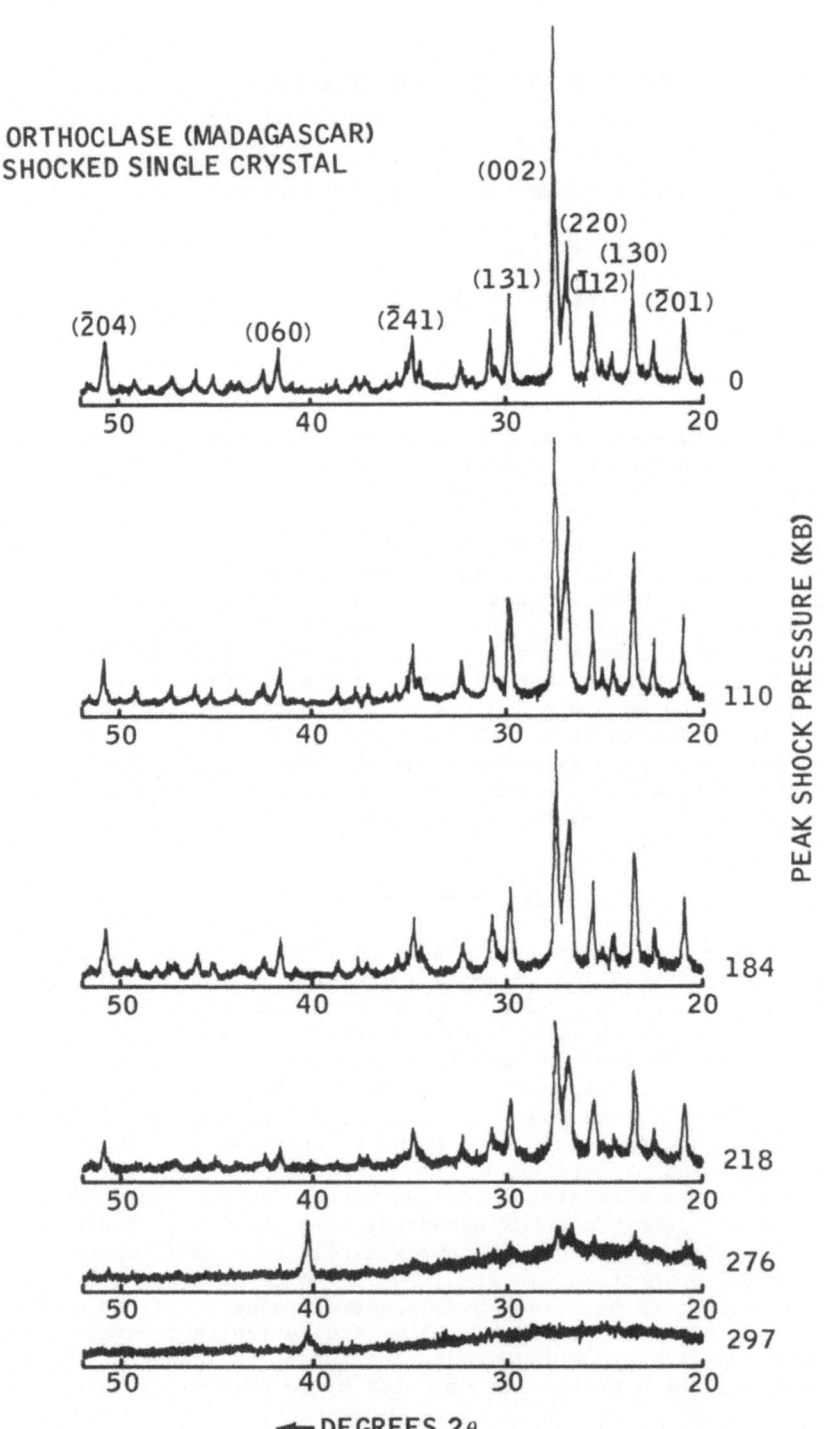

Figure 1.   Diffractometer patterns of orthoclase illustrating shock
            effects. (Anomalous 40° peak is from tungsten projectile.)

than 0.5% accuracy. The shock loaded targets are recovered intact and machined free from the sample holder. The peak impact pressure is then calculated from the velocity data, using equations of state for the single-crystal material, with an accuracy of ± 5%. Shock pressures ranging from 0-297 kb for orthoclase and 0-310 kb for quartz were utilized in this project.

SHOCK EFFECTS OBSERVED WITH X-RAY DIFFRACTION

X-ray diffraction studies of shocked materials have been mostly limited to Debye-Scherrer and Laue techniques. These studies are reviewed by Stöffler (1,2) and Hörz and Quaide (3). The latter investigators developed a series of standard Debye-Scherrer patterns showing increased streakiness, line broadening, and disappearance of reflections, for rotated single grains which had been exposed to increasing peak shock pressures. The length of time (4 or 5 hours) required for a Debye-Scherrer pattern plus the difficulty in quantitizing the diffraction line intensities led us to investigate the diffractometer as a potential rapid means of shock level determination.

Our diffractometer studies initially revealed a general decrease in the height of diffraction maxima with increasing shock pressure, until a certain pressure was reached, above which no diffraction maxima could be observed. Figure 1 shows the diffraction scans for samples prepared from single-crystal orthoclase shocked normal to the (001) plane. The decrease in height of the (002) diffraction maxima with increasing shock pressure values is shown in Table 1.

For detailed study of the behavior of a single diffraction peak with increasing shock pressure, we chose to work with quartz shocked normal to the (0001) plane, because it has a major peak (101) relatively isolated from the other quartz peaks, and because Debye-Scherrer studies have been performed on this mineral by Hörz and Quaide (3), Chao (4), Schneider and Hornemann (5), and others. The following sections describe this study of quartz.

SAMPLES AND DIFFRACTOMETER CONDITIONS

The recovered shocked samples usually consisted of 10-50 mg of clean material. The samples were ground to approximately 50 microns, and then mixed with 5 micron aluminum oxide as an internal standard. Aluminum oxide was chosen for this purpose because its diffraction pattern does not interfere with that of quartz in the two-theta region

TABLE 1:   ORTHOCLASE (002) PEAK AMPLITUDE VARIATION WITH INCREASING
           SHOCK PRESSURE.

PRESSURE (kb)	0	110	184	218	276	297
PEAK AMPLITUDE (counts/sec.)	1200	680	580	345	75	0

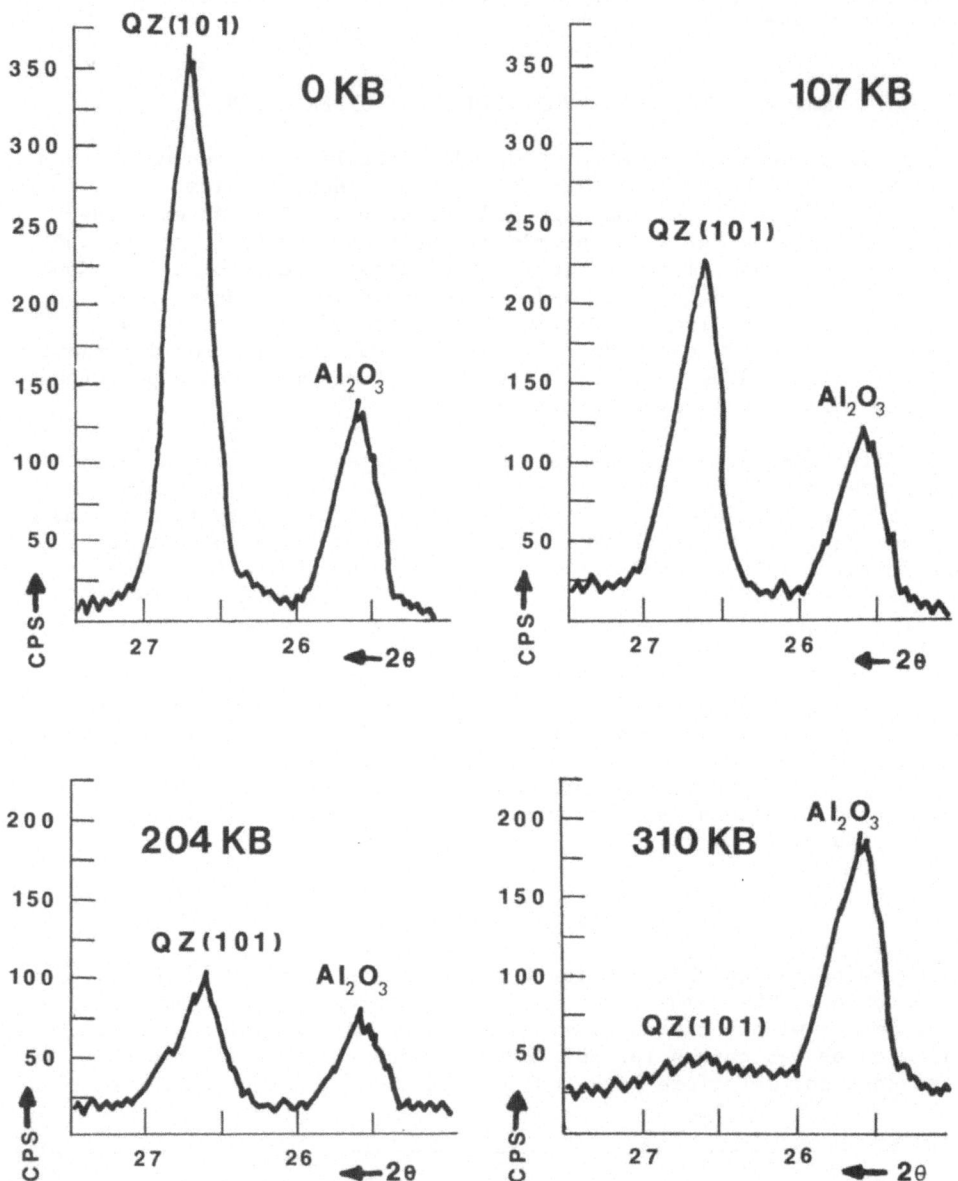

Figure 2.   Diffractometer patterns of the quartz (101) maxima illus-
            trating shock effects.

investigated. Since the unshocked quartz (101) and (100) diffraction maxima are much more intense than the nearby aluminum oxide (012) maxima, a 1:3 ratio of quartz to aluminum oxide by weight was used. This enabled us to utilize the small quantities of quartz available by dispersing it evenly in this mixture. The amount of sample used, when available, was 2 mg of quartz to 6 mg of aluminum oxide. Such small amounts of sample can introduce significant errors in weighing or in diffractometer results, but with reasonable care reproducible results were obtained.

The samples were mounted on two types of slides. The first consisted of a glass slide 2.5 x 1.5 cm upon which the sample was mounted using acetone within a rectangular area 1 x 1.5 cm. The use of this small glass slide was instigated by efforts to eliminate as much of the background radiation due to glass as possible. This effort eventually led to a second technique, in which a standard petrographic slide (27 x 47 mm) was covered with thin lead foil. A mask of lead foil containing a circular hole 7 mm in diameter was then aligned over the lead slide, pressed flat, and the sample placed in the resulting cavity and afixed to the surface using a mixture of half water and half methanol. After the sample had dried the mask was removed. This lead mounting decreased the background intensity by about 50%, and proved quite useful, especially as no lead diffraction maxima occurred in the two-theta region under study.

Diffractometer patterns of the shocked samples were obtained with a General Electric XRD-5 diffractometer equipped with a fine-focus copper-target tube, a 3° Soller slit, a graphite diffracted beam monochromator, and a xenon sealed-tube proportional counter. Two sets of operating conditions were used: 40kV/11mA and 50kV/5mA. In each case scans were made at 0.2° two-theta per minute.

VARIATION IN THE QUARTZ (101) AND (100) DIFFRACTION MAXIMA

In Figure 2 four representative quartz (101) diffraction maxima are shown, for shock pressures selected as close to 100 kb apart as possible, in order to illustrate the effects of increasing shock pressure on this diffraction maximum. In the first pattern, that of unshocked (0 kb) quartz, the height is 358 counts/second, while in the last, which has been shocked to 310 kb, the height is a barely detectable 20 counts/second. Associated with the decrease in height is an increase in the width (° two-theta) as measured at half the height (i.e., the "half-width"). The half-width varies from 0.29° two-theta in the case of the unshocked (0 kb) quartz, to 0.75° two-theta at 310 kb.

To describe this change occurring in the peak profiles in a quantitative fashion, the ratio of the height above background divided by the measured half-width was used.

A need exists for normalizing this raw peak height/half-width ratio since it is affected by different x-ray conditions and minor discrepancies in sample size, surface area, and surface texture. This need for normalization is emphasized by referring in Figure 2 to the 204 kb shocked sample. This sample contained less than 2 mg of quartz and aluminum oxide combined, resulting in the very low heights of both

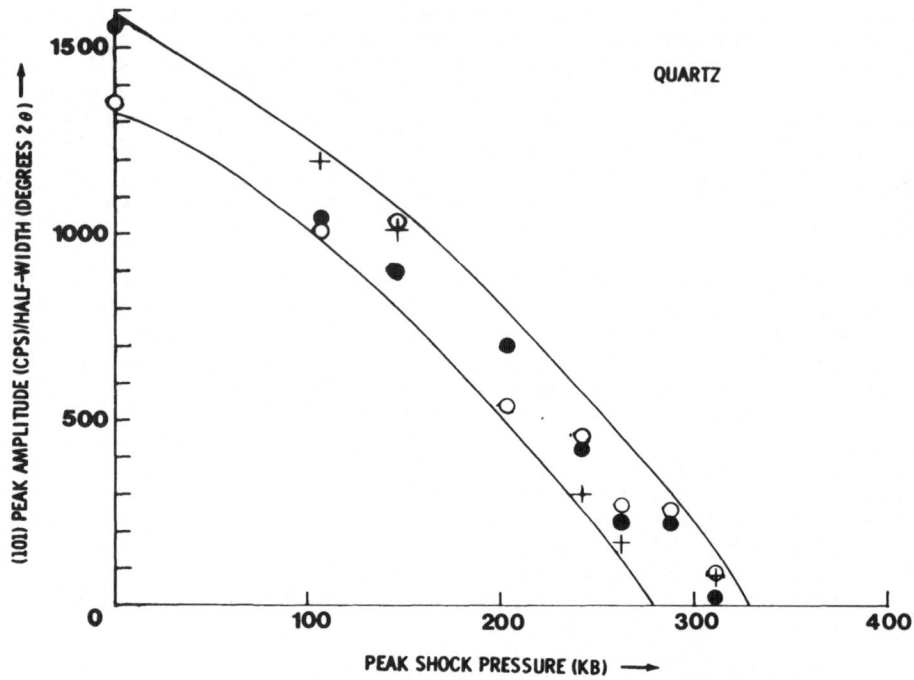

Figure 3.   Quartz (101) peak amplitude/half-width ratio as a function
            of peak shock pressure. (Solid circles-40kV/11mA, lead sub-
            strate; open circles-50kV/5mA, lead substrate; crosses-
            40kV/11mA, glass substrate.)

maxima. (It should be noted here that due to the small size of this
sample when compared to the other three in Figure 2, the increase of
half-width with respect to increased shock pressure is somewhat obscured
to the eye.) The normalization process chosen consists of dividing 500
by the aluminum oxide (012) peak-height/half width ratio, and multiply-
ing the raw quartz peak height/half-width ratio by this quotient to
obtain the normalized value. (The use of the fixed value 500 arose
simply because under our conditions the aluminum oxide (012) peak height/
half-width ratios tended to group about this value.)

    Figure 3 shows a graphic presentation of changes occurring in the
quartz (101) diffraction maxima. Here normalized peak height/half-width
ratios as a function of shock pressure are plotted for three sets of
data obtained under different conditions. This band provides an initial
calibration standard for an interpretation of peak shock pressure from
diffractometer data.

    An analysis of the quartz (100) diffraction maxima (20.8° two-theta)
for the same samples reveals a similar relationship between normalized
peak amplitude/half-width ratios and peak shock pressure. However, for
these diffraction maxima, there is a less steep slope for the data band
than in Figure 3 because of the less intense diffraction by the (100)
plane.

STRUCTURAL DEFORMATION IN SHOCKED SAMPLES

At each pressure level the shocked materials exhibit a range of
shock damage in spite of all experimental efforts to assure that the
target is traversed by a plane shock wave. Debye-Scherrer studies of
rotated single grains reveal this range, as do optical descriptions.
The diffractometer technique, which uses the equivalent of 10-20 single
grains of sample material, tends to reflect an average value of the
shock pressure experienced by the sample.

Several factors may cause the line broadening associated with the
decrease of the peak height/half-width ratio corresponding to increased
shock pressures. Internal strain in the polycrystalline aggregate produced by shock loading causes compressions and tensions at the interface
of crystallites, leading to a shortening or stretching of the theoretical d-spacing. The x-rays are thus diffracted by atomic planes having
a broad distribution of d-spacings about the mean values for unshocked
material. Since internal strain increases with pressure, line broadening is a function of pressure. In addition, line broadening is enhanced
by the decrease in domain size of the crystallites. The recovery products of the shocked single crystal are typically grains 1-2 mm in
diameter for low pressures (less than 100 kb) and somewhat finer aggre-
gates (0.5-1 mm) for higher pressures. The internal fragmentation of
such materials when shocked to successively higher pressures is evidenced in Debye-Scherrer patterns by increasing streakiness and the
fading out of first the back, then the front reflections. The ever-
decreasing size of the blocks finally reaches a level where it is beyond
coherent x-ray diffraction.

Hörz and Quaide (3) describe a proposed mechanism of 'fragmentation'
for quartz under shock conditions. This involves the generation of
relatively large blocks, somewhat heterogeneous in size, at pressures
up to about 100 kb. Above 120 kb, some of the SiO_4 tetrahedra are
rearranged during shock to high pressure phases (coesite, stishovite)
which relax after shock to highly-disordered, non-diffracting phases of
lower density than the original quartz. At pressures of about 200 kb,
the larger blocks break down into sizes too small for x-ray resolution
at the high-angle back reflections. The fragmentation and production
of high pressure phases increases until, at pressures above 300 kb, the
relaxed recovery products are x-ray amorphous and optically isotropic.
They are now a diaplectic glass.

DISCUSSION OF THE SIGNIFICANCE OF THE RESULTS

This study has been limited in detail to the major front-reflection
(low angle) diffraction peaks of quartz and orthoclase. No shocked
samples were available for the interval between 0 kb and 100 kb. Therefore, little information is provided here about the "fragmentation"
process or internal strains developing at pressures less than 100 kb,
except that it appears that some deformation effects begin in this
region and seem to continue, more or less linearly, up to about 300 kb
in both quartz and orthoclase. No samples of single-crystal quartz or
orthoclase shocked above 310 kb were available for diffractometer
studies, but scans were made of whole rock (granodiorite) samples

containing quartz and orthoclase and shocked in the JSC accelerator to
peak shock values of 320-480 kb. Some of these samples showed small
quartz (101) or orthoclase (002) maxima over a general non-diffracting
background. We interpret these small maxima at high peak shock pressures as expressing the fact that small parts of the polycrystalline
sample experienced pressures less than 300 kb.

However, the main value of this investigation, we believe, is the
development of a diffractometer technique to calibrate shock pressures,
using the peak height/half-width ratio as a parameter. This could be
of general use in the study of shocked crystalline materials (minerals
from terrestrial or lunar meteorite impact craters, nuclear explosion
sites, etc.) but several prerequisite studies are needed. These would
include the investigation of methods for adjusting data from one set of
x-ray experimental conditions to others, and the determination of standard calibration curves for each of the important rock-forming minerals.
If these studies are successful, the technique offers a rapid method
of shock level determination.

ACKNOWLEDGEMENT

This research was supported in part under NASA Grant NSG-9039.

REFERENCES

1. D. Stöffler, "Deformation and transformations of rock-forming
    minerals by natural and experimental shock processes. I. Behavior
    of minerals under shock compression," Fortschr. Miner. 49, 50-113
    (1972).

2. D. Stöffler, "Deformation and transformations of rock-forming
    minerals by natural and experimental shock processes. II. Physical
    properties of shocked minerals," Fortschr. Miner. 51, 2, 256-289
    (1974).

3. F. Hörz and W.L. Quaide, "Debye-Scherrer Investigations of Experi-
    mentally Shocked Silicates," The Moon 6, 45-82 (1973).

4. E.C.T. Chao, "Pressure and Temperature Histories of Impact Meta-
    morphosed Rocks-Based on Petrographic Observations," in B.M. French
    and N.M. Short, Editors, Shock Metamorphism of Natural Materials,
    Proceedings of the First Conference on Shock Metamorphism of Natural
    Materials, p. 135-158, Mono Book Corp. (1968).

5. H. Schneider and U. Hornemann, "X-ray Investigations on the Defor-
    mation of Experimentally Shock-Loaded Quartzes," Contrib. Mineral.
    Petrol. 55, 205-215 (1976).

A METHOD OF DETERMINING THE ELASTIC PROPERTIES OF ALLOYS IN SELECTED CRYSTALLOGRAPHIC DIRECTIONS FOR X-RAY DIFFRACTION RESIDUAL STRESS MEASUREMENT

Paul S. Prevey

Metcut Research Associates Inc.

Cincinnati, OH   45209

ABSTRACT

A technique and apparatus are described for obtaining the elastic constant $E/(1 + \nu)$ in selected crystallographic directions for the purpose of calibrating x-ray diffraction residual stress measurement methods. The preparation of a simple rectangular beam specimen with two active electrical resistance strain gages applied to the test surface is described. Samples are clamped in a diffractometer fixture designed to minimize displacement errors, and loaded in four-point bending to several stress levels below the proportional limit. A method is described for calculating $E/(1 + \nu)$ and an estimate of the experimental error.

Values of $E/(1 + \nu)$ obtained for several alloy-(hkl) combinations are presented. The results indicate that several alloys of current commercial interest exhibit significant elastic anisotropy.

INTRODUCTION

The strain in the direction defined by the angles ϕ and ψ in a sample of homogeneous material when under conditions of plane stress may be expressed in terms of the stress in the surface of the sample as:

$$\epsilon_{\phi\psi} = (\frac{1 + \nu}{E}) \, \sigma_\phi \, \sin^2\psi \, - \, \frac{\nu}{E} \, (\sigma_1 + \sigma_2) \quad . \tag{1}$$

In this expression, the quantities σ_1 and σ_2 are the principal stresses, σ_ϕ is the stress in the plane of the surface of the sample in the direction defined by the angle, ϕ, as shown in Figure 1; and ν and E are Poisson's ratio, and the elastic modulus of the material.

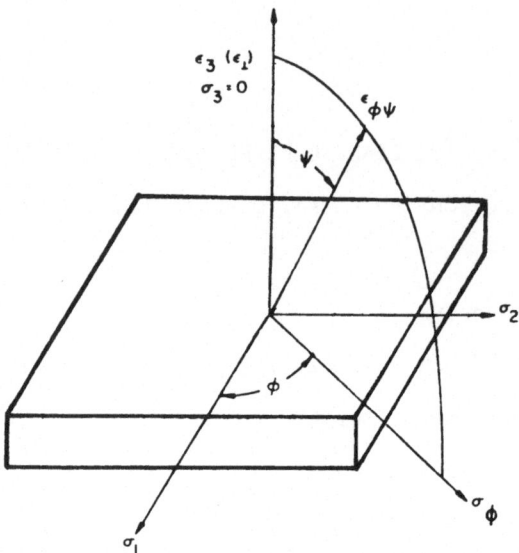

Figure 1.   Surface Under Plain Stress

Several x-ray diffraction techniques may be employed to solve equation (1) for the stress, σ_ϕ, which may be either applied or residual. If the strain, $\epsilon_{\phi\psi}$, is determined experimentally as the strain in the crystal lattice for at least two values of the angle ψ, σ_ϕ may be expressed in terms of the strain in the crystal lattice, as,

$$\sigma_\phi = (\frac{d_\psi - d_\perp}{d_\perp})(\frac{E}{1+\nu})\frac{1}{\sin^2\psi} \quad (2)$$

$$(1)$$

Equation (2) is the working equation for the "two-angle" technique , in which the lattice spacing is measured at $\psi = 0$ and $\psi = \psi$ to determine $d_\perp$ and d_ψ, respectively. The quantity $E/(1+\nu)$ is the elastic constant required to calculate the macroscopic stress, σ_ϕ, from the strain measured in a specific crystallographic direction. For an accurate calculation of σ_ϕ, $E/(1+\nu)$ must be determined in the direction normal to the lattice planes employed for stress measurement.

In the measurement of residual stresses, it is not uncommon to find mechanical values of E and ν being employed to reduce x-ray diffraction data. However, many of the alloys of current interest in the aerospace and nuclear industries are highly anisotropic. The use of a mechanically determined value of $E/(1+\nu)$ to reduce x-ray diffraction residual stress measurement data for these alloys can lead to errors as high as 80%.

It is the purpose of this paper to describe a technique for determining $E/(1+\nu)$ in four point bending which is used routinely in the author's laboratory. Specific elastic constant data obtained for several alloys is also presented to emphasize the importance of eliminating one of the major sources of systematic experimental error encountered in the measurement of residual stresses by x-ray diffraction techniques.

TECHNIQUE

Samples were prepared in the form of simple rectangular beams with
nominal dimensions of 4.0 x 0.750 x 0.060 inches as shown in Figure 2.
All surfaces of the samples were finish ground holding the thickness and
width of the beam to a tolerance of ±0.001 inch. Unless a specific heat
treatment of the alloy was to be investigated, test coupons were annealed
prior to grinding to the final dimensions. Fully annealed samples will
generally provide diffraction peaks in the high 2 θ range which are
sufficiently sharp to allow separation of the $K_{\alpha 1}$-$K_{\alpha 2}$ doublet. When
possible, diffraction data was taken using the $\bar{K}_{\alpha 1}$ peak. After final
grinding, a region 1 in. long in the center of one face of the sample
was electropolished to a depth of approximately 0.010 in. to remove the
plastically deformed layer produced by grinding.

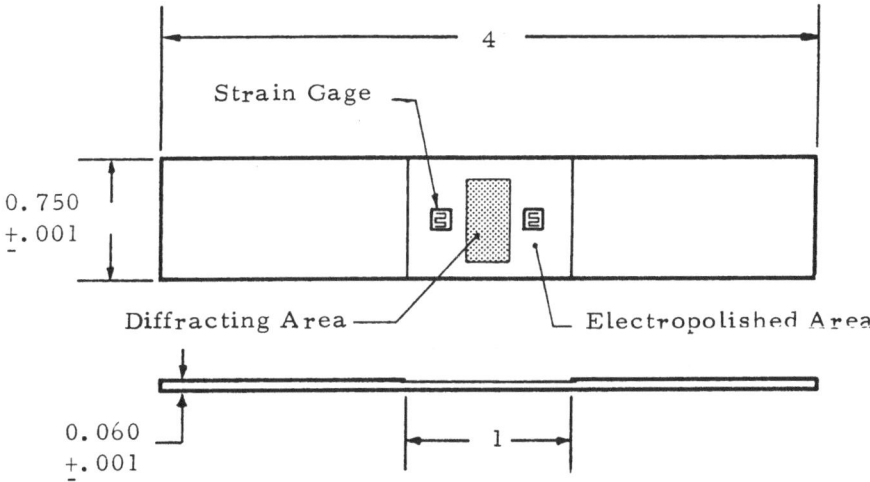

Figure 2   Four-Point Bending Sample

Two electrical resistance strain gages were applied to one face on
either side of the center of the sample in the electropolished region.
The gages were aligned to measure the outer fiber strain in the direction parallel to the longitudinal axis of the sample. The two gages
were positioned on either side of the center of the sample leaving a
bare region approximately 1/2 in. wide to be irradiated during the
determination of the elastic constants. The strain gages were bonded to
the electropolished surface using a furnace curing epoxy cement to
provide maximum stability of the strain gage bond. Room temperature
curing contact epoxies were found to be less stable, and appeared to
creep under sustained load.

The two active gages on the sample beam were wired with two identical
gages attached to a temperature compensating block of the same alloy to
form a full bridge circuit as shown in Figure 3. The circuit was
arranged so that the voltage across the bridge was proportional to the
sum of the strains measured by the two active gages. The strain in the

diffracting area between the two gages was assumed to be equal to the average strain measured by the two active gages. In this manner, any linear strain gradient along the length of the beam under four-point loading was eliminated. A protective coating was applied to the strain gages, and the entire assembly was allowed to cure at room temperature for at least 48 hour. prior to loading. After curing, the bridge circuit was attached to a strain indicator and the sample was flexed to approximately 80% of the yield strength several times until the gages would return to a reading of zero strain without hysteresis.

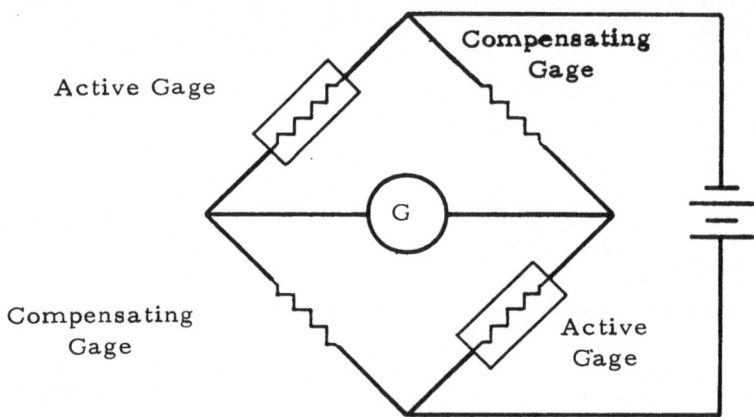

Figure 3. Strain Measurement Circuit

The strain gages were calibrated by placing the instrumented sample in a four-point bending fixture, and dead weight loading the sample to a known stress level. Knowing the applied load, the linear dimensions of the sample, and the moment arm of the four-point bending fixture, the actual stress in the outer fiber of the sample in four-point bending was calculated from the relation,

$$\sigma = 3Pa/bh^2 \tag{3}$$

where P is the applied load, a is the moment arm of the bending fixture, b is the width of the sample and h is the sample thickness. An effective elastic modulus for the sample was calculated from the ratio of the actual stress to the stress indicated by the strain gages using the strain gage manufacturer's supplied gage factor. Systematic errors due to misalignment of the strain gages and variations in the glue bond thickness were eliminated by using the effective modulus to calculate the applied stress when the sample was placed on the diffractometer.

The samples were placed on the diffractometer in a four point bending fixture designed to load the sample surface in tension while holding the sample rigidly over the center of the goniometer. The apparatus, shown in Figure 4, consists of a four-point bending fixture and a clamp, which is bolted to the ψ table and shimmed to position the diffracting volume of the sample over the ψ table-goniometer axis of rotation. The sample is held in the spring loaded clamp so that the bending fixture moves outward radially from the center of the ψ table as the sample surface is loaded in tension. The clamp minimizes displacement of the sample from the center of rotation of the goniometer as loads are applied.

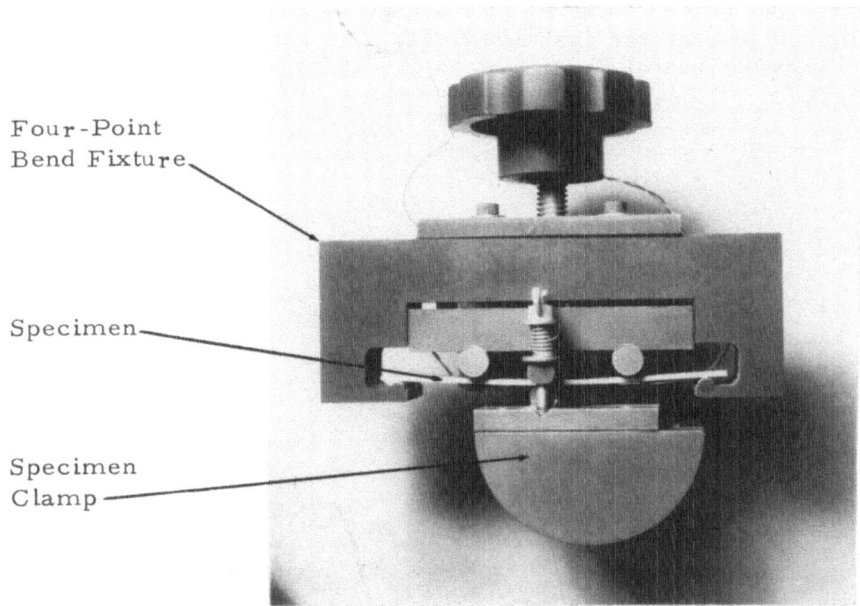

Four-Point
Bend Fixture

Specimen

Specimen
Clamp

Figure 4.   Four-Point Bending Apparatus

When positioned in the four-point bending apparatus, the samples were stressed to approximately 5%, 40%, and 75% of the yield strength of the alloy. At the highest and lowest levels of applied stress, calculated from the effective gage factor previously described, the lattice spacing of the selected set of planes was measured five or six times at ψ angles of 0.0 and 45.0 degrees. Two or three measurements were made at the intermediate stress level to assure that the lattice strain measured was linearly dependent upon the applied stress. A deviation from linearity indicates the failure of a strain gage bond or excessive loading beyond the yield strength of the material. The data were concentrated at the highest and lowest loads to minimize the uncertainty in the slope of a plot of applied stress versus the change in lattice spacing, Δd, between $\psi = 0.0$ and $\psi = 45.0$ degrees.

The lattice spacing at each load and ψ angle was determined using a parafocusing technique. The diffraction peak vertex was found using a least-squares parabolic regression procedure employing five points chosen in the top 15% of the diffraction peak after correction for a linearly sloping background intensity and for the Lorentz, polarization and adsorption factors. The inverse intensity at each of the five points chosen was determined by measuring the time required to obtain 100,000 counts. Calculation[2] of the systematic error due to the curvature of the samples under maximum load (9.0 in. minimum radius of curvature) at the lowest diffraction angle, 2θ, indicates a maximum error in 2θ of 0.015 degrees. As this error is on the order of the random error due to counting statistics, no correction was made for sample curvature. The shift in the lattice spacing, Δd, was then calculated between $\psi = 0.00$ and $\psi = 45.00$ degrees for each repeat measure-ment and plotted against the applied stress. Two sets of data taken for the (220) planes of Inconel 718 are shown in Figure 5.

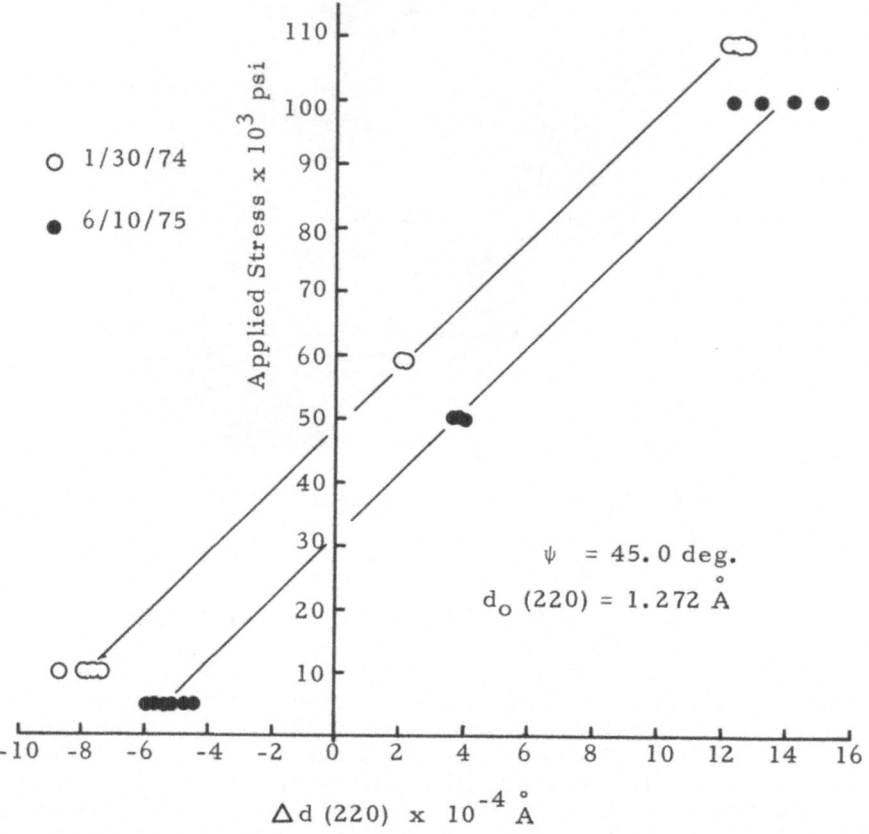

Figure 5. Change In d (220) Versus Applied Stress, Inconel 718, Annealed

A set of data was reduced by determining the slope of the applied stress as a function of the measured change in the lattice spacing, Δd, by linear least-squares regression. The uncertainty in the slope was obtained from the least-squares fit assuming all random error occurs in the determination of the change in the lattice spacing, Δd. The elastic constant is then calculated from the relation:

$$\frac{E}{(1 + \nu)_{(hkl)}} = m^* \sin^2\psi \; d_{o\,(hkl)} , \qquad (4)$$

where m^* is the partial derivative of applied stress with respect to the change in the lattice spacing, i.e., the slope of the data plot; d_o is the unstressed lattice spacing, taken to be the value for $\psi = 0$; and ψ is the range of the angle ψ, 45.0 degrees.

RESULTS AND DISCUSSION

The results obtained to date for iron, nickel, copper, aluminum, and titanium alloys are presented in Table I. The data have been grouped by base alloy and crystal structure. The commercial alloy name, (hkl), wave length, and approximate diffraction angle, 2 θ, are given. The elastic constant $E/(1 + \nu)$, determined in the direction normal to the (hkl) planes, is listed with an uncertainty equal to + one standard deviation. The published mechanically determined value of $E/(1 + \nu)$ is included for comparison, along with the percent difference between the (hkl) and mechanical values, $\Delta\%$. The quantity K_{45} giving the approximate stress required to produce a one degree, 2 θ, shift in the diffraction peak position for a ψ rotation of 45.00 degrees is calculated from the (hkl) value of $E/(1 + \nu)$ as:

$$K = \frac{E}{(1 + \nu)} \quad \frac{1}{\sin^2 \psi} \quad (\frac{\cot \theta_0}{2}) \quad \frac{\pi}{180} \quad . \tag{5}$$

The data in Table I are presented as empirical results, and no attempt shall be made to explain the origins of the anisotropy observed. Several observations can, however, be made concerning the data and technique. First, the most complex alloys, such as Inconels and Incoloys appear to be the most elastically anisotropic. The greatest variation between mechanical and (hkl) values of $E/(1 + \nu)$ occurs for Incoloy 903 and Inconel X750 which differ by 80.3 and 53.9%, respectively in the (220) direction. The large deviation for Incoloy 903 occurs because the elastic constant in the (220) direction agrees well with the values of Inconel 718, but the mechanical value is far lower than any other alloy of the Inconel or Incoloy series. Second, the degree of anisotropy appears to be highest for the lower order planes. Data obtained for Incoloy 800, Inconel 718, and Inconel 600 indicates a greater degree of anisotropy for the (220) direction than in the (331) or (420) directions.

Regarding the repeatability of the technique itself, repeat data taken on the same sample of Inconel 718 in both the (220) and (331) directions at intervals of approximately 18 months shows agreement within the estimated experimental error. These results are shown graphically in Figure 5, and are presented as separate entries in Table I. Measurements on two samples of 422 stainless steel with slight differences in hardness also agree within the estimated error. These repeat measurements on the same sample, and on separate but nearly identical samples appear to indicate that the random error in the determination of $E/(1 + \nu)$ is approximated well by the uncertainty in the calculation of m*.

The major source of systematic error is in the calculation of the applied stress. Care must be taken to determine the effective gage factor using a four-point bending apparatus which does not produce a tensile component. Ball bearing pivoted four-point bending grips of the type used for fatigue testing were found to give excellent repeatable results. Fixtures employing pins in sliding contact with the sample are generally not suitable. The determination of Δd is less susceptible to systematic error. The presence of a residual stress in the sample, or displacement

TABLE I

X-RAY AND MECHANICAL ELASTIC PROPERTIES

Material	(hkl)	λ, K$_\alpha$	2θ(deg.)	$E/(1+\nu)$ x 10^6 psi (hkl)	Mech.	Ref.	Δ%	K$_{45}$ (ksi/deg.)
Iron Base, BCT								
4340, 50 R$_c$	(211)	Cr	156.0	24.5 ± 0.4	22.7	3, 4	+ 7.9	89.3
410 SS, 22 R$_c$	(211)	Cr	155.1	25.6 ± 0.1	22.6	3, 4	+13.2	98.4
410 SS, 42 R$_c$	(211)	Cr	155.1	25.1 ± 0.2	22.6		+11.0	96.7
422 SS, 34 R$_c$	(211)	Cr	154.8	26.4 ± 0.2	22.7	3, 4	+16.3	103.2
422 SS, 39 R$_c$	(211)	Cr	154.8	26.1 ± 0.2	22.7		+14.9	103.4
Iron Base, FCC								
304 SS	(220)	Cr	129.0	20.2 ± 0.6	21.9	5, 6	- 7.7	170.0
Incoloy 903	(220)	Cr	128.0	31.2 ± 0.4	17.3	3, 4	+80.3	264.0
Incoloy 800	(220)	Cr	129.0	23.4 ± 0.6	21.4	4, 6	+ 9.3	196.0
	(420)	Cu	147.0	21.5 ± 0.4	21.4		+ 0.5	110.0
Nickel Base, FCC								
Inconel 718	(220)	Cr	128.0	31.2 ± 0.3	22.7	3, 4	+37.4	263.0
	(220)	Cr	128.0	31.4 ± 0.7	22.7		+38.3	265.0
	(331)	Cu	145.0	19.7 ± 1.0	22.7		-13.2	109.0
	(331)	Cu	145.0	20.3 ± 0.3	22.7		-10.6	112.0
Inconel X750	(220)	Cr	131.0	36.8 ± 1.2	23.9	3	+53.9	301.0

TABLE I (continued)

Material	(hkl)	λ, Kα	2θ(deg.)	$E/(1+\nu) \times 10^6$ psi (hkl)	Mech.	Ref.	Δ %	K$_{45}$ (ksi/deg.)
Inconel 600	(220)	Cr	131.0	21.1 ± 0.5	24.0	3, 4	-12.1	174.0
	(420)	Cu	151.0	23.1 ± 0.1	24.0	3, 4	- 3.7	105.0
Monel K500	(420)	Cu	150.0	21.0 ± 0.3	19.7	3, 4	+ 6.6	98.4
Copper Base, FCC								
85Cu-15Ni	(420)	Cu	146.0	18.6 ± 0.3	13.6	3	+36.8	98.8
Aluminum Base, FCC								
7075	(311)	Cr	139.0	8.83 ± 0.07	7.82	3	+12.9	56.9
Titanium Base, HCP								
Ti-6Al-4V	(213)	Cu	142.0	12.2 ± 0.07	12.3	7	- 0.8	74.0
Ti-6Al-2Sn-4Zr-2Mo	(213)	Cu	140.7	14.8 ± 0.2	12.5	3, 5	+18.4	92.3

of the sample from the center of rotation of the goniometer results in a shift in the intercept of the plot of applied stress versus Δd. The slope of plot as shown in Figure 5, and therefore $E/(1 + \nu)$, is not effected by either sample displacement or the presence of a residual stress.

CONCLUSIONS

The results indicate severe elastic anisotropy in several of the alloys of current interest in the aerospace and nuclear industries. Failure to determine the elastic constant $E/(1 + \nu)$ in the direction normal to the lattice planes employed for stress measurement can result in systematic errors in the measurement of residual stresses in these materials as large as 80%.

REFERENCES

1. Residual Stress Measurement by X-Ray Diffraction, SAE J784a, pp. 12-15, NY: Society of Automotive Engineers, Inc. (1971).

2. H. Zantopulos and C.F. Jatczak, "Systematic Errors in X-Ray Diffractometer Stress Measurements Due to Specimen Geometry and Beam Divergence," Advances in X-Ray Analysis, Vol. 14, pp. 260-376, 1971.

3. Alloy Digest, Upper Montclair, NJ: Engineering Alloys Digest, Inc.

4. Aerospace Structural Metals Handbook, AFML-TR-68-115, Traverse City, MI: Mechanical Properties Data Center, Belfour Stulen Inc., (1975).

5. Titanium Alloys Handbook, MCIC-HB-02, Columbus, Ohio: Metals and Ceramics Information Center, Battelle Columbus Laboratories (1972).

6. Handbook of Engineering Fundamentals, O. W. Eshbach, p. 1332, NY: John Wiley & Sons (1975).

7. Metal Progress Databook 1975, published as Metal Progress, Vol. 108, No. 1 (Mid-June 1975).

THE NEED FOR EXPERIMENTALLY DETERMINED X-RAY ELASTIC CONSTANTS[*]

R. H. Marion

Sandia Laboratories, Albuquerque, New Mexico 87115

J. B. Cohen

Northwestern University, Evanston, Illinois 60201

ABSTRACT

In order to convert residual strains measured by x-ray diffraction techniques into residual stresses, appropriate x-ray elastic constants have to be measured. Since these x-ray elastic constants may depend on the metallurgical state, deformation, and entire specimen history, errors in stress values may result if the constants are not measured for representative material states. In the present work, it is shown that in some cases these errors may be large.

The x-ray elastic constant, $S_2/2 = (1 + \nu)/E$, has been measured for the 211 CrK_α reflection from an Armco iron sample which had been previously deformed by rolling (69 pct. reduction in thickness) and for the 211 CrK_α and 310 CoK_α reflections from a 1045 steel specimen which had been previously elongated in tension. The measured elastic constant for the Armco iron specimen was 40 pct. lower than the value calculated from the average of the Reuss and Voigt values.

INTRODUCTION

In order to relate strains measured by x-ray diffraction techniques to a stress value, one must have appropriate values for the elastic constants. As has been pointed out many times in the literature, the elastic constants determined by mechanical means may not be applicable because of features inherent in the x-ray measurement (1-3). The x-ray technique is inherently selective in that the strain deduced from the change in position of an x-ray diffraction peak represents an average value in a given crystallographic direction for only those grains in the polycrystalline aggregate which are oriented to contribute to the particular x-ray reflection. Therefore, the effective values of E (Young's modulus) and ν (Poisson's ratio) in these orientations may differ from the overall average orientation, the latter being measured in a mechanical test.

[*] This work was supported by the Office of Naval Research and the U. S. Energy Research and Development Administration.

The x-ray elastic constants may be obtained theoretically or experimentally. A rigorous theoretical calculation of the x-ray elastic constants requires a complete theoretical solution of the influence of elastic anisotropy and grain interactions on x-ray strain measurements. Because such a solution has not yet been achieved, various assumptions of the nature of the coupling of the crystallites have been used. The most common are those of Voigt (4), which assumes equal strains in all crystallites, Reuss (5), which assumes equal stresses in all crystallites and Kroner's (6,7) "coupled crystallites" model. A common procedure used by a number of workers in this field is to take the arithmetic average of the Reuss and Voigt values calculated for a material with random crystallite orientation. X-ray elastic constants have also been calculated by using one or more of the above assumptions and considering the effects of nonrandom crystal orientation (8,9) and the effects of more than one phase (10,11).

The x-ray elastic constants calculated by any of the methods described above do not always agree with the experimentally determined values. The magnitude of the disagreement depends on the state of the materials and there is evidence that "constant" x-ray elastic constants may not exist. They may depend on composition and second phase components (2,12), grain size (3), microstructure (3), deformation (13-15), and heat treatment (16). The magnitude of these effects depends on the hkl reflection being considered.

The purpose of this paper is to report experimentally determined x-ray elastic constants which differ substantially from calculated values, to discuss their importance, and to discuss possible reasons for the large disagreements between calculated and measured values.

EXPERIMENTAL PROCEDURE

Specimen Preparation and Deformation

The materials employed in this study were Armco iron and 1045 steel. A 0.89 mm (.035 in) thick flat tensile specimen of 1045 steel (designated 1045-5) and a 0.89 mm (.035 in) thick block of Armco iron (designated Armco-9) were prepared in the manner outlined in (17). The final samples (before deformation) were in the annealed, furnace-cooled condition with dimensions also given in (17). The final annealing was done in a vacuum which was at worst 1.3 mPa (10^{-5} torr). Electropolishing was then performed in a phosphoric-sulfuric acid bath (18). Sample 1045-5 was subsequently elongated on an Instron tensile machine at a strain rate of $\approx 2.7 \times 10^{-4}$/sec to a true strain of 13 pct. It was deformed to the ultimate load with a true stress preceding unloading of 706.7 MPa (102,496 psi). Sample Armco-9 was subsequently reduced in thickness 69 pct. (final thickness = .267 mm (0.0105 in)) by rolling on a two-high mill driven at 31 rpm (roll diameter = 133.4 mm (5.25 in)). The reduction in thickness was 0.1 mm (0.004 in) per pass and the sample was reversed end for end after each pass. After the rolling deformation, a tensile specimen with the same dimensions as 1045-5 was carefully machined with excess coolant.

These materials and deformation history were chosen because a previous study (17) showed the rolled Armco iron to possess large oscillations in d_ψ vs. $\sin^2\psi$ (ψ and d_ψ are defined in the next section) and the tensile deformed 1045 steel did not. Therefore, these two specimens should provide information on the usefulness of the calculated x-ray elastic constants for samples which do or do not satisfy the classical linear d_ψ vs. $\sin^2\psi$ requirement.

Measurement of X-Ray Elastic Constants

The basic relation for x-ray stress analysis written for a uniaxial stress state, (surface stress = σ), is (2):

$$\epsilon_\psi = \frac{S_2}{2}\,\sigma\,\sin^2\psi + S_1\sigma = \left(\frac{\Delta d}{d}\right)_\psi = \frac{d_\psi - d_o}{d_o} \tag{1}$$

where ψ is the angle from the sample normal in the plane defined by the sample normal and the direction of the stress (it is the angle of tilt of the specimen away from the usual diffraction position for which the incident and diffracted beam make equal angles with the sample surface), ϵ_ψ is the strain in the direction defined by ψ, d_ψ is the lattice spacing in the direction defined by ψ, d_o is the lattice spacing in the unstressed state and S_1 and $S_2/2$ are elastic constants given by:

$$S_1 = -\frac{\nu}{E}\quad,\quad \frac{S_2}{2} = \frac{1+\nu}{E} \tag{2}$$

To measure the x-ray elastic constants, a uniaxial tensile test has to be performed within the elastic range (on the diffractometer). From measurements of d_ψ vs. $\sin^2\psi$ in the plane given by the sample normal and the applied load with different known values of the applied tensile stress $\sigma = \sigma_{app}$, one obtains m^* (given by Eq. (3)) and $d_{\psi=0}$ as a function of σ_{app}.

$$m^* \equiv \left(\frac{S_2}{2}\right)\sigma = \frac{\partial\epsilon_\psi}{\partial\sin^2\psi} = \frac{1}{d_o}\frac{\partial d_\psi}{\partial\sin^2\psi} \tag{3}$$

Applying these values in the partial differentiation of Eq. (3) and Eq. (1) written for $\psi = 0$, one obtains the x-ray elastic constants as:

$$\frac{S_2}{2} = \frac{\partial m^*}{\partial\sigma_{app}} \tag{4}$$

$$S_1 = \frac{1}{d_o}\frac{\partial d_{\psi=0}}{\partial\sigma_{app}} \tag{5}$$

Since d_ψ, $d_{\psi=0}$ and d_o rarely differ by more than 1 pct. (and since d_o in Equations 3-5 is a multiplier and not used in a difference), the value of d_o has been chosen equal to $d_{\psi=0}$ in this work.

All of the x-ray measurements were made with a Picker diffractometer
equipped with filtered radiation, a scintillation detector and a pulse-
height analyzer set for 90% acceptance of the K_α peak. The peak
position at a ψ value other than 0° was measured by the parafocus
method (18) in which the receiving slit is moved into the calculated
focal point. A fixed vertical slit at the stationary counter assured
that the same range of orientations was examined at each position.
The ω motion on the diffractometer was employed for the ψ rotation.
The method of positioning the sample to within ± .025 mm (± .001") of
the center of the goniometer is described in (17-19). The peak position
was determined to within ± .005° 2θ by performing a least squares fit
to a parabola for 10-20 data points with intensity greater than 85 pct.
of the maximum intensity (for each data point the time necessary to
accumulate 100,000 counts was measured).

Deformation on the X-Ray Unit

A small tensile unit for use on the diffractometer was constructed
and is shown in Figure 1. It can apply a uniaxial load on a tensile
specimen but the magnitude of the load had to be determined in some
other way. This was done as follows: A strain gage was applied to the
back of the tensile bar on which the measurements were to be made. As
the specimen was stressed, a value of strain could be recorded. The

Fig. 1. Tensile device for the x-ray diffractometer. (Shown in the
vertical position--the results presented in this study were obtained
with the load applying part rotated 90° about the normal to the specimen
surface.)

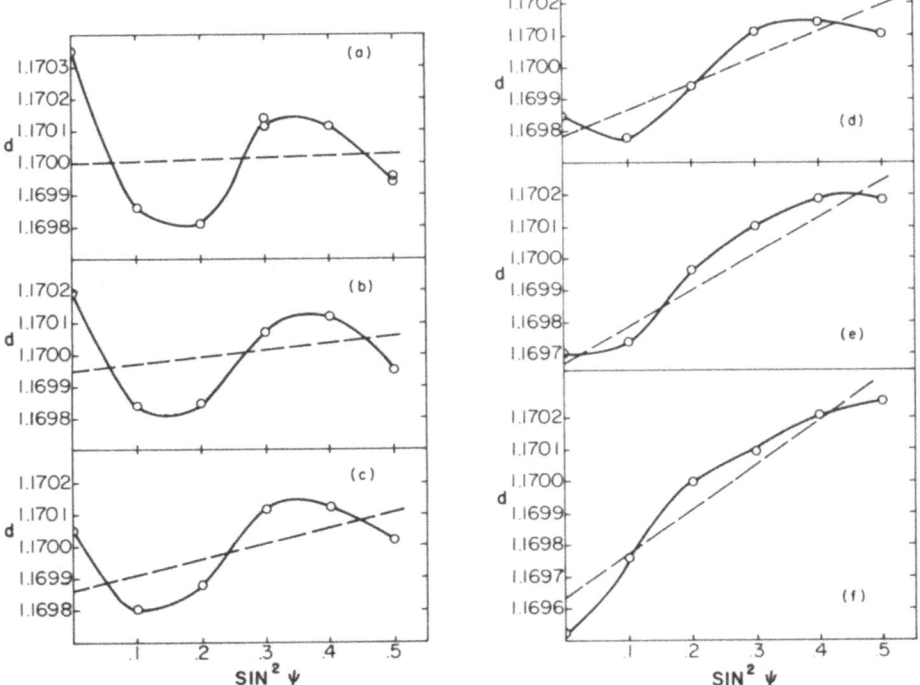

Fig. 2. The dependence of d (or d_ψ) on $\sin^2\psi$ for various applied loads for specimen Armco-9 previously reduced 69 pct. by rolling; 211 peak with CrK_α; ——— curve through experimental points; – – – represents slope obtained with the Marion-Cohen method described in Ref. (17). a) σ_{app} = 0; b) σ_{app} = 46.0 MPa (6667 psi); c) σ_{app} = 99.8 MPa (14,476 psi); d) σ_{app} = 191.8 MPa (27,810 psi); e) σ_{app} = 269.3 MPa (39,048 psi); f) σ_{app} = 375.7 MPa (54,476 psi).

load corresponding to this strain was determined by taking the same sample, strain gage and grips and loading it on an Instron testing machine until that value of strain was reached. The track in which the grips slide has been precisely machined to eliminate bending. The maximum bending strain was measured to be 25 μm/m. The gearing at the end of the unit makes it possible to apply the load very easily without deflecting the unit. The tensile unit is mounted so that the load applying rig can be rotated about the normal to the specimen surface. At the base of the support is a micrometer adjustment to allow for accurate specimen positioning.

RESULTS

The dependence of d_ψ on $\sin^2\psi$ for the 211 CrK_α reflection from sample Armco-9 is given as a function of applied load in Figures 2a-2f. A complete discussion of the oscillations in d_ψ vs. $\sin^2\psi$ is given in Ref. (17), where a method is presented for obtaining the true slope (or

m*), independent of the oscillations. This method was used to subtract
out the effects of the deviations from linearity and obtain the slopes
represented by the dashed lines in the figure. The resulting values of
m* can then be plotted vs. σ_{app} as is shown in Figure 3. All of the
points lie on a good straight line except for the last point. At this
load the sample was apparently no longer behaving in an elastic manner--
that is, it was microscopically plastic even though it was still below
the macroscopic yield stress of 703.9 MPa (102,095 psi). This is substantiated by the observation that the oscillations in d_ψ vs. $\sin^2\psi$
reversed. Therefore, this point was not included in the x-ray elastic
constant determination. A least-squares line was passed through the
first five points to obtain the value of $S_2/2$ given in Table I. The
standard deviation given is the standard deviation of the slope obtained
from the fit. The "texture-independent directions" approach of Hauk
et al. (20) to obtain m* was tried with these data. This method was
considered inadequate because it resulted in a nonlinear dependence of
m* on σ_{app} and $S_2/2$ determined from a best fit straight line was
excessively large.

A value of the other elastic constant, S_1, cannot be obtained
from these data because only the slope of the straight line (m*) can be
obtained from the techniques described in (17) and not its intercept.
Therefore, since $d_{\psi=0}$ cannot be obtained as a function of applied load,
S_1 cannot be determined from Eq. (5). This is not a severe limitation,
however, because $S_2/2$ is the only elastic constant needed in the commonly
used "$\sin^2\psi$-method".

The experimental results for d_ψ vs. $\sin^2\psi$ for sample 1045-5 are
given in Figures 4 and 5 for the 211 CrK$_\alpha$ and the 310 CoK$_\alpha$ reflections,
respectively. Since this sample had little or no oscillations in d_ψ vs.

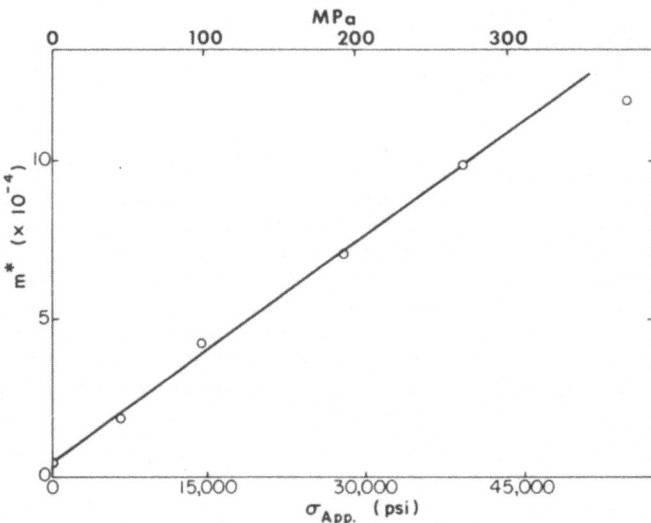

Fig. 3. m* vs. applied load (σ_{app}) for specimen Armco-9 previously
reduced 69 pct. by rolling; 211 peak with CrK$_\alpha$. ——— least-squares
line through the first five points.

Table I.   X-Ray Elastic Constants

| | hkl | Units | Experimentally Determined[+] | | Theoretical | | |
			ARMCO-9, Previously Reduced 69 pct. by Rolling	1045-5, Previously Elongated to a True Strain of 13 pct.	Reuss	Voigt	Average of Reuss and Voigt
$\frac{S_2}{2}$	211	10^{-8} MPa^{-1}	349.5 (± 10.2)	506.1 ± 16.0)	593.1	567.0	580.0
		10^{-8} psi^{-1}	2.41 (± .07)	3.49 (± .11)	4.09	3.91	4.00
	310	10^{-8} MPa^{-1}		723.6 (± 23.2)	904.8	567.0	736.6
		10^{-8} psi^{-1}		4.99 (± .16)	6.24	3.91	5.08
S_1	211	10^{-8} MPa^{-1}		-103.0 (± 7.3)	-117.5	-108.8	-113.1
		10^{-8} psi^{-1}		-0.71 (± .05)	-0.81	-0.75	-0.78
	310	10^{-8} MPa^{-1}		-158.1 (± 10.2)	-221.9	-108.8	-165.3
		10^{-8} psi^{-1}		-1.09 (± .07)	-1.53	-0.75	-1.14

[+]Numbers in parenthesis are standard deviation.

$\sin^2\psi$, it wasn't necessary to obtain d_ψ at as many ψ values as done previously. Even though the deviations from linearity were small in this sample, the Marion-Cohen method described in (17) was used to obtain m^* because more accurate values could be obtained. A plot of m^* vs. σ_{app} is given in Figure 6 for both 211 and 310 reflections. The lines drawn on the figure are a least-squares fit and from the slope of these lines the values given in Table I were obtained for the elastic constant $S_2/2$.

Since the deviations from linearity were not large in this sample, the other elastic constant, S_1, could be determined because the positioning of the straight line and the determination of the intercept ($d_{\psi=0}$) could be done relatively accurately. A plot of $d_{\psi=0}$ vs. σ_{app} is given in Figure 7 and it can be seen that deviations from a straight line are quite small. A least-squares fit was performed and the x-ray elastic constant, S_1, given in Table I was obtained from the slope of this line (see Eq. (5)).

DISCUSSION AND CONCLUSIONS

The difference between the measured and the theoretical x-ray elastic constants can be seen in Table I. Since the most commonly used theoretical x-ray elastic constants are the average of the Reuss and Voigt values, they will be used in this discussion. The measured x-ray elastic constant, $S_2/2$, for the 211 reflection from the heavily rolled Armco iron sample is 40 pct. lower than the average of the Reuss and Voigt values. The measured constants for the tensile deformed 1045 steel are closer to the calculated values: The experimental $S_2/2$ value for the 211 reflection does not lie within the Reuss and Voigt limits and is 13 pct. lower than the average of the two values; the value for the 310 reflection is only 2 pct. lower than the average calculated

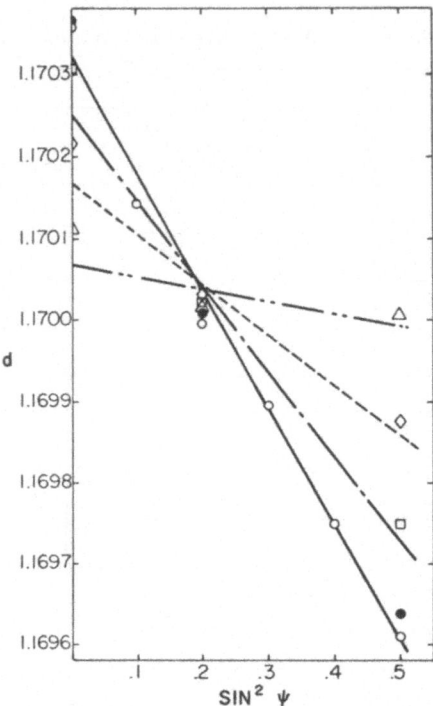

Fig. 4. The dependence of d (or d_ψ) on $\sin^2\psi$ for various applied loads for specimen 1045-5 previously elongated to a true strain of 13 pct.; 211 peak with CrK_α; the straight lines represent the slope obtained with the Marion-Cohen method described in Ref. (17).

● ——————————— σ_{app} = 0 (○ → Ref. (17))

□ ————— - ————— σ_{app} = 65.5 MPa (9500 psi)

◊ — — — — — — — — σ_{app} = 137.9 MPa (20,000 psi)

△ ————— - - ————— σ_{app} = 211.0 MPa (30,600 psi)

value and if the standard deviation is considered, there is no difference. The experimental results for the other x-ray elastic constant, S_1, behave in a manner similar to $S_2/2$.

The 40 pct. difference between the calculated and experimental value of $S_2/2$ for the rolled Armco iron sample may seem large but a number of large changes in the experimentally determined elastic constants have been reported in the literature. S. Taira et al. (13) have reported a decrease in $S_2/2$ with increasing plastic deformation in iron with 0.01 pct. carbon. At a plastic strain of 20 pct., $S_2/2$ for the 211 reflection had decreased by 36 pct. from its annealed value. Prümmer and Macherauch (14) found a 27 pct. decrease in $S_2/2$ (for the 211 reflection) with deformation for a uniaxially deformed sample of 0.86 pct. C steel. Esquivel (15) has reported a 30-45 pct. decrease

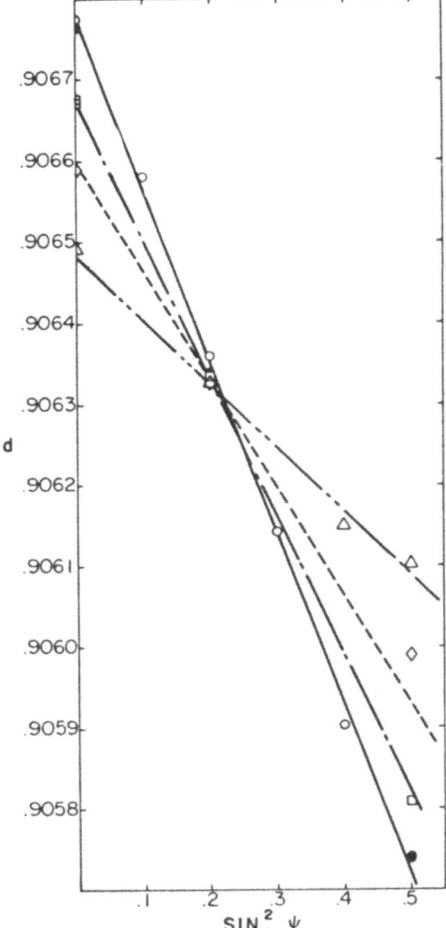

Fig. 5. The dependence of d (or d_ψ) on $\sin^2\psi$ for various applied loads for specimen 1045-5 previously elongated to a true strain of 13 pct.; 310 peak with CoK_α; the various straight lines represent the slope obtained from the Marion-Cohen method described in Ref. (17).

● ———————————— $\sigma_{app} = 0$ ($\bigcirc \rightarrow$ Ref. (17))

□ ——— — ——— $\sigma_{app} = 59.3$ MPa (8,600 psi)

◇ — — — — — — — — $\sigma_{app} = 128.3$ MPa (18,600 psi)

△ ——— — — ——— $\sigma_{app} = 201.4$ MPa (29,200 psi)

in the elastic constants measured after uniaxial plastic deformation for several hardened steels.

The rolled Armco iron sample used in this study was highly textured and probably had a highly deformed microstructure. Therefore, it

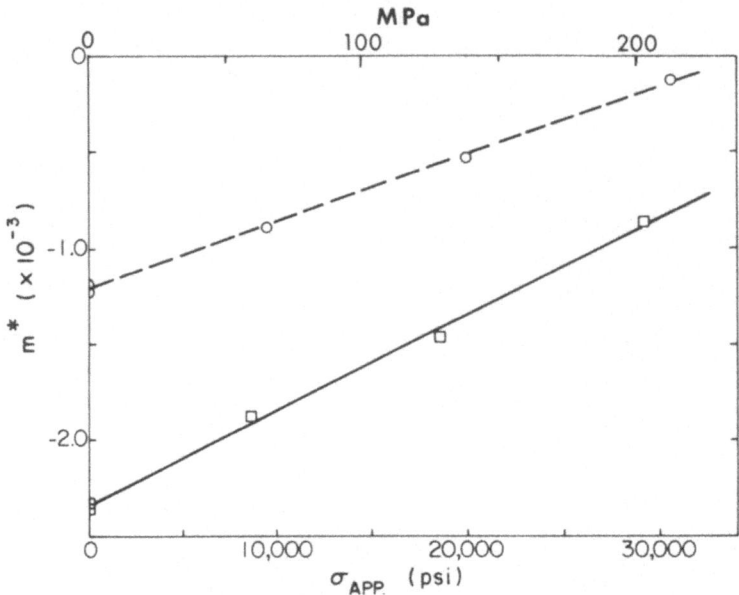

Fig. 6.   m* vs. applied load (σ_{app}) for specimen 1045-5 previously elongated to a true strain of 13 pct.

○ — — — — — — — —   211 peak with CrK$_\alpha$

□ ————————————   310 peak with CoK$_\alpha$

is quite likely that the x-ray elastic constants will decrease substantially from their annealed value because only certain regions (the coherently diffracting subgrain interior regions) contribute to the peak and they may be straining in a different manner than in the annealed state. (For a more complete discussion see Ref. (17).) Macherauch and Müller (21) have determined $S_2/2$ for the 211 reflection from Armco iron which had been reduced 75 pct. by cold rolling and subsequently annealed for 4 hours at 500°C. They obtained $S_2/2 = 594.5 \times 10^{-8}$ MPa^{-1} (4.10 x 10^{-8} in^2/lb). If one assumes that the values are suitable for comparison (similar grain size, etc.), it can be seen that the substructure has probably played a large role in the decrease observed in the results reported here. Additional support for the importance of the microstructural state is given by Fuks and Belozerov (22). They elastically loaded 15-20 μm thick condensed nickel films on the diffractometer and observed that as the size of the coherently diffracting regions increased, $S_2/2$ decreased from 938.8×10^{-8} MPa^{-1} to 591.8×10^{-8} MPa^{-1} for the 400 reflection and remained essentially constant for the 222 reflection. They also found that the anisotropy decreased as the size of the substructural elements increased.

The measured elastic constants for the tensile deformed 1045 steel are similar to those reported in the literature (2). The reason that

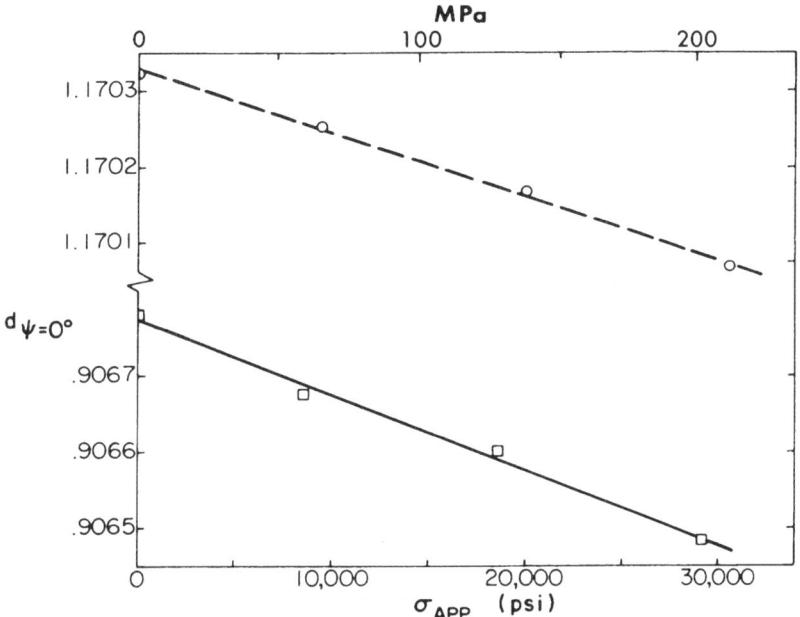

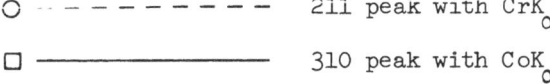

Fig. 7. $d_{\psi=0}$ vs. applied load (σ_{app}) for specimen 1045-5 previously elongated to a true strain of 13 pct.

○ — — — — — — —   211 peak with CrK$_\alpha$

□ ——————————   310 peak with CoK$_\alpha$

$S_2/2$ for the 211 reflection is farther below the average of the Reuss and Voigt values than the 310 value is probably because the 211 direction is a "softer" direction than the 310 direction and is affected more by deformation. This is substantiated by the experimental observation that residual strains measured with the 310 reflection are always higher than those measured with the 211 reflection (2,17).

Based on the results presented here, it is obvious that if one is measuring residual stresses which originate with deformation, care should be taken to use appropriate x-ray elastic constants. As mentioned previously, additional results in the literature demonstrate large potential errors due to other causes (Prümmer (16) has reported a 25-53 pct. difference between the experimentally measured x-ray elastic constants for the 211 reflection of hardened and annealed steels of the same composition; and, in plain carbon steels $S_2/2$ for the 211 reflection increases 30 pct. as the carbon increases from 0.03 pct. C to 1.0 pct. C (2).) All this experimental evidence demonstrates the necessity of having the experimental x-ray elastic constants for a specimen exactly the same (same composition, grain size, heat treatment, deformation history) as the material being studied. To date, theoretical calculations have not been able to explain these experimental results and are

therefore considered to be inaccurate. Residual stress errors as large
as 50 pct. may result if the wrong x-ray elastic constants are used.

ACKNOWLEDGEMENTS

This research was sponsored by the Office of Naval Research and
the U. S. Energy Research and Development Administration. Portions of
the work were submitted (by R. H. M.) in partial fulfillment of the
requirements for the Ph.D. degree at Northwestern University in August
1972. Mr. W. P. Evans, Caterpillar Tractor Co., kindly provided the
1045 steel specimens.

REFERENCES

1. C. S. Barrett and T. B. Massalski, Structure of Metals, 3rd Edition,
    McGraw-Hill, New York (1966).

2. E. Macherauch, "X-Ray Stress Analysis," Exp. Mech., 6, 140-153
    (1966).

3. G. Faninger, "Gitterdehnungen in Verformten Kubischen Metallen,"
    J. Soc. Mat. Sci., Japan, 19, 42-57 (1970).

4. W. Voigt, Lehrbuch der Kristallphysik, Teubner, Leipzig/Berlin,
    (1928).

5. A. Reuss, "Calculation of Flow Limits of Mixed Crystals on Basis
    of Plasticity of Single Crystals," Z. Angew. Math. Mech., 9,
    49-58 (1929).

6. E. Kroner, "Berechnung der Elastischen Konstanten des Vielkristalls
    aus den Konstanten des Einkristalls," Z. Phys., 151, 504-508,
    (1958).

7. F. Bollenrath, V. Hauk, and E. H. Müller, "Zur Berechnung der
    Vielkristallinen Elastizitätskonstanten aus den Werten der
    Einkristalle," Z. Metallkunde, 58, 76-82 (1967).

8. S. Taira and K. Hayashi, "X-Ray Investigation of Polycrystalline
    Metals (On the Effect of Fiber Texture on the Elastic Constants of
    α-Iron)," Proc. 13th Japan Cong. on Materials Research, 20-24
    (1970).

9. P. D. Evenschor and V. Hauk, "Röntgenographische Elastizitätskon-
    stanten und Netzebenenstandsverteilungen von Werkstoffen mit
    Texture," Z. Metallkunde, 66, 164-166 (1975).

10. J. Arima, N. Hosokawa and K. Honda, "Elastic Deformation Behavior
    of Two Phase Alloy," X-Ray Study on Strength and Deformation of
    Metals (Proceedings of the Seminar), 1-8, (1971), Tokyo, Japan,
    (The Society of Materials Science, Japan).

11. P. D. Evenschor and V. Hauk, "Berechnung der Röntgenographischen Elastizitätskonstanten von Mehrstoffsystemen," Z. Metallkunde, 66, 210-213 (1975).

12. R. Prümmer and E. Macherauch, "Der Verformungseinfluss auf die Röntgenographischen Elastischen Konstanten und die Oberflächeneigenspannungen un Legierter und Chromlegierter Stähle," Z. Naturforschg., 21A, 661-662 (1966).

13. S. Taira, K. Hayashi and Z. Watase, "X-Ray Investigation on the Deformation of Polycrystalline Metals (On the Change in X-Ray Elastic Constants by Plastic Deformation)," Proc. 12th Japan Cong. on Materials Research, 1-7 (1969).

14. R. Prümmer and E. Macherauch, "Zur Frage des Wellenlängeneinflusses auf die Röntgenographische Eigenspannungsbestimmung," Z. Naturforschg., 20A, 1369-1370 (1965).

15. A. L. Esquivel, "X-Ray Diffraction Study of the Effects of Uniaxial Plastic Deformation on Residual Stress Measurements," Adv. in X-Ray Analysis, 12, 269-298 (1969).

16. R. Prümmer, "Röntgenographische Eigenspannungsanalyze bei Gehärteten Stählen," Proc. 6th Int'l. Conf. on Nondestructive Testing, Hanover, Germany (1970).

17. R. H. Marion and J. B. Cohen, "Anomalies in Measurement of Residual Stress by X-Ray Diffraction," Adv. in X-Ray Analysis, 18, 466-501 (1974).

18. Society of Automotive Engineers, Residual Stress Measurement by X-Ray Diffraction, SAE J784a, (1971), (Society of Automotive Engineers, Inc., New York, NY).

19. R. H. Marion, Ph.D. Thesis, Northwestern University, Evanston, Illinois (1972).

20. V. Hauk, D. Herlach and H. Sesemann, "Über Nichtlineare Gitterebenenabstandsverteilungen in Stählen, ihre Entstehung, Berechnung und Berücksichtigung bei der Spannungsermittlung," Z. Metallkunde, 66, 734-737 (1975).

21. E. Macherauch and P. Müller, "Determination of the Röntgenographic Values of the Elastic Constants of Cold-Stretched Armco Iron and Chromium-Molybdenum Steel," Arch. Eisenhüttenwesen, 29, 257-260 (1958).

22. M. Ya. Fuks and V. V. Belozerov, "Crystallographic Anisotropy of Elastic Strain of Crystals in Polycrystalline Specimens," Fiz. Metal. Metalloved, 34, 107-113 (1972).

A MODIFIED DIFFRACTOMETER FOR X-RAY STRESS MEASUREMENTS

E. Macherauch and U. Wolfstieg

Institut für Werkstoffkunde I

Universität Karlsruhe (TH)

D-7500 Karlsruhe, West Germany

ABSTRACT

Usually, the classical Bragg-Brentano diffractometer arrangement is used for X-ray strain measurements. For measurements in different ψ-directions, the specimen is rotated clockwise or counter-clockwise around the diffractometer axis using a separate driving system (ω-diffractometer). During the measurements, primary and reflected X-ray beams are placed in the same plane as the normal of the specimen and the normals of the measured $\{h,k,l\}$ -planes. However, the Bragg-Brentano principle also can be applied to strain measurements in different ψ-directions with a number of advantages if the specimen is rotated around an axis lying parallel to the diffractometer plane (ψ-diffractometer) (1). The principle, construction and characteristics of the ψ-diffractometer are described and compared with those of the ω-diffractometer. An example of the application of the Karlsruhe-type of the ψ-diffractometer is given.

INTRODUCTION

X-ray stress analysis is based on the measurement of lattice strains, that is, measurements of the relative changes dD/D_0 of the lattice spacings D_0 of certain sets of $\{h,k,l\}$ - planes in special oriented surface grains of polycrystalline samples, due to the influence of internal and/or external stresses. In the simplest case of a uniaxial state of stress given by the principal stress σ, the theory of linear elasticity--as indicated in the left part of Figure 1--yields in a macroscopic plane fixed by the normal to the surface of the specimen, L, and the stress σ, the strain is given by

$$\varepsilon_\psi = [\frac{\nu + 1}{E} \sin^2\psi - \frac{\nu}{E}] \sigma \tag{1}$$

where ψ is the angle between L and the direction of strain, E is Young's modulus and ν Poisson's ratio. The engineering strains ε_ψ are dependent on $\sin^2\psi$ in a linear manner.

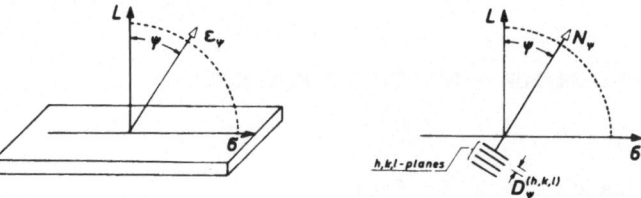

Engineering strain:

$$\varepsilon_\psi = \left[\frac{1+\nu}{E}\sin^2\psi - \frac{\nu}{E}\right]\cdot \delta$$

Lattice strain:

$$\frac{D_\psi^{(h,k,l)} - D_0^{(h,k,l)}}{D_0^{(h,k,l)}} = \left(\frac{dD}{D_0}\right)_\psi^{(h,k,l)} = -ctg\,\Theta_0\,(d\Theta)_\psi^{(h,k,l)}$$

$$\varepsilon_\psi \cong \left(\frac{dD}{D_0}\right)_\psi^{(h,k,l)}$$

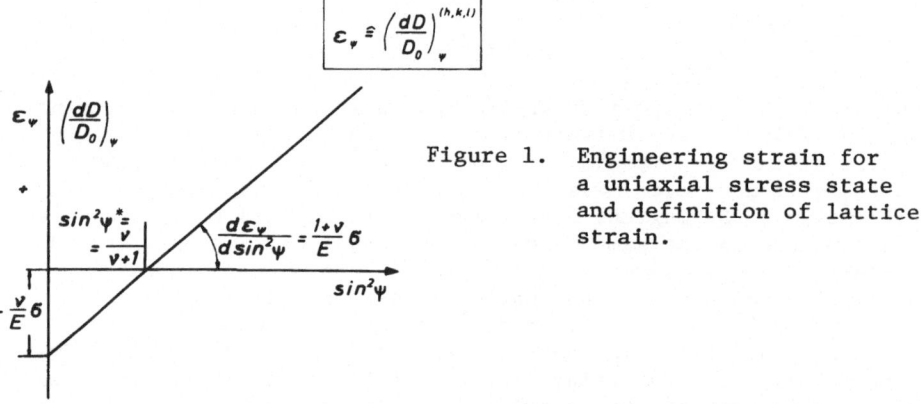

Figure 1. Engineering strain for a uniaxial stress state and definition of lattice strain.

As noted in Figure 1, lattice strains $(dD/D_0)_\psi^{(h,k,l)}$, defined by the equation

$$\left(\frac{dD}{D_0}\right)_\psi^{(h,k,l)} = \frac{D_\psi^{(h,k,l)} - D_0^{(h,k,l)}}{D_0^{(h,k,l)}} = -\cot\theta_0 \cdot (d\theta)_\psi^{(h,k,l)} \quad (2)$$

can easily be measured by means of X-rays, from the appropriate shifts $(d\theta)_\psi^{(h,k,l)}$ in the interference line (h,k,l) in directions $\psi \lesssim 60°$. $D_\psi^{(h,k,l)}$ is the lattice spacing of $\{h,k,l\}$ -planes in the stressed state, $D_0^{(h,k,l)}$ the lattice spacing and θ_0 the Bragg-angle in the stress-free state. In doing so, the spacings of the $\{h,k,l\}$-planes of some grains near the surface of the specimen under investigation are determined, whose normals N_ψ coincide with the direction ψ. Assuming the strains derived from the theory of elasticity (Eq. 1) are identical with the actual lattice strains (Eq. 2), the identity

$$\left(\frac{dD}{D_o}\right)^{(h,k,1)}_{\psi} = \varepsilon_{\psi} \qquad (3)$$

is obtained. As shown in the lower part of Figure 1, that means that
the lattice strains are also a linear function of $\sin^2\psi$ in the macroscopic plane that contains L and σ. Then the stress σ can be calculated from the slope $d\varepsilon/d \sin^2\psi$ of the linear relationship of ε_{ψ} vs.
$\sin^2\psi$.

$$\sigma = \frac{E}{1 + \nu} \cdot \frac{d\varepsilon_{\psi}}{d \sin^2\psi} \qquad (4)$$

This is the basis of the $\sin^2\psi$ method of X-ray stress analysis (2)
which has now gained worldwide application. It requires the strain
distribution in a cross-section plane of the strain ellipsoid to be
determined as exactly as possible. Consequently, strain measurements
are necessary in several directions ψ_i if sufficiently accurate information about the stress is to be obtained.

LATTICE STRAIN MEASUREMENTS

If X-ray diffractometer equipment is available, lattice strains
are usually determined with the classical Bragg-Brentano arrangement
(3) as sketched in Figure 2. The X-ray source or the entry slit F and
the radiation detector or the
detector slit D lie on a circle,
the diffractometer circle. The
specimen is mounted in the center
of the diffractometer. The
stress component of interest, σ,
and the normals of the reflecting
lattice planes N_{ψ} as well as the
entry slit F and the detector
slit D lie in the same macroscopic plane. In order to record
the interference line of the
{h,k,1} - planes, the specimen is
rotated at the angle speed ω and
the detector at 2ω around the
diffractometer axis which passes
through the point O in the sketch
and is perpendicular to the plane
of the sketch. The position of
the radiation source is held constant.

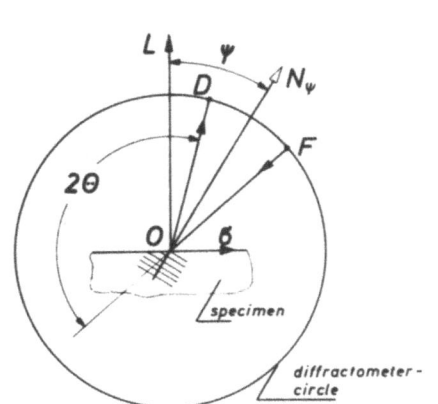

Figure 2.   Schematic illustration
of the Bragg-Brentano
arrangement used in
X-ray strain measurement .

GEOMETRICAL CONSIDERATIONS

It can be easily understood that the lattice strain ε_{ψ} in a direction ψ can also be measured by using other positions of the radiation
source F and the detector slit D. As shown in Figure 3, the lattice

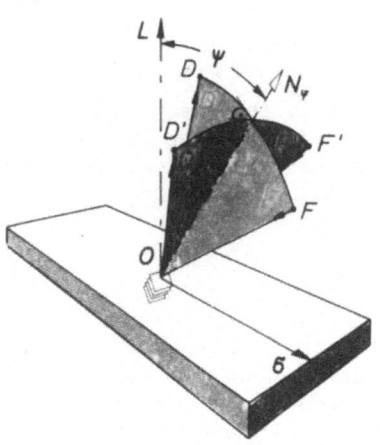

Figure 3. Different arrange-
    ments of radiation source
    F resp. F' and detector
    slit D resp. D' to measure
    the same lattice strain ε_ψ.

strain ε_ψ can also be determined, for
example, if the radiation source and
the detector slit are in positions
F' and D'. F'N$_\psi$D' lie in a macroscopic plane perpendicular to the
plane LDN$_\psi$F.

Instead of varying the positions
of F and D it is of course possible
to alter the specimen orientation
only, so as to give the equivalent
relationship between L, ψ, D', and
F'.

There are two convenient ways
for measuring the lattice plane spacings normal to N$_\psi$. One is as indicated in Fig. 2 and by F and D in
Fig. 3; the other is as indicated by
F' and D', Fig. 3. The former is
used in ordinary diffractometers,
with the ψ rotation of the specimen
around the diffractometer axis being
done with a separate drive called
the ω drive; one refers to these as ω-diffractometers. The other type,
using F' and D', Fig. 3, requires that the specimen be rotated by an
additional drive, called the ψ drive, about an axis perpendicular to
the usual diffractometer axis, the ψ axis. Therefore it has been called
the ψ-diffractometer.

COMPARISON BETWEEN ω- AND ψ-DIFFRACTOMETER

The ω- and ψ-diffractometer setups are schematically compared in
Figures 4a, b, c for lattice strain measurements in the directions
$\psi = 0$ (Figure 4a), $\psi = +45°$ (Figure 4b) and $\psi = -45°$ (Figure 4c). Obviously, there are characteristic differences between the two arrangements. Table I summarizes the main features of both diffractometers.
From this comparison it can be summarized that, compared with the
classical ω-diffractometer, the ψ-diffractometer setup is of considerable advantage. The following characteristics should especially be emphasized since they are quite important for various problems in X-ray
strain measurements:

(1) Symmetric path length of the incident and reflected X-rays in the
     specimen independent of θ, ψ and the vertical divergence.

(2) Small and symmetric defocusing effects; parafocusing movements of
     the detector not necessary.

(3) Soller slits not required if point source is used.

(4) Absorption factor independent of θ and ψ; no absorption corrections
     required.

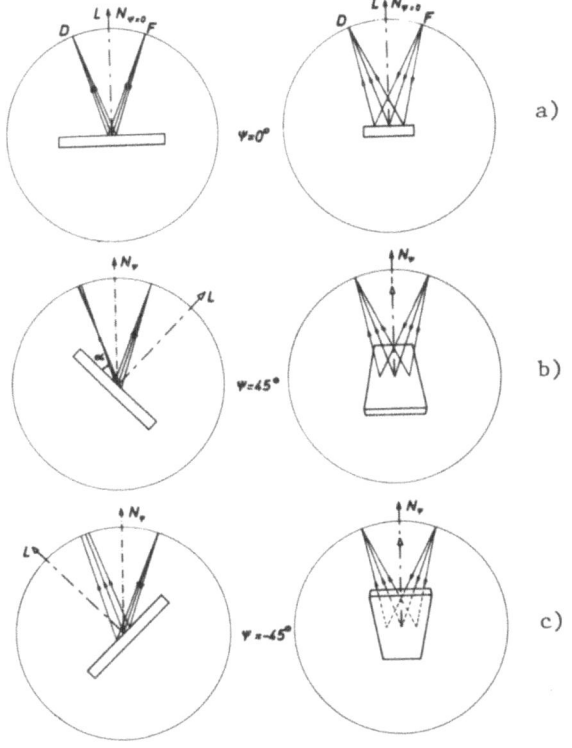

Figure 4. The geometrical
conditions for measur-
ing lattice strains
with ψ- and ω-diffrac-
tometers in the direc-
tions $\psi = 0°$ and $\psi = \pm 45°$.

(5) Ratio of penetration depths at $\psi = 0$ and $\psi = \psi_i$ independent of θ.

(6) Large usable θ-range without restrictions of the ψ-range.

THE KARLSRUHE TYPE ψ-DIFFRACTOMETER

Figure 5 shows an experimental setup which has been developed at Karlsruhe to realize the geometrical conditions discussed above. A high-load fine-focus X-ray tube (X) with an effective focus size of 0.4 · 0.8 mm is used. The entry slit system is built in the interior of a holding device T. The radiation is measured by scintillation counter C with an appropriate slit system S. Intensities are registered by either a ratemeter or a counter. Further radiation processing is done with customary recording instruments and by computer. The radius of the diffractometer circle is 286.5 mm. Thus, 5 mm in the circumference of the diffractometer circle corresponds to 1° of 2θ. With the specimen holder (H) or bending device holder the specimens' surface areas to be measured can be placed exactly on the optical center of the device. The ψ-variations around an axis parallel to the diffractometer plane are supplied with a step motor (M).

Attached to the ψ-diffractometer shown in Figure 5 is a double bending apparatus by means of which the surface area in the center of a cross-shaped specimen can be biaxially loaded by two lever arms. Measurements with any combination of applied tensile and compressive

Table I.   Characteristic features of the ω- and the ψ-diffractometers

Characteristics	ω-diffractometer	ψ-diffractometer
Plane of measurement	L,σ plane in the diffractometer plane	L,σ plane perpendicular to the diffractometer plane
Light path in the diffractometer plane	unsymmetrical	symmetrical
Irradiated area of specimen in the diffractometer plane	dependent on ψ	independent of ψ
Horizontal divergence	critical in respect to the irradiated surface during ψ variation	uncritical in respect to ψ-variation, can be chosen at random
Vertical divergence	can be chosen at random	must be kept slight
Defocusing	large, unsymmetrical, increases with increasing ψ	small, symmetrical increases with increasing ψ
ψ-rotation in clockwise and counterclockwise direction	unequal	equal
Ratio of penetration depths t_0/t_ψ in directions $\psi = 0$ and ψ	strongly dependent on ψ with decreasing θ	slightly dependent on ψ, independent of θ.
Absorption correction	mandatory	unnecessary
Soller slits	mandatory	unnecessary
Possible recording speeds	relatively small	relatively large
Maximum measurable Bragg-angle θ_{max}	81°	86.5°
Minimum measurable Bragg-angle θ_{min}	$\theta > \alpha + \psi \simeq 65°$	$\theta > \arc\sin \left(\dfrac{\sin \alpha}{\cos \psi}\right) \simeq 30°$

(where α is the angle between the specimen surface and the incident X-ray beam. The data are valid for $\alpha > 20°$ and $\psi < 45°$)

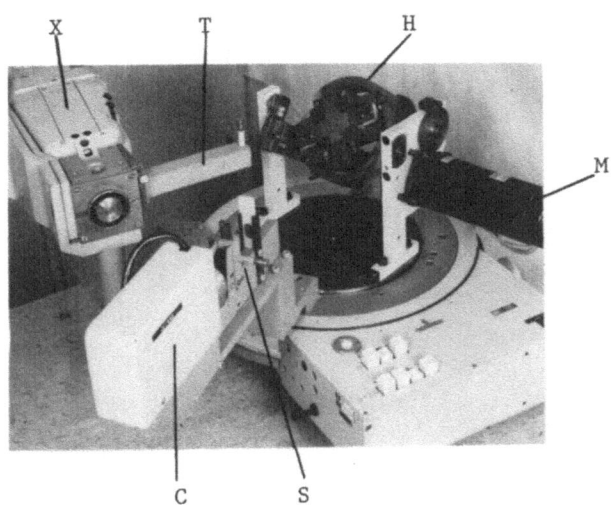

Figure 5. A Karlsruhe type ψ-diffractometer with a device for measuring biaxial bending stress states.

stresses in the surface layers of the specimen are possible. In addition, by rotating the bending device around an axis perpendicular to the surface of the specimen, the lattice strain distributions can be measured in randomly chosen cross-section planes of the strain ellipsoid.

SOME EXPERIMENTAL RESULTS

The usefulness of the ψ-diffractometer comes to light especially when using Kα-radiation with short wavelength for lattice strain measurements. As an example, Figure 6 shows the interference lines recorded for a plain carbon steel with 0.15 wt.% C at $\psi = 30°$ using molybdenum Kα-radiation. As can be seen, there is no problem in evaluating the interferences 521, 440, 530/433, 600/442, 611/532, 620, 541, 622, 631, 444, 710/550/543, 640, 721/633/552, 642, 730 and 732/651 with clear separation of the Kα-doublets. It is obvious, for example, that problems as to the influence of elastic anisotropy (4) and/or texture (5) on lattice strain distribution can be studied more thoroughly and advantageously this way than was previously the case.

As another example of the usefulness and applicability of the ψ-diffractometer, the complete measurement of a state of plane stress will be discussed. The study of this previously unexamined problem is of fundamental importance for the assessment of X-ray stress analysis. By means of the ψ-diffractometer described above the entire surface stress distribution of a biaxially bent plain steel specimen with a carbon content of 0.45 wt.% was determined. The measurements were performed with CrKα radiation using the $\sin^2\psi$-method (2) and measuring the lattice strain distributions in ϕ-intervals of 10°. ϕ is the azimuth angle between the first principle stress σ_1 and the stress component

$$\sigma_\phi = \sigma_1 \cos^2\phi + \sigma_2 \sin^2\phi , \qquad (5)$$

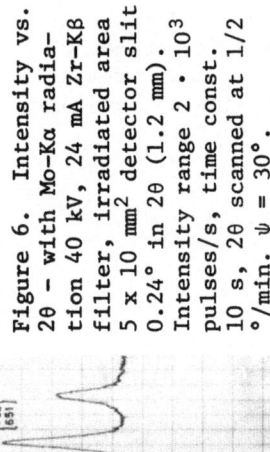

Figure 6. Intensity vs. 2θ – with Mo–Kα radiation 40 kV, 24 mA Zr–Kβ filter, irradiated area 5 x 10 mm² detector slit 0.24° in 2θ (1.2 mm). Intensity range 2 · 10³ pulses/s, time const. 10 s, 2θ scanned at 1/2 °/min. ψ = 30°.

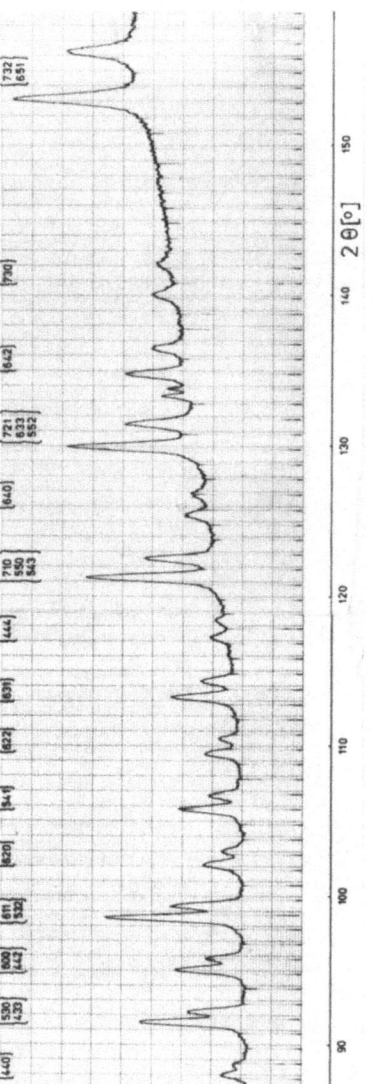

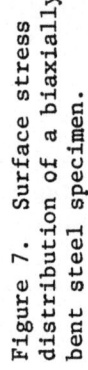

Figure 7. Surface stress distribution of a biaxially bent steel specimen.

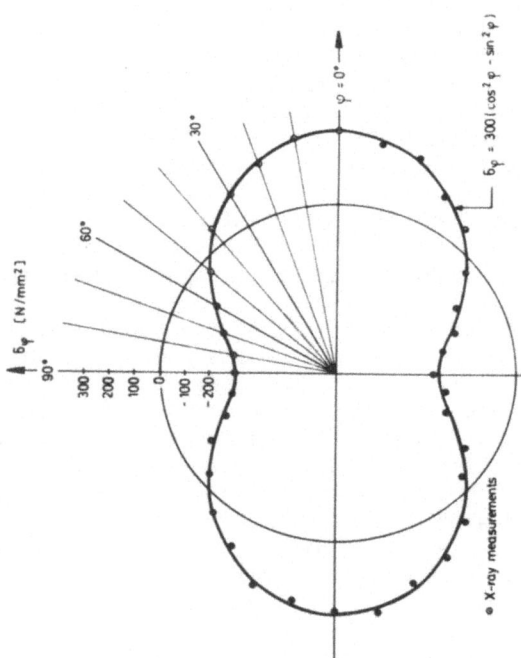

where σ_2 is the second principle stress. The influence of elastic
anisotropy on lattice strain distributions was taken into account, using the results of independently performed measurements of X-ray elastic constants according to the method proposed by (6). Figure 7 shows
the comparison of calculated and measured results for principle stresses
σ_1 = 300 N/mm^2 and σ_2 = -300 N/mm^2. The full line gives the theoretical stress distribution for different azimuths ϕ according to the
theory of elasticity. The circled dots represent measured surface
stresses. As can be seen there is an excellent correspondence between
theory and experiment.

REFERENCES

1. U. Wolfstieg, Arch. Eisenhüttenwes. 30 (1959), 447.
    U. Wolfstieg, Compt. Rend. du Colloque Rayons X et Matière, Monaco
    1973, 189.

2. E. Macherauch, III. Int. Col. Hochschule ET Ilmenau, 1958.
    E. Macherauch, P. Müller, Z. angew. Physik, 13 (1961), 305.

3. W. H. Bragg, Proc. Phys. Soc., London, 33 (1921), 211.
    J. Brentano, Proc. Phys. Soc., London, 37 (1925), 184.

4. V. Hauk, U. Wolfstieg, HTM, 31 (1976), 39.

5. G. Faninger, V. Hauk, HTM, 31 (1976), 98.

6. E. Macherauch, P. Müller, Arch. Eisenhüttenwes., 29 (1958), 257.

A DUAL DETECTOR DIFFRACTOMETER FOR MEASUREMENT OF RESIDUAL STRESS

C. M. Mitchell

Physical Metallurgy Research Laboratories

CANMET, Department of Energy, Mines and Resources, Ottawa

ABSTRACT

The diffractometer employs dual fixed detectors and a moving, vertically mounted X-ray tube. Lattice strain is measured in two specimen directions simultaneously. The drive has three concentric shafts; the outer driven by a cone drive worm and gear; a middle fixed shaft supporting the detectors and an inner shaft geared in a 1:2 ratio carrying the specimen mount. Scanning angles are measured to an accuracy of $\pm 0.001°$. A large range of specimen sizes can be accommodated.

Si(Li) semi-conductor detectors are used with single-channel analysers to provide monochromatic radiation. Coincidence of diffraction occurs for the interdetector angle β at

$$\beta = 2\phi_{HKL}$$

for an angle χ between diffraction normals

$$\chi = \phi_{HKL}$$

where ϕ_{HKL} is the supplemental Bragg angle. The coincidence setting has the geometry of the single-exposure film method.

The line position is measured by resolving the scattering distribution which is done by direct Fourier unfolding of instrumental and wavelength distributions.

Lattice strain measurement in the coincidence setting is independent of specimen eccentricity when the detector radii are equal.

Surface stress measurements have been carried out on a control-rolled HSLA steel.

INTRODUCTION

A dual detector diffractometer has been constructed for determination of the surface stress following the method of Gisen, Glocker and

Osswald(1) by simultaneous measurement of lattice strain in two directions in the material. This method has been extensively treated by
C.S. Barrett(2). The detectors are fixed at a chosen interdetector
angle and scanning is carried out by a moving, vertically mounted X-ray
tube.

The instrument is designed to measure specimens over a wide size
range. Specimens up to 50 lb are mounted on a specimen stage rotating
in the conventional manner at half the angular velocity of the scan,
but larger specimens can be measured without rotation. Solid state
Si(Li) detectors are used in the monochromatic mode.

The instrument has been calibrated using a diffraction line of a
standard specimen to determine the interdetector angle and the scanning
angle of the bisector which defines the diffractometer axis.

The instrument can be regarded as two diffractometers having a
common source and a common axis. The specimen setting errors are fixed
for the two observations and this can be shown to reduce and, under
certain conditions, eliminate the eccentricity error in the lattice
strain measurement.

The lattice spacing measurements have been carried out by resolving
the scattering distribution by direct Fourier unfolding of the
instrumental and wavelength distributions in order to correct for the
effects of deformation broadening and dispersion on the diffraction line
centers. The error in lattice spacing determination and in the surface
stress measurement has been examined.

Preliminary measurements have been carried out on a high-strength,
low-alloy steel in the as-rolled condition comparing the observed
diffractometer surface strain values with strain gauge values under
uniaxial load in bending. Stress measurements have also been carried
out to measure radial stress in the heat-affected zone of a weld. The
computed accuracy of the diffractometer stress measurements has been
compared with the observed deviations.

DIFFRACTOMETER DESIGN

The instrument is shown in Figure 1. The detectors and X-ray tube
are supported on columns above the drive unit providing 24 in. specimen
clearance below the equatorial plane.

The drive unit has a cone drive worm and gear with a 60:1 turn
ratio. The gear is mounted on the outer of three concentric shafts,
the middle shaft being fixed and the inner shaft geared in a ratio of
1:2 to the drive shaft. The three shafts are mounted on tapered roller
bearings and are concentric and parallel within ±0.0002 in. This
tolerance is required to insure concentricity of the rotations in the
equatorial plane.

An auxiliary 60:1 worm and gear is connected to the cone drive
worm and is driven by a slosyn stepping motor giving step increments of
0.001°. The specimen mount on the inner shaft has a translation table

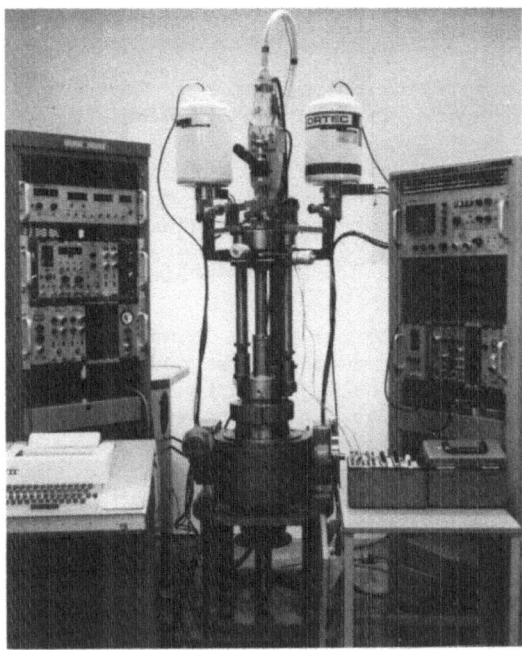

Figure 1. Stress diffractometer.

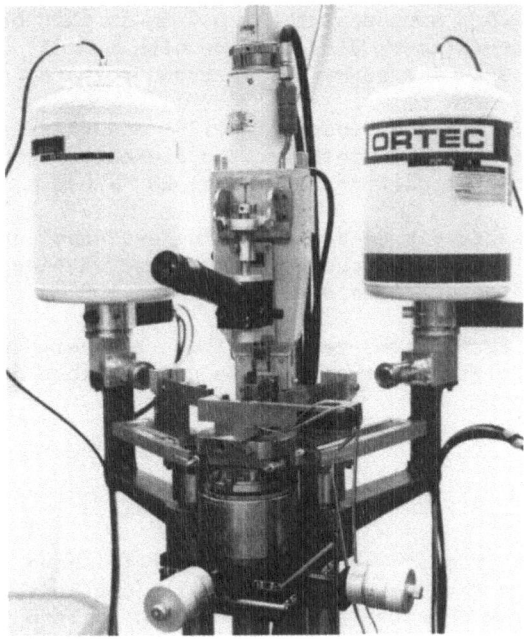

Figure 2. Equatorial plane of the instrument with specimen translation mount
and bending jig.

capable of supporting specimens up to 50 lb in weight and has an 18 in.
vertical height adjustment.

The mount and supporting column can be removed to accommodate
larger specimens which are fixed during scanning.

The detector slit and Soller slit systems are mounted on shafts
set in collars on the fixed middle shaft of the drive unit. In operation, the angle between the detectors is adjusted to give simultaneous
observation of one reflection for a given material and is fixed for the
duration of the experiment.

The diffractometer equatorial plane is shown in Figure 2. The
slit and Soller slit for each detector are mounted on a track so that
the detector slit radius can be adjusted to meet the focus condition
for an inclined planar specimen. A toolmaker's microscope is mounted
on the X-ray tube for centering small specimens. The detectors used
are Si(Li) semi-conducting detectors having an 80 sq mm sensitive area.
These are mounted independently on the support pillars so that the slit
adjustments can be made without moving the detectors.

The X-ray tube is mounted vertically in an inverted position above
the equatorial plane to provide clearance at high scanning angles. The
scan limit is $2\theta = 160°$ in the present arrangement.

The vertical mounting of the X-ray tube is a necessary compromise
since the square projection of the target must be used and the source
width is larger than with the line focus geometry. A Siemens fine-focus
tube is employed with a source width of 0.4 mm in the equatorial plane.
The resultant source-detector distribution with a 0.05° detector slit
is shown in Figure 3 as imaged by a 70-micron aperture.

The angular width of the source is 0.20° and this is less than the
$K\alpha_1$ characteristic line half width for the elements, chromium to copper
in the periodic table at diffraction angles $2\theta > 120°$.

The Si(Li) detector is used with a single-channel analyser to act
as a monochromator separating the characteristic wavelength by energy
dispersion. The detectors have a resolution of 175 ev.

The degree of resolution, measured from the energy dispersion
spectrum showed complete separation of $K\alpha$ and $K\beta$ lines and a high peak
to background ratio.

DIFFRACTOMETER CHARACTERISTICS

A schematic diagram of the diffractometer is shown in Figure 4.
The dual detectors A and B have scanning slits set at the Rowland focus
radii R_A and R_B. The detector slits subtend an angle β at the diffractometer axis. The scanning angle of the X-ray tube is ω and the
diffractometer axis lies on the bisector of the interdetector angle β
at scanning angle ω_o.

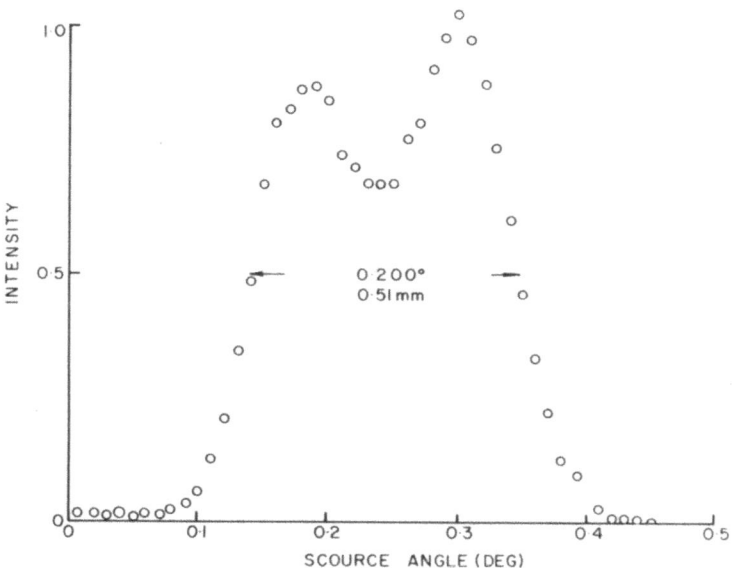

Figure 3.   Source distribution.   Siemens fine focus Fe target, 70 μm diaphragm

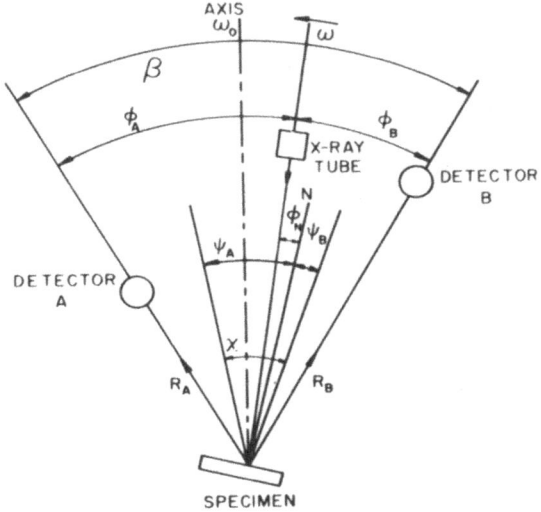

Figure 4.   Dual detector diffractometer

For a wavelength λ and diffraction from HKL planes of strain-free reciprocal lattice spacing s_{HKL} and supplemental diffraction angle ϕ_{HKL}, the reflection at detectors A and B occurs at scanning angles

$$\omega_A = \beta/2 - \phi_{HKL} + \omega_o$$

$$\omega_B = -\beta/2 + \phi_{HKL} + \omega_o \qquad (1)$$

where

$$\phi_{HKL} = 2 \cos^{-1}\left(\lambda s_{HKL}/2\right)$$

The included angle χ between diffraction plane normals for the two reflections

$$\chi = \phi_{HKL} + \omega_A - \omega_B \qquad (2)$$

In a strained lattice such that ϕ^1_{HKL} and ω_A^1 are the diffraction angle and scanning angle for detector A and ϕ^{11}_{HKL} and ω_B^{11} for detector B respectively, the relative displacement

$$\phi^1_{HKL} - \phi^{11}_{HKL} = 2\omega_o - (\omega_A^1 + \omega_B^{11}) \qquad (3)$$

The peak values of the diffraction lines will occur simultaneously when $\omega_A = \omega_B$. The peak coincidence will occur for the interdetector angle

$$\beta = 2\phi_{HKL}$$

and

$$\omega_o = \omega_A = \omega_B$$

The angle between the diffraction plane normals for coincidence $\qquad (4)$

$$\chi = \phi_{HKL} = \beta/2$$

The geometry of the coincidence arrangement corresponds to that of the single-exposure film method(2).

The values for the diffractometer axis scanning angle ω_o and the interdetector angle β for a given diffractometer setting can be determined using a standard of known lattice spacing s_s having a diffraction angle ϕ_s for the $K\alpha_1$ wavelength. Where ω_A^s and ω_B^s are the diffraction line centers of the standard, relation 1 gives

$$\omega_o = \left(\omega_A^s + \omega_B^s\right)/2$$

$$\beta = 2\phi_s + \omega_A^s - \omega_B^s \qquad (5)$$

ECCENTRICITY ERROR

In powder diffractometer measurements the systematic errors of the first order are the zero error, and the specimen eccentricity error. In the dual detector arrangement the zero error is eliminated by the axis calibration. At the coincidence setting the eccentricity error is balanced as illustrated in Figure 5. The eccentricity displacement Z of the specimen plane from the rotation axis produces equal parallel displacement of the diffracted beams. The eccentricity error ω_e in the strain displacement

$$\omega_e = Z \frac{\sin \phi_{HKL}}{\cos \gamma_n} \left(\frac{1}{R_A} - \frac{1}{R_B} \right) \qquad (6)$$

where γ_n is the angle between the specimen surface normal and the incident X-ray beam.

In the focussing arrangement the radii R_A and R_B are

$$R_A = R_o \cos \left(\phi_{HKL} + \gamma_n \right) / \cos \gamma_n$$

$$R_B = R_o \cos \left(\phi_{HKL} - \gamma_n \right) / \cos \gamma_n$$

where R_O is the source to specimen distance. The eccentricity error can be eliminated by relaxing the focus and making the detector distances equal. For a divergent beam this will produce defocussing and a resultant line broadening. The line broadening can be balanced by taking a common detector radius R_D the mean of R_A and R_B

STRESS DETERMINATION

At the surface the normal stress $\tau_z = 0$ and the principal stresses τ_x and τ_y lie in the surface plane. In an isotropic material the corresponding principal strains e_x, e_y and e_z can be expressed in terms of the partial strains ξ_x and ξ_y as

$$e_x = \xi_x - \upsilon \xi_y, \; e_y = \xi_y - \upsilon \xi_x, \; e_z = - \upsilon(\xi_x + \xi_y)$$

where ξ_x denotes the strain along the x axis when τ_x acts alone and ξ_y the strain along the y axis when τ_y acts alone.

$$\xi_x \equiv \tau_x/E, \qquad \xi_y \equiv \tau_y/E$$

and E and υ are Young's modulus and Poisson's ratio respectively.

Where the principal stress τ_x lies in the equatorial x z plane, a direction in this plane at an angle ψ_i to the surface normal has direction cosines $\sin \psi_i$, 0, $\cos \psi_i$ and the strain ε_i in this direction is, following Barrett(2),

$$\varepsilon_i = e_x \sin^2 \psi_i + e_z \cos^2 \psi_i$$

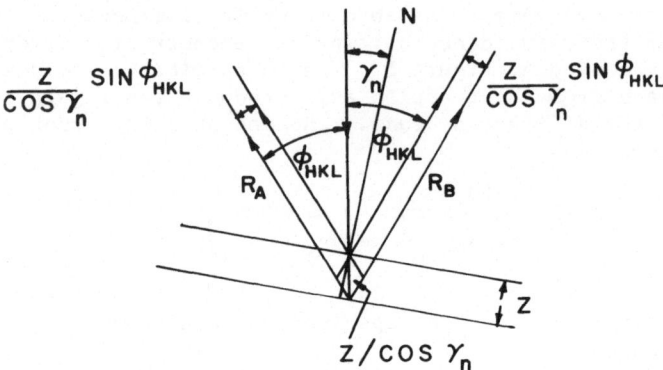

Figure 5.   Eccentricity error; coincidence setting

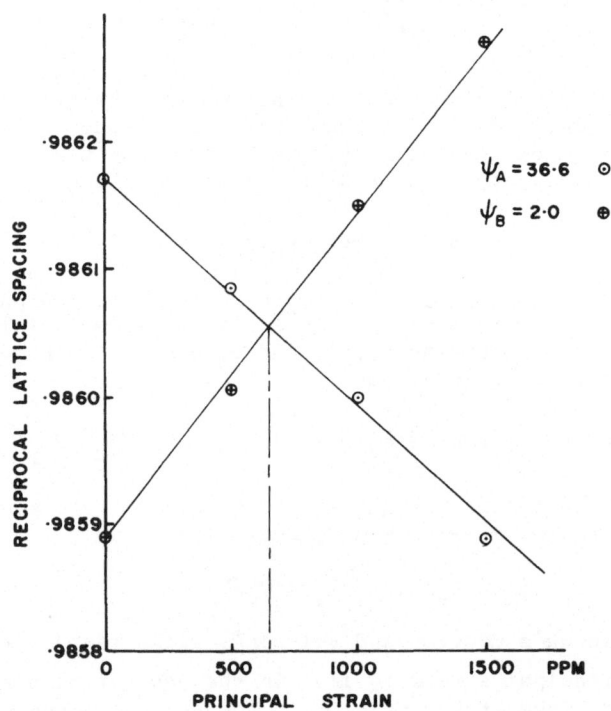

Figure 6.   Tapered bar in uniform strain, HSLA steel

In terms of the partial strains ξ_x and ξ_y the strain

$$\varepsilon_i = \xi_x (\sin^2 \psi_i - \upsilon \cos^2 \psi_i) - \upsilon \xi_y \qquad (7)$$

The coefficient of the partial strain ξ_x is

$$K_i = \sin^2 \psi_i - \upsilon \cos^2 \psi_i$$

Measurement in two directions ψ_A and ψ_B in the equatorial plane gives the partial strain

$$\xi_x = \left(\varepsilon_A - \varepsilon_B\right) / \left(K_A - K_B\right) \qquad (8)$$

which is directly proportional to the principal stress τ_x. The standard deviation σ_ξ of the partial strain ξ_x can be expressed in terms of the standard deviation σ_ε of the strain ε_i as

$$\sigma_\xi = \sqrt{2} \, \sigma_\varepsilon / (K_A - K_B)$$

Poisson's ratio is an undetermined constant in these relations. It can be obtained from strain measurements when uniaxial stress is applied. Stresses τ_x and $\tau_x{}^1$ applied holding τ_y constant produce strains ε_A, ε_B and $\varepsilon_A{}^1$, $\varepsilon_B{}^1$ respectively. Poisson's ratio is obtained as

$$\upsilon = \frac{\sin^2 \psi_A - r_{AB} \sin^2 \psi_B}{\cos^2 \psi_A - r_{AB} \cos^2 \psi_B} \qquad (9)$$

$$r_{AB} = \left(\varepsilon_A - \varepsilon_A{}^1\right) / \left(\varepsilon_B - \varepsilon_B{}^1\right)$$

STRAIN MEASUREMENT

The effect of instrumental and deformation broadening and of dispersion on the measured position of the diffraction line can be reduced by resolving the scattering distribution.

This has been done by the direct Fourier unfolding procedure which has been described previously(3). The treatment is similar to the Stokes method(4) but no deformation-free specimen of the material is required.

The source distribution was unfolded on the ϕ scale. The effect of dispersion was eliminated by the method of Mitchell and de Wolff(5), unfolding the wavelength on the scale $\ln (s/\hat{s})$ to resolve the scattering distribution.

Where the scattering distribution was symmetric the axis of the distribution has been located from the axis of the Fourier cosine series in transform space. The standard deviation of the axis position was also determined by the program.

Systematic errors due to specimen absorption and equatorial divergence which are a function of specimen tilt have been extracted using
the relations of Singh and Balasingh (6).

EXPERIMENT

The diffractometer was set for coincidence measurement on steel
specimens using Fe Kα radiation and the ferrite 220 line having the
diffraction angle $\phi_{220} = 34.6°$.

The instrument was calibrated using a Siemens standard gold specimen and the 400 diffraction line at $\phi = 36.6°$.

The diffractometer axis ω_0 and the interdetector angle β obtained
for three calibrations made over the course of the experiment were

$$\omega_0 = 9.016 \pm .009$$

$$\beta = 69.134 \pm .015$$

The standard deviation of the lattice spacing measurements

$$\sigma_L = \pm 25 \text{ ppm}$$

The experiment has been made on a control-rolled, high-strength, low-
alloy steel having a guaranteed yield of 60 ksi. This material is expected to be typical in preferred orientation and deformation to a wide
range of structural steels.

A comparison of the strain measured by diffractometer with strain
gauge values and a determination of Poisson's ratio was made using a
tapered beam in uniform bending.

A 3/-16 in. beam with its length parallel to the rolling direction
was tested. The irradiated area was electropolished using a probe to
remove deformation due to grinding. Strain gauges were mounted on
either side of the irradiated zone and on both sides of the beam.

The reciprocal lattice spacings are shown in Figure 6 for the two
diffraction directions in the material. The curves are linear and intersect at 650 ppm strain, indicating a residual stress in the electropolished surface. This was subsequently confirmed by the chemical
polishing and redetermining the surface strain for zero deflection, which
was found to be 150 ppm in tension. The value of Poisson's ratio obtained using relation (9) was

$$\nu = 0.273 \pm 0.005$$

The partial strain during bending ξ_x, calculated from equation 8,
has been treated as a superposition of the partial strain ξ_x' due to
bending alone and the partial deformation strain ξ_x^0 observed in the
tapered bar at zero deflection.

The observed partial strain and the partial bending strain are compared with the strain gauge values in Table I. The partial deformation strain is

$$\xi_x^o = -580 \text{ ppm}$$

TABLE I.   Surface Strain in Tapered Bar in Bending

Strain Gauge	Tilt		Factor	Observed Difference Strain	Partial Surface Strain	Partial Bending Strain
	ψ_B	ψ_A	$(K_A - K_B)$	$\varepsilon_A - \varepsilon_B$	ξ_x	ξ_x'
ppm	deg.	deg.	ppm	ppm	ppm	ppm
0	2.0	36.5	.4509	−262	−580	0
500	2.4	36.9	.4588	61	130	710
1000	2.8	37.3	.4666	151	330	910
1500	3.3	37.8	.4762	396	830	1410

The line center standard deviation and the corresponding partial strain standard deviation for the experiment were

$$\sigma_\varepsilon = \pm \, 39 \text{ ppm} \quad \text{and} \quad \sigma_\xi = \pm \, 130 \text{ ppm}$$

The partial bending strain values obtained by diffractometer and the surface strain measured by strain gauges are in agreement within the partial strain standard deviation.

The radial stress distribution over the heat affected zone of a weld was measured in a circular test plate of this steel. The plate was 12 in. in diameter and 1 in. thick and contained a concentric circular weld 4 in. in diameter.

The plate was mounted with its cylindrical axis in the equatorial plane of the diffractometer and the radius along which the surface intersected the equatorial plane defined the radial strain direction. From symmetry the radial strain direction was considered to be a principal stress axis.

The partial strain ξ_r and the corresponding stress τ_r in the radial direction were determined for a series of positions along the radius close to the weld edge on the exterior side of the weld. The results are shown in Table II. The radial stress is slowly varying and has a maximum at 0.25 in. from the weld of 42.2 ksi, which is close to the guaranteed yield stress of the material of 60 ksi.

TABLE II.   Radial Stress Distribution Over the Heat Affected Zone
of a Circular Weld in HSLA Steel

Weld Distance in.	Partial Radial Strain ppm	Radial Stress ksi	Weld Distance in.	Partial Radial Strain ppm	Radial Stress ksi
0.02	1181	35.5	0.40	1107	33.2
0.10	1292	38.8	0.60	1309	39.3
0.25	1406	42.4	0.80	981	29.4

DISCUSSION

The instrument was designed to enable stress measurements to be made on a large range of specimens while preserving the inherent accuracy of the diffractometer in angular measurement.

The design compromise to obtain dual detector operation was the vertical mounting of the X-ray tube. For a fine focus X-ray tube, the square source has been shown to produce only small instrumental broadening of the diffraction line over the range $\phi \leq 60°$.

The Si(Li) semi-conductor detectors have been found to be efficient monochromators, for the full range of targets, and to be stable and have high reproducibility in intensity measurement.

The tolerance in lattice strain measurement for stress measurements accurate to less than ± 2000 psi lies near the limiting accuracy of diffractometer measurement. The measurement of line position by Fourier unfolding corrects for doublet resolution and dispersion, and gives high accuracy, particularly in cases of large lattice deformation. It has the disadvantage that the full diffraction line intensity distribution is required.

The coincidence setting gives reduced eccentricity error in the focussing arrangement and negligible eccentricity error for equal detector radii.

Measurements on high-strength low-alloy steel specimens using the Fe 220 line at $\phi_{220} = 34.5°$ with Fe $K\alpha$ radiation show a computed standard deviation in lattice spacing of 30 ppm and a corresponding surface strain deviation $\sigma_\xi = \pm 130$ ppm in agreement with the observed variations in strain measurement.

ACKNOWLEDGEMENTS

The diffractometer drive unit was constructed by the Harrington Tool and Die Co. of Montreal, Canada, from the author's design.

The stress diffractometer was constructed by Mr. S. Samson of the CANMET development section, who also designed many of the components, including the alignment devices.

The Fourier unfolding program was developed in collaboration with Mr. K.S. Milliken.

Mr. E.J.-C. Cousineau assisted in the experiment and Mr. V. Chartrand carried out the electropolishing and chemical polishing of the specimens.

REFERENCES

1. F. Gisen, R. Glocker and E. Osswald, "Original Mitteilungen Einzelbestimmung von Elastischen Stannungen mit Röentgenstrahlen", Z. Tech. Physik <u>17</u>, 145 (1936).

2. C.S. Barrett, and T.B. Massalski, "Structure of Metals", 3 ed., Ch. 17, McGraw Hill, N.Y. (1966).

3. C.M. Mitchell, "Direct Determination of the Reciprocal Lattice and Radial Interference Distribution by the Fourier Method", <u>Advances in X-ray Analysis</u>, Vol. 12, p.352-354, Pergamon Press(1968).

4. A.R. Stokes, "Fourier Analysis for the Correction of Width and Shapes of Diffraction Lines", Phys. Soc., London, <u>61</u>, 328 (1948).

5. C.M. Mitchell and P.M. deWolff, "Elimination of the Dispersion Effect in the Analysis of Diffraction Line Profiles", Acta Cryst. <u>22</u>, 325 (1967).

6. A.K. Singh and C. Balasingh, "The Effect of X-ray Diffractometer Geometrical Factors on the Centroid Shift of a Diffraction Line for Stress Measurement", J. Appl. Phys. <u>42</u>, 5254 (1971).

X-RAY RESIDUAL STRESS MEASUREMENTS USING PARALLEL BEAM OPTICS

Richard M. Chrenko

General Electric Research and Development Center

Schenectady, New York 12301

ABSTRACT

X-ray residual stress measurements have been made with a commercial portable X-ray diffraction apparatus that uses parallel beam optics and that was specifically designed for residual stress measurements. This machine differs from X-ray diffraction units using the usual parafocusing geometry in several respects, most notably reduced sample placement errors and larger sample sizes that can be accommodated. Two special modes of operation are available and will be discussed. These are the ability to use the side inclining method for stress analysis and the ability to use an oscillating ψ motion, the latter mode being useful for examining large grain size materials.

The apparatus will be described as well as tests that compare the unit with those using parafocusing geometry. Some examples of the types of materials examined to date will also be described.

INTRODUCTION

Residual stress measurements using X-ray techniques are based on the determination of the interplanar spacings between certain atomic planes of a specimen. In other words, lattice strains are being determined. Many factors influence the accuracy and/or reproducibility of the residual stress results as well as the ease of obtaining the results. The major factors can be divided into: the sample or material; the type of equipment; and the technique and data reduction. Many subdivisions exist under each of these major divisions.

In this article the equipment factor will be emphasized with a portable Rigaku Strainflex X-ray residual stress apparatus using parallel beam optics being described. The points of emphasis will be the general features that are useful in certain applications. Also to be discussed will be how instrumental and technique factors affect the stresses obtained. Finally some examples of the types of materials examined will be included.

    While the apparatus has been used by us for several months, not
all the instrumental variables have yet been systematically studied.
Nor do we claim the highest possible accuracy and degree of reproducibility has yet been obtained from the machine. The main purpose of
this paper is to present data that will give a good indication of what
this type of parallel beam apparatus can do. Also both advantages and
disadvantages with respect to parafocusing apparatus will be mentioned.
A number of Japanese workers (1-3) have published articles on various
aspects of X-ray stress determinations using the parallel beam technique and these papers should be consulted for additional details.

                        APPARATUS DESCRIPTION

    One main feature of the Rigaku Strainflex is that the X-ray head
is on a movable arm so that stresses can be measured on samples that
are too large to fit on such diffractometers as the GE XRD-6. The entire apparatus can also be moved so that more massive structures can be
examined. The 2θ ranges available are 140° to 170° and 120° to 150°
depending on which of two positions the detector is located. The ψ
angles can be continuously varied.

    The apparatus is not designed to be used for wide 2θ scans such as
are used for sample analysis by X-ray diffraction techniques. To obtain
the different ψ angles the head is rotated, not the sample, which is an
advantage when large samples are measured. The correct sample position
is obtained with a calibrated pointer.

    Figure 1 [from Taira (3)] shows the differences in diffraction geo-
metries between the parallel beam method used in this apparatus and the
parafocusing method. The angle ψ has the same meaning for both methods.
However, the Rigaku ψ rotation is actually calibrated in terms of an
angle ψ_0, where ψ_0 is the angle between the incident X-ray beam and the
normal to the sample surface.

    The apparatus is also capable of performing residual stress
measurements using the "side inclining method." This technique is compared in Figure 2 [from Taira (3)] with the ordinary method. The major
difference is that in the ordinary method of obtaining residual stress
the ψ and 2θ rotations are co-planar whereas in the side inclining
method these motions are normal to each other. As can be seen from
Figure 2, if stresses are wanted in recessed areas, the side inclining
method has advantages in that at $\psi = 45°$ the diffracted beam is at a
larger angle with respect to the sample surface than with the ordinary
method. The data presented in this paper will be limited to that from
the ordinary method.

    The apparatus has an oscillating ψ feature which is useful for
obtaining stress data on large grain materials. In such materials the
Debye ring is spotty with the result that only a few grains contribute
to the diffracted intensity and accurate stress values representative
of the material are not obtained. This problem can be somewhat allevi-
ated by means of slight oscillations of the ψ angle about the mean ψ
angle, whereby more grains can contribute to the diffracted intensity
with the result that the stress value obtained is more representative

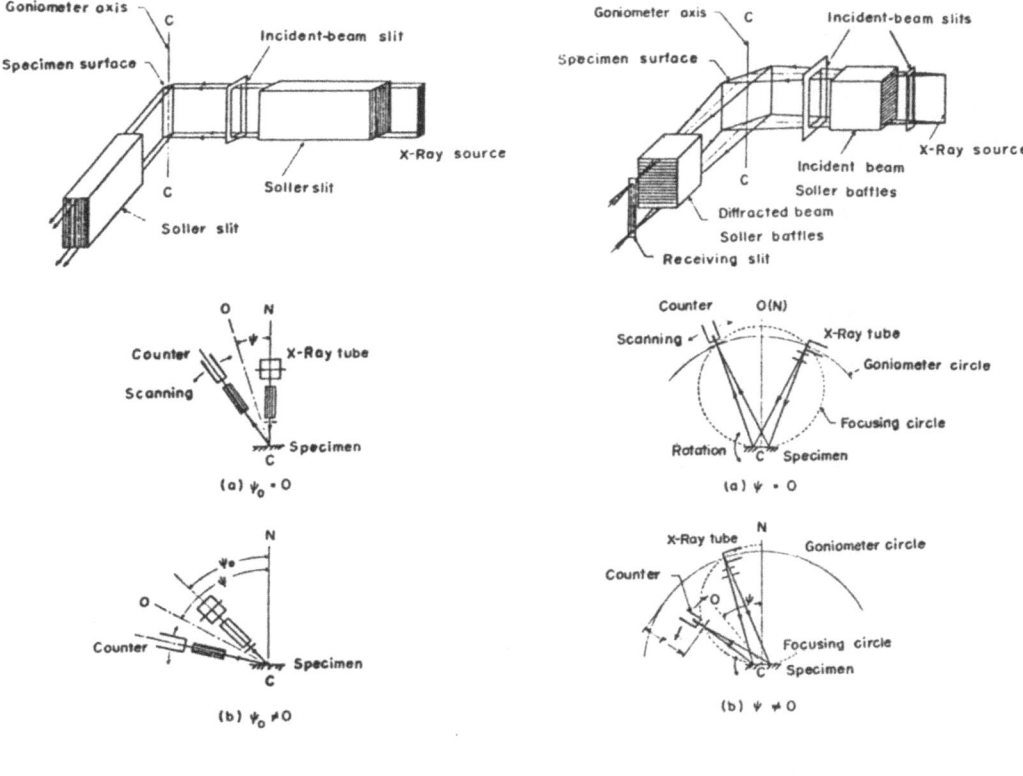

(a) Parallel-beam method   (b) Parafocusing method

Figure 1. Diffraction geometries of (a) parallel beam
method and (b) parafocusing method.

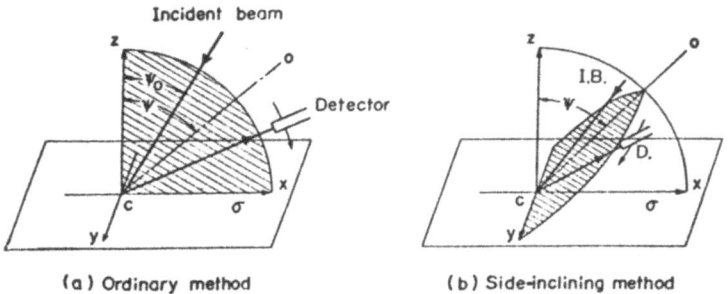

(a) Ordinary method   (b) Side-inclining method

Figure 2. Comparison of ordinary and side-inclining
methods of X-ray stress measurement.

of the sample. For this apparatus the X-ray head oscillates instead
of the sample, so that large samples with large grain size can be
conveniently measured.

The relative positions of the 2θ indicators and manual 2θ control
with respect to the other controls are quite inconvenient compared to
the GE XRD-6. The result is that the usual counting method for obtaining 2θ peak values is not as rapid.

Another feature allows one to obtain recorded diffraction profiles for up to four ψ angles automatically. This feature can be used to an advantage where multiple ψ readings are needed, but it is also useful for obtaining data at two ψ values.

TESTS

A number of tests have been made to determine the accuracy and reproducibility of the apparatus.

In order to obtain 2θ values the usual technique of counting and fitting to a three point parabola was first followed (4). Fitting the data to a five point parabola was not tried since previous experience had shown that, for the type of samples being examined, the added accuracy did not justify the extra time needed to obtain the data. The manipulations that must be done to set the 2θ goniometer to the desired value for counting even for a three point parabola fit made the counting method slow and cumbersome. Hence, a number of tests were made to compare the usual three point parabola method and the half value breadth method. In this latter method a chart scan is made and a line drawn parallel to the base at half maximum intensity. The peak 2θ is then defined as the center value of this drawn line on the 2θ scale. Comparison tests were made on typical samples. The $\Delta 2\theta$ values obtained from 2θ values at $\psi = 0°$ and $\psi = 45°$ using the counting and half value breadth methods were compared since these $\Delta 2\theta$ values were used to calculate σ: $\sigma = \Delta 2\theta \times K$. The results showed that the maximum difference between the two methods was $0.06°$ with the average difference $0.04°$. This maximum difference of $0.06°$ was within the error obtained using the three point parabola counting method on any one sample. Other authors also noted differences as large or larger on tests of reproducibility of residual stress data (5).

Most, but not all, of the residual stress values presented were obtained using the half value breadth method. A scanning speed of $1°$ 2θ per minute was used. The time constant can vary within certain limits. However, for the half value breadth method both 2θ scanning speed and time constant must be held constant for any one set of ψ measurements to obtain meaningful data.

The reproducibility of the half value breadth method for obtaining 2θ values was tested by making a number of runs using Soller slits of different divergences. Equivalent slits were used both for the divergent and receiving slits. The data were obtained on a typical sample, a ground bar of type 304 stainless steel, and are shown in Table 1. For any slit combination the sample was set in place, six scans were made at both $\psi = 0°$ and $\psi = 45°$, and the 2θ values were obtained by the half value breadth method. Then the slits were changed and the process was repeated. The sample area was defined by vinyl electricians tape and was 2.5 mm by 7.0 mm.

The reproducibility of the half value breadth method, which includes instrumental variations for any one set of slit and ψ conditions, is at worst $0.04°$ 2θ and usually less. As more divergent slits are used the $\Delta 2\theta$ values between $\psi = 0°$ and $\psi = 45°$ increase and the stress becomes

more tensile. Jatzczak and Boehm (5) also observed a tendency towards
more tensile stresses with increase in slit divergence in their study
of GE XRD-3 and Siemens-Halske diffractometers.

TABLE 1. Reproducibility of Half Value Breadth Method

Slit	.34°/.34°		.68°/.68°		1.02°/1.02°	
	2θ		2θ		2θ	
	$\psi = 0$	$\psi = 45$	$\psi = 0$	$\psi = 45$	$\psi = 0$	$\psi = 45$
	129.11	128.78	129.12	128.75	129.19	128.80
	129.13	128.78	129.12	128.76	129.18	128.81
	129.11	128.76	129.12	128.75	129.19	128.81
	129.13	128.77	129.13	128.75	129.20	128.79
	129.12	128.75	129.11	128.77	129.19	128.81
	129.10	128.74	129.13	128.73	129.19	128.81
Ave 2θ	129.117	128.763	129.122	128.752	129.190	128.805
Max. Diff.	.03	.04	.02	.03	.02	.02
Ave Δ2θ		.354		.370		.385

Radiation CrKα   I = 10 ma
Planes   {220}   V = 30 kv
Irradiated area 2.5 mm x 7.0 mm

The variation in residual stress values as a function of sample
location was examined. Previous data obtained on a GE XRD-6 are shown
in Figure 3, which plots Δ2θ ($\Delta 2\theta = 2\theta_{\psi=0} - 2\theta_{\psi=45}$) as a function of
sample location. The sample was a ground coupon of austenitic stain-
less steel and {220} planes were used at 2θ ∿ 129° with CrKα radiation.
The data indicates an average slope of 0.015° Δ2θ per 0.001 inch of
sample placement error. This is equivalent to ∿ 3 KSI per 0.001 inch
of placement error.

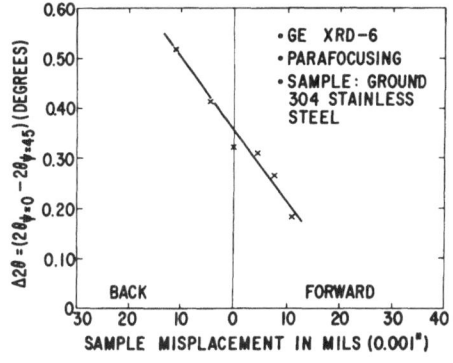

Figure 3. Variation of Δ2θ with
sample misplacement. Apparatus is
GE XRD-6 with parafocusing geometry. The sample is a ground
flat plate of 304 stainless steel.
X-ray beam parallel to grinding
lay.

The data for the Strainflex parallel beam apparatus for a fairly
smooth ground bar of 304 stainless steel are shown in Figure 4 on the
same scale as Figure 3. The standard deviation of the Δ2θ values is
<0.01, the slope <.0003° Δ2θ per 0.001 inch of sample placement error,
and the maximum difference in Δ2θ values for all the data is 0.03°.
This later value is roughly the reproducibility of the half value
breadth method. The conclusion is that there is little error in the
residual stress obtained for reasonable sample misplacements from the
correct location.

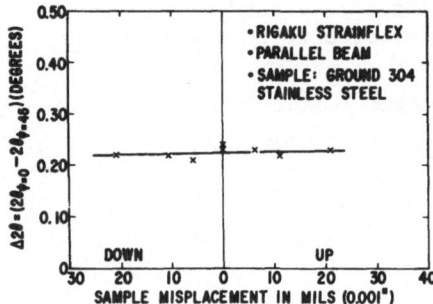

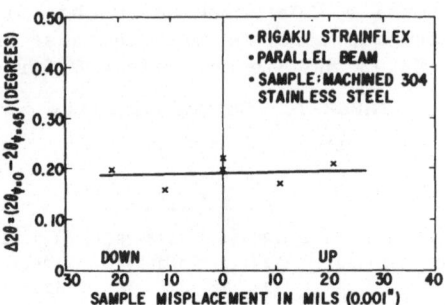

Figure 4. Variation of Δ2θ with
sample misplacement. Aparatus is
Rigaku Strainflex with parallel
beam geometry. The sample is a
ground flat plate of 304 stainless steel. X-ray beam parallel
to grinding lay.

Figure 5. Variation of Δ2θ with
sample misplacement. Apparatus is
Rigaku Strainflex with parallel
beam geometry. The sample is a
machined 304 stainless steel pipe.
Surface is ID; X-ray beam perpendicular to machining lay.

     Similar data for the Strainflex on a piece of machined 304 stain-
less steel with well defined machining grooves is shown in Figure 5.
In this case, the standard deviation of the Δ2θ values is 0.02°, the
slope is <0.0003° Δ2θ per 0.001 inch of sample placement error and the
maximum difference in Δ2θ values for all the data is 0.06°. Therefore,
even for the machined sample, there is little error in the residual
stresses obtained for reasonable sample misplacements from the correct
location. Also, the larger scatter in Δ2θ values is consistent with
other data that show larger scatter in Δ2θ values with rougher surfaces.

     X-ray residual stress constants were obtained for type 304 stain-
less steel using a four point bend device. Strain gauges were mounted
on either side of the area on which the X-ray strains were measured.
The divergence and receiving slits were both 0.68°. Using $CrK\alpha$ radiation and examining the {220} planes, a plot was made of applied stress
versus Δd. The beam was loaded and unloaded in a random fashion and at
each loading 2θ values were obtained at $\psi = 0°$ and $\psi = 45°$. From
$E/(1 + \nu) = $ slope x $\sin^2 \psi$ x d_o, $E/(1 + \nu) = 25.2$ x 10^3 KSI with an
index determination of 0.99. If one assumes a reasonable value of
$\nu = 0.29$, then $E = 32.5$ x 10^3 KSI for the X-ray value of Young's
modulus of the {220} planes. This value is higher than the bulk value
of $E = 28$ x 10^3 KSI. Conversations with others using this type of
machine in which the specimen does not rotate with 2θ indicate that the
deduced X-ray value for Young's modulus tends to be higher than for
diffractometers that have θ-2θ motion.

                                APPLICATIONS

     This parallel beam apparatus has been used to obtain residual
stresses on a number of different materials that had been prepared in
various ways. A few of these applications that will be described are:
1) residual stress distributions near welds in 304 stainless steel
pipes; 2) effects of nitriding on residual stress in 17-4 PH

precipitation hardened steel; and 3) use of the oscillating ψ feature in large grain Stellite 6B.

The first study concerns the roles of residual stress on certain material properties near welds in type 304 stainless steel pipes. Residual stresses have been measured on welded pipes of different diameters, in different directions, with distance beneath the surface, and with distance from the weld. Only surface residual stress data taken as a function of distance from the weld will be described here. Figure 6 shows such a stress distribution for the ID surface of a 10 inch diameter pipe. The stress direction is longitudinal, that is, perpendicular to the weld. The data in Figure 6 are for pieces cut from five different azimuths around the pipe and do not include corrections for the stress relaxed when the pieces were cut.

This apparatus has been especially convenient for measuring the stresses on these pipes for two reasons. First, sample placement is easy and sample adjustments due to different pipe curvatures are practically nil. Second, some of the pipe pieces were too large to measure on a usual diffractometer with stress attachments.

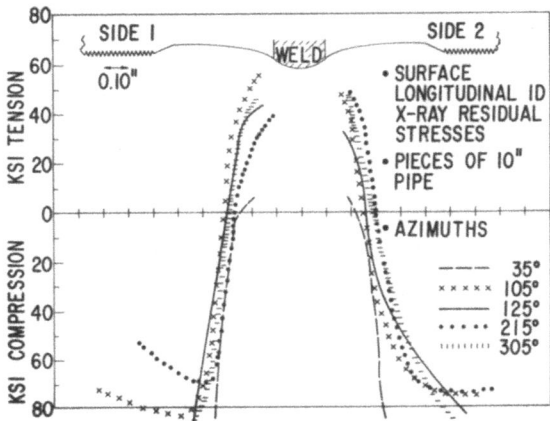

Figure 6.   X-ray surface longitudinal residual stresses as a function of distance from weld. For ID of 10" pipe. Data are shown for various azimuths.

The data of Figure 6 appear to show a remarkable similarity in that all azimuths give the characteristic bell-shaped stress distribution present near welds due to weld metal shrinkage. Despite the similarities, one can observe distinct differences with azimuth if one plots stress as a function of azimuth at a certain distance from the weld. For example, Figure 7 shows surface longitudinal ID stresses at 0.1" from the weld fusion line plotted against azimuth. Shown are both the X-ray stresses and total original stresses (X-ray stresses measured on the pieces added to the stresses relieved when the pieces for the X-ray measurements were cut out). Azimuthal stress variations of different magnitudes have been seen near all welds examined and have been partially correlated with welding procedures.

Another application has been the study of residual stress effects
due to nitriding 17-4 PH precipitation hardened stainless steel. The
nitriding was done to prepare a hard, wear resistant surface and
residual stress depth profiles were also needed. In Figure 8 are shown
such profiles of nitrided and non-nitrided samples. Another parameter
that can provide information is the half value breadth of the 2θ peaks,
obtained from chart scans. Such data is shown in the upper left corner
of Figure 8 for the nitrided sample. Both the half-breadth and
residual stress reached their non-nitrided values at ∿ 7-8 mils beneath
the surface, approximately where metallurgical cross sections show the
nitrided layer to end.

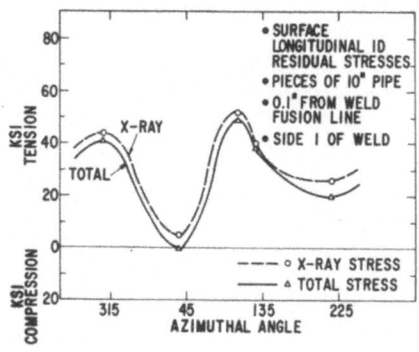

Figure 7. Azimuthal variation
of surface longitudinal ID
residual stresses at 0.1" from
weld fusion line. Data are
for a 10" diameter pipe.

Figure 8. Residual stress effects due
to nitriding 17-4 PH steel. Data are
not corrected for stress gradient and
layer removal. Also shown is half
value breadth data for nitrided
sample.

The oscillating ψ feature has proven useful for measuring large
grain size materials such as 304 stainless steel and Stellite 6B.
Both have face centered cubic structures. In Figure 9 for a 304
stainless steel sample are shown photos of sections of the Debye ring
at the detector slit at 2θ ∿ 129° both with no oscillation and ± 3°
oscillation of ψ. The ψ oscillation gives a more uniform intensity
ring.

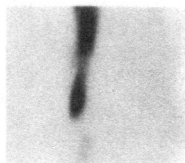

Figure 9. Effect of ψ
oscillation on spotty Debye
ring. Sample is large grain
304 stainless steel.

± 0° ψ
Oscillation

± 3° ψ
Oscillation

Similar exposures have been obtained on Stellite 6B, and a 2Θ - $\sin^2 \psi$ plot, using $\pm$ 3° oscillation of ψ is given in Figure 10.

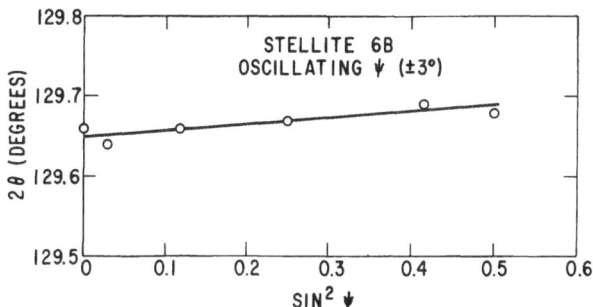

Figure 10.   Plot of 2Θ versus $\sin^2 \psi$ for Stellite 6B of large
             rain size.   Oscillating ψ of $\pm$ 3° used at each ψ.
             Debye ring of stationary sample was spotty.

SUMMARY AND CONCLUSIONS

X-ray residual stress measurements have been made with a commercial portable X-ray diffraction instrument that uses parallel beam optics and that was specifically designed for residual stress measurements. This machine differs from X-ray diffraction units using the usual parafocusing geometry in several respects, most notably reduced sample placement errors and larger sample sizes that can be accommodated.

Two special modes of operation are available with this machine. These are the ability to use the side inclining method for stress analysis and the ability to use an oscillating ψ motion. The side inclining method mode of operation enables one to measure stresses in samples having deep depressions where geometrical problems do not render the usual method feasible. The oscillating ψ feature enables one to obtain more representative residual stresses of polycrystalline metals that have large grain size and give spotty Debye rings.

The materials of principal interest to date have been commercial alloys, especially those used in power generation systems. Residual stresses have been obtained on welded samples as well as those prepared with different surface treatments.

Some general conclusions are:

1. For obtaining peak locations, the layout of controls makes the half value breadth method much easier to use than the usual method of counting at three 2Θ values and fitting the data to a parabola.
2. Both the half value breadth and counting methods give comparable results and reproducibilities.
3. For the half value breadth method both 2Θ scanning speed and time constant must be held constant for any one set of ψ measurements to obtain meaningful data.

4. The reproducibility of the half value breadth method is less than 0.04° 2θ.
5. For Soller slits of higher divergence the apparent stresses become more tensile.
6. Sample placement errors are low, at least in the range examined, 0.020 inches (0.5 mm) on either side of correct sample location.
7. An oscillatory ψ feature has been successfully used to obtain more representative values of residual stresses on large grain size materials.
8. Residual stress measurements can be made on large size samples that cannot be accommodated on usual diffractometers that have stress attachments.
9. The apparatus is designed for residual stress measurements with the result that wide 2θ scans such as are available on usual diffracto-meters are inconvenient or, in some cases, impossible to make.

ACKNOWLEDGMENTS

The author wishes to thank J. F. Fleischer for the careful assist-ance given in many aspects of the above work. Partial support by EPRI under contract RP 449-2 for the work on pipe welds is gratefully acknowledged.

REFERENCES

1. S. Aoyama, K. Satta, and M. Tada, "The Effect of Setup Errors on the Accuracy of Stress Measured with the Parallel Beam X-ray Diffractometer," J. Soc. Mat. Sci., Japan, 17, 1071 (1968).

2. J. Fukura and H. Fujiwara, "Evaluation of Various Techniques of X-ray Stress Measurement with the Diffractometer," J. Soc. Mat. Sci., Japan, 15, 825 (1966).

3. S. Taira, "X-ray Diffraction Approach for Studies on Fatigue and Creep," Experimental Mechanics, 13, 449 (1973).

4. "Residual Stress Measurement by X-ray Diffraction," SAE J 784a, Soc. Auto. Eng. (1971).

5. C. F. Jatczak and H. H. Boehm, "The Effects of X-ray Optics on Residual Stress Measurements in Steel," Adv. in X-ray Analysis, 17, 354, Plenum Press (1973).

PROTON-INDUCED X-RAY EMISSION ANALYSIS OF HUMAN AUTOPSY TISSUES*

R. D. Lear and H. A. Van Rinsvelt

Department of Physics and Astronomy

University of Florida

Gainesville, Florida 32611

and

W. R. Adams

Department of Pathology, General Rose Memorial Hospital

and University of Colorado Medical Center

Denver, Colorado 30220

ABSTRACT

The 3.8 MeV proton beam from the University of Florida Van de
Graaff accelerator has been used to perform trace element analysis of
approximately 1200 samples (mostly from autopsies) of human tissues by
proton-induced X-ray emission analysis (PIXE). Fifteen different or-
gans and a variety of diseases have been studied. Preliminary data
are presented indicating the variations of various elements in human
kidney as a function of age. Analysis of samples from infants also
indicate essential and non-essential elements in human kidney. On the
average twelve trace elements (with atomic number equal to or larger
than nineteen) are observed in each organ. Quantitative measurements
have been made on several elements including K, Ca, Mn, Fe, Cu, Zn,
Pb, Br, Rb, Sr, Cd, and Ba.

Mass normalization is accomplished by monitoring the elastically
scattered protons. This technique is as accurate as previously used
methods of weighing, spiking, or dissolving and pipetting each individual sample, and is much faster. This technique also eliminates many
beam handling and target integrity problems. Since mass information
comes only from the portion of the sample exposed to the beam, the
normalization is correct even if part of the target is missed or lost
during irradiation.

---

*Supported in part by grants GM20281-02 and GM20282-02 from the
National Institutes of Health.

Both dried and ashed samples have been examined and the results are compared. Standard samples have been prepared for quantization by serial spiking of selected tissues.

The ultimate goal of this investigation is the detection of possible correlation between human diseases and trace element level imbalances.

INTRODUCTION

Proton induced x-ray emission has become an effective means of doing trace element analysis. Sensitivities from 0.1 to 1.0 parts per million wet weight are easily obtained in biomedical samples. These sensitivities can be achieved in about 15 minutes using as little as 100 micrograms of material. This amount of material would make it possible to do elemental analysis of biopsies. PIXE analysis has a further advantage of simultaneous detection of several elements. Ten to fifteen elements are routinely detected in a human organ in a single analytical run.

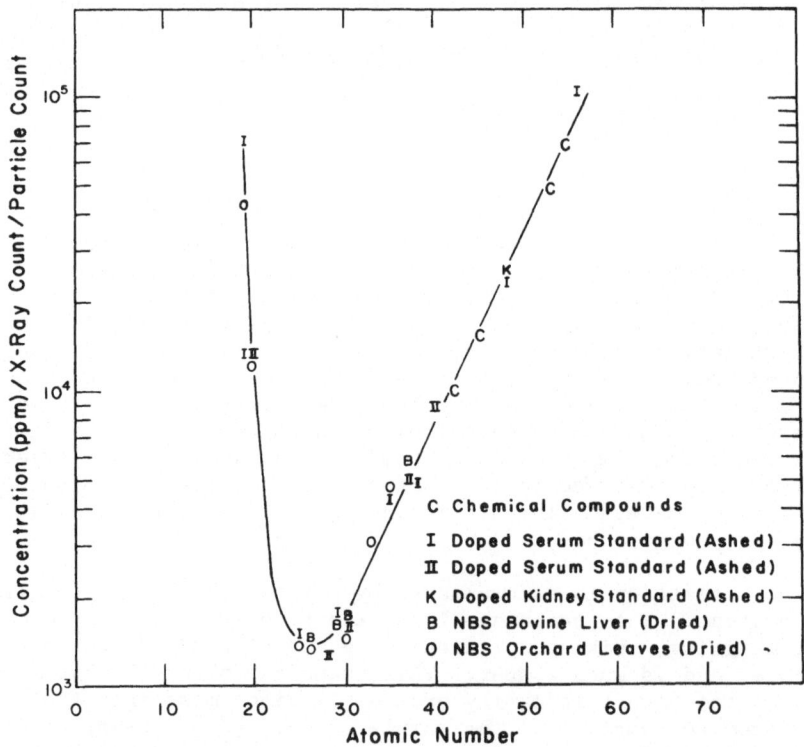

Figure 1. System calibration curve for K-shell x-ray emissions. The steep slope of the curve for low atomic number is a result of a 0.013 cm teflon absorber used to trap out low energy photons.

TARGET PREPARATION AND ANALYSIS

Because samples must be placed in a vacuum, target manufacturing can be somewhat of a problem. Biomedical samples must first be dried. This can be done by vacuum drying, low temperature drying or high temperature ashing. Once a technique is chosen a powdery residue remains which must be made into targets.

Targets are made by placing approximately 100 micrograms of material in a 3 mm circle on a thin formvar backing. Formvar was chosen because it makes low contributions to the overall background and contains no contaminants in the regions of interest. The sample is then encapsulated by placing another formvar cover over the backing and target material.

The mass of the material used to prepare the target has not been determined to this point. The problem now is to determine the amount of material exposed to the proton beam. Massing of each individual target before and after deposition of the target material on the formvar backing is time consuming and often inaccurate. Thus, this method has been avoided. Previous investigators have spiked each sample with a known amount of a non-interfering element which is then used as a mass indicator (1,2). This is also time consuming, especially when tissues must be dissolved before spiking.

The mass determination for experiments described here was done by monitoring back-scattered protons. A particle detector is placed at 135° to the beam direction in the vertical plane inside the target chamber. The x-rays are detected at 135° from the beam direction in the horizontal plane outside a 0.013 cm mylar vacuum window. Back scattered protons are then detected simultaneously with x-rays from the sample. Thus, mass determination and elemental determination are done simultaneously. This minimizes target problems as mass information and elemental information both come only from the portion of the target directly exposed to the beam.

The typical particle spectrum obtained from 3.8 MeV protons elastically scattered from a target of ashed biological tissue does not have sufficient resolution to allow extraction of any information about elemental constituents. However, integration of the particle spectrum yields a number which is proportional to the amount of the target

Table I. Reproducibility check in an ashed kidney. December 1975 to May 1976 (30 determinations)

ELEMENT	PPM WET WEIGHT	ONE STD. DEV. (%)
K	5149	8.7
Ca	193.3	12.6
Mn	3.9	12.8
Fe	44.3	7.3
Cu	1.4	11.4
Zn	18.7	7.3
Rb	0.7	19.2

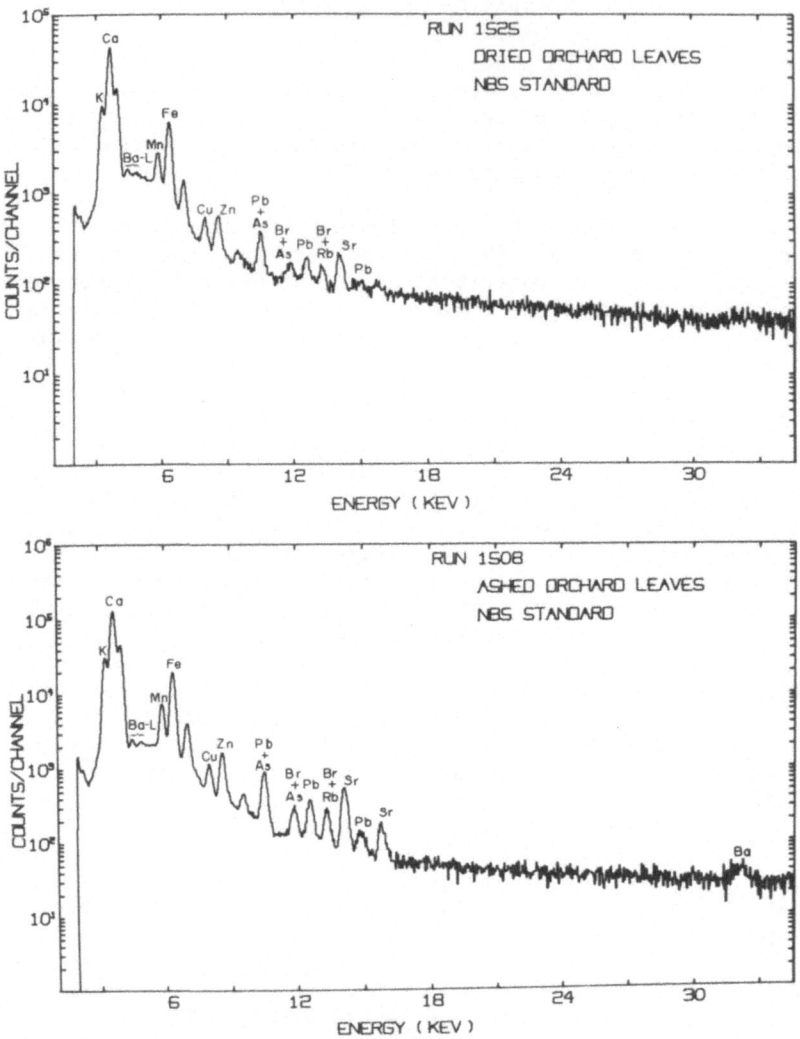

Figure 2. Comparison of NBS standard orchard leaves; unashed (dried) and ashed at 400°C. The sensitivity is enhanced for elements heavier than zinc when the material is ashed.

material exposed to the beam and to the integrated beam current (3). Thus, dividing the x-ray count by the particle count normalizes the target for both mass and integrated beam current.

Figure 1 shows the curve which relates the x-ray count per particle count for a given element to the concentration of that element in the sample. This curve was determined using five different standard materials. Three were serially doped biomedical standards and two were NBS standard reference materials. The biomedical samples were ashed at 500°C while the NBS standard materials (orchard leaves and bovine liver) were dried, but not ashed. Because of the strong dependence of the

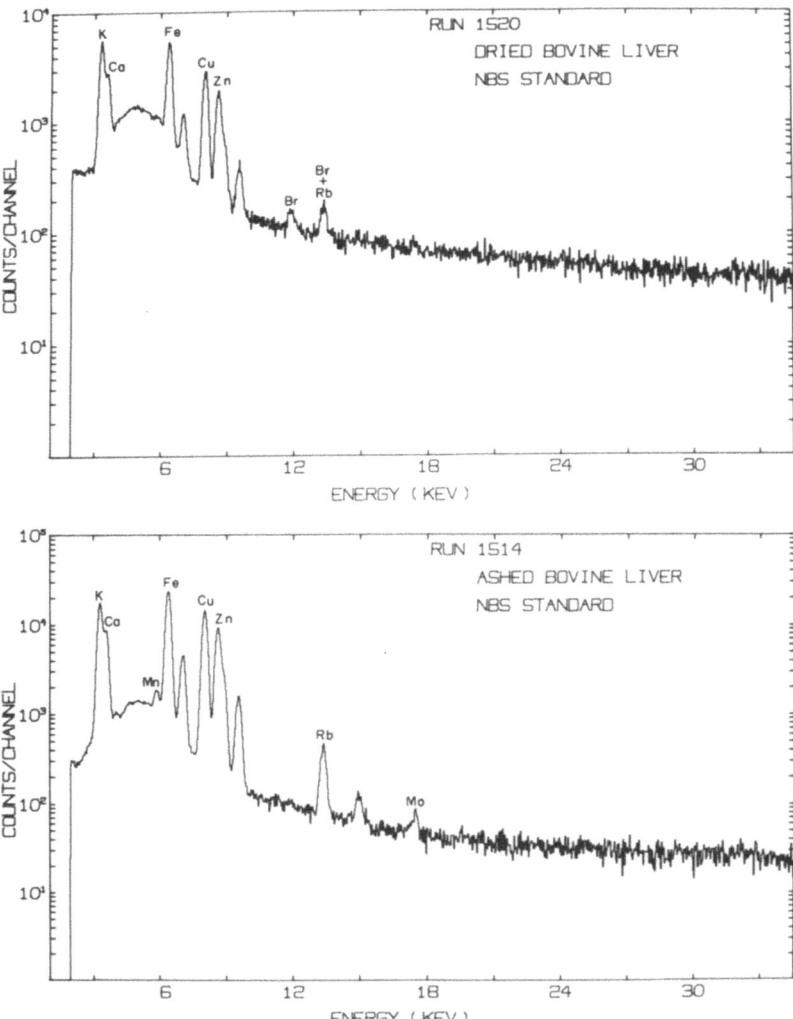

Figure 3. Comparison of NBS standard bovine liver, unashed (dried) and
ashed at 400°C. Sensitivity has been enhanced for Mn, Rb,
and Mo in the ashed sample. Bromine has been completely removed during high temperature ashing.

Rutherford cross section upon the atomic number of the target atoms,
one might expect a difference in normalization between ashed and unashed samples. The consistency of the data, however, shows this technique of normalization to scattered particles to be equally valid for
ashed or unashed material.

Table I shows a tabulation of data from a reproducibility check on
the system and the technique carried out over a six month period.
Thirty determinations of a single ashed kidney were made in this time
period. One standard deviation from the mean determination runs from

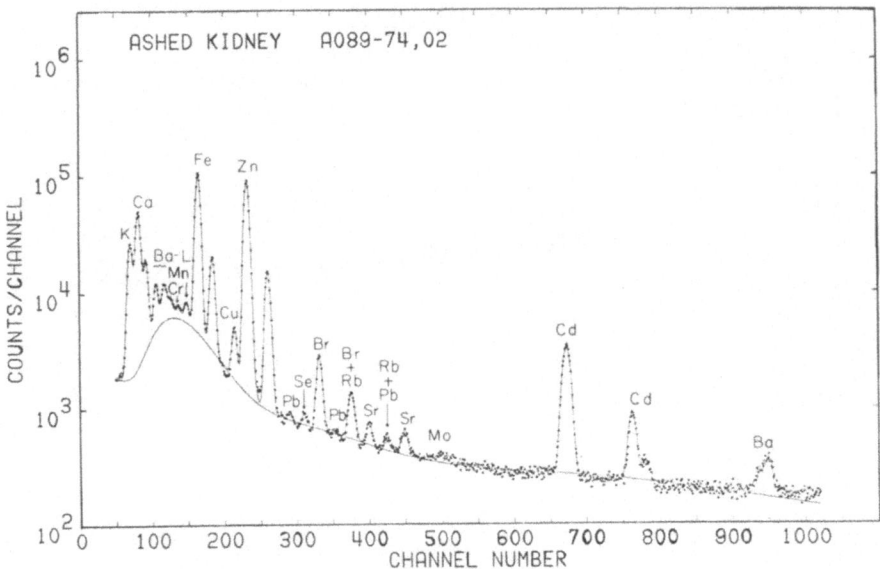

Figure 4. The spectrum from an ashed human kidney as modeled by the
           computer code REX. The quantitative analysis is given in
           Table II.

7.3% for the easily measured elements to 19.2% for the difficult to
measure elements. The overall reproducibility for one standard deviation from the mean is approximately 10%.

TYPICAL EXPERIMENTS

Once the technique has been established the question is whether to
dry or ash the samples. Figure 2 shows the results of ashing NBS
orchard leaves standard. The sensitivity for all elements is enhanced.
Arsenic, lead and strontium are much easier to detect and the barium K_α
lines have appeared. Thus it would appear to be advantageous to ash
samples.

However, figure 3 shows the result of ashing NBS bovine liver standard. While the sensitivity is greatly increased for manganese, rubidi-
um and molybdenum, bromine has been completely lost. Preliminary data
indicate most elements have lost 20-30% during ashing. Thus, care
must be exercised if ashing is to be used.

Figure 4 shows a spectrum from a human kidney which has been
modeled by the computer code REX (4). There are a few problems as the
background is too high in the region from lead to strontium and too low
under barium. The overall fit, however, is quite good. The wet weight
concentrations for the sample are listed in Table II.

Preliminary data show the concentration of various elements in the
kidney to change with age as follows: iron slowly increases until
approximately 55 years of age then slowly decreases; copper increases

Table II.   Elemental analysis of ashed human kidney (A089-74,02).

Element	Concentration (ppm wet weight)
K	235 ± 118
Ca	73.5 ± 35.1
Mn	0.58 ± 0.12
Fe	26.3 ± 2.6
Cu	1.11 ± 0.12
Zn	35.2 ± 3.5
Se	0.13 ± 0.03
Pb	0.12 ± 0.08
Br	1.84 ± 0.19
Rb	0.66 ± 0.10
Sr	0.28 ± 0.06
Zr	0.40 ± 0.07
Cd	20.0 ± 2.1
Ba	11.0 ± 1.2

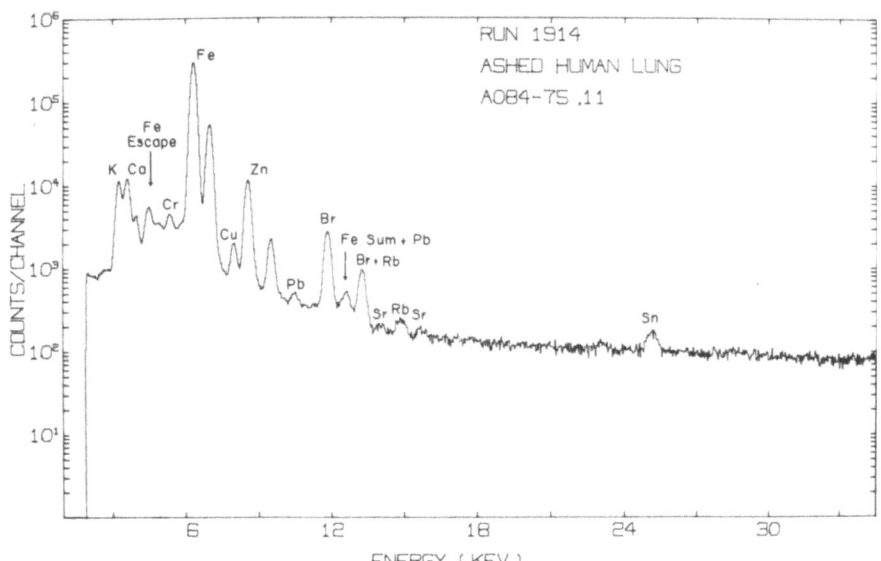

Figure 5.   Problems can arise when one very intense peak is present in
a spectrum.   The Fe escape peak interferes with the identification of Ti.   The Fe sum peak can result in improper quantization of Pb.

until approximately 65 years then decreases; zinc starts at a low concentration at birth and quickly increases to an approximately constant
amount; calcium continually increases,approximately tripling by age
85; cadmium starts at zero at birth, slowly increases until approximately 65, then decreases. The zero level of cadmium at birth would
indicate it is a non-essential element. All other elements mentioned
were non-zero at birth, indicating them to be essential.

Figure 5 demonstrates the problem which can arise if one very
intense line is present in the spectrum. The very intense iron line in
this spectrum has resulted in an iron escape peak and an iron sum peak.
The Fe escape peak interferes with the identification of Ti and the iron
sum peak can result in erroneous concentration for lead. Thus, samples
with one dominant element require careful handling.

SUMMARY

PIXE has been found to be a very useful method for doing trace
element analysis of biomedical samples. Using elastically scattered
protons as a mass monitor also greatly reduces target preparation time.
However, careful consideration must be given to whether samples are to
be dried at low temperature or ashed at high temperature.

REFERENCES

1. R. C. Bearse, D. A. Close, J. J. Malanify and C. J. Umbarger,
    "Elemental Analysis of Whole Blood Using Proton-Induced X-Ray
    Emission," Analytical Chemistry 46, 499-503 (1974).

2. V. Valkovic, R. B. Liebert, T. Zabel, H. T. Larson, D. Miljanic,
    R. M. Wheeler and G. C. Phillips, "Trace Element Analysis Using
    Proton-Induced X-ray Emission Spectroscopy," Nuclear Instruments
    and Methods 114, 573-579 (1974).

3. R. D. Lear, H. A. Van Rinsvelt and W. R. Adams, "An Investigation
    of the Correlation Between Human Diseases and Trace Element
    Levels by Proton-Induced X-Ray Emission Analysis," Advances in
    X-Ray Analysis 19, 521-532 (1975).

4. H. C. Kaufmann,Florida State University, private communication.

POLYMER FILMS AS CALIBRATION STANDARDS FOR X-RAY FLUORESCENCE ANALYSIS

T. G. Dzubay and P. J. Lamothe
U. S. Environmental Protection Agency
Environmental Sciences Research Laboratory
Research Triangle Park, NC 27711

H. Yasuda
Polymer Research Laboratory
Research Triangle Institute
Research Triangle Park, NC 27709

ABSTRACT

A new type of calibration standard has been developed for X-ray fluorescence analysis of thin samples. Each standard consists of a polymer film containing a single calibration element. The film is cast from a homogeneous solution containing known amounts of an organometallic compound and a polymer. Polymers of cellulose acetate-propionate and polystyrene are used because of their low moisture affinity. The films have a low ($2-4$ mg/cm^2) mass per unit area for minimal X-ray attenuation. Standards of V, Co, Ni, and Pb have been successfully made. Typical concentrations of 20 µg/cm^2 yield excellent spectra with a low background. The method shows promise for making standards of additional elements.

INTRODUCTION

As X-ray fluorescence (XRF) analysis becomes widely used for determining the elemental composition of environmental samples, large numbers of accurately known calibration standards will be needed. Standards consisting of vapor-deposited metal films, dried solution deposits, and nebulized multielement deposits have previously been used successfully by a number of laboratories (1,2,3,4,5). Because such standards are made individually, it is costly to replicate and certify them.

We have developed a new type of standard that can be accurately replicated in large quantities. For each element, the standard consists of a polymer film containing 0.5 to 1% by weight of a single calibration element. The film is cast from a homogeneous solution containing known amounts of an organometallic compound and the polymer. A sufficiently large quantity of the polymer film can be made in a single batch to meet the needs of a large number of laboratories employing XRF. Compared

411

with the type of standards consisting of deposits on membrane filters
made of cellulose esters, the polymer film standards are more rugged
and are less likely to become mechanically warped.

The most desirable feature of polymer film standards is the attainment of a high degree of homogeneity of the element in the polymer.
The method does not rely on casting films with an accurately known
thickness. Rather, the films are cut into discs of known area and
weighed to determine the total mass per unit area. Then from knowledge
of the weight fraction of the calibration element in the polymer film,
the mass per unit area of the calibration element is determined. By
using sufficiently thin films (2-4 mg/cm^2), the attenuation in the
standards can be a negligible effect for X-rays with energies above 8
keV. For X-rays in the 4- to 8-keV range, the attenuation can be reliably calculated because the chemical matrix of the standards is known.
For this, it is essential that the calibration elements do not occur as
particles or crystals in the film but rather as a homogeneous dispersion.

Standards were made using cellulose acetate-propionate (CAP) and
polystyrene (PS) as the polymers. Standards containing V, Fe, Co, and
Ni were made in CAP films; standards containing Cu and Pb were made in
PS films. For each element, sheets of polymer film were cut into 50
individual 33-mm-diameter discs. The discs were analyzed by gravimetric
and XRF methods to determine the homogeneity of each batch. The weight
percentages of the calibration elements were determined by XRF comparison with vapor deposited standards and by destructive atomic absorption
analysis.

APPROACH

To use a circular disc of polymer film as a standard of known mass
per unit area of a calibration element, one must evaluate the expression

$$m = M f/(\pi r^2) \qquad\qquad (1)$$

where:

  m = mass per unit area of the calibration element, μg/cm^2

  r = radius of calibration disc, cm

  M = total mass of calibration disc, μg

  f = fractional weight of calibration element in disc.

The mass, M, is easily and accurately determined by weighing on a
microbalance. Although the weight fraction, f, can be gravimetrically
determined at the time of fabrication of the films, as will be discussed in the next section, it is necessary to make at least one independent determination of f by using some other analytical procedure
such as atomic absorption or neutron activation analysis. For this, it
is essential that f be constant throughout the batch to allow measurements on a few standards to be representative of the entire batch.

If one wishes to use a polymer film to represent an infinitely thin standard of mass per unit area m', an attenuation correction must be applied to m:

$$m' = mA \qquad (2)$$

where the factor A describes the absorption of the emitted and exciting X-rays in the standard. For homogeneous films,

$$A = [1 - \exp (\mu_L X)]/\mu_L X \qquad (3)$$

where:

$$X = M/(\pi r^2)$$

$$\mu_L = \mu \sec \theta + \mu' \sec \theta'$$

θ and θ' are the mean angles between the normal to the standard and the exciting and observed radiation, respectively.

μ and μ' are the total mass absorption coefficients ($cm^2/\mu g$) of the incident and observed radiation, respectively.

For a multielement substance,

$$\mu = \sum \mu_i W_i \qquad (4)$$

where μ_i and W_i are the mass absorption coefficients and the fraction by weight, respectively, of the individual components (i.e., H, C, O, and calibration element).

EXPERIMENTAL

Preparation of Polymer Standards

Each standard was prepared by casting a film from a solution containing a known amount of an organometallic compound and a polymer. Special emphasis was placed on choosing a combination of a polymer and an organometallic compound that yields a clear film after the solvent is evaporated. The compatibility of the organometallic compound with the polymer is dependent on many factors such as the chemical structure of the polymer and the organometallic compound, the degree of polymerization, the viscosity of the solution, and the concentration of the organometallic compound in the polymer. Selection of combinations of organometallic compounds and polymers was based upon their solubilities in a common solvent; however, their compatibility cannot always be predicted from solubilities alone. If the compatibility is poor, the film contains crystals of the organometallic compound that cause a milky appearance of the film. Clear films do not contain crystalline material detectable by X-ray diffraction or by optical or scanning electron microscopy. Optical clarity of the film was used, therefore, as the criterion for determining the absence of crystals.

Table 1.   PREPARATION OF FILM STANDARDS

Element	Organometallic compound	Polymer	Solvent	Casting thick-ness,um	Film thick-ness,um	Percent metal in film
V	V-acetyl ace-tonate 1.557 g	CAP-20 30 g	THF 170 ml	280	26	1.00
Fe	Fe^{III}-acetyl ace-tonate 1.896 g	CAP-20 30 g	acetone 170 ml	200	20	0.94
Co	Co^{III}-acetyl ace-tonate 1.813 g	CAP-20 30 g	acetone 170 ml	200	19	0.94
Ni	Ni^{II}-acetyl ace-tonate 0.851 g	CAP-20 30 g	acetone 170 ml	200	17	0.64
Cu	Cu^{II}-oleate 2.280 g	PS 30 g	toluene 170 ml	280	46	0.77
Pb	tetraethyl lead 0.368 g	PS 30 g	toluene 170 ml	280	27	0.79

In the selection of polymers, the dimensional stability, moisture insensitivity, mechanical strength, and stiffness were considered. Standards used in this study were based on cellulose acetate propionate (Eastman CAP -20*) and polystyrene. Table 1 shows the amounts of the organometallic compound, polymer, and solvent used for each standard.

In the preparation of the film, the organometallic compound was first dissolved in 170 ml of the solvent, as specified in Table 1. Then 30 mg of polymer was added and allowed to dissolve under continuous stirring for at least 48 hours. The films were cast on a glass plate from the resulting solution; an adjustable casting blade was used to yield the desired film thickness. The film was cast in a room equipped for clean air with vertical laminar flow to avoid the contamination of film by airborne dust and also to facilitate uniform and fast drying of the film. The films were dried in the clean-air room overnight and were then peeled off from the glass plates. The films were cut to the size of approximately 10 by 10 cm, and their weights and areas were determined.

The metal content per unit weight of film was calculated from the amounts of polymer and organometallic compound shown in Table 1. For these calculations, it was assumed that all of the solvent was evaporated from the film and that the weight percent of the metal in the organometallic compound could be computed from the formula for the compound. The latter assumption is valid only if the organometallic com-

* Mention of commercial products or company names does not constitute endorsement by the U. S. Environmental Protection Agency.

pound contains no impurities. For the present feasibility study, the organometallic compounds were obtained from commercial sources and had no certification of purity.

The films for all of the elements except copper were optically clear, indicating that the organometallic compound did not crystallize. The cloudy condition of the film containing copper made it unsuitable as an XRF standard. Further investigation will be required to find a compatible polymer and organometallic compound containing copper.

For each element, ten sheets of 10 by 10 cm films were cast. An alloy steel punch was used to cut five discs of about 37-mm diameter from each sheet to yield a total 50 standards per element. Repetitive measurements indicated that the diameters were 36.47 (0.08)mm (standard deviation in parentheses); the area of the discs was 10.44 (0.05)cm^2.

Gravimetric Analysis

The accuracy of each calibration standard depends in part upon the determination of the total mass, M, as indicated in Equation (1). Each calibration disc was weighed to the nearest 10 μg using a Perkin Elmer Model AD-2* electrobalance. Before the discs were weighed, they were exposed briefly to a ^{210}Po radiation source of 2-mCi activity to reduce any electrostatic charge. A second similar radiation source was positioned inside the balance enclosure below the sample to further reduce any electrostatic charge.

Table 2. RESULTS OF GRAVIMETRIC DETERMINATIONS
OF MASS M OF CALIBRATION DISCS

Standard	Minimum mass[a],mg	Maximum mass[a],mg	Std dev[b] mg	Average mass/area, mg/cm^2
V–CAP	25.61	31.40	0.02	2.6
Fe–CAP	18.15	25.16	0.02	2.0
Co–CAP	14.49	23.21	0.01	1.9
Ni–CAP	13.76	20.05	0.01	1.7
Cu–PS	45.07	51.60	0.04	4.6
Pb–PS	27.84	30.49	0.02	2.7

[a]Analysis of 50 discs from sheets for all elements except copper for which 10 discs from 2 sheets were analyzed.

[b]Standard deviation for three determinations of weight of a single disc on three successive days.

For each standard, an average mass was determined based on measurements on three consecutive days at relative humidities ranging from 39 to 50%. Table 2 shows the range of the masses for each type of standard. As shown in Table 2, the standard deviations for repetitive weighings over a 3 day period ranged from 0.01 to 0.05 mg. When the mass measurements were repeated 30 days later at a relative humidity of 58%, the CAP standards had slightly gained in mass by 0.02 to 0.14 mg. After 30 days the polystyrene standards had lost about 1% of their initial mass.

X-Ray Fluorescence Analysis

An energy dispersive XRF spectrometer equipped with secondary fluorescers for excitation (6) was used to determine the weight fraction, f, and the uniformity of the f values for each batch of polymer standards. Figure 1 shows spectra for the V, Fe, and Ni standards observed using a molybdenum secondary fluorescer. To determine m, the thin film equivalent mass per unit area of a standard, the count rate in the region about the full width at half maximum for the $K\alpha$ line was determined. The absolute calibration in units of $\mu g/cm^2$ was provided by similar analysis of thin vacuum evaporated metal film standards obtained from MicroMatter Company*, Seattle, Washington. Using the measured total masses, the measured radius, r, and the calculated attenuation factors, the weight fractions, f, were calculated according to Equation (1) and are listed in Table 3. Each XRF determination was repeated three times. The precision (1σ) was deduced from the repetitive measurement and is shown in Table 3.

Table 3. X-RAY FLUORESCENCE ANALYSIS OF WEIGHT FRACTION, F, AND
CALCULATED ATTENUATION FACTOR, A, FOR POLYMER STANDARDS

Standard	A^a	$100f^b$	Uniformity of batch[c]	XRF accuracy[d]
V–CAP	0.955	0.92 (0.007)	0.007	0.07
Fe–CAP	0.983	1.07 (0.007)	0.046	0.05
Co–CAP	0.987	0.98 (0.007)	0.014	0.08
Ni–CAP	0.991	0.60 (0.005)	0.007	0.05
Cu–PS	0.987	0.54 (0.005)	0.005	0.03
Pb–PS	0.995	0.80 (0.008)	0.012	0.04

[a] Calculated attenuation factor for average mass per unit area shown in Table 2.

[b] Normalized using vacuum deposited films obtained from MicroMatter Company. The numbers in parenthesis represent 1σ measuring precision determined from repetitive XRF analysis of each standard.

[c] Uniformity of batch (including XRF measurement errors) expressed as measured standard deviation of 100f.

[d] Accuracy of 100f based upon the estimated accuracy of the vacuum evaporated films.

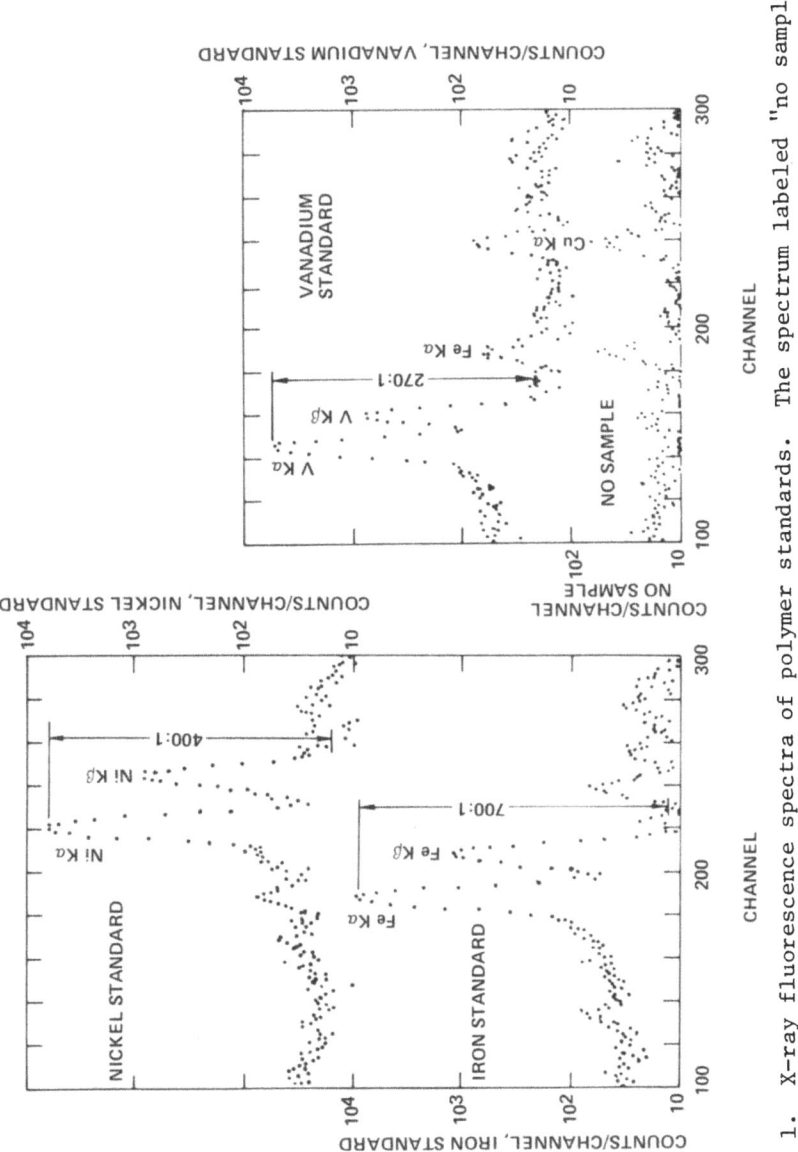

Figure 1. X-ray fluorescence spectra of polymer standards. The spectrum labeled "no sample" represents the background from the XRF spectrometer when no sample is present. All spectra were obtained using the energy dispersive spectrometer described in reference 6 operated with a molybdenum secondary fluorescer, an anode potential of 50kV, a current of 0.28 mA, and a counting time of 300 seconds.

The values of f in Table 3 represent the average for 10 randomly
selected samples from each batch. The standard deviation for the 10
samples represents the uniformity of the batch. Excellent uniformity
in the weight fraction, f, was observed for the films containing V, Co,
Ni, Cu, and Pb. For those elements, the measured standard deviations
amounted to less than 1.5%. Since this includes the XRF measuring precision of about 1%, there is no evidence of any significant nonuniformity.
For the film containing Fe, the measured uniformity for the batch is
4.3%. This is significantly larger than the measuring precision and
indicates a significant nonuniformity in that batch.

The overall accuracy in the XRF determination of the weight fraction depends primarily on the accuracy of the vacuum-deposited calibration films which were used as a reference. Although the manufacturer
does not state the 1σ accuracy of such films, we believe that it is
about 8%. For Co and Ni, only one vacuum-deposited reference film for
each element was used, and the uncertainty is therefore estimated to be
8%. For Fe, Cu, and Pb, the use of three such reference films resulted
in an uncertainty of 5%. Vacuum-deposited films of Ti and Cr were used
to determine vanadium value by interpolation.

In Table 4, the above XRF determinations of f are compared with the
XRF determinations of Dr. B. W. Loo of Lawrence Berkeley Laboratory and
Dr. J. R. Rhodes of Columbia Scientific Industries. The XRF spectrometer used by Dr. Loo was calibrated with a copper film known to be
accurate and with a set of standards containing copper and other elements
in known ratios (5). Dr. Rhodes used solution-deposited standards in
his calibration (1).

Table 4. RESULTS OF THE XRF DETERMINATIONS OF WEIGHT PERCENT
OF THE CALIBRATION ELEMENT IN POLYMER

Standard	Weight percent			
	Present work	Loo LBL	Rhodes CSI	Mean (σ)
V–CAP	0.92 (0.07)[a]	1.05 (0.03)[a]	1.10 (0.10)[a]	1.02 (0.09)[b]
Fe–CAP	1.07 (0.05)		1.15 (0.05)	1.11 (0.06)
Co–CAP	0.98 (0.08)	1.05 (0.03)	1.09 (0.05)	1.04 (0.06)
Ni–CAP	0.60 (0.05)	0.67 (0.02)	0.68 (0.02)	0.65 (0.04)
Cu–PS	0.54 (0.03)			
Pb–PS	0.80 (0.04)	0.75 (0.03)	0.74 (0.03)	0.76 (0.03)

[a] Estimated overall accuracy.

[b] Standard deviations for the three determinations.

Atomic Absorption Analysis

To achieve an independent determination of the elemental content of several polymer standards, samples were sent to the National Bureau of Standards (NBS) and to the DuPont Analytical and Physical Measurement Service for atomic absorption (AA) and Flame Emission Spectrometric (FES) analysis. Each laboratory received four samples of blank polymer films plus triplicate samples of the following elemental standards: V, Co, Ni, and Pb.

Samples of the standards in the CAP matrix (V, Co, Ni) were digested for 1 hour using 20 ml of hot concentrated nitric acid followed by dilution to a known volume using deionized water. Two different techniques were used for the digestion of the Pb in polystyrene samples. DuPont used a sodium carbonate fusion procedure whereas NBS dissolved the sample in benzene followed by nitric acid digestion of the solution.

Analyses of the digested samples were accomplished using background-corrected atomic absorption spectrometers. An air-acetylene flame was used for the Pb, Co, and Ni samples, and a nitrous oxide-acetylene flame was used for the vanadium determinations. Results of the analyses are given in Table 5. It is believed that the low Pb results were due to tetraethyl lead losses that occurred during the digestion processes.

DISCUSSION

The ultimate usefulness of the polymer films as XRF standards depends upon the accuracy with which the mass per unit area can be specified. From the previous discussion, this depends upon the films having a high degree of total mass stability and homogeneity and an accurately known weight fraction for the calibration element.

Table 5. COMPARISON OF DETERMINATIONS OF THE WEIGHT PERCENT

Standard	Weight percent			
	Prepared	XRF mean (σ)	AA Lab 1	AA Lab 2
V-CAP	1.00	1.02 (0.09)[a]	0.82 (0.04)[b]	0.89[c] (0.01)[b]
Fe-CAP	0.94	1.11 (0.06)		
Co-CAP	0.94	1.04 (0.06)	0.90 (0.01)	0.61 (0.02)
Ni-CAP	0.64	0.65 (0.04)	0.52 (0.01)	
Pb-PS	0.79	0.76 (0.03)	0.28 (0.04)	0.64 (0.01)

[a]Standard deviation for independent determinations by three different laboratories.

[b]Standard deviation of three replicate determinations

[c]The vanadium standard was analyzed by FES rather than AA.

The mass stability of the polystyrene and CAP films is very good. The polystyrene films exhibited no humidity dependence, but did lose about 1% of their mass in a 1-month period. This loss may be due to a slow rate of release of residual toluene that was used in the preparation of the films. This problem might be eliminated by using benzene, which would be expected to evaporate more completely soon after the films are cast. For the CAP films, the masses exhibited a slight humidity dependence, but the effect was less than 1% for relative humidity below 58%.

Table 3 indicates that the uniformity of the batches of V, Co, Ni, and Pb standards was better than 2%, an indication that large numbers of such standards could be replicated with good precision. On the other hand, the uniformity of the iron standards was an unacceptable 4.3%. The cause of this nonuniformity is presently unknown. For copper, the measured uniformity in the weight percent was excellent, but the cloudy films indicate a chemical incompatibility between the organometallic compound and polymer. Further development work is needed for both the iron and copper standards.

The XRF measurements provided the best determinations of the weight percent of the calibration element in each batch. The standard deviations for the determinations by three laboratories ranged from 4% for Pb to 9% for V. Because each XRF laboratory used an independent calibration procedure, any systematic errors in the mean XRF values are assumed to be no larger than these standard deviations.

There is a significant lack of agreement between the AA determinations shown in Table 5. Possible causes may be incomplete digestion of the sample matrix or loss of sample during digestion. Until the cause of this lack of agreement is determined, the present AA results are not usable for specifying the true elemental weight percents.

There is fair agreement between the XRF measured and gravimetrically prepared weight fractions for the V, Fe, Co, Ni, and Pb standards (2%, 18%, 11%, 2%, and - 4%, respectively). Because the organometallic compounds were of uncertified purity, the gravimetrically determined weight fractions are of unknown accuracy. We are presently attempting to purify the organometallic compounds prior to preparation of the standards.

The long-term stability of the calibration elements in the standards needs consideration. Although exposure to X-rays or sunlight may eventually make the polymer weak and brittle, such exposures would not necessarily cause loss of the calibration element. Present experience indicates that there is no deterioration over a 2-month period of storage at room temperature. Until long-term exposure studies are complete, it would be prudent to store the standards in a dry, opaque enclosure at temperatures below 22° C.

CONCLUSIONS

A set of XRF standards consisting of organometallic compounds that are homogeneously dispersed in a polymer matrix was developed for V, Co, Ni, and Pb. The standards are presently known to an accuracy ranging from 5-9% (1 σ), which is based upon three independent XRF determinations. If organometallic compounds of known composition could be obtained, then the composition of the standards could be gravimetrically determined with much better accuracy when the standards are prepared. Further development effort is expected to lead to the production of polymer standards for additional elements.

ACKNOWLEDGEMENTS

The authors are indebted to Mr. G. W. Ivey for performing the gravimetric and numerical analysis. Dr. B. W. Loo and Dr. J. R. Rhodes are to be thanked for providing the results of their XRF analyses.

REFERENCES

1. A. H. Pradzynski and J. R. Rhodes, "Development of Synthetic Standard Samples for Trace Analysis of Air Particulates," Publication 598, p.320-336, American Society for Testing and Materials (1976).

2. D. C. Camp, A. L. VanLehn, J. R. Rhodes, and A. H. Pradzynski, "Intercomparison of Trace Element Determinations in Simulated and Real Air Particulate Samples," X-Ray Spectrom. 4, 123-137 (1975).

3. R. M. Baum, W. F. Gutknecht, R. D. Willis, and R. L. Walter, "Preparation of Standard Targets for X-Ray Analysis," Anal. Chem. 47, 1727-1728 (1975).

4. A. R. Stiles, T. G. Dzubay, R. M. Baum, R. L. Walter, R. D. Willis, L. J. Moore, E. L. Garner, J. W. Gramlich, and L. A. Machlan, "Calibration of an Energy Dispersive X-Ray Fluorescence Spectrometer," in R. W. Gould, C. S. Barrett, J. B. Newkirk, and C. O. Ruud, Editors, Advances in X-Ray Analysis Vol. 19, p.473-486, Kendall/Hunt (1976).

5. R. D. Giauque, R. B. Garrett, and L. Y. Goda, "Calibration of Energy Dispersive X-Ray Spectrometers for Analysis of Thin Environmental Samples," in X-Ray Fluorescence Methods for Analysis of Environmental Samples, Chapter 11, Ann Arbor Science (1976).

6. J. M. Jaklevic, F. S. Goulding, B. V. Jarett, and J. D. Meng, "Applications of X-Ray Fluorescence Techniques to Measure Elemental Composition of the Atmosphere," in R. K. Stevens and W. F. Herget, Editors, Analytical Methods Applied to Air Pollution Measurements p.123-146, Ann Arbor Science (1974).

CHEMICAL ANALYSIS OF NICKEL ORES BY ENERGY DISPERSIVE X-RAY FLUORESCENCE

Bradner D. Wheeler, Daniel M. Bartell, and John A. Cooper

ORTEC, Incorporated*

Oak Ridge, TN   37830

INTRODUCTION

Chemical analysis of geological materials such as nickel ores has been accomplished by atomic absorption (1,3) x-ray fluorescence (11,14) and conventional wet methods (10). Procedures utilizing these techniques are capable of producing excellent results but are often difficult and time consuming.

Minerals often present serious problems in chemical analysis by wet methods. X-ray analysis can therefore offer the analyst considerable savings in time providing the obstacles which exist are understood and minimized or eliminated. The most serious problems to solve are absorption and enhancement effects, mineralogical differences among samples, sample preparation, and particle size effects which often influence the intensities of the analytical lines. In addition, the element of interest may be of low concentration in a variable and unknown matrix. The relative intensities of a standard and an unknown are often approximate measurements and not directly proportional to the concentration since the matrix, in addition to the concentration of the element, determines the intensity of the measured characteristic radiation. Matrix effects are generally considered as self absorption, interelement effects, mineralogical differences among samples, and inhomogeneity of the sample particles. Evaluation of these problems, thus providing a useful and workable method of analysis through x-ray fluorescence, has been approached by use of internal standards (4), comparison with standards approximate in composition to the unknowns (9), fusion and dilution with transparent materials (2,7,11) reduction of particle size by fine grinding (4,5,12, 13), and mathematical corrections (2,7,11,12). The method used by the authors utilizes the powder method, fine grinding, and empirical calculations for corrections due to interelement effects.

THEORY

X-ray fluorescence analysis of any material is best applied to samples where the compositional range of elements is reasonably small. Garnierite nickel ores fall into this category but several of the elements of interest are of low atomic number in a variable matrix.

---

*A wholly owned sudsidiary of EG&G Incorporated, Wellesley, MA.

In order to successfully apply an x-ray fluorescence technique, the characteristics affecting the reproducibility and accuracy must be identified and corrected. The factors affecting the analysis of granierite are particle size, mineralogy, and interelement effects due to varying chemical composition among samples.

PARTICLE SIZE

Accurate and reproducible analytical results by this method in the determination of mineralogical samples requires proper sample preparation in order to eliminate intensity fluctuations as a function of variations in particle size and distribution. A study by Burnstein (5) working with a common element in several minerals cites examples where fine grinding tends to reduce the intensity differences normally associated with mineralogical variations. A similar study by Campbell and Thatcher (6) measuring calcium in wolframite where the calcium may be present as a carbonate, tungstate, or phosphate supported Burnstein's work. Differences in intensities were observed for equal concentrations of calcium in the three chemical states when the particle size is large as compared to the depth of penetration. Extensive grinding illustrated the intensities from the different mineral forms approach a common value by reducing the absorption within the individual particles to a very small value.

INTERELEMENT EFFECTS

Quantitative analysis of any material by x-ray fluorescence requires that the measured intensity of a particular element is proportional to the percent composition. In complex geological materials, such as garnierite, the intensity of an element may not be directly proportional to the concentration due to the effect of another element in the sample. The nonlinearity in multielement systems is commonly referred to as the interelement effect and, as mentioned previously, may be in the form of enhancement of absorption.

Interelement effects were found to be significant in nickel silicate ores as illustrated in Figure 1. One of the most severe effects is the absorption of Ni K_α line which occurs just slightly on the high energy side of the iron K absorption edge. The same case is true of silica since the Si K_α line occurs near the aluminum K edge on the high energy side resulting in strong absorption of Si K_α by aluminum and enhancement of Al K_α by silica. The analysis of the low Z elements in this matrix is further complicated by the presence of the heavier elements such as Cr, Fe, and Ni which all act as absorbers for Mg, Al, and Si. Iron, for example, has a mass absorption coefficient in excess of 3500 at the energy levels of magnesium through silicon, in addition to chromium and nickel, acts as a strong absorber for these low energy elements.

Mathematical corrections can be readily applied in order to resolve these interelement effects. Although numerous methods have been proposed, the authors have utilized an exponential method known as Beer's law (14) as described by Hasler and Kemp (8). The generalized equation for interelement corrections for this application is as follows:

$$\%C = A + BI_c \exp (m_1I_1 + m_2I_2 + m_3I_3) + \ldots + m_nI_n) \qquad (1)$$

where

$\%C$ = present of element C present

A = the X intercept

B = the slope of the calibration curve

I_c = intensity of element c

m_1, m_2, m_3 = positive or negative constants of interfering elements 1, 2, and 3 derived from the calibration standards.

I_1, I_2, I_3 = intensities of interfering elements 1, 2, and 3.

Calibration of the standards using an iterative process with Equation 1 establishes the x intercept, slope, and proportionality constants. Unknown samples are analyzed in an identical manner utilizing the previously calculated equation derived from the standards. The interelement correction procedure as described is not designed as a general case and was not intended to correct for peak overlap. This technique, however, may be applied to other types of problems (8,12) providing major interelement effects are identified and treated in a similar manner.

SAMPLE PREPARATION

Particle size as previously discussed can affect the intensity of most elements to be determined in powder samples. In order to minimize particle size effects, six separate samples of one standard consisting of five grams of ore and 100 mg of Ivory Snow for a grinding aid were placed in a tungsten carbide rotary swing mill and ground for two to seven minutes. The resulting powder was then pressed into a pellet under 15 tons/square inch with boric acid as a backing material. Each pellet was analyzed for 200 seconds and the intensities of each element plotted as a function of grinding time. Examination of the grinding time vs. intensity illustrated that the intensities stabilized after a four minute grinding time. As a result, all the samples were ground for the minimum grinding time plus one minute or for a total of five minutes.

INSTRUMENTATION

An ORTEC 6110 Tube-Excited Fluorescence Analyzer (TEFA) was used for this analysis. The samples were analyzed in a vacuum under the following instrumental parameters:

Anode: Rh
Filter: None
Anode Voltage: 15 kV
Anode Current: 50µA
Analysis Time: 100 seconds

B. D. Wheeler, D. M. Bartell, and J. A. Cooper

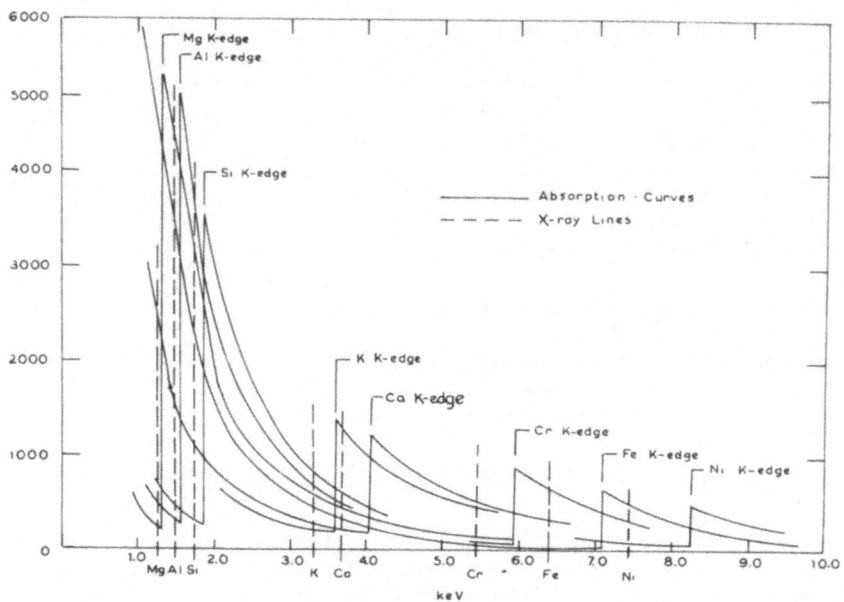

Figure 1

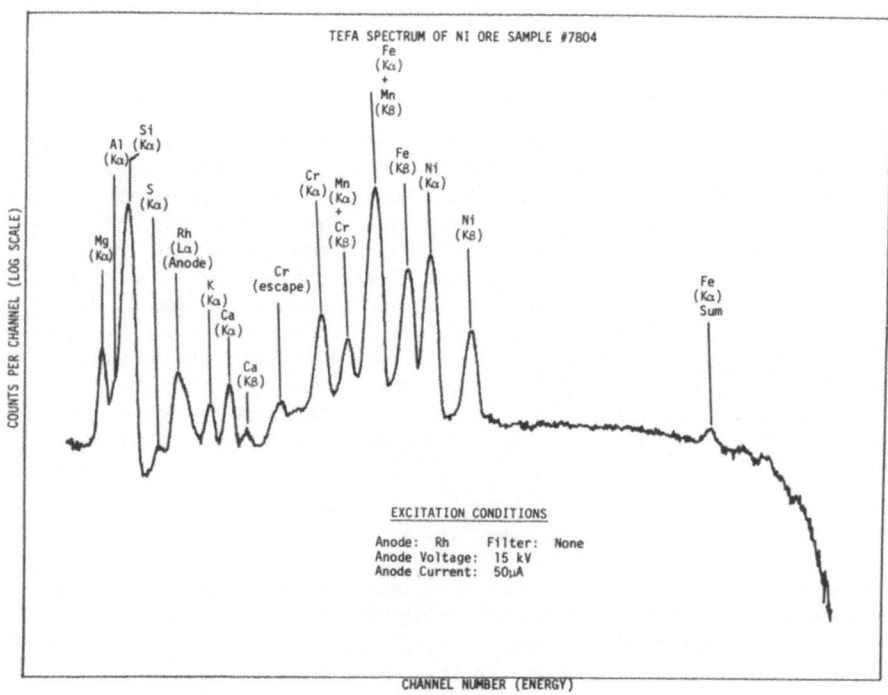

Figure 2

Representative spectra illustrating the resolution is illustrated in
Figure 2.

RESULTS AND DISCUSSION

The samples utilized in this study were twelve well characterized
garnierite nickel ores. Quantitative calculations and interelement
corrections were performed with a 16 bit, 16,000 word PDP 11/05 computer.

Nickel, which is the most important element in garnierite, exhibited
an absolute error of approximately 1% when utilizing a linear least
squares fitting program of concentration vs. uncorrected intensities.
Applying interelement corrections for the effect of chromium and iron
reduced this error to 0.03% absolute. Figure 3 illustrates the corrected and uncorrected calibration for nickel.

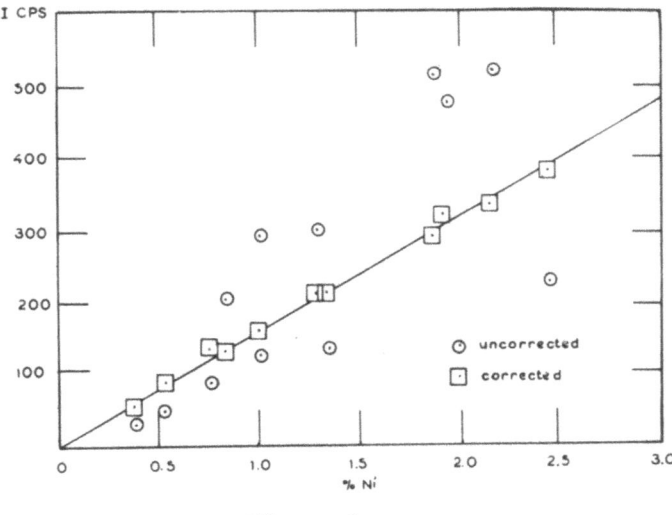

Figure 3

Analysis of the light elements in geological materials usually
require interelement correction due to severe enhancement and absorption
effects. Silicon K_α, as previously described, is strongly absorbed by
aluminum which in turn excites aluminum K_α. The problem is further
complicated by the fact that the mass absorption coefficients of nickel,
iron, and chromium are in excess of 3500 at the energy of Si K_α. Consequently, interelement corrections involving these four elements must be
applied in the analysis of silicon. Prior to correction, as illustrated
in Figure 4, the absolute error of silicon with uncorrected linear fitting
was 5% absolute. Following interelement corrections, the error was to
0.26% absolute. Another major element in this material, iron, exhibited
errors as large as 4% absolute without interelement corrections. Correcting iron for the influence of chromium and nickel, the errors were reduced
to 0.24% absolute as illustrated in Figure 5. The results of the total
analysis of the samples described are listed in Table 1.

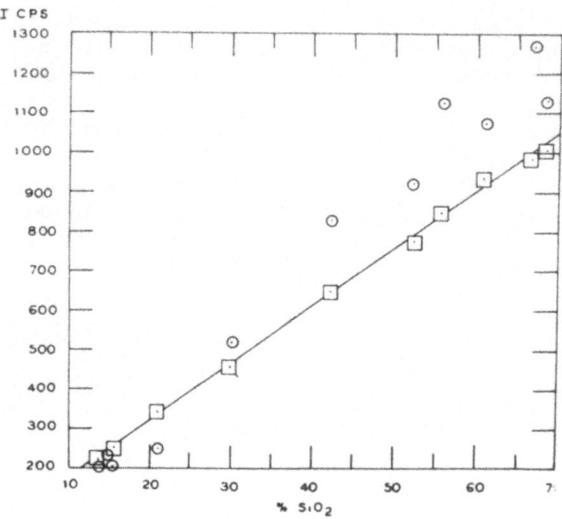

Figure 4

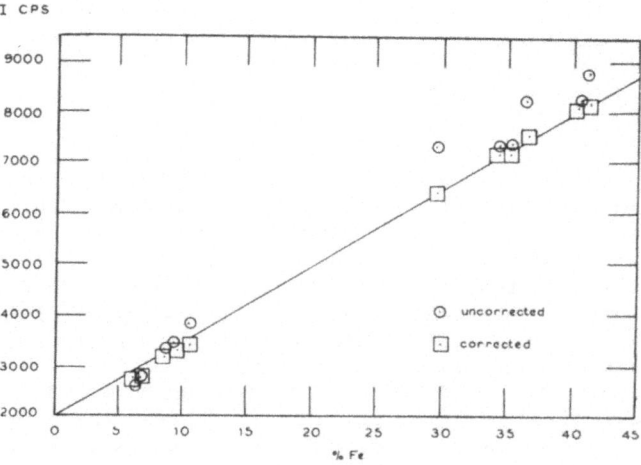

Figure 5

TABLE 1

SAMPLE I.D.	Ni		Fe		Mg		Al		SiO$_2$	
	LIST	TEFA	LIST	TEFA	LIST	TEFA	LIST	TEFA	LIST	TEFA
8071	1.94	2.01	10.5	10.3	9.6	9.7	1.6	1.70	42.7	42.8
7806	2.15	2.10	8.7	8.9	8.1	8.1	1.3	1.26	52.6	52.9
7798	1.04	1.02	29.6	29.7	2.6	2.6	2.2	2.20	30.2	29.9
7983	0.83	0.83	8.8	8.6	4.7	4.9	0.7	0.58	61.4	61.3
7981	1.00	1.03	6.5	6.4	5.1	5.1	1.4	1.35	67.3	67.0
21747	0.55	0.54	41.2	40.8	0.3	0.5	5.6	5.56	15.4	15.2
7804	1.87	1.83	6.8	6.9	6.7	6.4	0.5	0.69	55.8	55.5
7979	1.30	1.34	10.6	10.8	3.8	3.8	1.6	1.71	68.6	68.9
21720	0.77	0.84	36.6	36.6	0.9	0.7	7.4	7.34	21.0	21.3
21702	0.39	0.36	40.6	40.6	0.4	0.4	4.6	5.00	14.8	14.4
21722	1.34	1.37	34.5	34.8	4.7	4.5	1.9	1.91	21.3	21.1
7844	2.44	2.44	35.3	34.9	5.2	5.3	2.6	2.62	15.0	15.4
Avg Err	0.03		0.24		0.11		0.09		0.26	

SUMMARY

The application of energy dispersive x-ray fluorescence in the analysis of nickel silicate ores utilizing the powder method can provide the analyst with a rapid and accurate procedure for mineral assay. Many of the problems commonly termed matrix effects are due to particle size effects. Consequently, prior to application of interelement corrections, the samples must be properly prepared by fine grinding and pelletizing. Improvements in analytical results by interelement corrections are evident with the most significant being nickel, silicon, and iron. Analytical values obtained by this method are within the limits accepted by conventional techniques.

REFERENCES

1. S. Abbey, "Analysis of Rocks and Minerals by Atomic Absorption Spectroscopy", Geol. Survey, Canada Paper 67-37.

2. G. Anderman and J. O. Allen, "X-ray Emission Analysis of Finished Cements", Anal. Chem. 33(12), 1961.

3. C. B. Belt, "Partial Analysis of Silicate Rocks by Atomic Absorption", Anal. Chem. 39, 1967.

4.   O. E. Brown, "Use of X-Ray Emission Spectroscopy in Chemical Analysis of Cement, Raw Materials, and Raw Mix", ASTM, 66th Annual Meeting, 1963.

5.   F. Burnstein, "Particle Size and Mineralogical Effects in Mining Applications: 11th Annual Conference of Applications of X-Ray Analysis, Denver Research Institute, University of Denver, 1962.

6.   W. J. Campbell and J. W. Thatcher, Advances in X-Ray Analysis, Vol. 2., University of Denver, Plenum Press, New York, 1958.

7.   F. Claisse and C. Samson, Advances in X-Ray Analysis, Vol. 5, p 335, Plenum Press, New York, 1961.

8.   M. R. Hasler and J. W. Kemp, Amer. Soc. for Testing Materials, A.S.T.M. Committee E-2, Philadelphia, PA., 1957.

9.   B. Kester, AIEE Cement Industry Conference, Milwaukee, WI, 1960.

10. J. A. Maxwell, <u>Rock and Mineral Analysis</u>, Interscience Publishers, New York, 1968.

11. M. J. Rose, J. Adler, and F. J. Flanagan, "X-ray Fluorescence Analysis of Light Elements in Rock and Minerals", Applied Spectroscopy 17, 4, 1963.

12. B. D. Wheeler, "Cement Raw Mix Control Through X-Ray Emission Spectroscopy", Proceedings of Third Forum on Geology of Industrial Minerals, Special Distribution 34, University of Kansas, 76-96, 1967.

13. B. D. Wheeler, D. M. Bartell, and J. A. Cooper, "Chemical Analysis of Portland Cement by Energy Dispersive X-Ray Fluorescence", Pittsburgh Conference on Analytical Chemistry and Applied Spectroscopy, 1976.

14. P. D. Zemany, M. A. Liebhafsky, and H. G. Pfeiffer, <u>X-Ray Absorption and Emission in Analytical Chemistry</u>, John Wiley and Sons, Inc., New York, 1954.

DETERMINATION OF SULFUR, ASH, AND TRACE ELEMENT CONTENT OF COAL, COKE, AND FLY ASH USING MULTIELEMENT TUBE-EXCITED X-RAY FLUORESCENCE ANALYSIS

J. A. Cooper, B. D. Wheeler, G. J. Wolfe, D. M. Bartell, and D. B. Schlafke

ORTEC, Incorporated*

Oak Ridge, TN   37830

SUMMARY

A procedure using tube excited energy dispersive x-ray fluorescence analysis with interelement corrections has been developed for multielement analysis of major and trace elements and ash content of coal, coke, and fly ash. The procedure uses pressed pellets and an exponential correction for interelement effects. The average deviations ranged from about 0.0003% for V at an average concentration of about .003% to 0.1% for S at an average concentration of 4%. About 25 elements were measured and 100 second minimum detectable concentrations ranged from about one part per million for elements near arsenic to about one tenth of one percent for sodium.

INTRODUCTION

Coal is an important current energy source and will continue to play a major role in meeting our future energy needs. The quality and cost of coal currently depends on its sulfur and ash contents and may depend on other specific elemental parameters in the future as environmental requirements become more restrictive. This paper describes a procedure using tube excited energy dispersive x-ray fluorescence which has been developed to measure about 25 elements in coal, coke, and fly ash and to estimate the ash content of coal and coke.

EXPERIMENTAL CONDITIONS

An ORTEC 6110 TEFA (Tube-Excited Fluorescence Analyzer) System was used for this investigation. It included a PDP 11/05 Computer and dual drive floppy disk.

The samples analyzed in this study consisted of a variety of coals, cokes, and coal ash samples (1-4). They were ground for two minutes in a Spex Shatter Box (5) rotary swing mill using a variety of grinding aids and pelletized. The resulting pellets were loaded directly into the sample chamber which was then evacuated.

---

*A wholly owned subsidiary of EG&G Incorporated, Wellesley, MA.

The resulting spectral data were quantitatively analyzed using ORTEC's FLINT (6) software which provides a linear least squares fit to the inter-element corrected intensities. This program corrects the observed x-ray intensities for absorption and enhancement due to the presence of other elements. The concentration of the ith element is given by the equation:

$$C_i = A' + B'I_i \, [exp \, (-M_{ij} \, C_j)] \tag{1}$$

where

M_{ij} is the interaction coefficient for element j on element i and

C_j is the concentration of the jth element.

The interaction coefficients are determined by a nonlinear multiple least squares fit of the standards concentration-intensity data. This requires a minimum of n + 2 standards where n is the number of interfering elements.

Elemental concentrations in unknown samples were calculated with an iterative process using Equation 1 and interaction coefficients calculated from standards.

RESULTS AND DISCUSSION

Representative x-ray spectra excited under different excitation conditions are illustrated in Figures 1 and 2. Figure 1 shows the semilogarithmic plot of the low energy portion of the x-ray spectrum of a fly ash specimen. This spectrum was excited with direct bremsstrahlung radiation from a rhodium anode which maximizes the sensitivity for the light elements including sulfur. This excitation condition provides for the simultaneous analysis of the major and minor elements from sodium to iron. Although there is considerable peak overlap in the sodium to silicon region of the spectrum, simple "region-of-interest" peak integrations can be used since the interelement correction program will minimize both the spectral interference and concentration variation effects.

The optimum sensitivity for trace elements from iron to strontium, including elements near lead, is obtained by using molybdenum filtered radiation from a molybdenum anode as shown in the semilogarithmic plotted spectrum in Figure 2. Twenty elements are measurable in this single spectrum. Between twenty-five and thirty elements can be measured in these samples by using different excitation conditions. Upper limits can be set for many other elements based on the minimum detectable concentrations (MDC) shown in the 1000 second MDC plots in Figure 3.

The qualitative results for coal, fly ash, and coke are summarized in Tables 1 and 2. Table 1 compares the results obtained by tube-excited fluorescence analysis (TEFA) with those listed by the Illinois State Geological Survey (1). The TEFA results are not averages of replicate analyses but the results of a single analysis. Even so, the average deviations of about 0.02% represent relative accuracies of about 1%. These analyses, as well as the aluminum, sulfur and calcium, were corrected for interelement effects while the other elements did not require corrections.

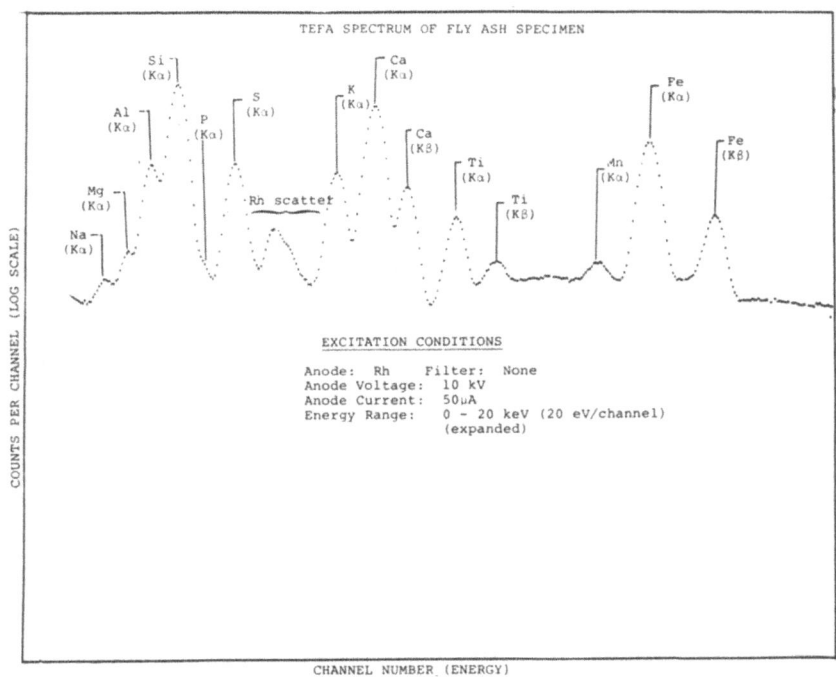

Figure 1. Logarithmic plot of the x-ray spectrum obtained from a
specimen of fly ash obtained with a Tube-Excited Fluorescence Analyzer.

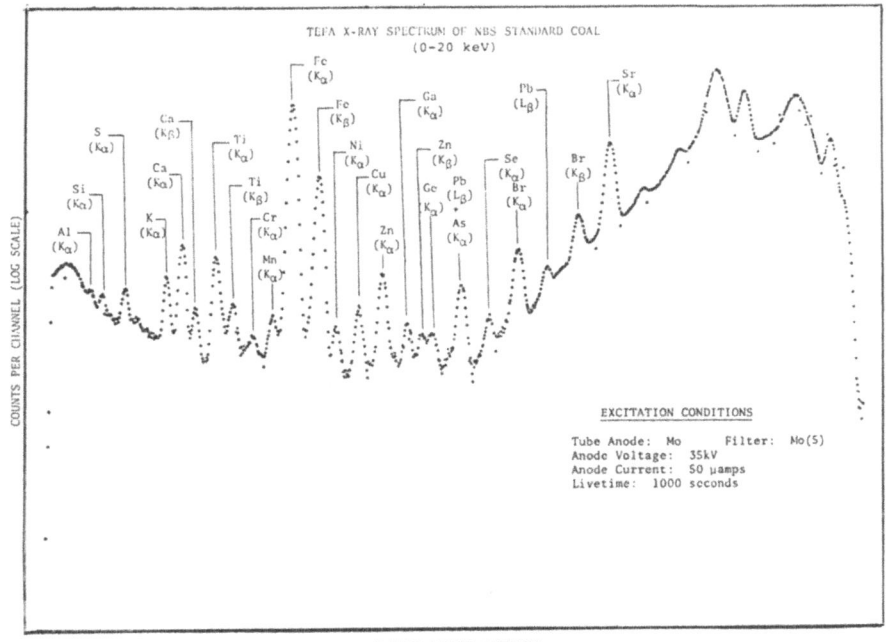

Figure 2. Logarithmic plot of the x-ray spectrum obtained from a
specimen of coal obtained with a Tube-Excited Fluorescence Analyzer.

TABLE 1. COMPARISON OF RESULTS OBTAINED FOR THE ANALYSIS OF ILLINOIS STATE GEOLOGICAL SURVEY COAL SAMPLES

	C15263		C15278		C16264		C16317		C16408		C16408*/C16317		C16264*/C16317		C15278*/C16264		Unknown		Avg. Deviation
	TEFA	LIST	TEFA	LIST	TEFA	LIST	TEFA	LIST	TEFA	LIST	TEFA	LIST	TEFA	LIST	TEFA	LIST	TEFA	LIST	
Percent																			
Na	.0300	.0140	.0300	.0480	.0300	.0510	.0300	.0170	.0300	.0070	.0300	.0120	.0300	.0370	.0300	.0050	.0300	.0310	.0200
Mg	.0400	.0400	.0400	.0500	.0400	.0400	.0400	.0500	.0400	.0300	.0400	.0400	.0400	.0450	.0400	.0450	.0400	.0450	.0050
Al	1.0100	1.0100	1.0500	1.1300	.9400	.9200	1.1300	1.1200	1.0000	1.0200	1.0900	1.0700	1.0300	1.0200	1.0600	1.0200	1.0800	1.0700	.0200
Si	1.6400	1.6500	2.1700	2.1100	1.8900	1.9200	2.4400	2.4800	1.3800	1.4100	2.0700	1.9400	2.1900	2.2000	2.0800	2.0400	2.0200	1.9100	.0400
P	.0030	.0020	.0060	.0070	.0120	.0110	.0090	.0020	.0140	.0200	.0110	.0110	.0100	.0066	.0040	.0090	.0020	.0050	.0030
S	3.0500	3.1600	3.5800	3.3500	4.3200	4.5200	3.5100	3.2500	5.0300	4.9900	4.0200	4.0800	3.8000	3.8800	3.7700	3.9400	3.0000	3.2600	.1500
Cl	.0600	.0200	.0500	.1100	.0600	.0100	.0300	.0200	.0600	.1000	.0400	.0600	.0400	.0150	.0600	.0600	.0600	.0550	.0300
K	.1500	.1400	.1700	.1700	.1600	.1500	.1500	.1700	.1400	.1300	.1500	.1500	.1600	.1600	.1500	.1600	.1500	.1600	.0080
Ca	.1400	.1000	.6100	.8200	.6700	.5600	.7400	.7300	.2100	.2300	.4800	.4800	.7200	.6400	.6700	.6900	.4600	.4600	.0600
Ti	.0600	.0500	.0600	.0600	.0600	.0500	.0600	.0700	.0500	.0500	.0600	.0600	.0600	.0600	.0500	.0550	.0500	.0550	.0050
V	.0029	.0023	.0027	.0027	.0026	.0022	.0027	.0032	.0029	.0031	.0028	.0032	.0027	.0027	.0025	.0024	.0026	.0025	.0003
Fe	2.6400	2.6300	1.6000	1.6500	2.0500	2.0500	1.6100	1.570	3.5200	3.5100	2.5000	2.5400	1.8300	1.8100	1.8600	1.8500	2.1900	2.1400*	.0200
Parts Per Million																			
Ni	42	40	14	8	16	22	23	30	24	26	26	28	25	26	21	15	30	24	.0004
Cu	40	44	7	8	5	10	22	20	15	16	21	18	19	15	11	9	24	26	.0003
Pb	106	96	30	9	43	51	53	72	41	40	52	56	52	62	37	30	46	52	.0010
Ash (%)	8.8	8.0	9.4	11.0	10.2	12.4	9.7	12.0	11.5	11.2	10.5	11.6	9.9	12.2	9.2	11.7	9.2	9.5	.0200

*These samples were composed of 50% of each of the samples listed.

TABLE 2. COMPARISON OF TEFA AND WET CHEMICAL ANALYSIS RESULTS FOR COAL ASH

ELEMENTAL CONCENTRATIONS (%)

SAMPLE ID	SiO2		Al2O3		Fe2O3		TiO2[f]		P2O5[f]		CaO		MgO		SO3		Na2O[f]		K2O		Sr		TOTAL	
	TEFA	WC	TEFA	WC	TEFA	WC	TEFA	WC	TEFA	WC	TEFA	WC	TEFA	WC	TEFA	WC	TEFA	WC	TEFA	WC	TEFA	WC	TEFA	WC
2819	47.76	48.15	14.72	15.70	3.08	2.86	.85	1.03	.26	.29	12.08	12.32	4.40	3.35	8.72	8.96	6.27	6.57	1.28	1.14	0.33	-	99.99	100.13
2822	53.00	53.50	16.73	17.95	4.42	3.52	.75	0.05	.25	.46	9.88	9.38	3.44	3.27	4.94	5.57	2.80	3.07	1.87	1.88	0.16	-	98.87	98.02
2840	39.60	39.79	25.68	25.80	6.89	7.01	.72	0.53	.36	.22	13.67	12.74	4.35	3.95	8.05	8.06	.19	0.22	.58	0.05	0.05	-	100.14	98.02
2843	40.67	39.59	28.35	27.75	5.25	5.38	.69	1.19	.37	.23	12.35	12.60	2.80	3.32	10.16	10.21	.19	0.15	.66	0.04	0.04	-	101.74	100.14
2844c	38.88	39.05	24.91	25.25	5.16	5.15	.69	0.89	.24	.19	17.18	17.70	4.39	4.32	6.24	6.35	.19	0.34	.80	0.87	0.04	-	98.83	98.85
2945	49.95	49.22	24.13	24.75	18.56	18.79	1.00	1.04	.42	.46	.90	0.56	1.58	1.43	.75	.79	2.97	3.06	2.54	2.52	0.01	-	101.03	101.10
2946	47.00	46.54	21.89	22.12	17.11	17.02	.85	1.00	.46	.42	4.83	4.90	.79	.75	4.82	5.12	.19	0.40	1.66	1.66	0.05	-	100.49	100.36
3003	42.33	43.67	27.43	26.80	20.12	20.02	1.09	1.10	.45	.52	2.72	3.25	.98	1.06	5.12	5.12	.38	0.11	1.79	1.45	0.02	-	101.64	101.52
3023-1	79.09	79.11	15.67	15.30	.78	.80	1.09	1.00	.33	.15	.17	.14	.14	.35	.40	.05	.38	.11	1.59	1.46	0.01	-	99.85	98.43
3048	48.69	49.22	27.70	28.00	3.91	3.43	.85	.88	.17	.17	8.17	8.40	.79	.91	6.31	6.24	.46	.11	1.30	1.20	0.07	-	99.04	98.43
3054	46.66	47.08	21.01	20.37	3.11	3.72	.87	.89	.21	.30	12.03	11.76	6.07	6.47	5.21	5.25	.19	.28	1.46	1.20	0.27	-	100.79	98.43
3055	47.90	46.54	18.54	18.25	3.76	4.00	.78	.89	.26	.33	11.20	11.02	5.67	6.22	6.70	7.10	4.78	4.62	1.10	1.15	0.30	-	100.92	100.90
3056	57.17	57.24	12.27	11.00	3.40	3.86	.70	.72	.19	.16	12.14	12.18	4.00	4.00	5.86	6.19	3.22	3.20	1.18	1.18	0.13	-	101.06	100.95
WC REPROD.		-		-		0.7		0.25		.15		0.32		0.5		0.23		0.3		0.3		-		6.60
TEFA AV DEV	0.55		0.59		0.30		0.13		.13		0.32		0.29		0.23		0.21		0.07		0.3		2.82	

f Ti, P, and Na were not corrected for interelement effects

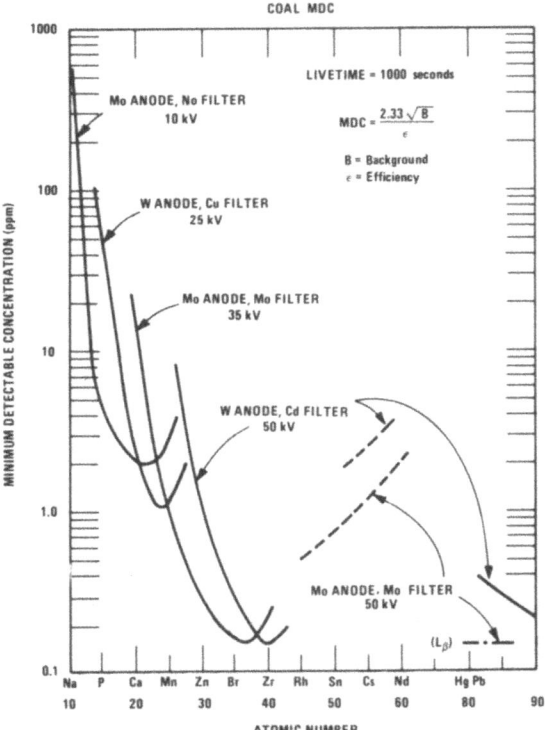

Figure 3. Minimum detectable concentration (MDC) in coal material.

The sums of the major element compositions are listed at the bottom of the table. The agreement with the listed ash values is relatively good despite the complex chemical changes taking place. The agreement is usually within 10% (relative) which is often adequate for estimation of the ash content. These coals, however, were quite similar in composition and a broader range of coal types should be studied before drawing general conclusions.

The quantitative results of the fly ash analyses are listed in Table 2. All of the elements except Na, P, Ti, and Sr were corrected for interelement effects. Although the agreement is not as good as might be expected, the average deviations are in all cases less than the estimated wet chemical reproducibilities listed at the bottom of the table.

CONCLUSION

Tube-excited energy dispersive x-ray fluorescence can provide rapid and accurate multielement analyses of about 25 elements in coal, coke, and fly ash ranging from sodium to lead. Accuracies approaching a tenth of a percent (absolute) can be achieved in the analysis of the major elements but require interelement corrections. One hundred second minimum detectable concentrations range from about one part per million for elements near arsenic to about one tenth of one percent for sodium.

ACKNOWLEDGMENTS

The authors thank John Kuhn of the Illinois State Geological Survey,
Joe Corpening of Monsanto Industrial Chemicals Company and Bill Montgomery
of Canadian Center for Mineral and Energy Technology for providing the
coal, coke, and fly ash samples used in this study. We also thank
David Smiley for his assistance with the analysis.

REFERENCES

1. R. R. Ruch, H. J. Gluskotev, and N. F. Shimp, "Occurrence and Dis-
    tribution of Potentially Volatile Trace Elements in Coal", Environ-
    mental Geology Notes, Report Number EGN-72, Aug. 1974.

2. The coke samples were provided by the Monsanto Industrial Chemicals
    Company, Columbia, Tennessee.

3. The fly ash samples were provided by the Canadian Center for Mineral
    and Energy Technology, Ottawa, Canada.

4. National Bureau of Standards standard reference material 1632 and 1633.

5. Spex Industries Incorporated, P. O. Box 798, Metuchen, New Jersey
    08840.

6. J. A. Cooper, B. D. Wheeler, D. M. Bartell, and D. A. Gedcke,
    "Analysis of Portland Cement, Clinker, Raw Mix, and Associated
    Ceramic Materials Using An Energy Dispersive X-Ray Fluorescence
    Analyzer With Interelement Corrections", Advances in X-Ray Analysis,
    <u>19</u> 213, 1976.

ADVANCES IN THE PRECONCENTRATION OF DISSOLVED IONS IN WATER SAMPLES

D. E. Leyden

Department of Chemistry, University of Denver

Denver, Colorado 80208

X-ray fluorescence is well established as an analytical method for the determination of multi-element systems. It is a technique which provides high accuracy and precision. However, it has serious limitations in the lower limit of detection for elemental analysis in environmental samples. In order to overcome this limitation some method of preconcentration or enrichment of the trace elements in water samples must be provided. In recent years many techniques have been developed for this purpose. Some of these techniques involve simple chemical or physical manipulations of the sample. In all cases, the technique should be rapid, simple, and not contaminate the sample. Because it is more convenient to use solid samples for X-ray fluor-escence analysis, the ideal method of preconcentration results in a solid sample. In some cases exhaustive recovery of the elemental com-position of the original sample is of importance. In other cases some selectivity of the ions to be recovered is desired. Frequently this selectivity can be obtained by using selective reagents, masking agents, pH adjustment, or control of the oxidation state of the ions to be recovered. The method of enrichment should require minimum sample pre-treatment.

Perhaps the simplest method of preconcentration of ions from solu-tion is evaporation of the solvent. This technique has been applied to X-ray spectrometry using conventional wavelength dispersive techniques and energy dispersive X-ray spectrometry. However, evaporation tech-niques have limited potential for trace-element determinations. Con-sidering a nominal detection limit of $0.1 - 0.01 \ \mu g/cm^2$ on a thin film sample, more than 10 ml of water must be evaporated in a confined spot to achieve precise analyses at the ppb level. However, in many cases of industrial effluent analysis, evaporation of 1 to 2 ml of water sample to dryness on filter paper may be used for analyses at the ppm level. In the case of X-ray methods applied to samples where one or more element is high in concentration but variable in concentration, evaporation would not be suitable because of the effect of the major element upon the intensity of emission from more minor constituents. An example where simple evaporation techniques have been used effective-ly is given by Rönicke in the analysis of rain water.[1] In this example special aluminum dishes were employed in which 2 ml of sample

were placed and evaporated. In this way, sulfur analysis in rainwater
was performed. One of the advantages of this simple physical method of
enrichment is that all constituents in the solution are retained with
the possible exception of those which may be volatile.

A method of enrichment which is of limited use to those techniques
requiring liquid samples, but of significant value when X-ray spectrometry is employed, is precipitation or co-precipitation. Precipitation
is one of the oldest chemical techniques for the separation and concentration of ions. A massive effort has been made over the years to
identify specific precipitation reagents and to minimize co-precipitation. This was because non-selective methods of quantitation were
normally employed. For trace-analysis using a technique such as X-ray
spectrometry, selective precipitation is usually not desirable. The
completeness of the precipitation of the ions of interest is much more
important. Many of the techniques which were used for classical gravimetric analysis are not applicable at the ppb level. For example,
nickel in this concentration range cannot be precipitated quantitatively with dimethylglyoxime. In many cases, quantitative precipitation
may be achieved by the addition of a co-precipitant. The application
of co-precipitation to analysis by X-ray spectrometry was first presented
extensively by Luke.[2] An extensive scheme for both general and selective precipitation of metal ions using a variety of reagents was
devised. The work presented by Luke was one of the first examples of the
application of chemical manipulation to the preparation of samples for
X-ray spectrometric analysis. One of the major advantages of precipitation methods is their simplicity. In addition, the precipitates may be
collected as uniform deposits on a filter substrate such as paper or
synthetic filter materials. Such a sample form is ideal for X-ray
spectrometry. However, when heavy metal ions are used as carriers for
co-precipitation, or when the precipitate contains a large and variable
amount of heavy metals such as iron, the data may not be completely free
of matrix effects and some caution is advised.

Evaporation and co-precipitation methods have been presented as
representative methods which have been used for some years for the
enrichment of samples for the purpose of X-ray spectrometric analysis.
Many of the newer techniques of enrichment of trace ions in aqueous
media are based upon ion-exchange reactions.[3-6] The major advantage
of most of the ion-exchange methods is that the enrichment process is
performed by the use of a functional group which is immobilized on some
type of solid substrate. This provides the potential to perform either
a batch extraction of ions from solution, or to use the ion exchange
material in a flow system such as a column. A simple form of ion-
exchange which has been used for X-ray spectrometric analysis of trace
elements is to deposit a thin film of an insoluble metal sulfide such
as silver sulfide on to a filter substrate. The aqueous sample to be
analyzed is then passed through the filter containing the sulfide deposit. Metal ions which also form insoluble sulfides exchange with the
silver and are retained on the filter. Conversely, if a deposit of
silver chloride or silver iodide is used on the filter, sulfide ions
in water may be determined by exchange with the halide. More commonly,
ion exchange techniques employ the use of commercially available ion
exchange resins either in a bead form in a column, or as ion-exchange
resin impregnated filter papers. The applications of these papers have

been briefly reviewed.[7] Most of the commercially available ion exchange resins consist of either strongly acidic or strongly basic
functional groups. These functional groups operate essentially on an
ion-pair formation basis and selectivity is frequently a problem. For
example, enrichment of transition metal ions from sea water is difficult because sodium ions compete well with the less concentrated transition metal ions. In cases such as these, the use of chelating functional groups offers an advantage.[8,9] These resins contain functional groups which form chelates with the metal ions. Unfortunately,
few commercial resins are available, although many varieties have been
synthesized.[10] One chelating ion exchange resin which is commercially available and which has proved very useful in preconcentration
of metal ions from solution, is Chelex-100. Application of this resin
has been reported in the literature.[6,11,12]

   Occasionally, some selectivity may be obtained by the use of
chemical parameters in conjunction with chelating ion exchange resins.
An example is the determination of Cr(III) and Cr(VI) using Chelex-
100.[12] Cr(III) is recovered from solution by Chelex-100, whereas
Cr(VI) is not. Therefore, Cr(III) may be determined directly by extraction with Chelex 100, and total chromium on a second alaquot
treated with a reducing agent to convert Cr(VI) to Cr(III). The most
serious disadvantages of Chelex-100 are the slow rate at which metal
ions are recovered from solution and the fact that alkaline earth ions
such as calcium and magnesium are also recovered. A second major limitation to chelating ion exchange resins is that few varieties are
commercially available. In some cases effective chelating resins may
be synthesized using simple procedures, for example, Siggia has synthesized polyamine resins using very simple single-step reactions.[13]
These resins offer a chemical advantage as well as simplicity in preparation. They are much more effective in chelating the "hard acid"
metal ions such as Ni(II), Cu(II) and Zn(II).

   An important modification of the ion exchange preconcentration
approach is the use of ion exchange impregnated materials. Two major
types of materials are used. Ion exchange impregnated papers have
found widest use. Campbell and co-workers have performed in-depth
studies of the distribution of ions when solutions are filtered through
paper discs impregnated with ion exchange resins.[14] These materials
are very easy to use, as the solution is simply filtered through (often
repeatedly) the filter disc which is held in a suitable holder. The
discs are well suited for X-ray spectrometric analyses. A few problems do exist with these materials and the importance of these limitations depends upon the specific applications. In cases in which low
energy X-rays are measured, the distribution gradient of the ions
through the thickness of the paper used in a filtration mode is an
important consideration. Soft X-rays are absorbed by the paper and
higher intensity is observed from the side of the disc to which the
sample was added. As long as this point is recognized, corrections
can be made. For higher energy X-rays, corrections may not be necessary. These papers have found wide use and over fifty publications
of application had appeared by 1974. Recently, the preparation of
ion-exchange resin impregnated papers on a continuous roll of paper
has been reported.[15] Using this material an automatic sequential
sampling system has been devised. The paper is automatically fed into

a filtration device. A controller allows the operator to program the
cycling of the sample solution through the ion exchange impregnated
filter paper. Once the cycle has been completed, the paper is advanced
and a new sample is treated. In this way, the filter paper strip contains spots upon which the aqueous sample has been concentrated. The
ions which are removed from the sample will depend upon the choice of
the ion-exchange material which has been impregnated into the paper.
The potential of the application of this device to the automated monitoring of rivers, industrial effluents, and other water sources is
obvious. A second form of impregnated material is the use of open-
cell polyurethene foam to adsorb chelating reagents.[16] This material
is easily prepared and appears to be effective as a preconcentration
tool. A major limitation is the very low capacity of approximately 25
micro-equivalents per gram. These materials are so recently investigated that a critical evaluation of their potential use in X-ray spectrometry has not been performed.

Because of the limited commercial availability of ion-exchange
materials suitable for the enrichment of samples for X-ray spectrometric
determination, many workers have sought readily available materials and
chemical reactions which may be performed routinely to provide immobilized functional groups for the purpose of enrichment of trace elements
from water samples. One such application has been reported by Van
Grieken and co-workers.[17] In this work oxine was used as a complexing reagent to form metal complexes in the aqueous media. Oxine forms
metal complexes with a wide variety of metal ions. This reagent has been
used as a complexing agent for gravimetric analysis for many years.
However, at trace levels the metal complexes of oxine are soluble. The
procedure reported by Van Grieken is to add oxine to a water sample,
filter any residual precipitate that occurs and then add to the sample
activated carbon. Although free metal ions are not quantitatively adsorbed onto the surface of activated carbon, the oxine metal complexes
are adsorbed. The carbon is then filtered from the solution and analyzed by X-ray spectrometry. Levels as low as 1 ppb enriched from 1
liter of water sample were detected.

A second approach has been presented by Going.[18] Soluble chelating reagents such as 2-(3'-sulfobenzoyl)-pyridine-2-pyridylhydrazone
have been used to form soluble anionic metal complexes. Using the
chemical selectivity of the reagent to form the complexes and commercially available anion-exchange resins provide for a method of enrichment of trace metal ions. Again, the procedure is to add to the sample
the chelating reagent and then to pass the mixture through a column
containing the anion-exchange resins. Going reports the technique to
provide concentration factors of several hundred.

The last two procedures mentioned above are attempts to employ
commercially available reagents and ion-exchange materials using chemical processes to accommodate the nature of the materials. In other
studies, reactions to immobilize functional groups on suitable substrates have been investigated. For example, a simple reaction which
may be conducted easily is the silylation of substrates such as glass.
These reactions have been extensively used to prepare columns for gas
chromatography and to immobilize enzymes on substrates. A variety of
silylation reagents are available. The reaction may be represented as

```
S
U
B    |—OH
S    |
T    |—OH  +  (CH₃O)₃SiCH₂CH₂CH₂-N̈-CH₂CH₂NH₂  →
R    |                              H
A    |
T    |—OH
E    |
```

$$(CH_3O)_3SiCH_2CH_2CH_2\overset{H}{N}CH_2CH_2NH_2$$

```
       |—OH  |
       |     O
       |     |
       |—O—  Si-CH₂CH₂CH₂N̈ CH₂CH₂NH₂ + 3CH₃OH
       |     |            H
       |     O
       |—OH  |
```

where the polyhydroxl species may be a molecule of a solid substrate
such as glass or silica. Some of the available silylation reagents contain chelating or coordinating functional groups and modifications of
these groups are possible. Several applications of the immobilized
functional groups on glass on controlled pore glass beads have been reported.[19-23] An interesting example of the effective use of chemistry for the enrichment and determination of trace elements by X-ray
fluorescence is illustrated by the following.[20] The functional group
illustrated in the reaction above contains an ethylenediamine terminal
group. It was observed that this functional group would effectively
abstract molybdate ions from solution at a pH of approximately 3. Phosphate in natural ground water was determined by taking advantage of
this observation. Excess sodium molybdate was added to a 50 ml sample
of ground waters. The molybdate forms the 12-MPA complex with phosphate. The complex was extracted into ethylacetate. Glass beads containing the ethylenediamine functional group were added to the ethylacetate, the molybdate was extracted on to the glass beads which were
filtered, and the molybdate determined by X-ray fluorescence. Because
of the enrichment factor of 12 molybdate atoms per phosphate ion and
the higher X-ray fluorescence sensitivity for molybdenum, the detection
limit of phosphate in natural ground waters by this procedure is approximately 7 ppb. The silylated glass beads have been used in a wide
variety of enrichment procedures for both cations and anions in aqueous
solutions. The simplicity of their preparation offers considerable
advantage as an enrichment tool. One limitation is that the silyl-ether
bond may hydrolyze in acid or base, but most use of the materials will
be in the pH 3-10 range. In this range the bonds are sufficiently
stable for preconcentration applications.

In summary, considerable effort is being expended to develop methods
of preconcentration of trace elements from aqueous solution. The capability for simultaneous multi-element determinations by X-ray fluorescence is very attractive. However, before this capability can be
fully realized in the analysis of trace elements in environmental and
industrial water samples, effective means of enrichment and sample
preparation must be devised. From the efforts being conducted at the
present, it would appear that one or more effective methods, perhaps

even an automated sampling method, may be developed for the purpose of
X-ray spectrometric analysis of trace elements in water. It is con-
ceivable that one of these methods may be incorporated directly into an
X-ray spectrometer such that the sampling and analysis can be conducted
in an on-line, automated method.

REFERENCES

1. Beitz, L. and Rönieke, G., Proceedings of the Third International
    Clean Air Congress, Düsseldorf, 1973, p. C32.

2. Luke, C. L., Anal. Chim. Acta, 41, 237 (1968).

3. Collin, R. L., Anal. Chem., 33, 605 (1961).

4. Van Nickerk, J. N., DeWet, J. F. and Wybenga, F. T., Anal. Chem.,
    33, 213 (1961).

5. Kashuba, A. T. and Hine, C. R., Anal. Chem., 43, 1758 (1971).

6. Blount, C. W., Leyden, D. E., Thomas, T. L. and Guill, S., Anal.
    Chem., 45, 1045 (1973).

7. Law, S. L. and Campbell, W. J., in "Advances in X-Ray Analysis,"
    Vol. 17, C. L. Grant, C. S. Barrett, J. B. Newkirk and C. O. Ruud,
    eds., Plenum Press, New York, 1974, p. 279.

8. Leyden, D. E., Patterson, T. A. and Alberts, J. J., Anal. Chem.,
    47, 733 (1975).

9. Alberts, J. J., Leyden, D. E., and Patterson, T. A., Marine Chem.,
    4, 51 (1976).

10. Helfferich, F., "Ion Exchange," McGraw-Hill Book Company, Inc.,
    New York, 1962, p. 47.

11. Leyden, D. E., in "Advances in X-Ray Analysis," Vol. 17, C. L.
    Grant, C. S. Barrett, J. B. Newkirk and C. O. Ruud, eds., Plenum
    Press, New York, 1974, p. 293.

12. Leyden, D. E. and Channell, R. E., Anal. Chem., 44, 607 (1972).

13. Dingman, J., Siggia, S., Barton, C. and Hiscock, K. B., Anal.
    Chem. 44, 1351 (1972).

14. Campbell, W. J., Spano, E. F. and Green, T. E., Anal. Chem., 38
    987 (1966).

15. Carlton, T. C., EPA Symposium on X-Ray Analysis of Environmental
    Samples, Chapel Hill, N.C., January 1975.

16. Braun, T. and Farag, A. B., Talanta, 22, 699 (1975).

17. Vanderborght, B., Verbeeck, J. and Van Grieken, R., Bull. des Soc. Chim. Belgium, in press.

18. Going, J. E., Wesenberger, G., and Andrejat, G., Anal. Chim. Acta, 81, 349 (1976).

19. Leyden, D. E., Luttrell, G. H. and Patterson, T. A., "Silica Gel with Immobilized Chelating Groups as an Analytical Sampling Tool," Anal. Lett., 8, 51-56 (1975).

20. Leyden, D. E., Nonidez, W. K., and Carr, P. W., "Determination of PPB Phosphate in Natural Waters Using X-Ray Fluorescence," Anal. Chem., 47, 1449-1452 (1975).

21. Leyden, D. E. and Luttrell, G. H., "Preconcentration of Trace Metals Using Chelating Groups Immobilized Via Silylation," Anal. Chem., 47, 1612-1617 (1975).

22. Leyden, D. E., Luttrell, G. H., Nonidez, W. K., and Werho, D. B., "Preconcentration of Certain Anions Using Reagents Immobilized Via Silylation," Anal. Chem., 48, 67-70 (1970).

23. Leyden, D. E., Luttrell, G. H., Sloan, A. E., and DeAngelis, N. J., "Characterization and Application of Silylated Substrates for the Preconcentration of Cations," Anal. Chim. Acta, 84, 97-108 (1976).

CONCENTRATION OF U AND Np FROM Pu AND Pu ALLOYS FOR DETER-

MINATION BY X-RAY FLUORESCENCE

J. M. Hansel, Jr., C. J. Martell, G. B. Nelson,

and E. A. Hakkila

University of California

Los Alamos Scientific Laboratory

Los Alamos, NM   87545

ABSTRACT

Methods are presented for the determination of uranium, or uranium and neptunium, in plutonium metal and plutonium alloys. Anion exchange or a combination of anion exchange - solvent extraction is used to concentrate the elements for x-ray fluorescence analysis, depending upon the impurities present and the elements to be determined. The precision for determining between 3 and 250 µg of uranium or neptunium ranges between 30 and 2%.

INTRODUCTION

Rapid analytical procedures were required for the determination of uranium or both uranium and neptunium in plutonium metal and in various plutonium alloys. Methods exist for the determination of uranium [1,2,3] or neptunium [4] in high purity plutonium metal. The x-ray fluorescence method [1] for determining uranium is time consuming, requiring three ion exchange separations, and the spectrophotometric methods are susceptible to interference from neptunium or from any plutonium that is not removed during the separation. Likewise, the spectrophotometric method for determining neptunium suffers from interference by uranium, and uranium and neptunium cannot be measured simultaneously. X-ray fluorescence is capable of providing rapid analyses with good precision and selectivity, and the final measurement of both uranium and neptunium can tolerate some contamination by plutonium. However, a separation from the major portion of the sample is needed for good sensitivity. Anion exchange, or a combination of anion exchange and solvent extraction were chosen for these separations. Several separation schemes were developed, dependent upon the alloy composition, the elements to be determined, and the following analytical requirements:

445

(1) a sensitivity of 1 µg or less for measuring uranium and neptunium, (2) reproducible recoveries of uranium and neptunium, and (3) carryover of less than 10 µg of plutonium with the uranium and neptunium to eliminate plutonium spectral interference with the x-ray measurement of neptunium.

SEPARATION PROCEDURES

For the simultaneous determination of uranium and neptunium in high purity plutonium metal, a sample containing between 10 and 250 µg each of uranium and neptunium is dissolved in 5 ml of 12 $\underline{M}$ HCl. Approximately 150 mg of ascorbic acid are added to reduce the plutonium to the cationic trivalent state. Uranium and neptunium remain as anions in the hexavalent and tetravalent states, respectively, and are adsorbed on a column of Dowex 1-x4 (50 to 100 mesh) anion exchange resin which has been conditioned with 12 $\underline{M}$ HCl. The plutonium is washed through the column with 20 ml more of 12 $\underline{M}$ HCl. The adsorbed uranium and neptunium are then eluted from the column with 20 ml of 0.01 $\underline{M}$ HCl. The eluate is evaporated to near dryness, the residue is dissolved in 0.5 ml of water, and approximately 75 mg of ascorbic acid are added. Fifteen ml of 12 $\underline{M}$ HCl are added, and a second separation and concentration for x-ray analysis is performed by passing the solution six times through a filter paper impregnated with anion exchange resin. The intensities of the Lα_1 x-rays for uranium, neptunium, and plutonium and background intensities are measured. Corrections for background and for overlapping x-ray lines are applied and the corrected intensities are compared to intensities obtained for standards prepared in the same manner as the samples. The recoveries for neptunium and uranium are only approximately 35%, but they are reproducible.

For determining only uranium in plutonium alloys the dissolved sample is evaporated to near dryness and the residue is dissolved in 5 ml of 6 $\underline{M}$ HCl. Samples containing zirconium or other elements that form cations in 6 $\underline{M}$ HCl also can be dissolved in 5 ml of 6 $\underline{M}$ HCl. Approximately 150 mg of ascorbic acid are added to reduce plutonium to the cationic trivalent state. At this acidity Zr(IV) and Pu(III) remain as cations and U(VI) remain as an anion. This solution is passed through a column containing Dowex 1-x4 anion exchange resin which has been conditioned with 6 $\underline{M}$ HCl. The plutonium, zirconium, and other cations (including neptunium) are washed through the column with 20 ml more of 6 $\underline{M}$ HCl. The adsorbed uranium is eluted with 20 ml of 0.01 $\underline{M}$ HCl which is then evaporated to a volume of approximately 200 µl. The solution is transferred dropwise using a micropipet onto a filter paper impregnated with anion exchange resin, and the paper is dried, leaving uranium highly concentrated for x-ray analysis. Only a few micrograms of zirconium and plutonium are present and do not interfere. This method is more sensitive for determining uranium than the previously described method because all of the uranium is collected on the filter paper. However, neptunium is not adsorbed quantitatively on anion exchange resin from 6 M HCℓ and cannot be determined.

Gallium, if present in amounts greater than 1 mg, interferes with the determination of uranium and must be separated before the anion exchange step. After initial dissolution of the sample and evaporation to near dryness, the residue is dissolved in 5 mℓ of 8 M HCℓ and is transferred to a separatory funnel with an additional 5 mℓ of 8 M HCℓ. Twenty mℓ of isopropyl ether are added to the separatory funnel which is then shaken for 2 min. More than 99% of the gallium is extracted into the ether phase; the aqueous phase, containing the uranium and plutonium, is drained into a beaker and evaporated to near dryness. The uranium is then separated from plutonium using 6 M HCℓ as described above, or if both uranium and neptunium are to be determined, an anion exchange separation is performed from 12 M HCℓ.

ANALYTICAL RESULTS AND DISCUSSION

The precision of the method for determining both uranium and neptunium was calculated from data obtained by analyzing one-gram plutonium samples containing various amounts of uranium and neptunium. Relative standard deviations for uranium (Table I) and neptunium (Table II) range between 29 and 7% for determining 3 to 120 ppm of the measured elements.

TABLE I

PRECISION OF X-RAY FLUORESCENCE MEASUREMENT OF URANIUM

(12 M HCℓ Ion Exchange Separation; 1-g Samples)

U added, μg	No. of determinations	Standard deviation, μg	Relative Standard deviation, %
0	9	0.7	--
4	11	0.8	20
10	10	1.5	15
20	7	2.2	11
40	9	2.9	7
100	10	10	10

TABLE II

PRECISION OF NEPTUNIUM X-RAY FLUORESCENCE

(12 $\underline{M}$ HCℓ Ion Exchange Separation; 1-g Samples)

Np added, μg	No. of determinations	Standard deviation, μg	Relative standard deviation. %
0	9	0.9	--
3.1	10	0.9	29
10.9	4	1.8	17
20.2	9	2.6	13
40.3	9	3.2	8
116	10	8.6	7.4

The precision for determining uranium by ion exchange separation from 6 $\underline{M}$ HCℓ (Table III) or by combined ion exchange - solvent extraction (Table IV) is better because of the direct pipetting of the final sample onto the ion exchange discs with nearly 100% recovery of the uranium.

TABLE III

PRECISION OF X-RAY FLUORESCENCE DETERMINATION OF URANIUM

(6 $\underline{M}$ HCℓ Ion Exchange Separation, 0.5-g Samples)

No. of determinations	U, μg	Standard deviation, μg	Relative standard deviation, %
7	0	0.2	--
8	5	0.5	10
9	17	1.0	6
10	75	4.8	6
10	100	5.2	5
7	250	15	6

TABLE IV

PRECISION OF X-RAY FLUORESCENCE DETERMINATION OF URANIUM

(Combined ion Exchange - Solvent Extraction; 0.5-g Samples)

No. of determinations	U, μg	Standard deviation, μg	Relative standard deviation, %
6	0	0.1	--
6	10	0.4	4
4	75	0.9	1
]2	100	5.5	6
6	150	10	7

The main sources of error in this method arise from
the presence of elements having high absorption coefficients
for anion exchange resin in hydrochloric acid, thus competing
with uranium and neptunium for the active sites on the ion
exchange paper. For example both zirconium and gallium are
strongly absorbed from 12 $\underline{M}$ HCl onto anion exchange papers.
The interference from zirconium is eliminated by reducing
the acid molarity to 6 $\underline{M}$ so that zirconium behaves as a cation. For elements such as gallium, that behave as anions at
acid molarities required for adsorption of uranium, other
separation schemes, such as solvent extraction, can effectively eliminate interference.

A second source of potential interference results from
overlapping x-ray lines of elements that may be present in
microgram amounts with the uranium and neptunium on the final
filter paper. Potential interference in this category includes plutonium in the neptunium determination and neptunium
in the uranium determination, as shown in Table V.

TABLE V

$L\alpha_1$ X-RAY WAVELENGTHS FOR URANIUM, NEPTUNIUM, AND PLUTONIUM

Element	X-Ray	Wavelength, A	Degrees 2θ (LiF)
U	$L\alpha_2$	0.923	26.48
	$L\alpha_1$	0.911	26.13
Np	$L\alpha_2$	0.901	25.85
	$L\alpha_1$	0.889	25.50
Pu	$L\alpha_2$	0.880	25.24
	$1\alpha_1$	0.869	24.92

By selecting fine collimators for the x-ray spectrograph to
provide line widths at one-half maximum of 0.40° 2θ, serious
overlap is eliminated. However, a correction to the neptunium x-ray intensity of 0.034 c/s per 1 c/s of plutonium intensity is required. Also a correction of 0.029 c/s of uranium x-ray intensity is required per 1 c/s of neptunium x-ray
intensity. This overlap is not serious in either case,
amounting to 0.17 c/s of neptunium x-ray intensity per microgram of plutonium, and 0.14 c/s of uranium x-ray intensity
per microgram of neptunium.

The accuracy of the ion exchange method from 12 $\underline{M}$ HCl
for determining uranium and neptunium was tested by determining uranium and neptunium in four metal samples which also
were analyzed by spectrophotometric procedures. The values
for neptunium (Table VI) are in good agreement by the two
methods, but values for uranium (Table VII) are significantly

higher using the spectrophotometric procedure. To test the source of this bias, three solutions were prepared and analyzed by the two procedures, and these data also are included in Tables VI and VII. Solution 1 was prepared from high purity plutonium metal (200 ppm total impurities) from which uranium and neptunium were removed by an anion exchange resin separation. Solution 2 consisted of solution 1 with 107 and 115 ppm of added uranium and neptunium, respectively, and solution 3 consisted of solution 1 with 426 ppm of added uranium, 173 ppm of added neptunium, and between 20 and 600 ppm each of aluminum, copper, chromium, iron, manganese, molybdenum, nickel, silicon, tin, vanadium, and tungsten to simulate an impure plutonium sample. Results for these analyses also are in good agreement between the two methods for determining neptunium, but recoveries for uranium are biased high on the spectrophotometric method for samples 1 and 2. The reason for the high bias is not known, but may be due to carryover of plutonium when uranium is separated from the plutonium.

TABLE VI

RESULTS OF SPECTROPHOTOMETRIC AND X-RAY FLUORESCENCE
MEASUREMENT OF Np

Sample	Spectrophotometric		X-Ray Fluorescence	
--------	No. of anal.	Np found, ppm	No. of anal.	Np found, ppm
Metal 1	2	< 15	2	7
Metal 2	2	40	2	38
Metal 3	2	39	2	37
Metal 4	2	< 15	2	6
Soln. 1	3	< 5	10	< 3
Soln. 2	5	113 ± 2	10	124 ± 19
Soln. 3	6	186 ± 5	24	202 ± 17

TABLE VII

RESULTS OF SPECTROPHOTOMETRIC AND X-RAY FLUORESCENCE

ANALYSES FOR U IN Pu

Sample	Spectrophotometric		X-Ray Fluorescence	
	No. of anal.	U found, ppm	No. of anal.	U found, ppm
Metal 1	2	30	2	16
Metal 2	2	36	2	15
Metal 3	2	35	2	13
Metal 4	2	24	2	17
Soln. 1	2	14	10	< 2
Soln. 2	2	144	10	100 ± 19
Soln. 3	3	440 ± 25	24	382 ± 37

SUMMARY

Uranium, or uranium and neptunium, may be determined rapidly in high purity plutonium metal or in a variety of plutonium alloys by one of several separation methods and subsequent x-ray fluorescence analysis.

Anion exchange procedures are used to concentrate the elements to be measured. Both uranium and neptunium are concentrated from plutonium using a 12 $\underline{M}$ HCℓ system. If only uranium is to be determined, a 6 $\underline{M}$ HCℓ system is used to effect a cleaner separation of uranium from plutonium, and also to separate elements such as zirconium that would interfere if the separation were performed from the higher acidity solutions. Neptunium is not recovered in 6 $\underline{M}$ HCℓ. Gallium can be removed by isopropyl ether extraction. The detection limits for measuring uranium and neptunium together are of the order of 3 ppm, or, for measuring only uranium, 1 ppm. One analyst can analyze six to eight samples daily for both elements.

ACKNOWLEDGEMENTS

The authors are indebted to R. B. Brooks, G. Coriz, and P. J. Hakkila for performing many of the analyses for this study, and to G. R. Waterbury for aid in manuscript review.

REFERENCES

1. J. A. Hayden "Determination of Uranium in Electro-Refined
    Plutonium by a Combined Ion-Exchange and X-Ray Fluores-
    cence Technique," Talanta, 14, 721-729 (1967).

2. R. D. Gardner and W. H. Ashley, "The Spectrophotometric
    Determination of Trace Amounts of Uranium in High Purity
    Plutonium Metal," U. S. Atomic Energy Comm. Rept. LA-3551
    (1966).

3. N. L. Koski, R. D. Gardner, and G. R. Waterbury, "Spectro-
    photometric Measurement of Uranium in Plutonium Oxide,"
    U. S. Atomic Energy Comm. Rept. LA-5490 (1974).

4. R. G. Bryan and G. R. Waterbury, "The Spectrophotometric
    Determination of Neptunium," U. S. Atomic Energy Comm.
    Rept. LA-4061 (1969).

PRECONCENTRATION OF URANIUM IN NATURAL WATERS FOR X-RAY FLUORESCENCE

ANALYSIS

L. R. Hathaway and G. W. James

Kansas Geological Survey

University of Kansas

Lawrence, Kansas 66044

ABSTRACT

Use of Chelex-100 impregnated filter membranes for preconcentration of uranium in ground-water samples for XRF analysis leads to lower processing time, sample size, and detection limit than achieved in a batch extraction process. Data is presented for potential effects of iron and organic matter in natural waters upon the recovery of uranium when using Chelex-100.

INTRODUCTION

Hydrogeochemical investigations represent an important technique in regional uranium potential evaluation(1). However, the ppb levels of uranium in most natural waters necessitate the use of a preanalysis concentration step when using X-ray fluorescence methods in these studies.

The feasibility of batch extraction using a chelating ion-exchange resin, Chelex-100, and direct determination of uranium on the pelletized resin has been demonstrated for ground waters of the High Plains region of western Kansas(2). This paper will investigate the application of Chelex-100 impregnated filter membranes to the determination of uranium in natural waters and compare results from membrane and batch extraction studies.

EXPERIMENTAL

Apparatus and Operating Conditions

Instrumentation and instrumental parameters for XRF analysis were the same as listed previously(2).

Reagents

Chelex-100 impregnated filter membranes (Gelman Instrument Company) were converted to the hydrogen form with 0.1 N HCl, rinsed thoroughly with distilled water, air dried, and stored for subsequent use.

Test Solutions

Test solutions were distilled water or ground water (preacidified to pH 1.5 with HCl) which were spiked to about a 50 ppb uranium level using a uranyl nitrate solution. Chemical interference studies involved spiking ground-water test solutions with a ferric chloride solution to produce a 10 ppm Fe level, or addition of 5 ml of "humic" extract (at pH 5--derived from leaching of a black shale) to produce an estimated 5 ppm "humic" level, based upon combustion of the "humic" residue.

Procedures

The pH of all solutions was adjusted by meter using dilute NaOH or HCl solutions. Sample volumes of 250 ml were used in the membrane studies and one liter in the batch extractions (three hour extraction times). The chelating membranes were loaded in the filtration apparatus with the "smooth" side toward the solution and a 0.2 μ prefilter membrane was stacked directly on top of the chelating membrane. After filtration was completed, the chelating membranes were removed, and air dried about five minutes before being analyzed.

Duplicate sets of Fe and "humic" spiked samples were adjusted to pH 4 and run as part of the chemical interference study. Solid sodium carbonate was added to another duplicate set of Fe-spiked samples until a pH of 9.3 was reached. The resultant precipitates were removed by filtration and the solutions were readjusted to pH 4 and run. A second duplicate set of "humic" spiked samples were adjusted to pH 0.5 with concentrated nitric acid and gently boiled for 30 minutes. The samples were then readjusted to pH 4 and run.

A standard additions-calibration curve was prepared by running triplicate sets of unspiked, 20 ppb- and 60 ppb-spiked portions of the ground-water test solution.

DISCUSSION

The data in Table 1 suggest that some retardation of the filtration rate during vacuum filtration is essential for achieving reproducibility in the filtration-extraction process. The stacked assembly of a 0.2 μ prefilter and a Chelex-100 membrane appears adequate in this respect and also serves to keep the surface of the chelating membrane free of particulate material which might be present in the sample.

Table 1. Effects of Filtration Rate
Upon Uranium Recovery

	Single Membrane	Stacked Membranes
Net Counts/Second	159 180 134	141 136 144
Filtration Rate (ml/min.)	∿250	∿63
Sample	Distilled Water	Ground Water

Dependency of the extraction efficiency upon pH is shown in
Figure 1. The greater variation noted in the pH 4-6 range for the
membranes as compared to batch extractions for preacidified samples
probably reflects lack of complete equilibration of the solutions
with the chelating membranes during the filtration process.

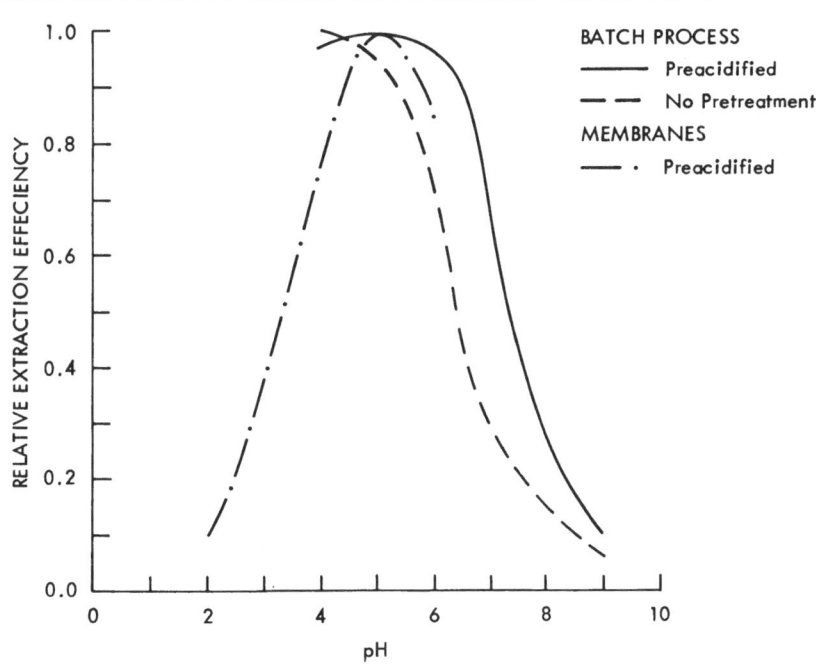

Figure 1. Extraction efficiency vs. pH for batch process and membranes with spiked ground-water samples.

Garrels(3) has noted that carbonate ions have a stabilizing influence upon uranium in solution, and results from batch extraction studies of non-preacidified waters suggest that uranium recovery by Chelex-100 is reduced by the presence of carbonate ions. Thus, a working pH of 4 has been selected in all studies employing Chelex-100 in order to minimize effects due to dissolved carbon dioxide related species. Accordingly, a three hour batch extraction appears to remove more than 90% of the recoverable uranium and multiple filtration studies with individual chelating membranes indicate about 55% and 80% recoveries of extractable uranium with single and double filtration steps, respectively, using vacuum filtration. An overall processing time of about 15 minutes for pH adjustment through XRF analysis is achieved for 250 ml samples using the chelating membranes.

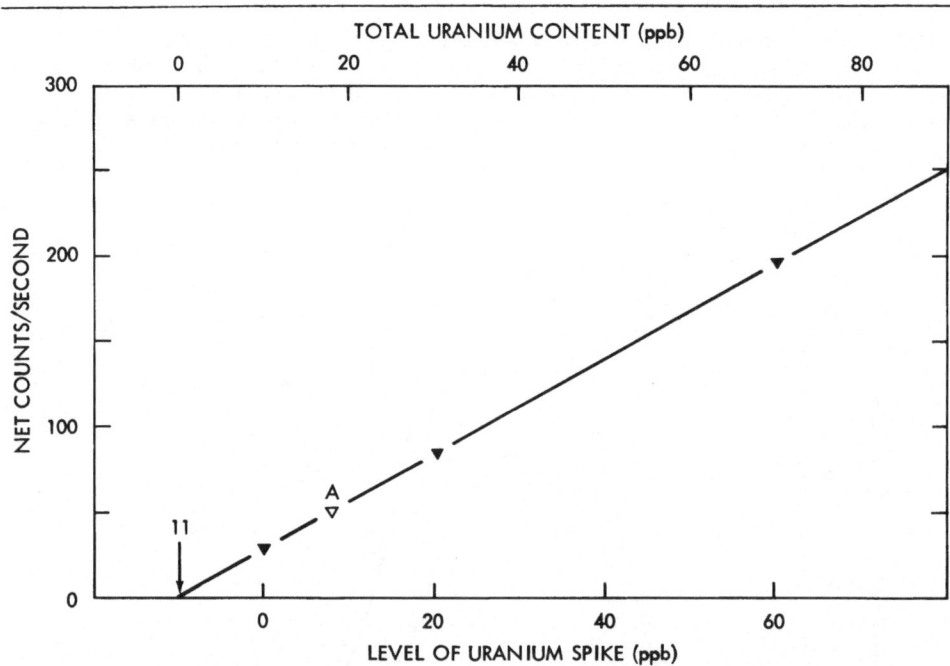

Figure 2. Membrane standard additions – calibration curve for alkaline earth–bicarbonate waters.

Figure 2 represents the standard additions-calibration curve obtained using the membrane concentration method for a preacidified alkaline earth-bicarbonate type ground-water sample (377 ppm total synthetic dissolved solids). Point A is a synthetic alkaline earth-sulfate type water (3280 ppm total dissolved solids) which has been spiked to a 20 ppb uranium level. A response of 11 cps/µg uranium in solution is obtained from the calibration curve. Similar studies using the batch extraction process yield a response of 2.2 cps/µg uranium in solution(2). The Chelex-100 content of the chelating membranes and the sample size are each essentially one-fourth the value used in the batch process, thus a 5-fold increase in response is realized in using the chelating membranes to preconcentrate uranium. A root-mean-square standard deviation of $\pm$ 3.8 cps is obtained from three 100 second countings of each of the triplicate samples of each of the three concentration sets of the calibration curve. The background for this group of sample determinations of 15.2 $\pm$ 0.5 cps, yielding a detection limit of 0.5 ppb (0.13 µg) uranium in a 250 ml sample which represents the concentration of uranium giving a net signal equal to three times the standard deviation of the background signals. This compares very favorably to the 2 ppb detection limit obtained with liter size samples by the batch process(2). Recovery from distilled water samples is found to be somewhat greater than from ground-water samples, thus the calibration curve should be determined using actual water samples. Improvement of the detection limit in membrane work appears feasible through use of slower filtration rates and/or multiple filtration steps and the use of larger sample sizes.

The chemical interference studies are related to potential problems in the application of Chelex-100 to preconcentration of uranium in surface waters, acid waters, and acid leaches of stream sediments. Most surface waters contain particulate ferric hydroxide and acidic waters and acid leaches of stream sediment may contain appreciable amounts of iron(4). In addition, organic matter is found in surface waters in a concentration range of 1-50 mg carbon per liter(5). The data of Table 2 show that the presence of either iron or organic matter can cause serious problems in the recovery of uranium using Chelex-100.

Application of a sodium carbonate treatment to remove excess iron in acidic samples prior to preconcentration of uranium appears to be an effective technique for this type of sample. The nitric acid treatment used by Florence and Batley(6) to liberate trace elements associated with a "bound" fraction, presumably associated with organic matter, in sea water for recovery by Chelex-100 is found to be ineffective for the "humic" spiked uranium solutions.

The satisfactory application of Chelex-100 to the preconcentration of uranium in samples containing organic matter appears to require additional work and evaluation of the extent of this problem in actual surface water samples. The partitioning of uranium which is possible in surface waters should influence the sampling and sample handling procedures as well as the interpretation which is placed on the final hydrogeochemical data.

Table 2. Chemical Interference of Iron and
"Humic" Material Upon Uranium
Recovery

Iron

Sample	Net Counts/Second
No Fe addition	133
10 ppm Fe	86
10 ppm Fe Na$_2$CO$_3$ Treatment	149

"Humic"

No "Humic" addition	133
<5 ppm> "Humic"	70
<5 ppm> "Humic" HNO$_3$ Treatment	68

<   > Estimate based upon combustion of "Humic" residue.

REFERENCES

1. H. Fauth, "Hydrogeochemical Reconnaissance Prospecting," Uranium Exploration Methods, p. 209-218, International Atomic Energy Agency, Vienna (1972).

2. L. R. Hathaway and G. W. James, "Use of Chelating Ion-Exchange Resin in the Determination of Uranium in Ground Water by X-Ray Fluorescence," Anal. Chem. 47, 2035-2037 (1975).

3. R. M. Garrels, Mineral Equilibria, p. 186, Harper (1960).

4. J. D. Hem, Study and Interpretation of the Chemical Characteristics of Natural Water, Geological Survey Water-Supply Paper 1473, 2 nd. ed., p. 114-126 (1970).

5. W. Stumm and J. J. Morgan, Aquatic Chemistry, p. 347, Wiley-Interscience (1970).

6. T. M. Florence and G. E. Batley, "Trace Element Species in Sea-Water--I," Talanta 23, 179-186 (1976).

"LOSS ON IGNITION" IN FUSED GLASS BUTTONS

R. LeHouillier, S. Turmel

Department of Natural Resources
Québec, Canada

F. Claisse

Department of Mining and Metallurgy
Université Laval
Québec, Canada. G1K 7P4

ABSTRACT

When a sample is fused with a flux to produce a glass button for X-ray fluorescence analysis, some of the constituents are lost by decomposition (H_2O, CO_2) or by evaporation (Br, Cl, S). Therefore the glass button is smaller than if no "Loss on Ignition" had occurred, and the X-ray intensities appear stronger. It is shown that "L.O.I." can be handled as a regular element and that corrections can be made by means of influence coefficients.

Since "L.O.I." cannot be measured by X-ray fluorescence, one should normally know the "L.O.I." to make the appropriate corrections, but under certain conditions, X-ray intensities yield both composition and "L.O.I." as shown by Tertian and as confirmed by our experiments.

INTRODUCTION

Natural samples often contain carbonates, hydration water, organic matter, volatile constituents which evaporate during fusion in the preparation of fused glass buttons. It is sometimes assumed that the effect on calculated concentrations is negligible when the loss on ignition (represented by L.O.I. or L) is small. But what should be done when it is large?

The most obvious way is to calcine the sample, determine the L.O.I., analyse the calcined product and recalculate the initial concentration.

Another way,which is the object of the first part of this paper, is to analyse the sample without calcination, and to make a mathematical correction to account for the L.O.I. which must be known or estimated by some other means.

A more elegant way which is also the object of this paper is to
start with the material in the as-received condition and to make
appropriate mathematical calculations to yield both the L.O.I. and the
correct analysis.

The principle on which this paper is based has been enunciated by
Tertian (1) who has applied it to his "self-consistent calibration
method". The fact that Tertian obtained good results on bauxite samples
is a good argument in favor of his approach.

It is desirable to make a more systematic study of that principle
and if it is correct, to apply it to the Lachance-Traill (2) and
Claisse-Quintin (3) numerical methods of analysis.

Corrections for L.O.I. through α coefficients have also been proposed by Anderson, Mander and Leitner (4) but, in fact, such corrections are for interelement effects of CO_2 on the analyte, and not for
the effect of an element which has escaped from the sample after the
latter was weighed and before the x-ray intensity was measured.

THEORY

To simplify the theory let us assume for a while that the
Lachance-Traill relation (2) is valid.

$$C_A = R_A(1 + \alpha_B C_B) \tag{1}$$

with $$\alpha_B = \frac{\mu_B^* - \mu_A^*}{\mu_A^* + D} \tag{2}$$

where   A = analysed element
        B = influencing element
        R_A = normalised intensity
        μ^* = effective mass absorption coefficent
        D = a function to account for the diluent

Equation (1) is usually applicable to fused glass buttons because
a nearly constant effective wavelength exists in diluted samples. Since
"mass" absorption coefficients are involved in Equation (2), this takes
into account the difference of x-ray absorption between the two elements
and also the difference of density of the same elements.

When a fraction of the sample for example CO_2, is lost during
fusion, the glass disk is smaller than if no loss had occurred, the
concentrations of the remaining elements in the glass appear higher and
the x-ray intensities do not represent the initial concentrations unless an appropriate correction is made.

To understand the physical meaning of the correction that will be
proposed let us imagine that the specimen contains an element B which
does not absorb x-rays and which occupies no space; then $\mu_B^* = 0$.

(Hydrogen is not very different from B). Obviously an influence coefficient α_B can be determined for the influence of B on each element of the sample according to Equation (2).

Now, this description of element B applies exactly to loss on ignition. L.O.I. is present in the original sample at a given concentration C_L just as B is at a concentration C_B; the effect of the absence of volatiles after fusion is the same as the effect of element B because they both occupy no space and they both do not absorb x-rays. As a result, matrix corrections for L.O.I. can be made through α coefficients just as for regular elements.

Since L.O.I. represents material that escapes during fusion, its chemical composition is immaterial. Glass disks with artificial L.O.I. can be prepared by using smaller samples than usual and considering the missing fraction of the sample as the L.O.I.; however, the amount of flux should remain the same as usual.

SAMPLE PREPARATION AND X-RAY MEASUREMENTS

Glass disks 30 mm diam. were prepared by fusing together 0.8 g sample and 4.8 g Li Tetraborate with 25 mg KBr as a non-wetting agent. The mixtures were fused for 10 minutes on a Claisse Fluxer.

Binary samples of several compositions were required to determine the α coefficients accurately. To prepare, for example, the CaO-L.O.I. mixtures, quantities of 0.16 g, 0.32 g, 0.6 g, etc. CaO were fused with a constant quantity of flux, 4.8 g. Therefore these samples represented respectively 20%, 40%, 75%, etc, CaO in the binary system.

Because the glass disks produced are of high quality, one disk only was prepared for each sample. Two 20-sec. readings were taken for each element and the count rate was kept below 10,000 c.p.s. to minimize dead-time corrections.

Line intensities were corrected for dead time and for background.

Concentrations were calculated using the iteration procedure.

RESULTS

Binary Systems

When L.O.I. is considered as an element, a binary system contains one real element only; for example pure $CaCO_3$ contains 56% CaO and 44% L.O.I. Experiments on two binary systems only, CaO with L.O.I. and ZnO with L.O.I. are reported in this paper.

Figure 1 represents the measured x-ray fluorescence intensities of Ca and Zn as a function of their concentrations in each binary mixture. As usual, the intensities are normalised so as to be equal to

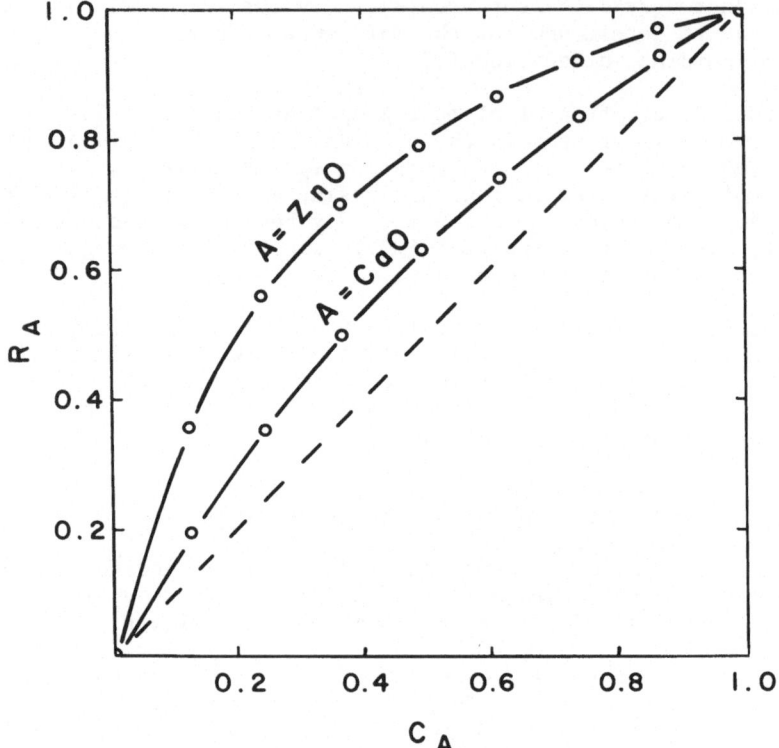

Fig. 1 - Normalised x-ray fluorescence intensities as
a function of concentration in the binary
systems "CaO plus Loss on Ignition" and "ZnO
plus Loss on Ignition". (Glass buttons)

one when the original sample is 100% pure oxide, that is, with no
L.O.I. The general shape of the curves is similar to what is observed
for binary mixtures of two elements, and this alone is sufficient to
suggest that alpha coefficients can be attributed to L.O.I. just as for
any real element. The deviation of the two curves from the straight
line is quite pronounced which means that the matrix effect in these
systems is high. It is more important for Zn than for Ca because Zn
is associated with shorter wavelengths which are less absorbed in the
flux.

In order to obtain the most accurate corrections for the L.O.I. we
will use the Claisse-Quintin (3) relation which contains one parameter
more than the Lachance-Traill relation. For a binary system we have

$$C_A = R_A(1 + \alpha_B C_B + \alpha_{BB} C_B^2) \tag{3}$$

or,

$$C_A = R_A \left[1 + (\alpha_B + \alpha_{BB} C_B) C_B \right] \tag{4}$$

or

$$Y = \frac{1}{C_B}\left(\frac{C_A}{R_A} - 1\right) = \alpha_B + \alpha_{BB} C_B \tag{5}$$

Equation (5) indicates that a plot of Y as a function of C_B is a straight line, the ordinate of which at the origin is α_B and the slope of which is α_{BB}. Such a plot is shown in Figure 2 where B is the L.O.I. represented by the letter L. Also comparing Equations (1), (4) and (5), we see that Y is equivalent to the α coefficient of Lachance-Traill and shows to what extent the latter varies with composition.

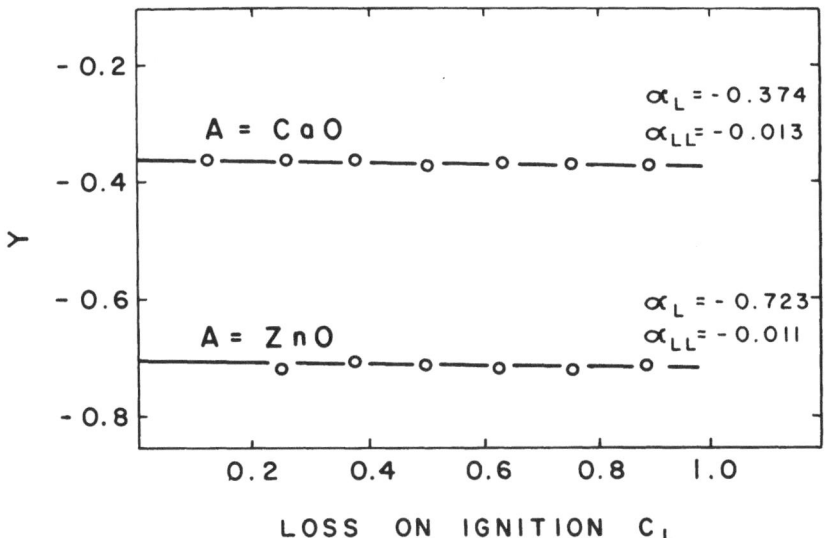

Fig. 2 - Determination of α_L and α_{LL} coefficients for the influence of Loss on Ignition (L) on CaO and ZnO in glass buttons.

It is observed that the slope is nearly zero in the two cases which means that the Lachance-Traill relation could be applicable.

The α values obtained for the analysis of zinc and calcium are respectively about -0.7 and -0.4. The minimum value of α_L would be 0 for no-matrix effect and the maximum value would be -1 for the largest effect in a non-diluted sample. In the present case where the dilution is medium, one part sample in seven parts mixture, the value of -0.7 observed for Zn represents a rather large effect. Consequently, corrections for L.O.I. seem to be far from negligible.

TERNARY SYSTEMS

These systems comprise two elements, for example ZnO and CaO, plus L.O.I. as the third constituent. The Claisse-Quintin relation for element A in the system of A, B and L is

$$C_A = R_A (1 + \alpha_B C_B + \alpha_{BB} C_B^2 +$$
$$+ \alpha_L C_L + \alpha_{LL} C_L^2 +$$
$$+ \alpha_{BL} C_B C_L) \qquad (6)$$

and a similar relation exists for element B; but there is no such relation for L since L.O.I. is not detected by x-ray fluorescence. The mixed-subscript coefficients such as α_{BL} can be neglected in diluted glass buttons. The coefficients with subscript L or LL are already known. The other coefficients are usual influence coefficients which are determined in the same way as for binary samples of A and L. For ZnO and CaO mixtures the fluorescence intensities are shown in Figure 3. Here, the intensities are nearly proportional to the concentration

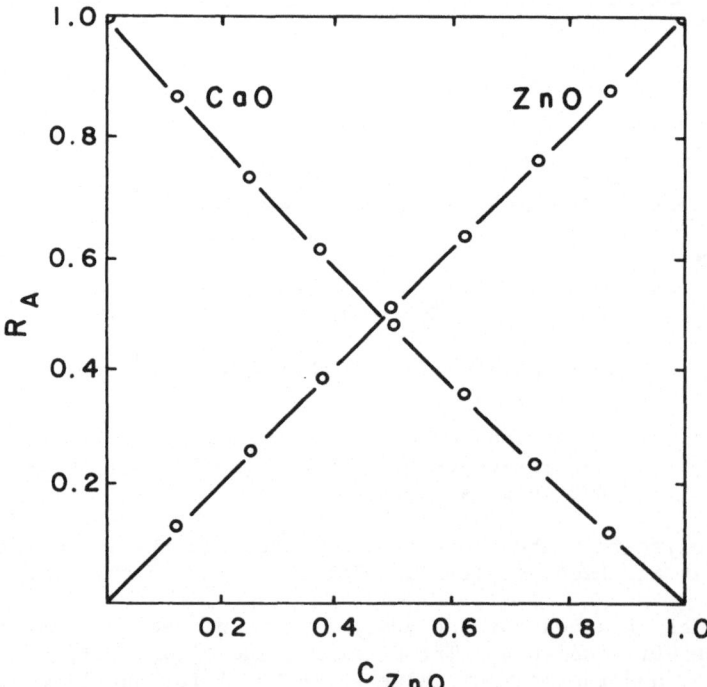

Fig. 3 - Normalised x-ray fluorescence intensities as a function of composition in the system CaO-ZnO (glass buttons).

of each element. This represents a situation with nearly no mutual effects between Zn and Ca and the α coefficients should be small. As expected, Figure 4 shows that Y is equal to about 0.0 and 0.1; however it depends slightly on concentration. Therefore the Claisse-Quintin relation with α_B and α_{BB} terms should be used to make more accurate corrections.

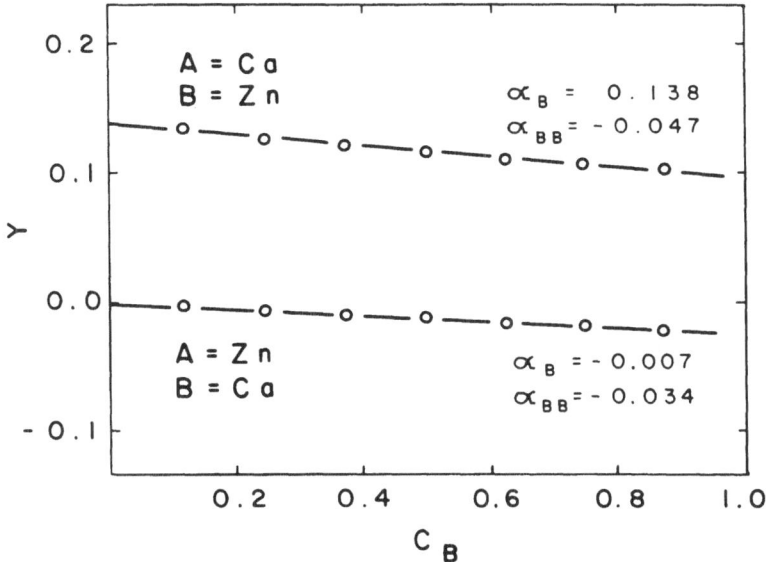

Fig. 4 – Determination of α_B and α_{BB} coefficients for the mutual influence of CaO and ZnO in glass buttons.

The method of correction for L.O.I. using influence coefficients has been applied to several samples containing ZnO, CaO and volatile constituents.

A first series of samples was prepared from pure ZnO and pure CaO. Loss on ignition was artificially produced as explained above. The values of L.O.I. were known and were used in conjunction with the appropriate α coefficients (Figures 2 and 4 and Equation 5) to make the desired corrections. A few results are given in Table I. Although the L.O.I. varies from as low as 6% to as high as 90%, the average deviation is 0.2% only and one result only has a deviation larger than 0.4% absolute.

A second series of samples was prepared from ZnO, CaO and compounds that are susceptible to lose weight during fusion: $CaCO_3$, $Ca(OH)_2$, $ZnCO_3$ and $Zn(OH)_2$. Here, no artificial L.O.I. was present and all the samples weighed 0.8 g when mixed with the flux. The L.O.I. was known from the composition of the salts used to make the samples.

R. LeHouillier, S. Turmel, and F. Claisse

TABLE I

Analysis of ZnO-CaO specimens, using KNOWN values
of loss on ignition (L.O.I.)

True composition (%)			Calculated composition (%)	
ZnO	CaO	L.O.I.	ZnO	CaO
81.3	12.5	6.2	81.5	12.6
68.7	2.50	28.8	68.7	2.58
18.8	50.0	31.2	18.8	50.4
31.2	18.8	50.0	31.2	19.1
3.75	6.25	90.0	3.85	6.39
12.5	25.0	62.5	12.7	25.2
12.5	2.50	85.0	12.5	2.55
12.5	62.5	25.0	12.7	62.9

Results of x-ray calculated CaO and ZnO concentrations are given in
Table II; the absolute average deviation is 0.3%.

TABLE II

Analysis of mixtures of CaO, $CaCO_3$, $Ca(OH)_2$, ZnO, $ZnCO_3$
and $Zn(OH)_2$ using KNOWN values of loss on ignition (L.O.I.)

True composition (%)			Calculated Composition (%)	
ZnO	CaO	L.O.I.	ZnO	CaO
12.5	29.7	37.8	12.6	50.2
62.5	28.0	9.5	62.5	28.0
0.0	56.8	43.2	0.0	57.5
0.0	74.7	25.3	0.0	75.1
71.3	0.0	28.6	70.9	0.0
96.5	0.0	3.47	96.2	0.0
42.8	50.0	7.2	42.3	50.5
94.1	2.50	3.36	93.8	2.63
6.03	53.2	40.8	6.03	54.2
73.3	16.9	9.9	73.1	17.2
2.67	95.6	1.73	2.68	95.9
98.8	0.70	0.54	99.0	0.80

We went one step further and attempted to determine the L.O.I. from x-ray results only as suggested by Tertian (1) by ignoring the true value of the L.O.I. and equating it to the difference between 1 and the sum of the concentrations of all the elements:

$$C_L = 1 - C_A - C_B \qquad (7)$$

The value of C_L is calculated at each iteration so that C_L becomes better and better known as the iteration process goes on. The results for the first series of samples, those with artificial L.O.I. are given in Table III. These figures were obtained from the same x-ray measurement as for Table I. The average deviation is 0.2% and is equal to the deviation observed in Table I for the same samples when the true values of L.O.I. were used in the calculations.

TABLE III

Analysis of ZnO-CaO specimens with UNKNOWN values
of loss on ignition (L.O.I.)

True composition (%)			Calculated Composition (%)		
ZnO	CaO	L.O.I.	ZnO	CaO	L.O.I.
81.3	12.5	6.2	81.9	12.5	5.5
68.7	2.50	28.8	68.6	2.56	28.8
18.8	50.0	31.2	18.9	50.0	381.1
31.2	18.8	50.0	31.3	18.9	49.8
3.75	6.25	90.0	3.87	6.38	89.7
12.5	25.0	62.5	12.7	25.1	62.1
12.5	2.50	85.0	12.5	2.53	85.0
12.5	62.5	25.0	12.7	62.7	24.6

The results for the second series of samples, those with natural L.O.I., are given in Table IV. The average absolute error in this case is less than 0.4% for the two elements as compared to 0.3% when the true values of L.O.I. are used.

CONCLUSION

It has been shown that the glass fusion method of analysis can be applied to samples that lose weight during fusion. When the L.O.I. is known a Lachance-Traill or Claisse-Quintin correction can be applied in the same way as for any chemical element.

It has also been shown that under certain conditions corrections for L.O.I. can be done even if L.O.I. is not known. To do that, L.O.I. has to be estimated by stating that the sum of the analysed elements

R. LeHouillier, S. Turmel, and F. Claisse

TABLE IV

Analysis of mixtures of CaO, $CaCO_3$, $Ca(OH)_2$, ZnO, $ZnCO_3$ and $Zn(OH)_2$ with UNKNOWN values of loss on ignition (L.O.I.)

True composition (%)			Calculated Composition (%)		
ZnO	CaO	L.O.I.	ZnO	CaO	L.O.I.
12.5	49.7	37.8	12.7	49.8	37.6
62.5	28.0	9.5	62.1	27.8	10.1
0.0	56.8	43.2	0.0	57.8	42.1
0.0	74.7	25.3	0.0	75.4	24.6
71.3	0.0	28.6	70.0	0.0	30.0
96.5	0.0	3.47	95.2	0.0	4.75
42.8	50.0	7.2	42.0	49.8	8.2
94.1	2.50	3.36	93.3	2.62	4.11
6.03	53.2	40.8	6.11	54.4	39.5
73.3	16.9	9.9	73.1	17.1	9.7
2.67	95.6	1.73	2.69	95.8	1.5
98.8	0.70	0.54	99.9	0.81	-0.8

plus the L.O.I. makes a total of 100%; this means that all the elements must be analysed or that some of their concentrations must already be known. In doing the calculations for unknown L.O.I. we have found that the iteration procedure is a slow one. In some cases up to ten iterations were necessary to reach the final result as compared to one or two iterations only when L.O.I. was known or absent. Perhaps a direct solution of the system of equations would be faster.

This research on L.O.I. has encouraged the authors to extend their work towards the development of a method of analysis in which no weighing would be necessary, neither the sample nor the flux. The first results are very encouraging and indicate that this goal is not utopia.

ACKNOWLEDGEMENTS

The authors are grateful to the Quebec Department of Natural Resources and to the Canadian National Research Council (Grant A-931) for their contribution and financial support.

REFERENCES

(1) R. Tertian, "A Self-Consistent Method for Industrial X-Ray Spectrometric Analysis". X-Ray Spectrom. 4 52-61 (1975).

(2) G.R. Lachance and G.R. Traill, "A Potential Solution to the Matrix Problem in X-Ray Analysis", Can. Spect. 11 43-48 (1966)

3. F. Claisse, M. Quintin, "Generalization of the Lachance-Traill Method for the Correction of the Matric Effect in X-Ray Fluorescence Analysis", Can. Spect. 12, 129-133 and 146 (1967).

4. C. H. Anderson, J.E. Mander and J.W. Leitner, "X-Ray Fluorescence Analysis of Portland Cement through the Use of Experimentally Determined Correction Factors". in Advances in X-Ray Analysis, Vol. 17, p. 214-224, Plenum Press (1974).

MEASUREMENT OF "CHEMICAL SHIFT" BY AN AUTOMATED COMMERCIAL X-RAY FLUORESCENCE SPECTROMETER

L. G. Dowell, J. M. Bennett, D. E. Passoja

Union Carbide Corporation

Tarrytown, New York 10591

The measurement of "chemical shift," that is, the change in energy of an element's x-ray emission lines with the state of its chemical combination, has been carried out for some years. Of the three major aspects of the technique, two have received major attention. Nagel (1) has an excellent treatise on the interpretation of valence band x-ray spectra, while such workers as Fischer (2) and Koffman and Moll (3) have attempted to correlate the data with structure. The third area, convenient data collection, has not been so well investigated. Much, but not all, of the effort has been toward direct electron excitation with its attendant problems of sample damage due to high vacuum and electron bombardment effects. It is the purpose of this paper to describe a practical x-ray fluorescent spectroscopic method to collect spectral data in digital form, to show examples of computer processing of these data to present the spectral composition of the emission lines, and to show one example of the application of this technique.

Figure 1 depicts the x-ray system. It is basically a fully automated Siemens system consisting of a K-4 generator, SRS sequential spectrometer, and associated electronics panel. The dedicated computer system consists of a PDP-11/10 20 K computer with tu60 dual cassettes and an ASR-33 teletype. The spectrometer is automatic vacuum or helium path and equipped with a 10-position sample changer, .15^0 Collimator and a "piggy-back" air conditioner. Software was developed by us to step-

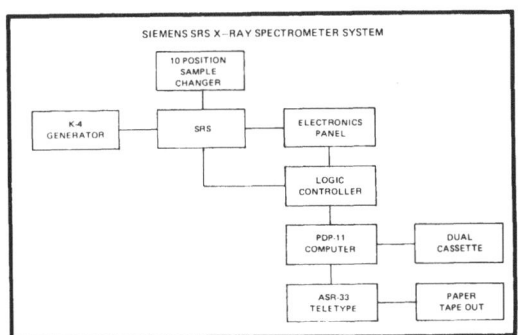

Figure 1

scan any desired angular range at optimal step sizes and counting times
and to output on punched paper tape the data pairs of counts vs.
diffraction angle. These data are then processed on our corporation
time-sharing computer system. Figure 2 is a typical automatically
plotted SiKβ spectrum from a sodium alumino-silicate molecular sieve
and shows the main Kβ peak and its satellite. A pentaerythritol
(P. E. T.) crystal was used.

All of our work has been with powders, and very little sample
preparation was required. Approximately one gram of powder was poured
into a standard liquid cell fitted with 0.00015 inch (four micrometer)
polyester bottom window. To reduce evacuation time for high surface
area materials prior degassing was sometimes done in a separate vacuum

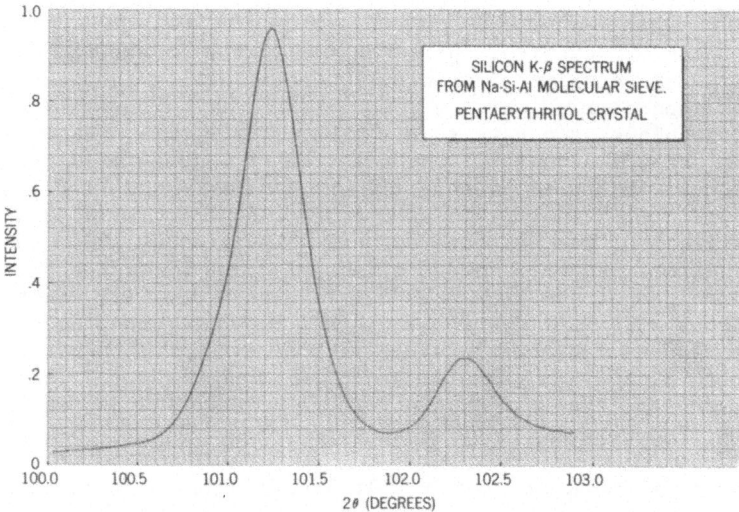

Figure 2

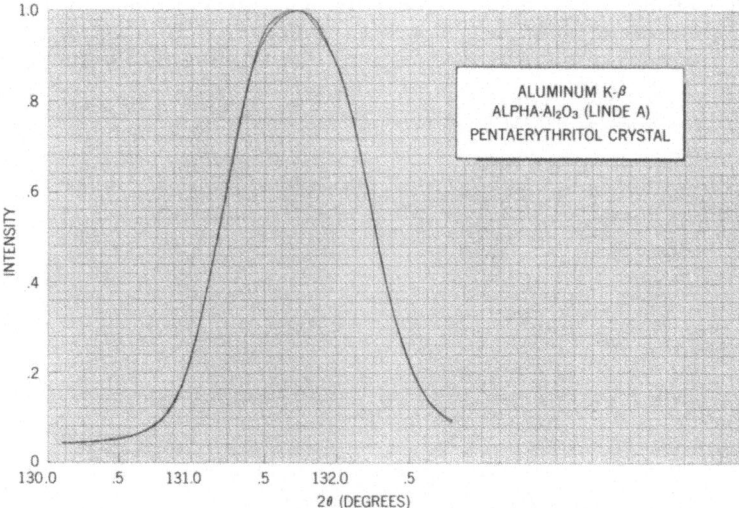

Figure 3

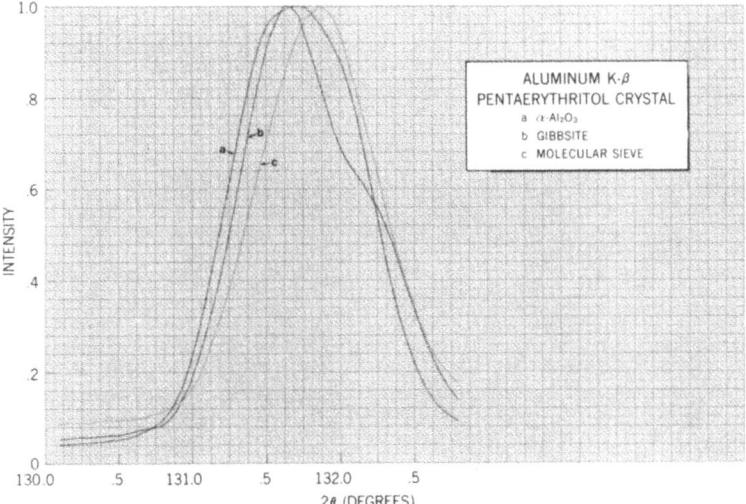

Figure 4

$$I = \frac{1}{1 + c^2 \epsilon^2}$$

where

$$c = \frac{FWHM}{2}$$

$$\epsilon = \text{displacement} = 2\theta \, (i) - 2\theta \, (\text{center})$$

$$I(OBS) = \frac{R}{1 + (c(1) \times \epsilon(1))^2} + \frac{1 - R}{1 + (c(2) \times \epsilon(2))^2}$$

5 PARAMETERS   2 FWHM
                   2 PEAK CENTERS
                   1 PEAK RATIO

Figure 5

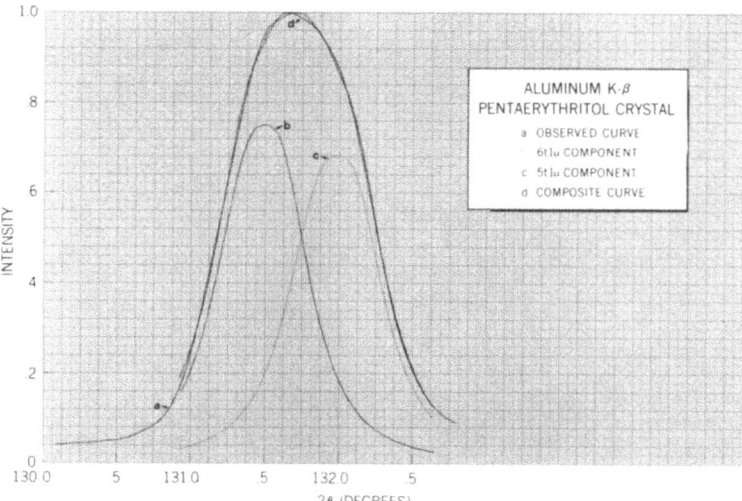

Figure 6

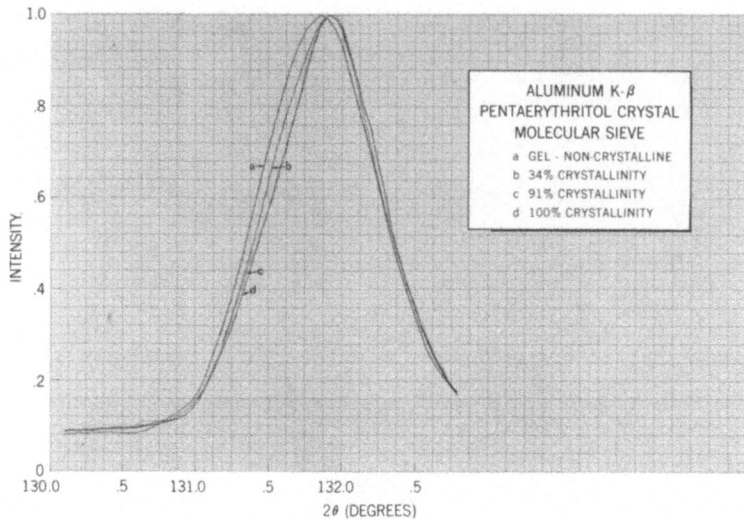

Figure 7

system. For a continual check of system stability two or three of the
ten sample positions are always occupied by an appropriate stable reference standard. All of our major activity has been with aluminum and
silicon, and our reference materials are Linde A (0.3 micrometer α-Al_2O_3
in corundum form) and SiO_2 (α-Quartz) for aluminum and silicon respectively.

Normally, measurements are made unattended over a weekend, during
which time up to three cycles of the ten samples are completed, yielding
stability and precision data. The system reproducibility is visually
expressed in Figure 3, which shows four measurements of the aluminum
standard obtained during one run. Figure 4 visually indicates the

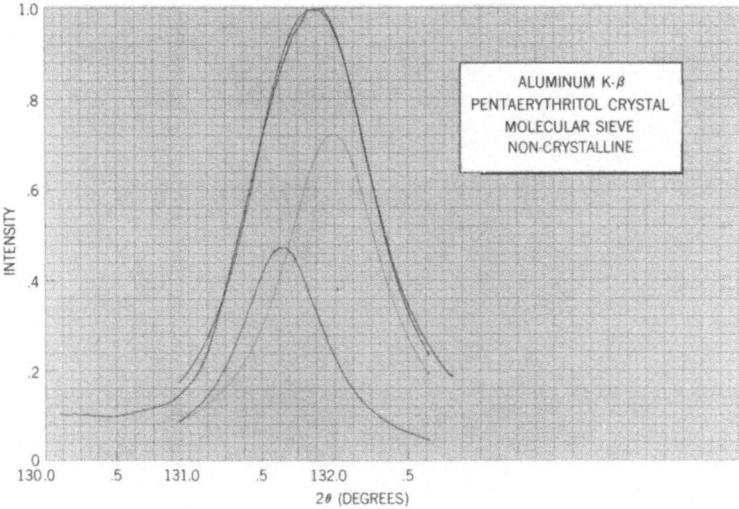

Figure 8

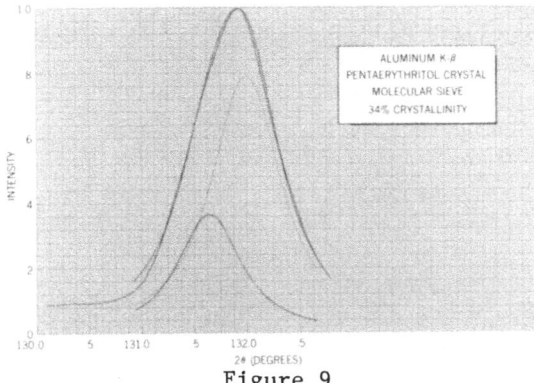

Figure 9

"chemical shift" of the aluminum Kβ peak from three materials, Linde A
(α-Al$_2$O$_3$), gibbsite (α-Al$_2$O$_3$·3H$_2$O) and a sodium alumino-silicate molecular sieve.

    Figure 4 illustrates the difficulty in correlating the shift in
line position (energy) with sample chemistry. Usually the peak center
is measured and the energy (or energy shift from some standard) calculated and correlated with some chemical factor, such as aluminum content
in alumina-silicates as reported by Kuhl (4). It is obvious that the
three spectra in Figure 4 have different shapes and half widths as well
as positions and that one cannot meaningfully refer to the shift of the
main Kβ peak. It is interesting to note that although Nagel (1), page
202, refers to "the shoulder in the SiKβ spectrum" which seems "to be
due to further level splittings" and Fischer (2), page 177, discusses
at some length the complexity of the main Kβ peak, researchers have
persisted in measuring the position of the entire main peak. Tossel (5)
uses molecular orbital theories to explain the Kβ spectra and his
assignment of orbitals to individual components will be used in the
results to be shown later. Assuming that there are always two components, we have developed a computer program that determines the centers,
half widths and magnitude ratio of each component. Figure 5 illustrates the plan of our mathematical analysis. We assume a Cauchy shape,
as used by Klug and Alexander (6), which has the form $I_i = 1/(1 + c^2 \varepsilon^2_i)$
where $c = FWHM/2$ and ε_i is displacement from peak center and is equivalent
to $2\theta(i) - 2\theta(center)$. Since the observed curve is the sum of the two

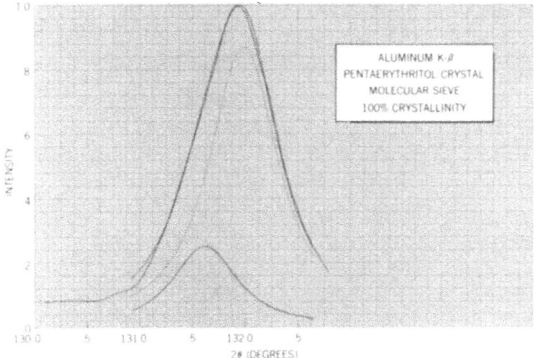

Figure 10

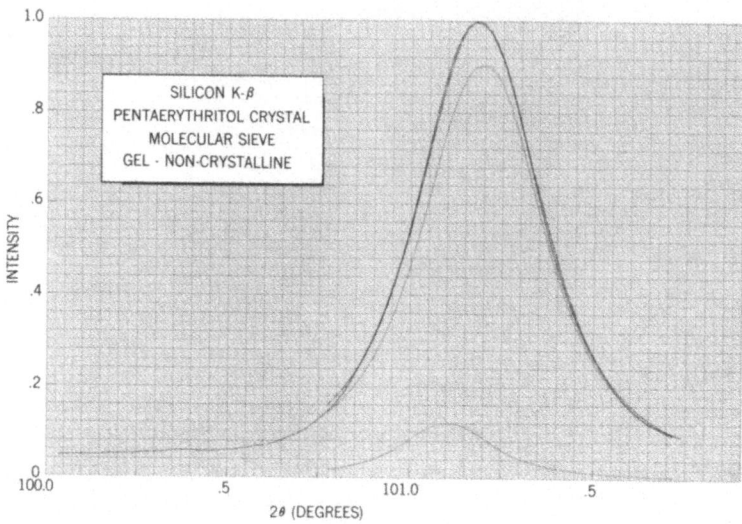

Figure 11

components, the equation becomes $I = R/(1+c_1^2\epsilon_1^2)+(1-R)/(1+c_2^2\epsilon_2^2)$, where R is
the fractional contribution of the high energy component. The optimal
values of the five parameters, c_1, c_2, 2θ(center)$_1$, 2θ(center)$_2$ and R,
are found which minimize the RMS difference between observed and calcu-
lated intensities. Using 30-50 points RMS values as low as 0.44% have
been obtained. Figure 6 shows the results of the foregoing treatment on
the spectrum of α-Al$_2$O$_3$. Using the molecular orbital treatment of
Tossel, the left-hand high energy and the right-hand low energy components
are consistent with the assignments 6t1u and 5t1u, respectively. In fact,
our observations of the angular split between the two peak centers translates to an energy difference of 3.08 ± 0.05 e.v. as compared to a calculated value of 3.1 e.v. by Tossel. Repeated observations of our two
reference materials have consistently shown that, within each batch of

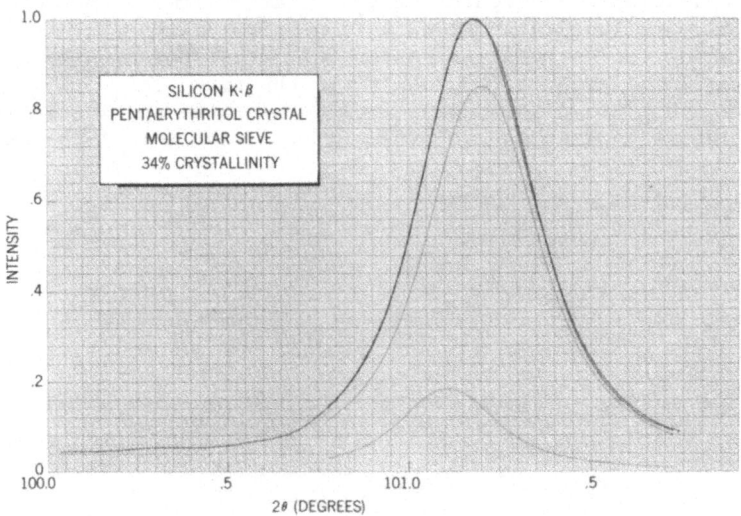

Figure 12

sample measurements, the angular positions of the stronger components can
be determined within ± 0.002 degrees (2θ) and, therefore, angular shifts
of a sample relative to its reference, to ± 0.004°, which translates to
an energy shift precision of at worst ± 0.05 e.v.

Generally known for their drying characteristics, molecular sieves
have a wide variety of applications, especially catalysis. The nature
of the silicon and aluminum bonds is of very considerable importance in
these materials. Our early work confirmed reports that the shift of the
silicon Kβ band center is closely related to the amount of aluminum in
the structures, but variances existed that were outside our error limits.
It is now apparent that, since the Kβ band is a composite of two peaks,
the term "chemical shift" as currently used is not valid, but the individual band components should be compared. In the work to be described

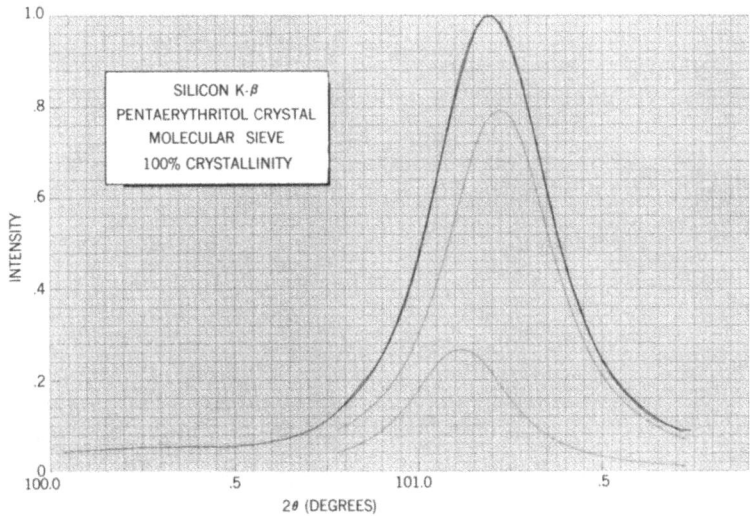

Figure 13

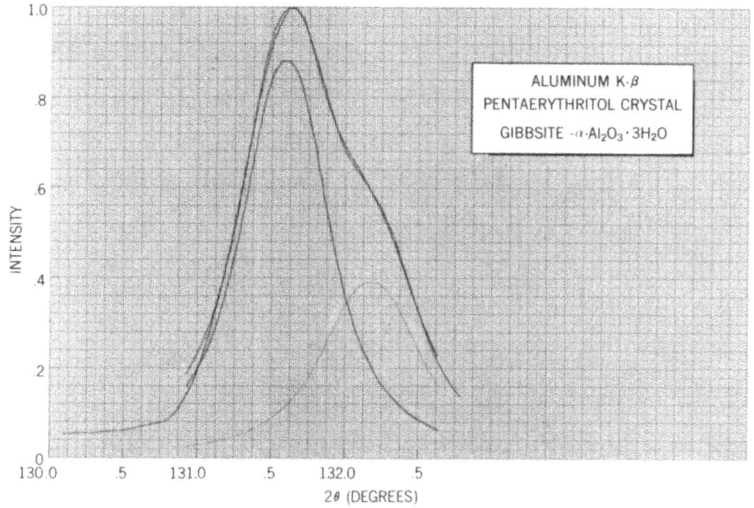

Figure 14

now, the energy shift of each molecular orbital is relative to the same
orbital in the reference material, α–quartz or α–Al₂O₃.

The procedure as described has been applied to follow the synthesis
of a molecular sieve. In outline, the starting material is in the form
of a non–crystalline gel which is subsequently heated. After a time the
crystalline molecular sieve product forms, soon reaching a maximum in
concentration. Samples of the solid portion are taken at various times
and analyzed by both x–ray diffraction and fluorescent chemical shift
techniques. The following figures show a portion of the results.
Figure 7 is a composite of the AlKβ main peak spectra in which very
obvious differences are seen.   Figure 8 is the AlKβ spectrum of a single
sample, the gel. Note particularly the higher energy 6tlu orbital and
its decreasing contribution as we progress through 34% (Figure 9), and
100% (Figure 10) crystallinity. Conversely, the silicon 6tlu orbital

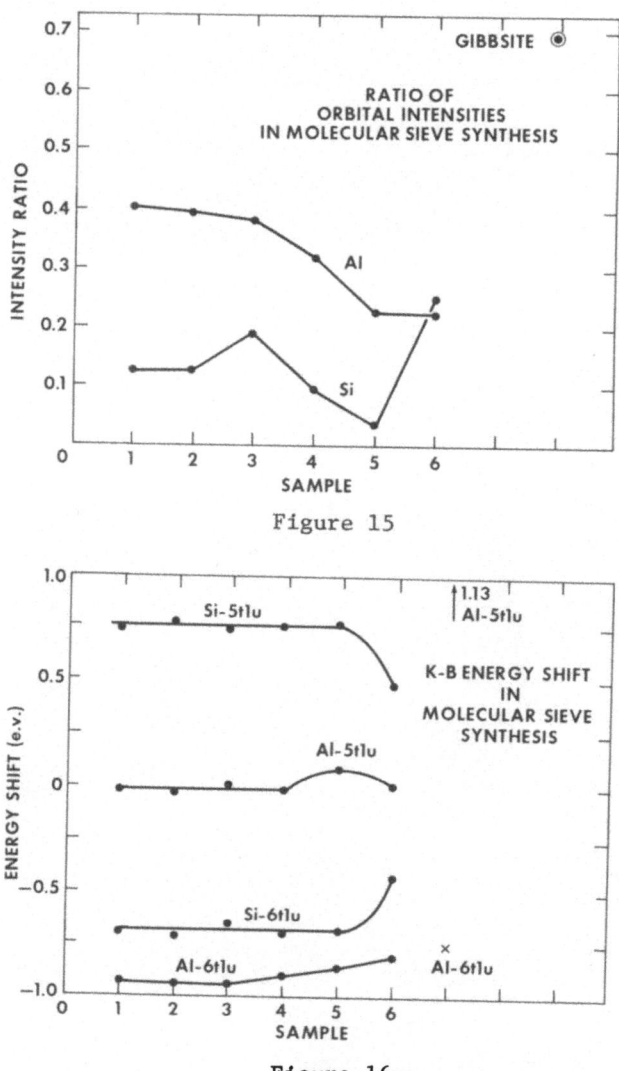

Figure 15

Figure 16

increases in this series as seen in 0% (Figure 11), 34% (Figure 12) and 100% (Figure 13) crystallinity. In contrast gibbsite (Figure 14) shows a complete reversal in intensities. Figure 15 summarizes the intensity data for the whole series. However, the energy shift of the components follows a different pattern and is summarized in Figure 16. The silicon curves show remarkably constant energy until at 100% crystallinity both show sharp decrease in shift from α-quartz. The high energy Al5t1u orbital is quite constant with a small change at 91% crystallinity, and the lower energy Al6t1u shows a small steady change.

The interpretation of these data are outside the scope of this conference and will be reserved for later communications. However, it seems clear from this work that a commercial x-ray fluorescence apparatus is capable of very precise measurement of line profiles. Although the work presented here deals solely with aluminum and silicon, the technique can be applied to any emission line of any element that falls within the range of the particular spectrometer. Using the digital output as a starting point, data treatment and presentation can then be tailored to each specific application.

REFERENCES

1. D. L. Nagel, "Interpretation of Valence Band X-Ray Spectra", in W. M. Mueller, Editor, Advances in X-Ray Analysis, Vol. 13, p. 182-236, Plenum Press (1969).

2. D. W. Fischer, "Chemical Bonding and Valence State -- Nonmetals", ibid, p. 159-181, Plenum Press (1969).

3. D. M. Koffman and S. H. Moll, "The Effect of Chemical Combination on the X-Ray Spectra of Silicon", in W. M. Mueller, Editor, Advances in X-Ray Analysis, Vol. 9, p. 323-328, Plenum Press (1965).

4. G. H. Kuhl, "A Study of Aluminum Coordination in Zeolites Using the Kβ Line", in Proceedings of Third International Conference on Molecular Sieves", p. 227, Leuven Press (1973).

5. J. A. Tossel, "The Electronic Structures of Mg, Al and Si in Octahedral Coordination with Oxygen from SCF Xα MO Calculations", J. Phys. Chem., Solids 36, 1273-1280 (1975).

6. H. P. Klug and L. E. Alexander, X-Ray Diffraction Procedures, Second Edition, p. 635, John Wiley and Sons (1974).

LOW ENERGY MASS ABSORPTION COEFFICIENTS FROM PROTON INDUCED X-RAY

SPECTROSCOPY

A. Lurio, W. Reuter and J. Keller

IBM Thomas J. Watson Research Center

Yorktown Heights, New York 10598

ABSTRACT

We describe a new and reliable experimental technique for the measurement of mass absorption coefficients in the 0.1 to 1 keV energy range. In this technique, the absorbing film is supported directly on a substrate which under proton bombardment will generate the x-rays whose absorption will be measured. Results are given for thirteen different metals at the C K_α (277 eV) line.

INTRODUCTION

At the present time there exists very little experimental data on mass absorption coefficients (μ) in the wavelength range 10-100 Å. Of the existing data, reliable measurements are principally available only for those elements which can be studied in the gaseous state (1). The reason for this circumstance is that previous techniques applicable to solids required very thin, large area, self-supporting films of the absorber. Difficulty in making uniform films and in accurately measuring their areal density and purity (e.g. surface oxides) have limited reliable measurements to a few special cases. For this reason, researchers needing reliable mass absorption coefficients in the soft x-ray region are presently using Henke's tabulation (1) which is based on interpolations from the calculated results of Veigele (2) which agree well with the measured mass absorption coefficients of gases.

We shall present in this paper a new and reliable technique for measuring the mass absorption coefficients of solids in the soft x-ray region. In this technique the absorbing film is supported directly on a substrate which, under proton bombardment, will generate the characteristic x-rays whose absorption will be measured.

EXPERIMENTAL

Basic Techniques

Figure 1 shows a schematic diagram of our experimental method. A

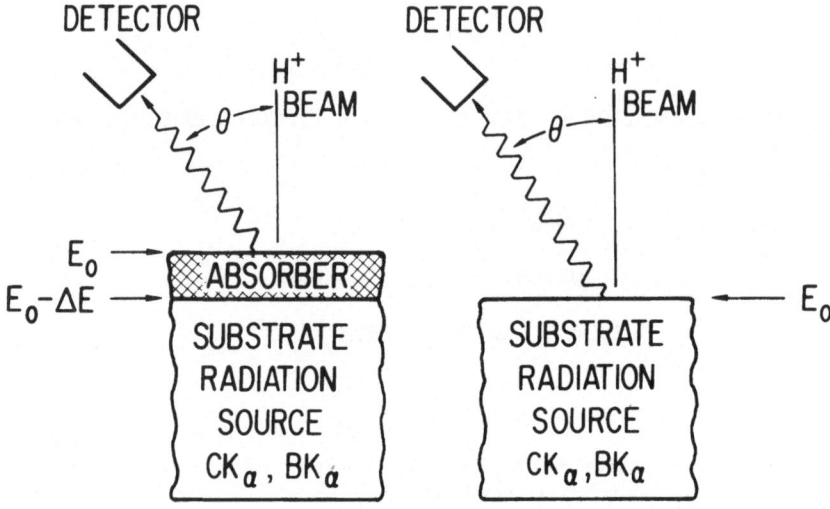

Figure 1.   Schematic illustration of the experimental technique.

beam of protons from our 3 MeV Van de Graaf accelerator is incident onto
a carbon target exciting thereby C K_α radiation.   This radiation is dispersed by a crystal spectrometer, using an LSD crystal, and is detected
by a gas flow proportional counter.   Other details of our vacuum chamber
and spectrometer have been reported previously (3).

We compare the intensity of the C K_α radiation produced from the
carbon substrate with the intensity produced from a similar substrate
having the absorbing film deposited onto it.

Substrate and Film Preparation

During the course of our work we have used substrates of Cu, Ni, Al,
C and BN.   In all cases it is necessary to have a substrate which is
smooth locally (50-500 Å) but which may have undulations on a large scale
(0.2 - 1μ).   This is required in order that the deposited films in
following the local topography have a uniform thickness for x-ray absorption.   In the case of carbon, we found that commercial vitreous carbon
squares (4) 25 x 25 x 3 mm could be cut into 8 x 8 x 3 mm squares which
were final polished with 1μ Al_2O_3 powder achieving a mirror-like finish.
The metal absorber films in the thickness range 300 to 2500 Å evaporated
onto these substrates also had mirror-like quality.   The base pressure
in the e-beam evaporation system was 2-3 x 10^{-7} Torr and did not exceed
10^{-6} Torr during deposition at rates from 5 to 20 Å/sec.

Measurement of Areal Density

The important quality for mass absorption coefficient measurements is the number of atoms/cm^2, ρ_A, in the x-ray path, not the film thickness. We measured ρ_A by two different techniques. For the non-destructive measurements we used Rutherford backscattering (RBS) of 2 MeV He$^+$ ions (5). In this method ρ_A is determined by counting the number of He ions backscattered from the surface film into a known solid angle for a measured number of incident ions. We estimate the overall accuracy of this technique to be $\pm$ 5%. Each film was sampled in several places with a beam spot of about 1 mm^2. Variations of less than $\pm$ 2% from the mean were typical. Local surface roughness of 100 Å or greater can be detected by RBS and were not observed for our carbon substrates.

After this non-destructive measurement, the films were dissolved and analyzed wet chemically. Atomic absorption spectroscopy with an estimated accuracy of $\pm$ 3% was used for the determination of Ag, Au, Ni and Co. Colorimetric methods were used for all other elements with an accuracy of $\pm$ 5% except for Hf where we claim an accuracy of only $\pm$ 10%.

Film Impurities and Uniformity

In addition to measuring the film areal density, RBS also yields a quantitative estimate of the oxygen content and impurities present in the films. In most cases we observe two small oxygen peaks in the RBS spectrum corresponding to oxygen in the front and back sides of the film. The largest oxygen content occurred for the 2100 Å Ti film where in contrast to all other metal films the oxygen was distributed throughout the entire film. We calculated that 1.98 μgm/cm^2 of oxygen was present but since μ_{oxygen} (44.7 Å) = 6044 this would cause less than 1.5% transmission loss. The next worst case was V which had 1.17 μgm/cm^2 of oxygen. The Hf film had a significant impurity of Nb, 1.53 μgm/cm^2, whereas the silver film showed no oxygen but 0.19 μgm/cm^2 sulfur in the surface region. The Nb impurity in Hf contributed a 5% transmission loss and was corrected for.

Surface roughness can be estimated from the sharpness of the rise and fall in the RBS film peaks. The sharpness of the high energy edge is determined by the resolution of the surface barrier detector (15 keV) while the low energy edge has additional broadening due to the straggle of the He ions. Taking the Bohr theory for straggling, we calculated the sharpness of our low energy edges for a perfectly uniform film thickness. We compared the theory with experimental observation and ascribed any broadening in the low energy edge to film non-uniformity. The technique is most sensitive for single isotope elements so we made the comparison for Al, Co Nb and Au. Our results indicated no non-uniformity within the precision of the technique, approximately 100 Å.

The thick target x-ray yield for protons on carbon goes through a rather flat maximum near 950 keV, (6) and in this region a $\pm$ 100 keV energy change of the proton beam causes a change in the emitted x-ray intensity of approximately 1%. The greatest energy loss for protons passing through our films was 14.5 keV for the case of a 2400 Å thick Ti film. In view of this we took all our data using a 2 MeV H$_2^+$ ion beam, which is equivalent to a 1 MeV H^+ beam as the molecule breaks up within 10 Å of the surface. In our system H_2^+ can be generated with higher intensity and better stability than the H^+ beam. The number of H_2^+ ions incident on the sample was obtained from a current integrator connected to a rotating-vane beam monitor. Not having to locate the samples in a Faraday cup greatly simplifies alignment, sample changing, and obtaining a clear path to the spectrometer crystal.

For each sample we counted the number of C K_α x-rays for the same number of protons incident onto the reference and film coated carbon substrates at a number of different positions on the film. Typical variations were ± 3% from the mean while samples having greater than 8% variations were discarded. Some of the metal films were checked for uniformity of the mass desposition using an electron probe microanalyzer operated with a beam energy of 29 keV and a beam size of 0.5 μm. Those tested were found to be homogeneous within ± 2%.

RESULTS

As previously explained, we may ignore the change in x-ray production due to the energy loss in the film. Let I_o be the detected x-ray intensity generated in the carbon substrate and I_F be the intensity from the film coated substrate. We may write therefore:

$$I_F = I_o \, e^{-\mu \rho_A / \cos\theta}$$

or

$$\mu = \frac{\cos\theta}{\rho_A} \, \ln \frac{I_o}{I_F}$$

The error in the determination of μ at given counting statistics has a minimum at $I_o/I_F = 2.3$, but changes insignificantly in the I_o/I_F range from 10 to 1.4 (7). We selected absorber thicknesses yielding I_o/I_F in the range from 10 to 2.5 permitting the use of relatively thick films without an appreciable increase in error. This minimized problems associated with the substrate topography and with the analytical determination of the absorber mass.

In our previous publication (8) we estimated an experimental uncertainty of ± 5%. The same considerations apply here but in the present work we have the additional problem of positioning identically the sample and reference targets on the Rowland circle of the crystal spectrometer. Also the question of surface topography as discussed previously can influence the results. For these reasons we give the conservative estimate of ± 10% to our results.

In Table 1 we list for all samples used in our experiments, the areal density ρ_A, the experimental value of μ and Veigele's theoretical value for comparison. In most cases the agreement is quite good. Two cases showed significant disagreement with Veigele. These are Mo and Nb. Since rather thin films of these elements, of the order of 300 Å were used in the first measurement we suspected possible non-uniformity on a 100 Å scale as the cause for the observed discrepancy between experiment and theory. We therefore repeated the measurements using films two to three times thicker.

Table 1

Element	ρ_A (μgm/cm^2) RBS	ρ_A (μgm/cm^2) Wet Chemistry	μ(cm^2 gm) Present exp	μ(cm^2/gm) Veigele	% difference columns 4 & 5
Al	35.4	34.8	28800	30200	− 4.7
Ti	104.3	106	8056	8094	− .5
V	56.3	60.2	10600	8840	+ 18
Co	58.9	57.4	16300	14730	+ 10
Ni	72.3	75.7	19100	17270	+ 10
Zr	31.0	31	26100	31300	− 18
	66.0		28200		− 10
Nb	23.7	26	22400	33900	− 41
	63.8		25000		− 30
Mo	27	27.4	17900	32420	− 58
	72.1		19200		− 51
Ag	98.9	106	6610	5507	+ 18
Hf	46.1	49+5	18600	18030	+ 3
Ta	45.6	43	18100	18390	− 2
W	42.2	47	16200	18750	− 14
Au	65.8	62.3	13800	15200	− 9
	114		14000		

average 9%

Table Caption – Experimental results for the mass absorption coefficient μ at C K$_\alpha$ (44 Å).

The second set of results shown in Table 1 for Zr, Nb, and Mo were obtained from these films. Consistently higher mass absorption coefficients were found for the thicker films, which may indicate some nonuniformity problems leading to erroneously low values if the mass deposition is less than 30 $\mu gm/cm^2$.

The very large difference in μ between theory and experiment for C K_α in Mo and also to a lesser extent in Nb is far outside our experimental uncertainty and must be attributed to the presence of the M V absorption edge in Mo (227 eV) and Nb (205 eV) close to the energy of the radiation source (277 eV). The omission of solid state effects in the free atom absorption theory becomes critical in such cases and explains the large observed deviation found for Mo.

ACKNOWLEDGMENTS

We wish to express our thanks to B. L. Gilbert and B. L. Olson for their careful wet chemical analysis of our samples and to S. Ruffini for his meticulous preparation of the evaporated samples used in the experiments. We also thank G. Pigey for polishing the carbon substrates.

REFERENCES

1. B. L. Henke and E. S. Ebisu, University of Hawaii, AFOSR Report 72-2174; B. L. Henke and M. L. Schattenburg, Adv. X-ray Anal. 19, 749 (1976).

2. W. J. Veigele, Atomic Data Tables 5, 51 (1973).

3. W. Reuter, A. Lurio and J. F. Ziegler, J. Appl. Phys. 46, 3194 (1975).

4. Obtained from Beckwith Carbon Corporation, 16140 Raymer Street, Van Nuys, California 91406.

5. See "Ion Beam Surface Layer Analysis", ed. by J. W. Mayer and J. F. Ziegler, Elsevier Co., Lausanne, Switzerland (1974).

6. J. M. Khan, D. L. Potter and R. D. Worley, Phys. Rev. 139 A1735 (1965); L. H. Toburen, Phys. Rev. A5, 2482 (1972).

7. I. M. Kolthoff and E. B. Sandell "Quantitative Inorganic Analysis," 3rd Ed., McMillan Company, New York, 632 (1952).

8. A. Lurio and W. Reuter, Appl. Phys. Letters 27, 704 (1975).

PROCESSING OF ENERGY DISPERSIVE X-RAY SPECTRA

J. C. Russ

EDAX Laboratories

Prairie View, IL   60069

Much attention has been given to the treatment of energy-dispersive X-ray spectra. Descriptions of mathematical approaches (and their implementation in computer programs) have dealt with the major areas of separating peaks from background and from each other. There has been an unspoken but widely accepted underlying assumption that the more complex the approach, and the more general it seems to be, the better. This assumption appeals to the mathematicians and programmers but should be sharply challenged by users, whose focus is more properly the analytical results that can be obtained in specific circumstances.

Background Subtraction

Methods in use for separating peaks from background in energy dispersive spectra can be generally summarized by the list below, in approximate decreasing order of complexity (at least in terms of their application if not the initial computation):

1.   Shape fitting.
2.   Interpolation.
3.   Blank Subtraction.
4.   Regressed Constant.

Shape Fitting (or recognition) covers both the frequency transform or digital filter approach, and the calculation of background. The first of these operates on the assumption that the peaks have a shape (for instance, their rate of rise and fall) that does not occur in the background and can be distinguished from it. The simplest such method, cross-correlation of the spectrum with a peak shape, serves quite well to locate peaks but does not preserve their areas (1). More elaborate methods overcome this problem for most peaks but the shape of the filter used to separate peaks from background does not work equally well for the range of peak heights and widths that may be encountered (2,3).

This general approach recognizes peaks by the change in slope, compared to the background. This gives rise to several problems. When peak overlaps produce a single broad peak, the "background" tends to rise beneath it, because the rate of rise and fall of the composite peak is less than an isolated peak, and it contains some of the frequencies associated with the background.   If the weighting factors are empirically chosen to give accurate results for small peaks, as are normally produced by electron excitation, and in the case of either electron or photon excitation are of interest in trace analysis, then very high peaks with small backgrounds tend to have satellite false peaks ("ringing") next to them, which can interfere with the detection of real peaks in that region.

Also, since the peaks in energy dispersive X-ray spectra are basically Gaussian, the tails are clipped by this procedure and this can alter the results for small peaks next to large ones (4). Finally, the highly sloped background and large discontinuities at absorption edges in the 1-3 KeV region are hard to distinguish from peaks and the largest errors are thus typically found for the light elements. Nevertheless, this method appears to be the most generally useful for spectra that cannot be handled by one of the following more specific techniques.   The significant attractive feature of this method is that it locates peaks with no input required from the user, and is thus especially valuable for qualitative analysis or reproducible detection of trace peaks.

Calculation of the actual background using semi-theoretical equations for generation and absorption has been shown to be effective for bulk inorganic samples with electron excitation (5,6).   It offers good accuracy for peaks above about 3 KeV but at lower energies (especially around 1 KeV) the equations in use do not exactly match observed results, possibly because the details of the depth of continuum generation in the samples are ignored. The method presumes the elements present are known and their compositions are being quantitatively calculated with the background correction included in the iteration. This is much too slow if the entire background is computed, and calculating only single values under the peak centroids is incompatible with methods for stripping of peak overlaps by fitting of either stored or generated peaks. The calculation method has not yet been adapted for low atomic number matrices, or thin samples, or X-ray photon excitation. A method to calculate the scattered background at any energy in X-ray fluorescence analysis spectra relative to a reference point on the spectrum background, as a function of composition as reflected in total matrix absorption coefficient, has recently been proposed (7).   This method could become useful, especially for heavy elements, as it is more fully developed.

Interpolation has been used extensively in systems having limited computing capability. For isolated major peaks in the region above about 4 KeV, it is often adequate, but it cannot be used where peak interferences are present, or at low energies where very nonlinear backgrounds are encountered. For many specific applications in analysis of high energy lines, particularly in X-ray fluorescence analysis where a few specific elements are known to be present (with isolated peaks) it is the method of choice. This includes cases where the background is the continuously sloping tail of the Compton-scattered peak in monoenergetically excited X-ray fluorescence analysis.

Comparison of the frequency filter, calculation, and interpolation method is very revealing. Table 1 shows net intensities measured on a series of pure elements in a scanning electron microscope. For lines at relatively high energy, where the background changes smoothly and gradually, the net intensity values obtained by each method are virtually identical. Even at

low energy the values are very close. The results for the silver L lines
show the greatest disparity. The frequency filter method gives the lowest
net intensity (-2.2%) because of the tendency, shown before, for the
filtered background to rise under broad overlapped regions. The interpolations method gives the highest net intensity (+2.2%) because it includes
in the peak the background just below the quite large absorption edges.

Table 1
Intensities - Pure Elements

Element	Line	Total Intensity*	Calc.	Frequency	Interpolation
Al	K	114468	113180	113874	113892
Fe	K	134038	130754	130059	129704
Mo	K	14419	10261	10296	10152
Mo	L	304427	266681	269165	278816
Ag	L	119368	108570	106229	110901
Au	L	29094	24540	24491	24411

*   All measured at 25 kV, normal incidence, 32° takeoff angle,
      but with different beam currents.

The error in net intensities on these pure elements is relatively small,
because the peaks are large and the background relatively low.   For trace
peaks the errors can be expected to be larger. To investigate the effects
on minor elements further, analyses were carried out on two complex
samples, a metal alloy and a mineral.

The metal alloy, a nickel-based permalloy, gave the the intensity results
listed in Table 2. The large nickel and iron peaks give the same results
by any of the three methods. The molybdenum L peak shows a significant
difference (greater than three sigma) for the linear interpolation
method. The other minor peaks, manganese and silicon, do not differ by a
significant amount. These results suggest that, in many cases the choice
of background subtraction method is not critical.

A more severe test is presented by the mineral sample, an amphibole
containing appreciable concentrations of several low energy elements with
partially overlapped peaks.   For the elements sodium thru silicon the
difference between the three different background methods becomes
greater.

From these results we may conclude that although the exact calculation of
background gives the best results for samples to which it can be applied -
bulk samples excited by an electron beam - the loss in accuracy associated
with the frequency filter and interpolation methods is frequently small.
These methods are especially useful since they do not require knowledge of
the exact composition, and thus can also be used for semiquantitative or
qualitative analysis. The frequency filter method is further attractive
in that no user input is required. The user should examine the results
with any method to verify that the results are reasonable. In this way

Table 2

Ni-base alloy (Permalloy)
25 kV, 32° tilt, 41° takeoff angle, 200 sec.

Element	Line	Conc.	Total Counts	Freq. Filter	Interpolation	Calc. BG.	3σ Std Error
Ni	K	80.11%	227210	216604	216391	217476	2001
Fe	K	14.84	78090	66290	66509	66511	1141
Mo	L	4.29	27894	10895	11871	10968	591
Mn	K	0.49	9397	1928	1973	2072	321
Si	K	0.25	7057	1282	1234	1424	276

Table 3

Mineral (Amphibole)
12 kV, 30° tilt, 39° takeoff angle, 200 sec.

Element	Conc.	Total Counts	Freq. Filter	Interpolation	Calc. BG.	3σ Std Error
Na	1.9	2340	825	658	595	192
Mg	7.7	7200	4542	4340	3865	308
Al	7.7	22966	19909	19967	19268	490
Si	18.4	63605	58402	59037	58374	787
K	1.7	6592	3457	3487	3791	291
Ca	7.9	13740	10021	9923	10373	392
Ti	3.0	4538	2610	2237	2674	240
Mn	0.1	1478	85	127	72	161
Fe	7.8	7281	5552	5560	5540	285
O	43.8	not analyzed				

the errors from the more approximate methods can frequently be spotted and avoided.

Subtraction of a blank spectrum is preferred when the specimen is analyzed on a substrate and is itself so thin or sparse a deposit that it produces little background and causes negligible absorption of the X-rays from the substrate. X-ray fluorescence analysis of deposits on filter papers, or transmission electron microscope analyses of tissue sections on a grid, are often in this category. The subtraction of the blank spectrum measured on the bare substrate is preferable to the use of constant factors for background since the blank spectra can be easily re-measured when a new batch of filters, for example, are used.

For scanning electron microscope analysis of bulk samples, the blank technique cannot be applied directly. However, it has been shown to be possible in some cases to use a spectrum measured on a specimen containing no elements that produce spectral peaks(e.g.diamond) as a starting point in the calculation of background (8).

The regressed constant for background under each element is quite adequate when a calibration curve approach is used. For X-ray fluorescence analysis of narrow concentration ranges, as in the case of quality control of metal alloys, cements, etc., this is often the case. The constant, expressed in either intensity or concentration units, is the intercept of the best fit calibration curve, determined with standards. The intercept of the calibration curve is calculated as part of the least-squares fit of the data, and the analysis of an unknown then requires only a simple subtraction. It is important to remember that the "background" value, even if it is expressed in intensity units, may not correspond to the actual background underlying the peak in the spectrum. This is particularly likely if the element is present in moderately high concentration. The calibration curve is intended to fit only a narrow concentration range, and its extrapolation to zero concentration ignores the non-linear effects that are generally present. The intercept value is thus a useful background value for analyses of samples within the "calibrated" concentration range, but not outside the range.

Peak overlaps

Methods for peak stripping can be classed into three categories:

1. Least squares fitting of library spectra.
2. Fitting of generated peak shapes.
3. A stored matrix of overlap factors.

Library Spectra can be added together using a multiple-linear regression fit to match a measured spectrum. This fitting can only be properly done after background removal in both the measured spectrum and the library spectra, since backgrounds are not linearly additive. The equations used in the calculated background method described above show that the absorption of the generated background depends on the total matrix absorption coefficient. This is a linear sum of the individual elemental absorption coefficients in proportion to their concentration, but it enters into the absorption calculation in an exponential term.

The chief appeal of the library method is its intuitive simplicity and the

apparent assurance that the stored library spectra include all peaks
present and any irregularities in peak shapes (9). On the other hand, the
library requires the user to obtain and analyze numerous standards under
various excitation conditions and requires extensive storage. The
fitting procedure is very sensitive to shifts in spectrum gain or zero
calibration and to possible changes in peak shape (for instance, due to
count rate effects), or more often, peak ratios. These latter changes can
occur either due to changes in excitation or matrix absorption. It is
also essential that the user include all elements present in the set of
spectra to be fit.

The burden of judging the goodness of fit is ultimately the user's
responsibility, and is frequently hard to determine. The chi-squared
factor describes in mathematical terms the "goodness-of-fit" but is not
always consistently interpretable. The errors may be due to poor counting
statistics in one case, to peak shifts in another, or to missing an entire
element in a third, and yet give identical chi-squared values. Since the
method is identical to the method of overlap factors, in that it has no
ability to adjust peak ratios for matrix effects, there seems to be no
justification for its added complexity and inflexibility.

Adding together reference, or library spectra will sometimes give a very
good fit to the measured spectrum; indeed, for relatively major elements
with peaks at fairly high energies, even including the background in the
addition does not introduce significant errors. The problems with the
method become more apparent when fitting is attempted over broad energy
ranges. Since the different shell absorption edges have different
excitation energies, their relative excitation depends on the other
elements present in the matrix, which compete for the energy of the
incident electrons. Also, the generated X-rays (both characteristic and
background) are absorbed, and the matrix absorption coefficient depends
on composition. As a result, while the ratio of K_β to K_α peaks varies
little for a given element (unless a large amount of a matrix element with
absorption edge energy lying between the K_α and K_β is introduced), the
ratio of L lines (which lie farther apart, and have three different
absorption edges), and even more significantly the ratios of K to L or L
to M lines can vary greatly.

Fitting of generated peaks can be used to overcome some of the limitations
of the library approach: peak positions and peak ratios can be allowed to
fluctuate within specified ranges to obtain a best fit (10). The
principal drawback of the method is that the generated shapes may not
adequately describe the detector response. This is particularly the case
for small peaks,where statistical fluctuations are large, and for low
energy peaks,where a low energy tail can be present due to incomplete
charge collection in the detector. A further drawback is that a peak must
be generated and fit for every line energy in the element's spectrum,
including minor lines and doublets. There is a tendency to use too few
lines by ignoring the ones with a few percent intensity, and by averaging,
for instance the $K_\alpha 1$ and $K_\alpha 2$. This leads to errors, particularly in
the more complex L-spectra.

The generated peak method is generally the best approach when dealing with
elements for which standards with a similar matrix have not previously
been measured with similar excitation conditions. This is often the case

when complete unknowns are encountered in qualitative analyses. It is an appealing method because the spectrum can be examined with some peaks removed, for user interpretation, but good judgment is required since artifacts of stripping can be mistaken for uncovered peaks.

There are several methods that can be used to determine the height of the peak to be stripped; each has some drawbacks. The three principal methods are 1) using the greatest peak height that will fit without exceeding the overlapped spectrum; 2) judging the height from the ratio of two points on the exposed side of the peak; 3) simultaneously fitting two or more peaks using a least squares technique.

The first method seems to be logical in that, obviously, no single peak can be higher than the composite. A Gaussian peak is generated with the proper width for the detector resolution (as a function of energy), and is centered at an energy either entered by the user or taken from a table to correspond to the element and line of interest. The height is adjusted so that no point is higher than the corresponding point in the overlapped set of peaks. Then it is subtracted. The remaining peak or peaks can then be identified, integrated, or subjected to further stripping operations. Since this procedure is carried out one peak at a time, the user can identify one element, remove its peaks, and look for the next. The problem comes from the fact that the centroid energy chosen may not correspond to the actual peak centroid, either because of a slight analyzer miscalibration or because the "table" was condensed into an equation and generated inexact values. Equations that calculate peak energies to within 10 eV or even 20 eV are good enough to position lines on the display to identify peaks. They are definitely not good enough for peak stripping.

To illustrate the sensitivity of stripping to choice of centroid energy, a composite iron-cobalt spectrum was made by adding the pure element spectra together. The spectra were measured with the system exactly calibrated to correspond to the centroid energies which the computer used (in fact, a computer-assisted calibration program was used). When "automatic" stripping of the cobalt K peaks was performed, excellent results were obtained as shown in Table 4.

However, if the centroid energy is wrong, for any of the reasons previously mentioned, this method introduces large errors in intensity (Table 4, Section II).

The second method is much less "unstable" in its results when small errors in peak centroid energy are present. In this case the ratio of the height at the peak centroid to the height at the full-width at tenth-maximum point is used to estimate peak height. In principle any other point could be used as well. It is clear, of course, that this method can only be applied to the outside peak (lowest or highest energy) in a set of overlapped peaks. Also, while the intensity values shown in Table 4, Section III vary much less than the method described before, the errors particularly for the iron $K\beta$ are far from negligible. This emphasizes the need for great care in system calibration and the use of proper energies. Finally, this method works best on large peaks, because small ones have sizeable statistical fluctuations that can affect the height ratio. To a greater or lesser degree, this last limitation affects all

Table 4

Intensities obtained by stripping CoKα-FeKβ overlap

CoKα centroid energy	CoKα counts	FeKβ counts

I. "True" intensities on original spectra before adding.

	76914	10506

II. Stripping of CoKα with height not exceeding composite.

CoKα centroid energy	CoKα counts	FeKβ counts
6.929	77629	10441
6.920	49642	37778
6.924	58511	28949
6.928	66405	20840
6.932	77096	9421
6.936	81437	6607

III. Stripping of CoKα with height determined by ratio of centroid to FWTM value.

CoKα centroid energy	CoKα counts	FeKβ counts
6.920	75159	12154
6.924	76743	11382
6.928	77737	10487
6.932	77298	9566
6.936	76465	7113

IV. Simultaneous least squares fitting of both peaks with FeKβ fixed at 7.060 keV.

CoKα centroid energy	CoKα counts	FeKβ counts
6.920	75261	14038
6.924	76121	12063
6.928	76872	10011
6.932	77518	7881
6.936	78065	5672

V. Simultaneous least squares fitting of both peaks keeping FeKβ 130 eV above CoKα.

CoKα centroid energy	CoKα counts	FeKβ counts
6.920	75595	14678
6.924	76736	12263
6.928	77613	9934
6.932	78221	7862
6.936	76834	5741

stripping methods. The total statistical error in a net intensity value N after subtracting B background counts and stripping S counts from an overlapping peak is given by,

$$\sigma = \frac{\sqrt{N+2B+2S}}{N}$$

and this will often represent the major limitation on the ultimate analytical accuracy that can be achieved by stripping.

Simultaneous least squares fitting of both peaks gives results for intensity values that are comparable to this method. Two sets of data are shown in Table 4. Section IV moves one peak while the other stays fixed. This is the kind of error that results from inexact energy values for the various lines. Part V shifts both peaks together, as occurs when there is a slight miscalibration. The errors are similar in magnitude. This method's advantage is that it fits to all of the channels in the peaks - not just one or two - and can handle any number of peaks at once. The quality of the matrix inversion in the program shown in Appendix D to apply this method represents the principal limit on the number of peaks that can be included. The disadvantage is that the user must know ahead of time what peaks are present (and their energies). No visual feedback is supplied unless the user generates and adds the peaks back together to examine the agreement with the original.

For qualitative work the use of the methods that remove a single peak at a time (by either method of determing peak height) are usually best because they allow the user to judge and interpret the results. For routine quantitative analysis the simultaneous fitting method is usually better. In either case, one more reminder of the importance of exact calibration is worthwhile.

Overlap factors, describing the added intensity in an energy window or region of integration for the peak of the element of interest due to each of the other elements present, are identical in principal to the use of library spectra and hence share many of the limitations of that method. The factors can be used with little computation and require modest storage, however. It is possible from a table of line energies to predict which elements will have non-zero overlap factors, but in general the factors themselves cannot be calculated with acceptable accuracy. They must be measured for a given system, as they depend on detector efficiency and resolution, and for samples with the same general range of concentrations, and excitation conditions, as will be used for the unknowns. In general this is readily done for quantitative analysis.

Reference (11) shows such a matrix of factors, determined for electron probe analysis of oxide (mineral) samples at constant voltage and specimen geometry. In many cases the overlap factors can be determined as the pure element or compound standards are run. In principle it is possible to adjust the factors for the change in peak ratios that occur when the matrix composition changes, although this correction would have to be carried out as part of the overall procedure of iterative quantitative analyses. Most users of this method simply assume that the ratios do not change over the range of concentrations they encounter.

In X-ray fluorescence analysis the overlap factors can be included in the general regression on standard samples by using an equation of the form,

$$C_1 = K_i I_i (1 + \Sigma \alpha_{ij} C_j) + B_i (1 + \Sigma \xi_{ij} I_j)$$

In this case K is the reciprocal of the general slope of the calibration curve, α the correction for interelement absorption or enhancement, B the constant background term, and ξ the overlap factors.

The overlap factors are not necessarily the same values that would be obtained from an actual stripping operation. Like the "background," or intercept term discussed above, they include more than just peak overlaps - for instance any change in total matrix scattering power with composition that raises or lowers the general background level. This equation is widely used in both energy dispersive and conventional wavelength dispersive analysis, and is well known to give useful results provided that the "calibration" samples adequately cover the concentration ranges of the elements.

Conclusion

Each of the methods described for separating peak intensity from background and other overlapping peaks has validity and applicability for some purposes. Since no single approach is suitable for all cases, it must be the user's responsibility to select the method(s) for his needs,and the manufacturer's responsibility to make all methods available.

References

1. A. L. Connelly and W. W. Black, "Automatic Location of Area Determination of Photopeaks", Nuc. Instr. and Meth., 82, p. 141, 1970.
2. J. C. Russ, " Background Subtraction for Energy Dispersive X-ray Spectra", Proceedings 7th Nat'l. Conference on Electron Probe Analysis, San Francisco, California, p. 76A-76C, July, 1972.
3. T. D. Kirkendall, "Comprehensive Qualitative and Quantitative Analysis of Energy Dispersive X-ray Spectra", Tutorial and Proceedings 9th Annual Conference, Microbeam Analysis Society, Ottawa, Canada, p. 24A-24G, July, 1974.
5. S. J. B. Reed and N. G. Ware, "Quantitative Electron Microprobe Analysis using a Lithium Drifted Silicon Detector", X-ray Spectrometry, 2, 1973.
6. C. E. Fiori, R. L. Myklebust, and K. F. J. Heinrich, "Prediction of Continuum Intensity in Energy-Dispersive X-ray Microanalysis", Analytical Chemistry, 48, No. 1., January, 1976.
7. D. G. W. Smith, C. M. Gold, D. A. Tomlinson, "The Atomic Number Dependence of the X-ray Continuum Intensity and the Practical Calculation of Background in Energy Dispersive Electron Microprobe Analysis", X-ray Spectrom. 4, (1975), p. 149-156.
8. C. E. Feather, J. P. Willis, "A Simple Method for Background and Matrix Correction of Spectral Peaks in Trace Element Determination by X-ray Fluorescence Spectrometry", X-ray Spectrom. 5, (1976), p. 41-48.
9. F. H. Schamber, "A New Technique for Deconvolution of Complex X-ray Energy Spectra", Proceedings 8th Nat'l. Conference on Electron Probe Analysis, New Orleans, Louisiana, p. 85A-85D, August, 1973.
10. R. H. Geiss and T. C. Huang, "Quantitative X-ray Energy Dispersive Analysis with the Transmission Electron Microscope", X-ray Spectrometry, 4, No. 4., p. 196-201, October, 1975.
11. D. G. W. Smith, C. M. Gold, " A Scheme for Fully Quantitative Energy Dispersive Energy Dispersive Microprobe Analysis", Advances in X-ray Analysis, 19, (Gould, et al, ed.), 1976, p. 191-201.

USE OF X-RAY SCATTERING IN ABSORPTION CORRECTIONS FOR X-RAY FLUORESCENCE

ANALYSIS OF AEROSOL LOADED FILTERS

K. K. Nielson and S. R. Garcia

Battelle Pacific Northwest Laboratories

Richland, Washington   99352

ABSTRACT

Two methods are described for computing multielement x-ray absorption corrections for aerosol samples collected in IPC-1478 and Whatman
41 filters. The first relies on scatter peak intensities and scattering cross sections to estimate the mass of light elements ($Z < 14$) in
the sample. This mass is used with the measured heavy element ($Z \geq 14$)
masses to iteratively compute sample absorption corrections. The second method utilizes a linear function of $\ln(\mu)$ vs. $\ln(E)$ determined
from the scatter peak ratios and estimates sample mass from the scatter
peak intensities. Both methods assume a homogeneous depth distribution
of aerosol in a fraction of the front of the filters, and the assumption is evaluated with respect to an exponential aerosol depth distribution. Penetration depths for various real, synthetic and liquid aerosols were measured. Aerosol penetration appeared constant over a 1.1
mg/cm^2 range of sample loading for IPC filters, while absorption corrections for Si and S varied by a factor of two over the same loading
range. Corrections computed by the two methods were compared with measured absorption corrections and with atomic absorption analyses of the
same samples.

INTRODUCTION

The use of x-ray fluorescence methods for the analysis of aerosol
loaded filter samples has been extensive in recent years and has generally proved both reliable and inexpensive. Sample absorption corrections, when required, depend on the x-ray energy, the aerosol particle
size, and depth of penetration into the filter, the total aerosol loading on the filter, and the compositions of the aerosol and filter materials. Since many membrane-type filters are suitable for collecting
only small sample loads and do not permit appreciable sample penetration,
absorption only occurs in the aerosol particles and is often insignificant for elements as light as sulfur. Various fibrous filter matrices
such as Whatman 41 or IPC 1478 permit considerable particle penetration
and can potentially collect much greater masses of particulate material.
In such cases, the filter material is usually the predominant source of
x-ray attenuation, while absorption by the aerosol itself arises in

cases of heavy particulate loads or agglomeration of the particulates.

Our objective in this study has been to quantitatively analyze for such light elements as sulfur and potassium in IPC filters, in addition to a suite of fifteen heavier elements for which absorption corrections are small or negligible. The use of Zr exciting radiation gives reasonable sensitivity for all the required peaks in a single analysis, thus reducing the problem to one of determining appropriate absorption corrections for the light elements. We have found that the scattered exciting radiation, being proportional to both sample loading and atomic number, provides a convenient basis for computing the required absorption corrections. Scattered radiation has previously been used in absorption corrections for homogeneous, uniform or thick samples (1-4). Since IPC filters do not fit these criteria, we have examined the average deposition depth of aerosols, the change in this depth with sample loading, and the change in absorption corrections with sample loading. We have examined two separate methods of computing absorption corrections for the filters, both utilizing scattered radiation, and compared the results with those of atomic absorption analyses and measured absorption correction factors.

THEORY

In order to appropriately correct for absorption by the filter matrix, a definition of where the aerosol is located in the matrix is required. Although an exponential filter depth distribution of aerosol material has frequently been used for this purpose, we have found that an equivalent homogeneous distribution in some fraction of the front of the filter closely approximates the resulting correction factors. The two distributions are illustrated in Figure 1. The exponential distribution has been discussed by Dzubay and Nelson (5), and is defined by

$$C(x) = \exp\left[x \ln(P)/t\right], \tag{1}$$

where $C(x)$ is the aerosol concentration in the filter, x is the depth into the filter, P is the fractional penetration through the entire filter, and t is the total filter thickness (mass/cm^2). Absorption corrections (f) based on this exponential model take the form

$$f = \frac{(1-P)\ (1-\mu t/\ln P)}{1-P\ \exp(-\mu t)} \tag{2}$$

where μ is the sample mass absorption coefficient for the given line.

The equivalent homogeneous distribution shown in Figure 1 is defined as

$$\begin{align}
C(x) &= k \qquad 0 \le X \le 2D \\
C(x) &= o \qquad \quad X > 2D
\end{align} \tag{3}$$

where k is a constant and $2D$ represents the fraction of the filter containing the sample (twice the mean aerosol penetration depth). Absorption corrections based on Equation 3 thus take the form

$$f = \frac{2\mu D}{1-\exp(-2\mu D)} \ . \tag{4}$$

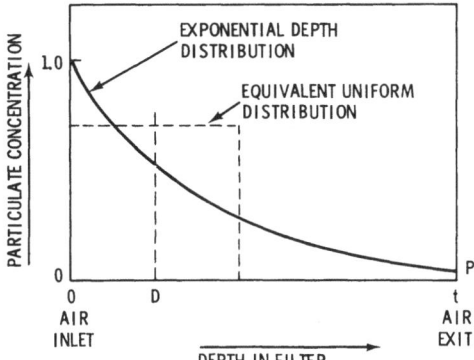

Figure 1. Assumed Aerosol Distributions in Filter Material.

The error which would result from corrections utilizing Equation (4) relative to corrections from Equation (2) can be estimated by integrating the exponential curve (Equation 1) in Figure 1 from zero to D and from D to t, and equating the resulting areas. Solving the resulting equation for the mean aerosol penetration depth, D, relative to the total filter thickness, t, gives

$$D/t = \frac{\ln (\frac{P+1}{2})}{\ln P} \quad . \tag{5}$$

Using Equation (5), the depth D which is equivalent to various P values can be readily used to evaluate corrections using Equations (2) and (4), which are compared in Figure 2 for various values of the multiple μX. To relate to real aerosol samples, the following μX multiples are given respectively for IPC and Whatman 41 filters: Si, 8, 3.5; S, 4, 2; Ca, 1, < 1. It can be seen from Figure 2 that the error in using Equation (3) will be less than 10% for IPC or Whatman filters.

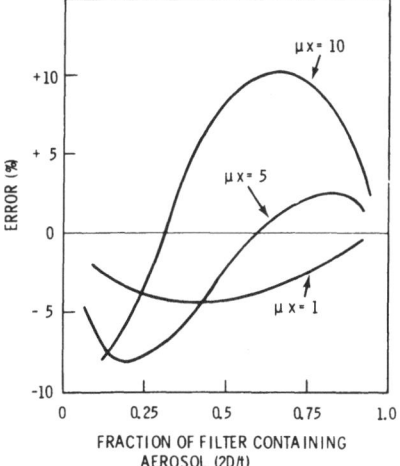

Figure 2. Correction Errors in Assuming a Uniform, Partial Aerosol Penetration [Equation (3)], Relative to an Exponential Penetration Distribution [Equation (1)].

EXPERIMENTAL

Real aerosol samples were collected on IPC 1478 filters (Knowlton Brothers, Watertown, New York) at Quillayute, Washington, over 24 hr intervals at 80 cfm through a 34 cm^2 filter area. Real aerosols collected on Whatman 41 filters were sampled in St. Louis, Missouri, and Colstrip, Montana.

Synthetic aerosols used in estimating depth distribution and scatter peak parameters were prepared by a method similar to that of Giauque, et al, for rock specimens (6). The aerosols consisted of 2μ ZnS particles (U. S. Radium Co., Morristown, New Jersey), and CuCl (J. T. Baker, analyzed reagent grade), ground in a Tungsten Carbide ball mill (Spex Industries, Inc., Metuchen, New Jersey). Wet aerosol samples were prepared as the dry synthetic ones except that the sample material was first dissolved in aqueous solution and sprayed through an atomizing nozzle.

X-ray fluorescence analyses were performed under vacuum using a zirconium-filtered zirconium secondary fluorescer excited by a tungsten anode x-ray tube (Model 810 excitation system, Kevex, Inc., Burlingame, California). X-ray spectra were collected with an 80 mm^2 Si(Li) diode having 200 eV resolution FWHM at 6.4 KeV (Kevex, Inc.). Data accumulation utilized a computer-based analyzer system (Model 4410, Nuclear Data, Palatine, Illinois), and data reduction utilized an off-line, 24 K PDP-15 computer.

Atomic absorption measurements on the real aerosol samples were accomplished by digesting a known fraction of the IPC filter sample for 12 hr in 10N sulfuric acid, followed by fusion of the resulting mixture with lithium metaborate for 15 min at 950° C and dissolution with 3N HCl prior to analysis. Standards were prepared from BCR-1 in a similar fashion, and appropriate blanks were used.

Measurements of the penetration depth, D, were made by analyzing the air filter from the front and back sides, in addition to measurement of the transmitted radiation through the filter by the method of Giauque, et al (6). The penetration depth was then computed as

$$D/t = 1/2 - \frac{\ln(I_f/I_b)}{2 \ln(\frac{I_o}{I-I_f})} \tag{6}$$

where I_f and I_b are the intensities of a given line as analyzed from the front and back sides, respectively. I is the line intensity of the sample with an elemental standard behind it, and I_o is the intensity of the standard without the filter sample. This relative depth is comparable to that measured by Adams and Van Grieken (7). The measurements of I, I_o, I_f, and I_b also permit one to compute an absorption correction factor for the given line, assuming the distribution in Equation 3

$$f = \ln A/(1-A), \tag{7}$$

where

$$A = \frac{I_f}{I_b} \cdot \frac{I - I_f}{I_o} \tag{8}$$

Absorption correction factors based on scatter peaks utilized two separate methods. The method used in Program 1 is illustrated in the flow chart in Figure 3. The contributions to scatter peak areas from various elements were determined from x-ray scattering cross sections for Zr Kα radiation, computed from the data of McMaster, et al (8). The fraction of the light element mass used in the absorption correction was defined as 2D/t. A more complete description of this computer program is given separately (9).

A more empirical approach was used by Program 2, which assumed the following: (a) the aerosol was uniformly deposited in a filter region of thickness 2D (mg/cm^2); (b) absorption edge jumps are not significant in the x-ray energy region of interest; and (c) ln(μ) is linearly related to ln(E) such that

$$\mu = E^{(a)} \exp(b), \qquad (9)$$

where a and b are the slope and intercept of the ln(μ) vs ln(E) plot. The slope a is adjusted for variation in sample atomic number using the relationship

$$a = k_1 + k_2 (\ln \frac{I}{C}) + k_3 (\ln \frac{I}{C})^2, \qquad (10)$$

where I and C are the incoherent and coherent scatter intensities and k_1, k_2, and k_3 are the quadratic coefficients of the curve shown in Figure 4. This curve was determined experimentally from several pure standard materials. The effective sample thickness was estimated from a linear function of the scatter peak parameter, C^2/I, and resulted in

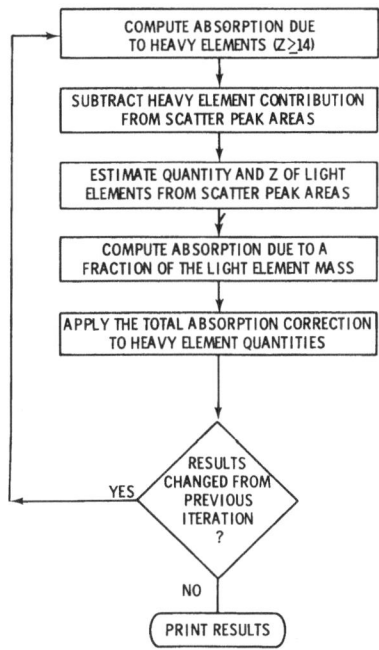

Figure 3. Flowchart for Filter Absorption Corrections Using X-ray Scattering Cross Sections (Program 1).

$$2\mu D \approx E^{k_1 + k_2(\ln \frac{I}{C}) + k_3(\ln \frac{I}{C})^2} [k_4(C^2/I)+k_5], \qquad (11)$$

from which absorption corrections were computed using Equation (4).
The constants k_4 and k_5 were determined experimentally as illustrated
in Figure 5. Data points used in the calibration represent both real
and synthetic aerosol samples. Data points for both Whatman 41 and IPC
filters are shown in Figure 5.

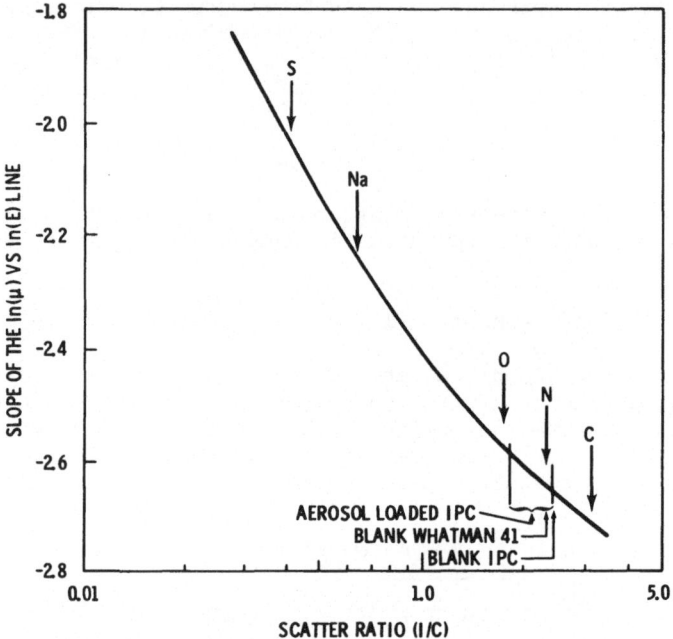

Figure 4.   Change in X-Ray Energy-Dependence of μ with the Observed
Scatter Peak Ratio

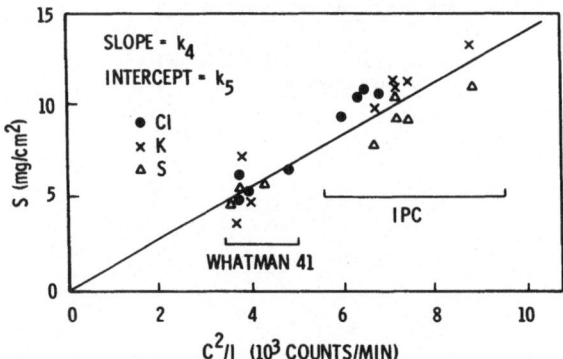

Figure 5.   Calibration of the Scatter Peak Parameter, C^2/I, Using Real
and Synthetic Aerosol Samples.

RESULTS AND DISCUSSIONS

    Experimentally measured sample depths were greater for sulfur than
for calcium, as illustrated in Figure 6a. This should be expected, considering the small sizes usually observed for sulfur-rich particles from
gaseous effluent streams relative to the larger, calcium-rich natural
terrestrial aerosols. The respective relative depths of calcium and
sulfur, 0.10 ± 0.05 and 0.27 ± 0.11 in Whatman 41 filters are in excellent agreement with reported depths (7) of 0.11 and 0.29, indicated by
arrowheads in Figure 6a. Corresponding depths of 0.35 ± 0.04 and 0.40 ±
0.14 in IPC filters have not previously been reported. Sample depths in
Millipore and Nuclepore filters were determined from only two samples
and are not conclusive. Synthetic aerosols failed to penetrate as deep-
ly as real samples, presumably due to larger particle sizes. Liquid
aerosols were nearly uniformly distributed through the filter matrix as
indicated in Figure 6b.

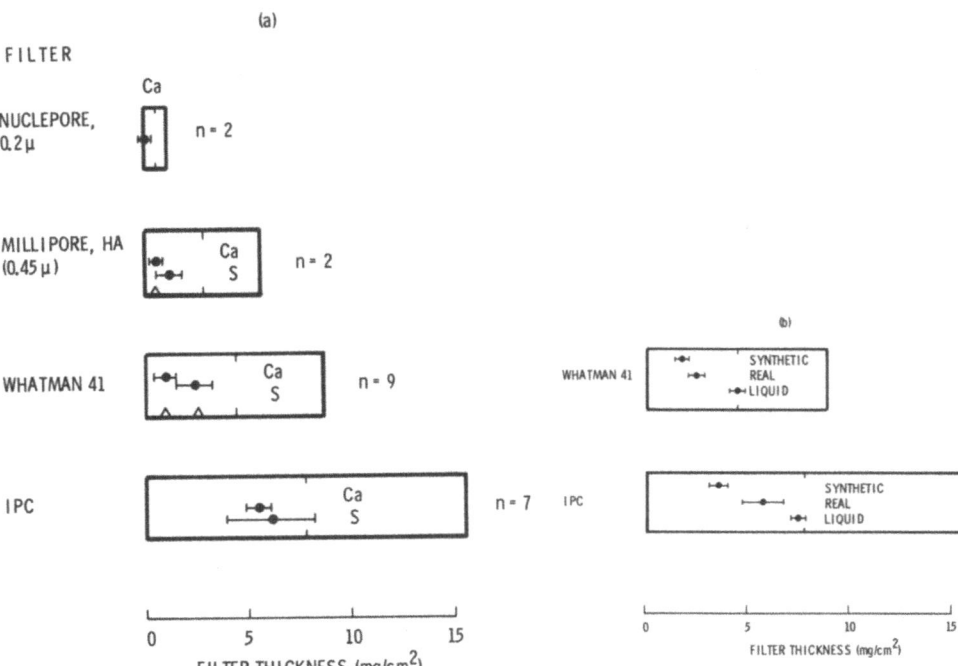

Figure 6. Comparison of Aerosol Penetration Depths in Various Filters
           for Real Aerosols (a) and Synthetic Aerosols Compared with
           Real (b).

    It was further shown that sample penetration did not vary with the
particulate loading over a 1.1 mg/cm^2 loading range for IPC filters, as
shown in Figure 7. The single high point for one of the real sample
penetrations appropriately fell into the region observed for wet aerosol
samples. This sample was moistened during the sampling process by rain-
fall. The invariance of sample penetration with loading by real, wet
and synthetic samples was also noted for Whatman 41 filters, but the
small range of loading considered (approximately 0.14 mg/cm^2) precludes
any conclusions for typical samples.

The need for absorption correction factors which vary with sample
loading is illustrated in Figure 8. Correction factors for sulfur and
silicon varied by over a factor of 2 for the sample loading range indicated.

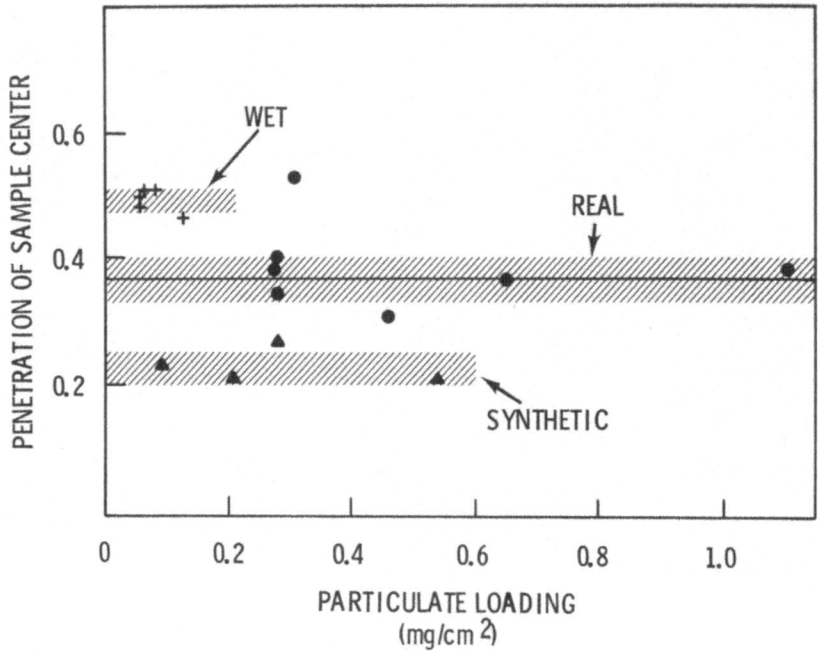

Figure 7. Penetration of Varying Aerosol Loads into IPC Filters.

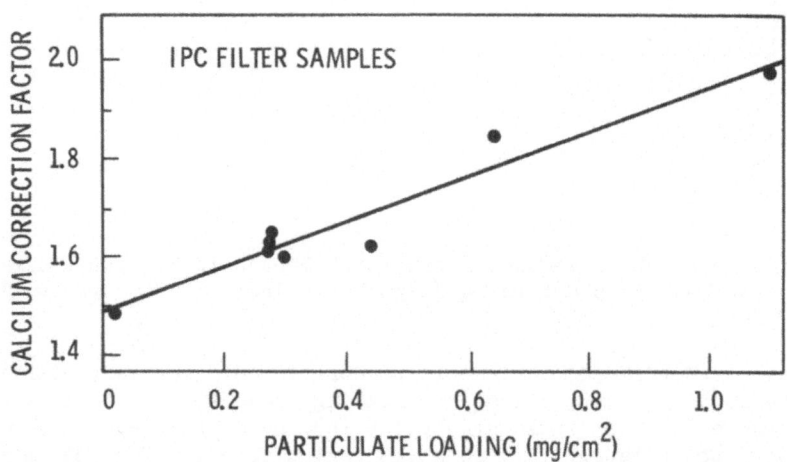

Figure 8. Variation in IPC Filter Corrections with Sample Loading.

Typical absorption correction factors as computed by Program 2 for
Whatman 41 filters are reported in Table I for calcium and sulfur and
compared with reported values. X-ray fluorescence data from IPC filter
analyses using Programs 1 and 2 are compared in Table II with atomic absorption measurements on the same samples and, where possible, with data
using measured absorption correction factors. Differences between the
x-ray data and the atomic absorption measurements may arise from a variety of possible errors in either of the analytical methods or in the
sampling procedure. The fact that the two methods show no systematic
difference common to all four elements for any sample indicates that
much of the error must be attributed to sources other than the absorption correction methods.

TABLE I

ABSORPTION CORRECTION FACTORS FOR WHATMAN 41
(as compared by Program 2)

	Ca	S
Ours	1.20 ± .02 n = 9	1.99 ± .08 n = 9
Reported[a]	1.18	2.18

[a]See Reference 7.

TABLE II

COMPARISON OF IPC FILTER ANALYSES
$(\mu g/cm^2)$

	Si			K				Ca				Fe			
	AA	Pgm1	Pgm2	AA	Pgm1	Pgm2	Meas	AA	Pgm1	Pgm2	Meas	AA	Pgm1	Pgm2	Meas
1	659	818	652	44	53	38	38	40	47	33	33	107	112	105	114
2	236	306	214	12	15	12	11	31	33	26	25	63	62	61	60
3	160	183	131	18	19	14	16	32	41	30	36	50	48	46	48
4	< 9	< 5	< 4	10	6	5	7	12	11	8	8	< 3	0.5	0.7	0.8
5	52	138	90	10	16	12	15	16	13	10	12	16	13	13	13
6	85	72	46	14	12	9	9	16	16	13	13	22	23	23	24
7	< 9	< 5	< 5	8	7	5	7	14	9	7	8	< 3	0.9	1.1	1.8

From the data presented in Table II, both methods of utilizing
scatter peaks for absorption corrections give good results. The tendency of Program 1 to give slightly high corrections is likely due to a
poor choice of light elements by the program in computing the absorption
correction. The assumption of a uniform distribution of aerosol in the
front part of the filters appears to be a good approximation of an exponential distribution as it is used here. We have shown that although

the absorption factor for real aerosols in IPC filters varies with loading, the average relative depth of penetration does not. This permits
the assumption of uniform sample deposition in a constant fraction of
the total filter thickness, which is necessary to both of the correction
methods which we considered. Scattered x-radiation can thus provide a
rapid means of estimating air filter absorption corrections without re-
sorting to separate absorption measurements on each sample.

ACKNOWLEDGMENTS

The authors appreciate the work of R. W. Sanders in performing the
atomic absorption analyses.

REFERENCES

1. C. E. Feather and J. P. Willis, "A Simple Method for Background and
    Matrix Correction of Spectral Peaks in Trace Element Determination by
    X-Ray Fluorescence Spectrometry," X-Ray Spectrometry $\underline{5}$, 41-48 (1976).

2. F. Bazan and N. A. Bonner, "Absorption Corrections for X-Ray Fluor-
    escence Analysis of Environmental Samples," Advances in X-Ray Analy-
    sis $\underline{19}$, 381-391 (1976).

3. G. Andermann and J. W. Kemp, "Scattered X-Rays as Internal Standards
    in X-Ray Emission Spectroscopy," Anal. Chem. $\underline{30}$, 1306-1309 (1958).

4. Z. H. Kalman and L. Heller, "Theoretical Study of X-Ray Fluorescent
    Determination of Traces of Heavy Elements in a Light Matrix," Anal.
    Chem. $\underline{34}$, 946-951 (1962).

5. T. G. Dzubay and R. O. Nelson, "Self Absorption Corrections for X-Ray
    Fluorescence Analysis of Aerosols," Advances in X-Ray Analysis $\underline{18}$,
    619-631 (1975).

6. R. D. Giauque, F. S. Goulding, J. M. Jaklevic and R. H. Pehl, "Trace
    Element Determination with Semiconductor Detector X-Ray Spectrome-
    ters," Anal. Chem. $\underline{45}$, 671-681 (1973).

7. F. C. Adams and R. E. Van Grieken, "Absorption Corrections for X-Ray
    Fluorescence Analysis of Aerosol Loaded Filters," Anal. Chem. $\underline{47}$,
    1767-1773 (1975).

8. W. H. McMaster, N. K. Del Grande, J. H. Mallett and J. H. Hubbell,
    "Compilation of X-ray Cross Sections," Sec. II, Rev. I, Univ. of
    Calif. Lawrence Radiation Laboratory Report, UCRL 50174 (1969).

9. K. K. Nielson, "Matrix Corrections for X-Ray Fluorescence Analysis
    Using Scattered X-Rays," Battelle-Northwest Laboratory Report
    BNWL-SA-5785, to be published (1976).

AN INTERACTIVE PROGRAM FOR THE CONTROL OF THE X-RAY SPECTROMETER, FOR

DATA COLLECTION AND DATA MANIPULATION - USE IN QUALITATIVE ANALYSIS

R. Jenkins, D. Myers and F. R. Paolini
Philips Electronic Instruments, Inc.
Mount Vernon, New York

Introduction

Computer controlled X-ray spectrometers have been available since the mid 1950's and there are many hundreds of these systems in use today. The degree of sophistication of these machines has increased gradually over the past several years, but in all cases the flexibility of the analytical system is directly relatable to the available computer hardware. In the design of any computer controlled X-ray spectrometer, it is highly desirable to keep the cost of the computer, plus its associated interfaces and software, to within a reasonable fraction of the total cost of the whole system. A figure of perhaps 30% or so is a good figure to aim for in this context. A compromise must, therefore, always be sought between cost and computational power--which generally means flexibility.

The evolution of the hardware used in these systems started with early machines in the mid 1960's that typically used 4-8K direct access core memories with mini-computers and with paper tape, had access times of some minutes, and used assembler programming language. The early 1970's saw the development of 8-16K memories, cassettes, assembler/high level language and with access times of some seconds. The mid 1970's brought in machines with 32K memories employing the floppy-disc, high level language, with access times in milliseconds.

Programs were almost exclusively written in assembler code and nearly all manufacturers offered relatively inflexible "canned" programs. By the early 1970's, a new level was reached with availability of magnetic cassette oriented systems and in many cases, use of real time cassette monitors allowed the use of high level language programs --typically FORTRAN or BASIC. The more recent advent of the floppy disc system now allows us to move to yet another plateau and this paper discusses certain aspects of a series of highly interactive, real-time, floppy disc oriented computer programs for the control of the Philips Semi-Automatic X-ray Spectrometer, for data collection and subsequent data manipulation. These programs are part of the AXS project which involves automation of a semi-automatic X-ray spectrometer using a 32K NOVA 3/12 computer, floppy discs, graphics terminal and X/Y plotter. The programs run in a real time, multi-tasking environment, and offer a very high degree of flexibility, both for qualitative and quantitative X-ray spectrometry. All data manipulation programs are written in FORTRAN IV with calls to assembler routines for the control of the spectrometer hardware.

A wide choice of correction models is available to the operator in the quantitative mode and data evaluation procedures are greatly simplified by the use of a high speed graphics terminal which can be used for graphical presentation and interpretation of data.

Since the total system has been described in detail elsewhere,[1] it is the intention to concentrate in this paper on one specific area of the system, namely its use for qualitative work, and to use this as an example of what can be done with a modern interactive spectrometer system.

Traditional Qualitative Analysis Techniques

Qualitative analysis using a wavelength dispersive X-ray spectrometer has traditionally been done by scanning the goniometer over a selected 2θ range, at the same time integrating the pulses from the detector in a ratemeter circuit and displaying the resulting analog signal on an X/t recorder. This process is relatively time consuming, particularly in those cases where data on the whole atomic number range are required.

TABLE 1
Typical Data Acquisition Times with a Wavelength, Dispersive Spectrometer

Element Range	Crystal	2θ Range and Scanning Time
F plus Na	TAP	2 x 3° = 6° at 1°/min = 6 min
Mg	ADP	3° at 1°/min = 3 min
Al to K	PE	45–145°=100° at 2°/min=50 min
Ca to U	LiF(200)	10–120°=110° at 2°/min=55 min

Total Scanning Time = 114 min
Resetting of Conditions = 6 min
Total = 120 min

Table 1 gives typical data acquisition times for a wavelength dispersive spectrometer and it will be seen that where the whole range of elements from fluorine upwards is required, about 2 hours of total scanning time is necessary, with 4 or 5 changes of instrumental conditions. This latter point means, in turn, that the operator must be in the vicinity to make the required parameter changes when necessary. In addition to the time problem, the automation of such a system is wieldy, particularly where multiple sample handling is desirable. The direct synchronisation of the goniometer and X/t recorder, both of which are scanning at a fixed rate, also leads to several major drawbacks. For example, due to the fact that a fixed $d\theta/dt$ rate is being used on the goniometer, a plot is obtained in which λ varies as the Sin of θ. Also, since λ varies as approximately $1/Z^2$, the atomic number scale in the resulting spectrogram is a complex function of the form as follows:

Bragg's Law: $\lambda = \dfrac{2d}{n} \cdot \sin\theta$

Moseley's Law: $\dfrac{1}{\lambda} = K\ (Z - \delta)^2$

From which
$$Z = \left(\frac{1}{K} \cdot \frac{n}{2d} \cdot \frac{1}{\sin\theta} \right)^{1/2} + \delta \qquad \text{------------------} \quad (1)$$

in which K and δ are the spectral series constants
    d the interplanar spacing of the analysing crystal
    θ the Bragg angle and n an integer

This, in turn, means that look-up tables or charts must be used in order to identify atomic numbers from the diffraction maxima in the spectrogram. A further consequence of equation 1 is that for a fixed spectral series (eg, $K\alpha$ lines) and a fixed (n/2d), Z varies as $1/\sqrt{\sin\theta}$, hence spectra will tend to be cluttered at low 2θ values and widely spread at high 2θ values. Table 2 illustrates this effect for elements giving $K\alpha$ and $L\alpha$ lines over the 2θ range $10-130^\circ$ using a LiF(200) analysing crystal.

TABLE 2

Distribution of $K\alpha$ and $L\alpha$ Lines as a Function of 2θ for the LiF(200) Crystal

Angular Range (°2θ)	Elements giving $K\alpha$ Lines		Elements giving $L\alpha$ Lines	
	Range of Z	Total	Range of Z	Total
10 – 30	58 – 35	24	92 – 87	6
30 – 50	34 – 28	7	86 – 70	17
50 – 70	27 – 24	4	69 – 61	9
70 – 90	23 – 22	2	60 – 56	5
90 –110	21	1	55 – 52	4
110 –130	20	1	51 – 50	2
10 –130	58 – 20	39	92 – 50	43

Splitting the whole angular range into 20° segments indicates that for example, 24 elements give $K\alpha$ lines within $10-30^\circ 2\theta$ but only 1 element gives a $K\alpha$ line between $90-110^\circ 2\theta$ or $110-130^\circ 2\theta$. Since the goniometer is slewing at a constant rate, some compromise has to be sought, which generally means that one scans too quickly at low angles and too slowly at high angles.

In view of all of the problems outlined above, one perhaps wonders why we have remained with this system for 25 years or more. Initially, the technique represented such a tremendous advance over what was then available, we probably didn't even consider the drawbacks to be as such. Over the intervening years, X-ray spectroscopists have become so used to the routine that it becomes almost second nature. However, the more recent advent of Energy Dispersive Spectrometry with its associated high speed, wide range atomic number data acquisition, and convenience of presentation on a video terminal, has prompted us to reconsider the "classical means of qualitative analysis.

A New Approach to Qualitative Analysis

In order to improve the traditional system of qualitative analysis, it is clear that a new hardware basis is required and, if possible, this should include

a) Independent control of goniometer and recorder speeds.
b) Variable (and computer selectable) scan rates on goniometer and x axis of recorder.
c) Closer control of the y axis of the recorder.
d) Form feed option on recorder to allow multiple specimens.
e) Automatic selection of optimum analytical conditions for different portions of the atomic number range.

In the AXS system we have direct computer control over all hardware functions including goniometer, crystal, collimator, etc., and, for qualitative work, include a 300 Baud Diablo 1550 X/Y Plotter. This gives the required hardware listed above. The Diablo X/Y Plotter uses a series of single characters – typically dots – in its graphics mode and has a maximum resolution of 120 dots per inch in the X direction and 48 dots per inch in the Y direction. Since we shall be using the dot sequence to define a profile, the profile definition will be dependent upon the dot frequency as illustrated in Fig. 1.

Definition of a Profile as a Function
of the Dot Frequency

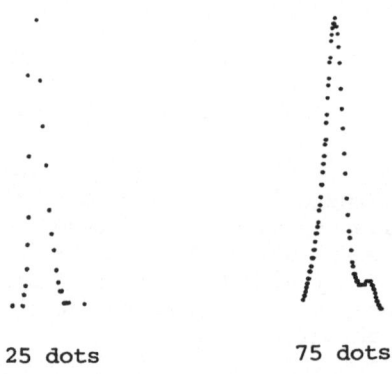

25 dots   75 dots

Here, a scan has been made over the CaKα multiplet using first 25 dots, then 75 dots, to define the profile. It is apparent that somewhere in this 25-75 dot range would yield satisfactory profile definition. The Diablo utilises a carriage width of 15 inches, so in one width of paper about 1800 dots or points are, in principle, available for spectral display.

Another point to remember in defining a spectral scan over
a wide 2θ range is that the profiles become much broader at higher
Bragg angles. This is partially due to the optics of the system
and is partially contrived because of the tandem type detector
combination commonly employed in modern X-ray spectrometers. This
means that a line occurring at a high 2θ value could be compressed
somewhat without loss of definition.

In defining an improved qualitative system, it is useful to
devote a minimum time to scanning over (generally) irrelevant portions of the spectral region. This is particularly the case in
the Low Z region F(9) to Ca(20) where the optimum situation is
probably to slew rapidly to the vicinity of the strongest line of
the element, step slowly over this line, slew to the next element,
and so on. By selecting the optimum measuring conditions for each
element during each slew cycle, an output of the type shown in
Fig. 2 can be obtained.

Profile Acquisition by a Slew/Step Sequence, Selecting
Optimum Conditions for Each Element between Each Step

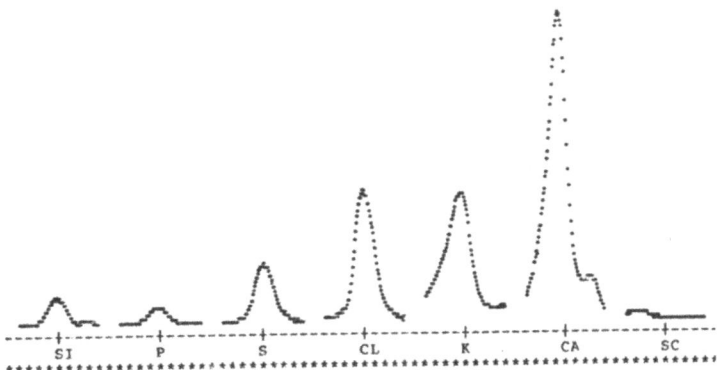

Here, a specimen containing Si, P, S, Cl, K and Ca has been run
in this mode and the peaks for each of these elements displayed.
Not only is the data acquired extremely quickly, about 1-1/2
minutes per element in this instance, and under optimum conditions
for each element, but also the data interpretation is trivial since
a scale can be automatically outputted directly on the plotter.
This is one of the techniques we use in the new qualitative analysis approach and we allow the operator to pre-select individual conditions for each element between F(9) and Ti(22). Since the whole
system runs in a multi-tasking environment, the time required to
select parameters such as 2θ angle, crystal, excitation conditions,
etc., is reduced to a minimum. Another feature of this output is
that we allow the intensity to be displayed as $\sqrt{I}$ rather than I,
which allows reasonable profiles to be obtained for both low and
high concentrations.

For the middle and high atomic number range, this slew/step technique is not the best since there are multiple series lines present
which may also be useful in interpretation. In this region, we use a
rather novel alternative approach in which the X axis of the plotter
is incremated at each goniometer step increment, but in which the
goniometer step increment is varied over the 2θ range. This allows us
to obtain a spectrogram which is linear in Z, giving the dual advantages
of data acquisition at an optimum rate, plus an output which is very
easily interpreted. Again, we have the √I option for the Y axis, if
required.

Fig. 3 shows the actual output of the new qualitative mode. As will
be seen, a single page is used for the complete output, this page being
divided roughly into two major segments.

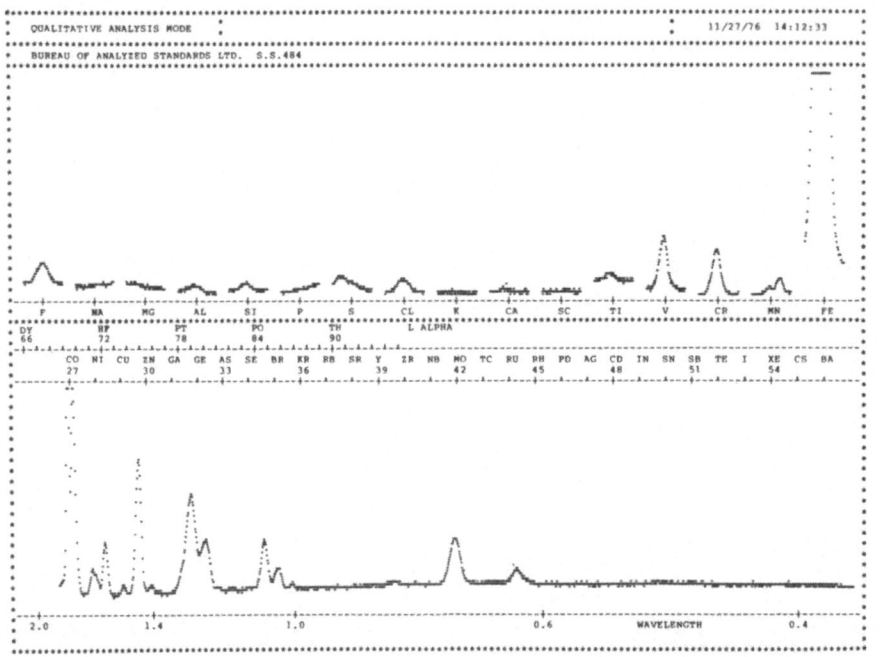

The top portion records all elements between F(9) and Ti(22) using the
slew/step approach, and the lower portion uses the asynchronous scanning
technique to cover the remaining elements above Ti in atomic number.
The top of the page contains an identification plus data and time of
day, the latter being derived from the real-time clock of the computer.
As was previously mentioned, the operator can pre-enter 13 different
sets of analytical conditions, 12 for the slew/step low Z portion and
one for the asynchronous scan, middle/high Z portion. Certain operating
parameters can also be changed automatically during the course of the
asynchronous scan. In addition to the change in the Δ 2θ increment, it
may be useful to vary the time per step, the collimator, the excitation
conditions, or the detector combination, in order to control the sensitivity of the spectrometer over the wavelength range covered. For example, in the scan shown in Fig. 3, only the step increment was changed

but by changing to the coarse collimator at 1.3Å, more sensitivity could be obtained for the elements Zn(30) through V(23). Again, by automatically changing the kV on the X-ray tube from 80 kV at Z = 60, to 60 kV at Z = 40, and to 50 kV at Z = 20, almost constant excitation conditions can be obtained. Obviously there are many additional programmable combinations of possibilities.

The time required to obtain this complete output was 37 minutes, this being broken down in 2 minutes for the print-out of the grid and scales, 15 minutes for the low Z segment and 20 minutes for the medium/high Z segment. When using the program, the operator can call for either or both segments and need only define the range of atomic numbers required. For example, if a scan from Z = 19 was called for, the program would automatically ignore all elements below potassium and start by choosing the correct conditions for this element.

Conclusion

It has been demonstrated that the development of a highly interactive, state-of-the-art computer controlled spectrometer leads to advantages and applications far beyond the usual quantitative use of the X-ray spectrometer. Rapidly acquired, easily interpretable qualitative scans can be obtained under optimum analytical conditions in which the operator can also control the sensitivity of the spectrometer over different portions of the wavelength range.

Reference

1) Jenkins, R., Myers, D., and Paolini, F.R., Norelco Reporter 23 (1976) in press.

LAMA I - A GENERAL FORTRAN PROGRAM FOR QUANTITATIVE

X-RAY FLUORESCENCE ANALYSIS

Daniel Laguitton and Michael Mantler

IBM Research Laboratory

San Jose, California 95193

ABSTRACT

A comprehensive Fortran IV program designed to perform the matrix correction in x-ray fluorescence analysis is described. Specimens and standards can be in bulk or film form. All necessary fundamental parameters are provided by internal routines thereby requiring a minimum of input data.

INTRODUCTION

The basic equations relating x-ray fluorescence intensity from a sample and its composition were derived several years ago in a few theoretical papers (1,2,3,4). The large number of parameters involved, and especially the polychromaticity of the primary beam used to excite the fluorescence make these formulae quite complex and difficult to handle. As a consequence, it is not surprising that most of the literature devoted to quantitative x-ray fluorescence analysis has dealt with methods which aimed at avoiding the use of the complete equations. This resulted in the development of methods using calibration standards and the concept of equivalent wavelengths. However, the increasing availability of computers in research laboratories is very likely to change this way of solving the problem, and this paper presents the first comprehensive computer program performing the complete matrix correction for quantitative x-ray fluorescence analysis, relying on the best known values of fundamental parameters and energy distribution of the primary beam.

THEORY

The expressions used in the program to express the intensity of the fluorescent radiation emitted by an element i in a bulk matrix containing one or several elements j can be written as follows and the meanings of the symbols are listed in Table 1.

$$R_i = R_{Pi} + R_{Si}$$

$$R_{Pi} = K\, P_i w_i \int_{\lambda min}^{\lambda edge_i} \frac{\tau_i(\lambda)}{X \sin \psi_1}\; I_0(\lambda)d\lambda$$

$$R_{Si} = \sum_j 0.5\, K(P_i w_i)(P_j w_j) \int_{\lambda min}^{\lambda edge_j} \frac{\tau_i(\lambda_j)\tau_j(\lambda)}{X \sin \psi_1} \;\Big[\quad\Big]\; I_0(\lambda)d\lambda$$

with: $\Big[\quad\Big] = \dfrac{\sin \psi_1}{\mu(\lambda)} \ln \left(1 + \dfrac{\mu(\lambda)}{\mu(\lambda_j)\sin \psi_1}\right) + \dfrac{\sin \psi_2}{\mu(\lambda_i)} \ln \left(1 + \dfrac{\mu(\lambda_i)}{\mu(\lambda_j)\sin \psi_2}\right)$

$$K = S\frac{\Omega}{4\pi}\kappa(\lambda_i)$$

$$X = \frac{\mu(\lambda)}{\sin \psi_1} + \frac{\mu(\lambda_i)}{\sin \psi_2}$$

$$P_i = \frac{r_i - 1}{r_i}\, p_i \omega_i \quad i = i, j \dots$$

Table 1. Definition of Symbols

D	Thickness of sample (cm)
exp	Exponential function
$I_0(\lambda)d\lambda$	Spectral distribution of primary radiation over interval $d\lambda$
i	Analyzed element
j	Interfering element
ln	Natural logarithm
p_i	Transition probability for considered line of element i
r_i	Jump ratio of element i for edge of considered line
R_i	Total counting rate for line of element i
R_{Pi}	Counting rate due to primary excitation for line of element i
R_{Si}	Counting rate due to secondary excitation for line of element i
S	Irradiated area of sample
w_i	Weight fraction of element i
θ	Integration angle, no direct physical meaning
$\kappa(\lambda)$	Efficiency of the counting system for wavelength λ
λ	Wavelength (subscript i indicates wavelength of a line of element i)
λ min.	Minimum wavelength of the primary radiation
λ edge$_i$	Wavelength of the considered edge of element i
$\mu(\lambda)$	Mass absorption coefficient of the sample for wavelength λ (cm^2/g)
π	3.14159...
ρ	Density of sample (g/cm^3)
$\tau_i(\lambda)$	Mass photoabsorption coefficient of element i for wavelength λ (cm^2/g)
ψ_1	Angle between central beam of tube radiation and surface of the sample
ψ_2	Take-off angle of measured radiation from specimen
ω_i	Fluorescence yield for line of element i
Ω	Detection solid angle

For a thin film, these relations become:

$$R_i = R_{P_i} + R_{S_i}$$

$$R_{P_i} = K\,P_i w_i \int_{\lambda min}^{\lambda edge_i} \frac{\tau_i(\lambda)}{X \sin \psi_1} \left[1-\exp(-XD\rho)\right]\ I_0(\lambda)d\lambda$$

$$R_{S_i} = \sum_j 0.5\,K(P_i w_i)(P_j w_j) \int_{\lambda min}^{\lambda edge_j} \frac{\tau_i(\lambda_j)\,\tau_j(\lambda)}{\sin \psi_1}\ \left\{\ \right\}\ I_0(\lambda)d\lambda$$

$$\text{with:}\ \left\{\ \right\} = \begin{cases} \displaystyle\int_0^{\frac{\pi}{2}} \tan\theta\, d\theta \left[\frac{1-\exp(-X_1 D\rho)}{X_1 X_2} - \frac{1-\exp(-XD\rho)}{X_2 X}\right] + \\[2em] \displaystyle\int_{\frac{\pi}{2}}^{\pi} \tan\theta\, d\theta \left[\frac{\exp(-X_2 D\rho)-\exp(-XD\rho)}{X_1 X_2} - \frac{1-\exp(-XD\rho)}{X_2 X}\right] \end{cases}$$

$$K = S \frac{\Omega}{4\pi} \kappa(\lambda_i) \qquad\qquad X = \frac{\mu(\lambda)}{\sin \psi_1} + \frac{\mu(\lambda_i)}{\sin \psi_2}$$

$$P_i = \frac{r_i-1}{r_i}\, p_i \omega_i \qquad i = i, j \ldots \qquad X_1 = \frac{\mu(\lambda_i)}{\sin \psi_2} + \frac{\mu(\lambda_j)}{\cos \theta}$$

$$X_2 = \frac{\mu(\lambda)}{\sin \psi_1} - \frac{\mu(\lambda_j)}{\cos \theta}$$

Because of the difficulty of determining certain experimental factors such as the efficiency of the detector and the reflectivity of the analyzing crystal, the intensities are expressed as the ratio of intensities measured for the selected line from the specimen and a standard of known composition. Even in that form, it can be seen from the above equations that the relation between intensity and weight fraction is quite complex and that its manual calculation would be almost impossible. Furthermore, since these relations express intensities as a function of weight fractions and are not reversible, the calculation of the weight fractions from the values of intensity ratios must be iterative, which adds to the complexity of the problem. Nevertheless, all the parameters involved in these equations are known to a good approximation as well as some spectral distributions of the x-ray tube for specific experimental conditions. It is thus possible to use a computer to solve these equations in a manner which is similar to the computer correction methods developed for electron microprobe analysis (5,6). The main difference between the two calculation techniques is the necessity of integrating the formulae over the range of wavelength of interest, and this is probably the main reason for the small number of previously published attempts of using the computer for complete quantitative x-ray fluorescence analyses (7,8,9).

DESCRIPTION OF THE PROGRAM

The program, named LAMA after the names of its authors, has been
written in FORTRAN IV and designed to process the widest variety of
problems with the minimum input requirement. Its main possibilities
are summarized in Table 2. A general flowsheet of the program is
presented in Figure 1. The structure of the program is modular, each
specific calculation being made in a special subroutine or function
described in the following.

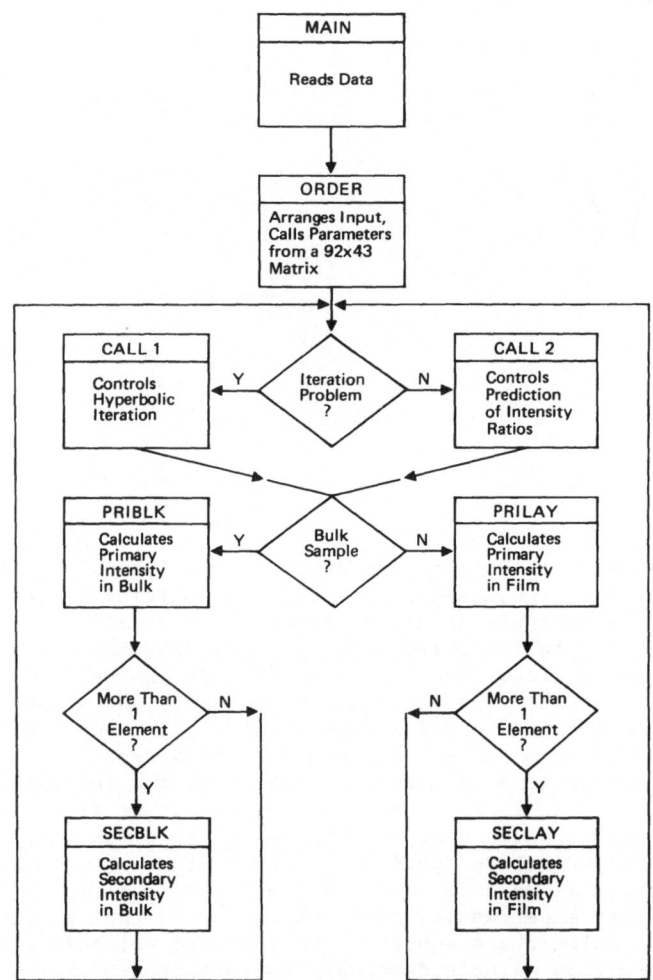

Figure 1. General flowsheet of the LAMA program.

Table 2. Main Characteristics of the LAMA Program

- Analysis of bulk specimen with bulk standards
- Analysis of thin specimen with thin standards
- Analysis of thin specimen with bulk standards
- Calculation of weight fractions from intensity ratios
- Calculation of intensity ratios from weight fractions
- Use of any stored experimental energy distribution of the primary beam or a description of the primary radiation by the Kramer's law.
- The specimen can contain up to 10 elements
- Each standard can contain up to 7 elements
- One element can be analyzed by difference, and several elements determined if their atomic or weight proportion is known / ex: PO_4, SO_4 ...
- Input data can be read on formatted cards or tape, or interactively on a time-sharing terminal
- Printed or displayed results can be optionally brief or detailed
- All input concentrations can be optionally expressed as weight fraction or as number of of atoms
- Core size for a complete execution is about 90K
- CPU time for a complete interaction problem on a three elements system 6.6 s, on an IBM 360/195
- CPU time for a prediction problem on a three elements system: 1.2 s.

MAIN Program

The MAIN program is an input program which reads the minimum amount of data essential to the execution of the calculation. Two different MAIN programs can be used depending on whether the data have been formatted on cards or in a stored data set, or the input is made from a time-sharing terminal on an interactive mode.

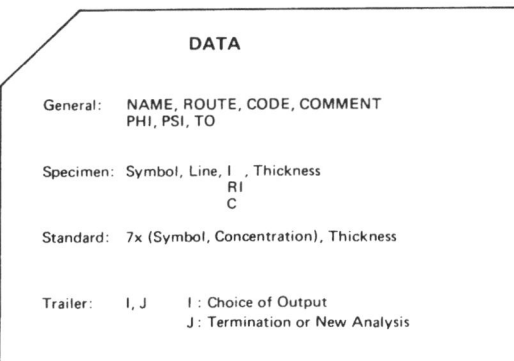

Figure 2.   Formatted input data for the MAIN program.

Four types of data are required for the MAIN program as illustrated in Figure 2 and described below.

a). <u>General data</u>: The name of the operator, the comment that one wants to associate with the output for identification purposes, and two essential keywords have to be coded. The first keyword is an integer variable which indicates the route followed by the calculation: a value of 1 shunts the calculation to the iteration subroutines whereas a value of 2 indicates that a direct prediction of intensity ratios is desired. The second key word is a three letter word which indicates the nature of the specimen and standards currently analyzed. It can take three different spellings, namely BLK, LAY, and LBS as shown in Table 3. The remaining general data consist of the incidence angle of the primary beam onto the specimen, the emergence angle of the fluorescent radiation from the specimen, and the dead time of the counting electronics. These three parameters have default values of 45°, 45° and 2 microseconds, respectively.

b). <u>Specimen data</u>: A single card or line of data per element is necessary for the data related to the specimen. It must contain successively the symbol of the element, the symbol of the line (KA, KB, LA, LB, LG for $K\alpha$, $K\beta$, $L\alpha$, $L\beta$, $L\gamma$, respectively), the experimental intensities of the selected line in the specimen, and in the standard, including the background intensity, the peak intensity and the counting time. If the intensity ratio has already been calculated, its value can be coded instead of using the measured intensities, and if the problem is a prediction problem, the weight fraction or the number of atoms of the element in the specimen replaces the intensity ratio. The last value given on the line, and only for the first element, is the thickness of the specimen if it is a film. If one element is analyzed by difference, its line contains only its symbol without a line symbol, and if several elements are being analyzed by difference, their weight or atomic proportion follows their symbol. The maximum permissible number of elements in the specimen is 10, and a blank card must follow the data of the last element if this maximum is not reached.

c). <u>Standards data</u>: Data related to the standards follow this card. If pure element standards are used for the analysis of a bulk specimen, one blank card is the only necessary information about the standards. If complex or film standards are used, their compositions and thicknesses must be given. The maximum permissible number of elements per standard is 7.

Table 3. The CODE Parameter

CODE	BLK	LAY	LBS
SPECIMEN	BULK	THIN	THIN
STANDARD	BULK	THIN	BULK

d). <u>Trailer card</u>: The data is now complete and must be closed
by a trailer card containing two integers. The first of these two
numbers is used to select the type of output required: 1 indicates
that a brief printout is requested, 2 will select a detailed output.
The second number on the trailer card can take the value 0 if no other
data follows, 1 if data related to a new system has to be used, 2 if
a new set of experimental results (or concentrations) for the same
elements using the same standards will be following. In this latter
case, the only necessary data is one card or line per element, followed
by a new trailer card. An example of the complete data required for
the analysis of a three element system is given in Figure 3 (next page).

The ORDER Subroutine

This subroutine has the role of ordering the complete set of data
necessary for the subsequent calculation. The input data read by the
main program is checked and, if necessary, completed by replacing
blanks by their default values, transforming compositions given in
number of atoms into weight fractions, performing dead time correction
and calculating intensity ratios. In order to do so, the program uses
a 92×43 matrix which contains all the necessary parameters as described
in Table 4. The subroutines adapted to the problem being solved are
then called and take control of the subsequent operations.

The CALL Subroutines

The CALL1 and CALL2 subroutines are the actual control programs
being used according to the value of the of the first keyword in the
general data.

Table 4. Content of the General Matrix

LAMATRIX (92, 43)

92 Lines: Elements H to U

43 Columns: Symbol
            Atomic number
            Atomic weight
            Density
            Edges energy   (K, L2, L3)
            Lines energy   ($K\alpha$, $K\beta_1$, $L\alpha_1$, $L\beta_1$, $L\gamma_1$)
            Jump ratios
            Fluorescence yields
            Line probabilities
            McMaster fit coef. ⎧ M Photoabsorption
                               ⎪ L Photoabsorption
                               ⎨ K Photoabsorption
                               ⎪ Coherent Scattering
                               ⎩ Incoherent Scattering

INPUT SAMPLE

```
    D.LAGUITTON      1BLK  FE-CR-NI ALLOY. NBS SRM5074 (ANAL.CHEM.46,1974,86)
    45.       63.00     33.00
  FEKA                                                          .4511
  NIKA                                                          .0203
  CRKA                                                          .3258


    2  0
```

OUTPUT SAMPLE

```
**************************************************************************
*  JOB SUBMITTED BY:  D.LAGUITTON                                        *
*                                                                        *
*  TYPE OF ANALYSIS:BULK SPEC. WITH BULK STD.                            *
*  COMMENT:  FE-CR-NI ALLOY. NBS SRM5074 (ANAL.CHEM.46,1974,86)          *
*  EXPERIMENTAL CONDITIONS                                               *
*  TUBE VOLTAGE (KV):45.                                                 *
*  DEAD TIME (E-6 S): 2.                                                 *
*  INCIDENCE ANGLE (DEG.):63.                                            *
*  EMERGENCE ANGLE (DEG.):33.                                            *
**********                                                      **********
*              DATA RELATED TO THE SPECIMEN                              *
*  ELEMENT-LINE    EDGE(KV)    LINE(KV)       INT.RATIO    WGT.FRACT.     *
*     FE-KA         7.1120      6.3980         0.4511       0.6863        *
*     NI-KA         8.3390      7.4710         0.0203       0.0474        *
*     CR-KA         5.9870      5.4110         0.3258       0.2572        *
*  DETAILS OF THE CALCULATION                                            *
*  4 ITERATIONS WERE NECESSARY                                           *
*  ELEMENT-LINE    IP        IS        IPS        ISS                    *
*     FE-KA     272.730     4.331    604.386      0.0                    *
*     NI-KA      15.545     0.0      758.601      0.0                    *
*     CR-KA     100.585    43.327    438.146      0.0                    *
**********                                                      **********
*              DATA RELATED TO STANDARDS                                 *
*ELEMENT (D)    COMPOSITION OF STANDARD                                  *
*  FE (     0.)FE1.0000                                                  *
*  NI (     0.)NI1.0000                                                  *
*  CR (     0.)CR1.0000                                                  *
**************************************************************************
```

IP: Primary intensity in the specimen
IS: Secondary intensity in the specimen
IPS: Primary intensity in the standard
ISS: Secondary intensity in the standard

Figure 3.   Example of data required for analysis of a three element
            specimen (above) and the corresponding output (below).

CALL1 controls the iterative process by the hyperbolic convergence method commonly used in electron probe microanalysis (10). Some modifications have been introduced to cover the cases where elements are analyzed by difference in thin films and to accelerate the calculation. They will be described in a separate publication.

CALL2 is the control program for the direct calculation of intensity ratios.

Other Subroutines and Functions

A large number of other subroutines and functions are used to calculate specific formulae or parameters. The main formulae for the primary and secondary intensity in a bulk or thin specimen are calculated by the PRIBLK, SECBLK, PRILAY, and SECLAY subroutines, respectively. As an option, these subroutines use a description of the primary energy distribution by the Kramer's law, according to the relation:

$$I_0(\lambda)d\lambda = k \left(\frac{\lambda}{\lambda_{min.}} - 1 \right) \frac{1}{\lambda^2} d\lambda$$

or any experimental energy distribution stored in the memory, such as that of Gilfrich and Birks (11). The total mass absorption and mass photo-absorption coefficients are calculated by special functions using the McMaster fit coefficients (12) stored in the general matrix. The result is printed or displayed by the subroutine RESOUT, and its size is controlled by the value of the first parameter given on the trailer card. The general organization of the program as described above is summarized in Fig. 4.

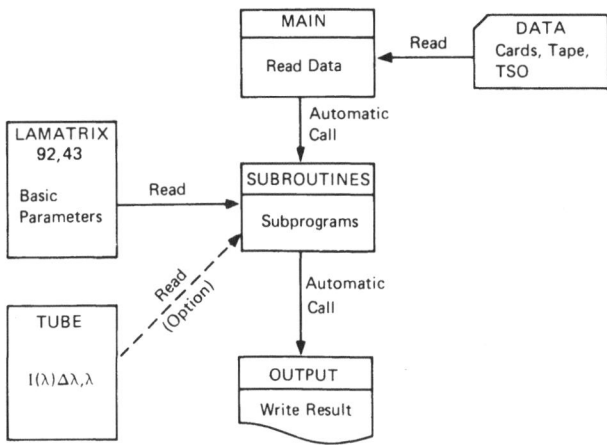

Figure 4. General diagram of the LAMA program.

TEST OF THE PROGRAM AND RESULTS

In order to test the proper functioning of the program, two sets
of experimental data have been processed.

1. Bulk Specimen

Rasberry and Heinrich (13) have recently published a series of
experimental results on several NBS standard reference materials
consisting of Fe-Ni, Fe-Cr, Ni-Cr, and Fe-Ni-Cr alloys. Since these
authors have published the intensity ratios, the certified composition,
and the experimental conditions of the measurements, it is possible
to perform the matrix effect correction on these data and to compare
the calculated concentrations with the NBS certified values which are
assumed to be the true concentrations. The results are presented in
Table 5 and the frequency histograms representing the error distribution
are shown in Fig. 5. The calculations were made with the energy
distribution of a tungsten tube operated at 45 kV (11). The calculated
values are in very good agreement with the true values even though
the corrections in this system are of exceptional magnitude (sometimes
more than 100%) due to the strong interaction of the analyzed
fluorescent lines with the specimen. The histograms are also almost
free from bias and the average absolute error is much less than 0.01
in weight fraction. The relative error is also quite small even though
it depends on the range of concentrations used, which for chromium
was always less than 0.4.

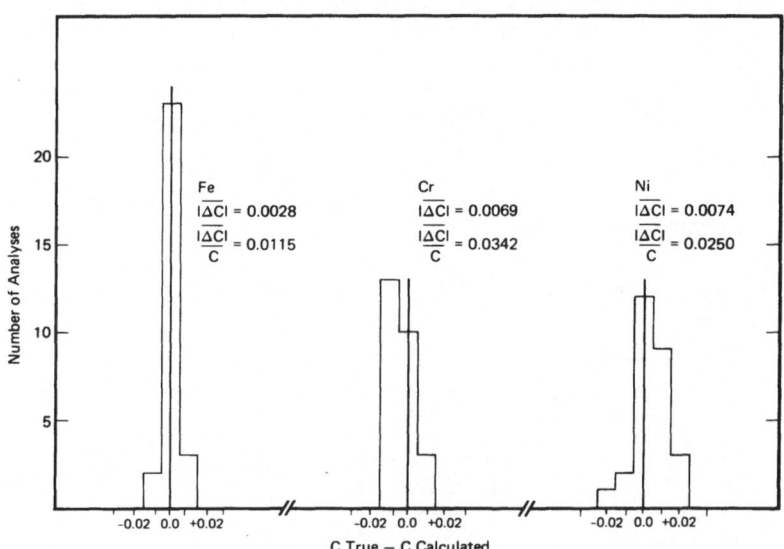

Figure 5. Result of matrix effect correction on Fe-Cr-Ni alloys.

Table 5. Intensity Ratios, True and Calculated Intensities for Fe-Ni-Cr Alloys
(Experimental Data from Ref. 13)

Specimen	Relative Intensity, R			True Weight Fraction, C^a			Calculated Weight Fraction, C^b		
	Fe	Ni	Cr	Fe	Ni	Cr	Fe	Ni	Cr
971	0.0789	0.8782	...	0.0462	0.9516	...	0.0459	0.9509	...
972	0.1104	0.8321	...	0.0659	0.9322	...	0.0662	0.9298	...
974	0.1621	0.7595	...	0.1018	0.8964	...	0.1019	0.8935	...
983	0.3172	0.5483	...	0.2263	0.7711	...	0.2275	0.7599	...
986	0.4007	0.4515	...	0.3067	0.6931	...	0.3062	0.6805	...
987	0.4373	0.4073	...	0.3431	0.6552	...	0.3430	0.6394	...
1159	0.5907	0.2553	...	0.5100	0.4820	...	0.5096	0.4667	...
126B	0.6958	0.1720	...	0.6315	0.3599	...	0.6322	0.3453	...
809B	0.9659	0.0125	...	0.9549	0.0329	...	0.9588	0.0308	...
4061	0.8970	...	0.0617	0.9627	...	0.0353	0.9645	...	0.0373
4062	0.8270	...	0.1004	0.9372	...	0.0608	0.9369	...	0.0632
4065	0.6974	...	0.1817	0.8766	...	0.1214	0.8764	...	0.1237
4173	0.5739	...	0.2587	0.8080	...	0.1900	0.8043	...	0.1886
4181	0.4748	...	0.3326	0.7477	...	0.2503	0.7328	...	0.2572
4183	0.4048	...	0.4023	0.6786	...	0.3194	0.6725	...	0.3265
4184	0.3579	...	0.4476	0.6322	...	0.3658	0.6266	...	0.3739
3995	...	0.6998	0.1906	...	0.8446	0.1425	...	0.8367	0.1564
4002	...	0.6556	0.2240	...	0.8041	0.1897	...	0.8064	0.1874
4003	...	0.6515	0.2304	...	0.7858	0.2069	...	0.8035	0.1934
4004	...	0.6224	0.2492	...	0.7837	0.2104	...	0.7824	0.2114
4011	...	0.5543	0.3085	...	0.7343	0.2621	...	0.7295	0.2698
4012	...	0.5392	0.3174	...	0.7192	0.2765	...	0.7171	0.2789
4013	...	0.4684	0.3840	...	0.6591	0.3360	...	0.6549	0.3475
4014	...	0.4119	0.4305	...	0.6064	0.3883	...	0.6004	0.3971
5074	0.4511	0.0203	0.3258	0.6838	0.0498	0.2525	0.6863	0.0474	0.2572
5181	0.4971	0.0416	0.2651	0.6945	0.0996	0.1988	0.6939	0.0951	0.2049
5324	0.3529	0.0821	0.3311	0.5280	0.1927	0.2696	0.5329	0.1754	0.2789
5321	0.4343	0.0898	0.2582	0.5919	0.2002	0.1988	0.5924	0.1924	0.2083
7271	0.5298	0.0343	0.2536	0.7159	0.0829	0.1879	0.7266	0.0794	0.1925
161	0.1460	0.4367	0.2072	0.1501	0.6429	0.1688	0.1556	0.6459	0.1796
1189	0.0125	0.5630	0.2263	0.0140	0.7260	0.2030	0.0124	0.7398	0.1915
3987	0.4377	0.4110	...	0.3431	0.6552	0.0000	0.3433	0.6429	0.0000
5054	0.4689	0.0006	0.3348	0.7250	0.0015	0.2577	0.7272	0.0014	0.2596
5202	0.4480	0.0642	0.2784	0.6303	0.1480	0.2130	0.6313	0.1419	0.2225
5364	0.3179	0.1115	0.3361	0.4721	0.2357	0.2784	0.4758	0.2291	0.2894
1188	0.0667	0.5534	0.1740	0.0660	0.7265	0.1540	0.0621	0.7410	0.1460

[a] Given by chemical analysis.   [b] Using LAMA

2. Thin Specimens

Another set of experimental results is presented in Table 6. The x-ray intensity ratios for pure element films and pure element bulk standards have been measured using a Philips universal vacuum spectrometer, a Philips tungsten target tube, and a Rigaku x-ray generator. The incidence and emergence angles were 66° and 35°, respectively, and the tube was operated at 45 kV. The measured intensity ratios were plotted as a function of thickness, along with the theoretically calculated values obtained from the LAMA program. Fig. 6, 7, and 8 show an example of the results for CuKα, FeKβ and GdLα lines, respectively. The good agreement between the two sets of values is a direct verification of the correctness of the equations used. The slight departure of experimental point from the theoretical curve in the case of Gd films can be attributed to some uncertainty in the mass absorption coefficients of the GdLα line or to some lack of precision in the interferometric measurement of the thickness.

CONCLUSION

The LAMA 1 program has been proven to perform with a minimum of input data, the complete matrix effect correction for bulk or thin specimen and standards. It constitutes the first comprehensive program to perform the complete calculation for bulk and thin specimen using the best known values of fundamental parameters. It will undoubtedly provide a very powerful tool for further improvement or simplification of the method, such as checking of new experimental energy

Table 6. Experimental Intensity Ratios Measured on Films,
Using Pure Element Bulk Standards

Fe Films on Si		Gd Films on SiO_2		Cu Films on SiO_2	
Thickness (Å)	Intensity Ratio (Kβ)	Thickness (Å)	Intensity Ratio (Lα)	Thickness (Å)	Intensity Ratio (Kα)
1,000	0.0244	1,265	0.0573	535	0.0152
2,730	0.0661	1,995	0.1113	1,010	0.0258
4,210	0.0944	2,510	0.1345	1,355	0.0277
7,050	0.1624	4,220	0.2427	2,280	0.0555
8,990	0.1872	7,265	0.3428	2,620	0.0674
12,260	0.2517	9,710	0.4263	3,845	0.0902
		12,100	0.5162	5,185	0.1257
				7,640	0.1833
				10,710	0.2291
				13,770	0.2905
				16,730	0.3322

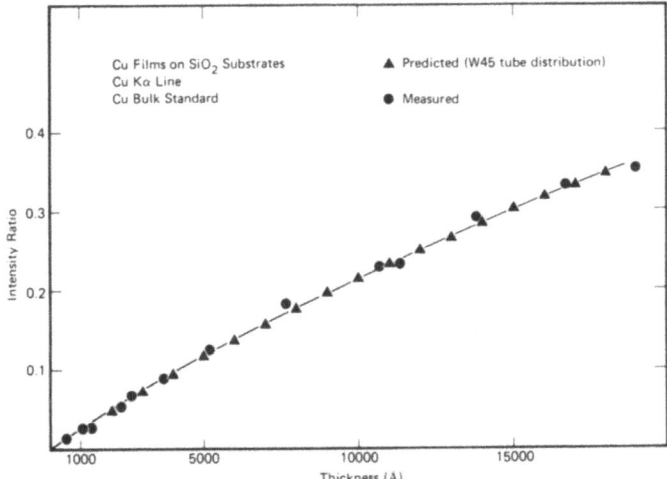

Figure 6. Intensity ratio vs. thickness for Cu films.

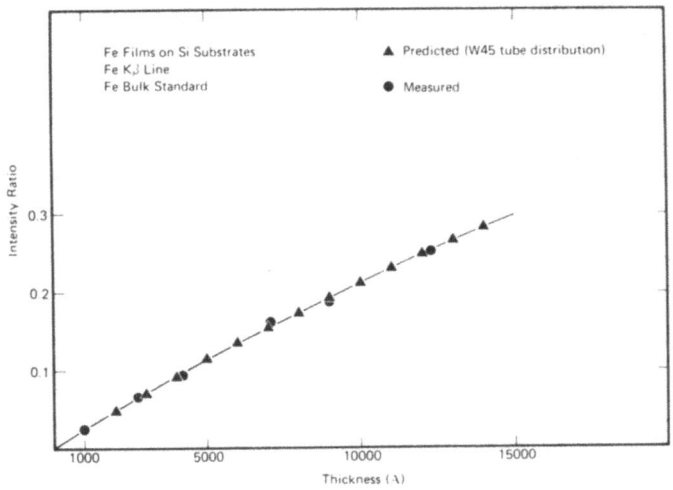

Figure 7. Intensity ratio vs. thickness for Fe films.

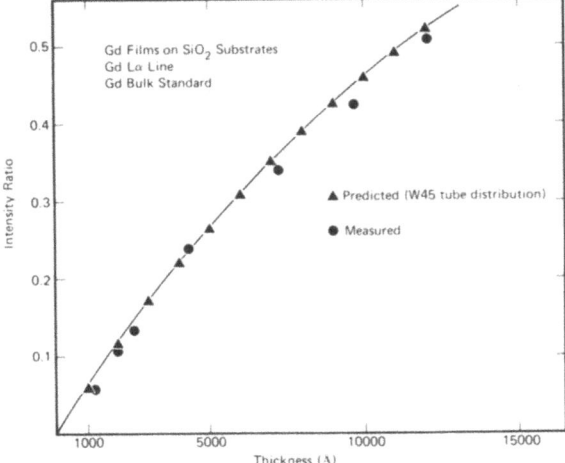

Figure 8. Intensity ratio vs. thickness for Gd films.

distributions of the primary radiation or of equations describing it, testing of new proposed values of experimental parameters, and ultimately opening the way to a generation of much smaller on-line computer programs.

ACKNOWLEDGMENT

The authors wish to thank Dr. William Parrish for assistance in the realization of this work.

REFERENCES

1. J. Sherman, "The Theoretical Derivation of X-ray Intensities from Mixtures," Spectrochimica Acta, 7, 283 (1955).
2. J. Sherman, "Simplication Formula in the Correlation of Fluorescent X-Ray Intensities from Mixtures," Spectrochimica Acta, 14, 466 (1959).
3. T. Shiraiwa and N. Fujino, "Theoretical Calculation of Fluorescent X-ray Intensities in Fluorescent X-Ray Spectrochemical Analysis," Japanese Journal of Applied Physics, 5, 10, 886 (1966).
4. G. Pollai, M. Mantler, and H. Ebel, "Die Sekundaranregung Bei der Rontgenfluoreszenzanalyse Ebener Dunner Schichten," Spectrochimica Acta, 26B, 747 (1971).
5. D. R. Beaman and J. A. Isasi, "A Critical Examination of Computer Programs Used in Quantitative Electron Microprobe Analysis," Anal. Chem., 42, 13, 1540 (1970).
6. D. Laguitton, Y. Bérubé, and F. Claisse, "Un Nouveau Programme de Calcul en Language APL pour L'analyse Quantitative par Microsonde Electronique," Can. J. Spectroscopy, 19, 3, 100 (1974).
7. J. W. Criss and L. S. Birks, "Calculation Methods for Fluorescent X-Ray Spectrometry. Empirical Coefficients vs. Fundamental Parameters," Anal. Chem., 40, 7, 1080 (1968).
8. Donald A. Stephenson, "Theoretical Analysis of Quantitative X-ray Emission Data: Glasses, Rocks, and Metals," Anal. Chem., 43, 13, 1761 (1971).
9. C. J. Everett and E. D. Cashwell, "MCP Code Fluorescence Routine Revision," Nuclear Science Abstracts, 29, 1, 230 (1974).
10. K. F. J. Heinrich, "Errors in Theoretical Correction Systems in Quantitative Electron Probe Microanalysis - A Synopsis," Anal. Chem. 44, 350 (1972).
11. J. V. Gilfrich and L. S. Birks, "Spectral Distribution of X-ray Tubes for Quantitative X-ray Fluorescence Analysis," Anal. Chem., 40, 7, 1077 (1968).
12. W. H. McMaster, et al. Compilation of X-Ray Cross Sections, Document UCRL 5074, National Technical Information Service U.S. Dept. of Commerce, Springfield, Va. 22161, (1969).
13. S. D. Rasberry and K. F. J. Heinrich, "Calibration for Interelement Effects in X-Ray Fluorescence Analysis," Anal. Chem., 46, 81 (1974).

A NOVEL X-RAY POWDER DIFFRACTOMETER DETECTOR SYSTEM

Susan K. Byram, Bui Han, G. B. Rothbart,

Roger N. Samdahl, and Robert A. Sparks

Syntex Analytical Instruments, Inc.

Cupertino, California 95014

ABSTRACT

A new proportional counter x-ray detector with application for powder diffractometry has been developed. The new detector collects virtually the entire diffraction spectrum simultaneously with good efficiency and angular resolution. Thus, the powder spectrum of a small sample can be obtained much faster than with film or a conventional powder diffractometer. The detector read-out is digital and is interfaced to a dedicated minicomputer. A description of the detector system is discussed and preliminary results are presented.

INTRODUCTION

Powder diffractometry is a useful and powerful tool in sample identification, and in favorable cases (where reflections are intense and indices can be assigned) makes possible crystal structure determination (1). The powder method requires small crystallites of random orientation, so that the technique is typified by relatively weak, often crowded, reflections. Presently available cameras and diffractometers using conventional x-ray film strips, or collimated moving counter detectors, respectively, do nothing to offset poor intensities. What is desired is a large angle acceptance detector with high eff[i]cency to x-ray radiation at the energy of interest, and angular resolution at least comparable to the limits imposed by finite beam collimation, beam energy spread, sample size, etc. Such a device should ideally have an electronic read-out method to enable a computer to acquire, store, process, and display the data gathered.

THE PROPORTIONAL WIRE CHAMBER APPLIED TO X-RAY DETECTION

We have developed a position sensitive single-wire proportional chamber which matches the detector specifications mentioned above. This chamber is a prototype of the Syntex XID 2000 electronic powder camera product.

Proportional wire detectors have been rather extensively used in the fields of high-energy physics and nuclear medicine. In the former, charged particle tracks (or converted gamma tracks) associated with elementary particle interactions are detected. The detection process is sensitive to Landau ionization of the traversing charged particle. In the latter, interaction of soft x-rays in biological matter enable sub-millimeter imaging of a spatial distribution of mass density and/or atomic number. Detection is achieved by sensing a photo-ionization event initiated by interaction of the x-ray in the proportional gas mixture.

From the broad range of experience with proportional chambers in these areas, it is possible to outline the general features of these detectors as follows:

i)   Energy resolution. Proportional gases, even at very high pressures, are total energy absorbers only for x-rays. X-ray energy resolution is limited by the finite statistics of the ion pair number, and fluctuations in gain about the anode wire. Several authors (2, 3) have reported full-width-at-half-maximum (FWHM) responses to monochromatic beams as 86% at 250 eV declining with energy to 10% at 17.4 KeV. Thus, the better applications of proportional chambers are those for which only crude information is to be extracted from the beam energy; the energy-role of the detector is to verify events with a known incident beam energy, and to discriminate against events with an appreciably differing energy.

ii) Position resolution. X-ray localization is limited in principle by signal-to-noise ratio (S/N) and the finite range of the photo-electron. At energies near the K-edge of the gas, S/N is poor while the range is small. At high energies S/N is good, but the range is large. It is clear that there exists an energy for which these two effects are optimally traded off and position resolution is a minimum. Figure 1 presents the results of a calculation incorporating these effects. S/N is taken to be 18.5 per KeV of photo-ejected kinetic energy, inferred from observations of our detectors.
Electron ranges are due to Tabatu, et. al. (4), where resolution depends on the isotropically oriented range projected onto the axis of the anode wire. Figure 1 refers to Argon and Xenon gas at two atmospheres pressure. These curves suggest that the best energies for position sensitive applications of proportional chambers are in the range of 6-10 KeV. In these situations, resolutions of 100 μm to 200 μm may be achieved. For instance, a FWHM resolution of 183 μm in high pressure Argon is reported by James, et. al. (5).

iii) Efficiency. The efficiency of proportional chambers is virtually 100% to the passage of charged particles, while for x-rays efficiencies in general tend to be small due to the small mass thickness of absorber involved in a real chamber. For low energy x-rays, and particularly those near the K- or L-edge of a high atomic number gas, efficiencies can be rather good. For example, three atmospheres of Argon 25 mm thick is sufficient to convert 90% of an 8 KeV incident beam.

iv) Time resolution and count rate. About 7% of the charge of an event is read-out in the first 10 nsec (6). Pulse development beyond this is usually determined by amplifier characteristics. In principle then, the proportional chamber is capable of high count rate. Read-out schemes designed for good position resolu-

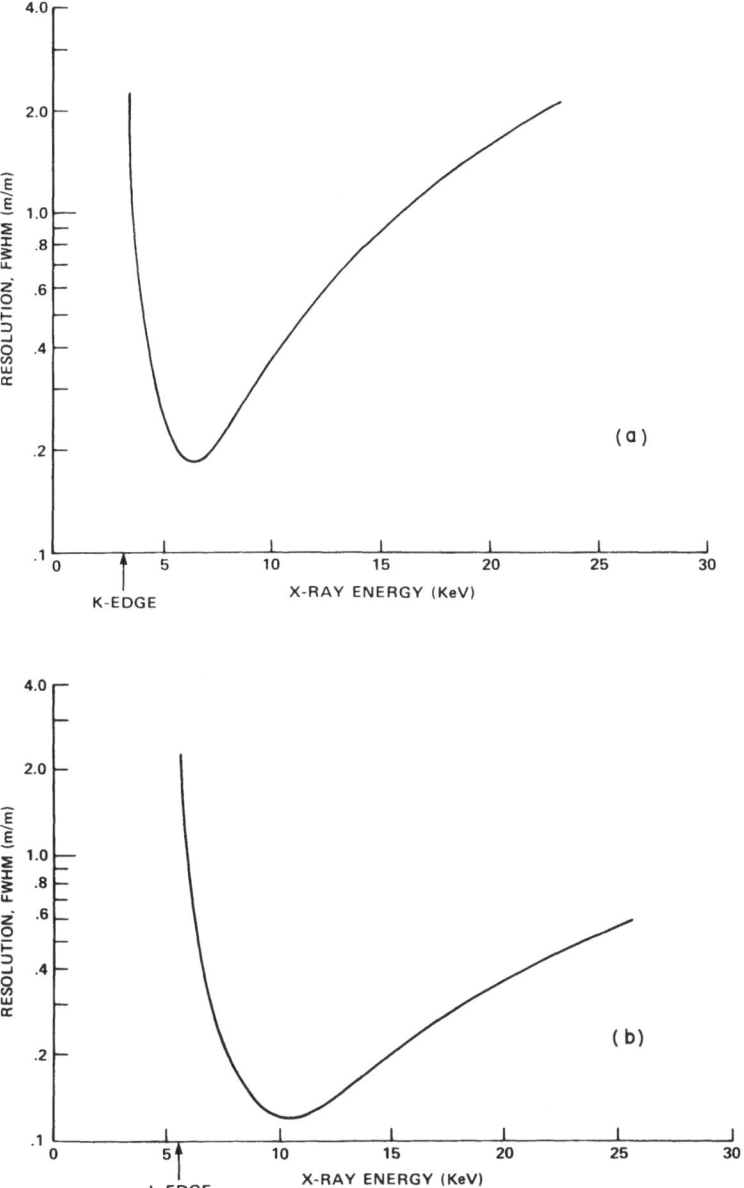

Figure 1. Ideal position resolution for x-rays in a position sensitive proportional detector where the signal-to-noise ratio and photoelectron range are the dominant effects. (a) for 2 atm Argon; (b) for 2 atm Xenon. Optimum resolution of 100-200 µm is obtained in the range of 6-10 KeV.

tion, however, usually require much larger encoding times.
Practical read-out methods encode the order of 1 mm per 10 nsec,
so that a 10 cm detector has a characteristic read-out time of
about 1 μsec. Since multi-events are difficult to read-out,
the best applications are those for which event rates are below
1 MHz.

v) <u>Size</u>. To the extent that anode wires can be supported, proportional chambers can be fabricated to be fairly large. One and
even two meter multi-wire chambers have been used in high energy
experiments where large solid angle acceptance is important (7).

   Consideration of i) through v) point to applications where a large
solid angle x-ray field of 6-10 KeV energy at rates less than 1 MHz is
to be detected to extract precise position (or angle) information, with
good efficiency, and where energy information is not of particular
interest. An ideal application, and quite surprisingly one which until
recently (5, 8) has been neglected, is x-ray powder diffractometry. The
principle problem in detector design for this application has been geome-
try--the preferred imaging surface is not a plane but rather a right
cylinder orthogonal to the beam axis, and centered on the sample. In
the scattering plane, the coordinate of the powder rings is a linear
measure of the reflection angle, 2θ.

   The unique aspect of the proportional chamber we have developed
is its cylindrical geometry, i.e., the anode wire and read-out plane are
curved into an arc in the scattering plane. This affords the advantage
of read-out that is strictly linear in 2θ, where the detector is neither
collimated nor mechanically scanned.

   A useful detector system must have an overall resolution on the
order of .15° FWHM in 2θ, or a resolution of .07° FWHM at the chamber
output. For Cu Kα radiation, nearly optimum position resolution is
possible, about 150 μm. This dictates the radius of curvature of the
detector, 115 mm. For purposes of fabrication, our diffractometer
evolved with a 135 mm radius. Nonetheless, it is evident from this that
a compact, high resolution wire chamber type diffractometer is in principle achievable.

                      PROPORTIONAL CHAMBER DESIGN

   Ideally, the wire chamber should cover 180° in 2θ. Our method of
conforming the anode wire does not easily admit this possibility. We
have empirically determined the optimum angular span of the chamber at
several radii for various wire thicknesses, to be about 70° at a radius
of 135 mm. We have taken advantage of this span and the symmetry inherent in a diffraction pattern through the use of an array of three
separate and independent detectors installed around the perimeter of a
vacuum chamber centered on the powder sample in the Debye-Scherrer geometry
(see Figure 2). In this way, very nearly 180° of spectrum are accessible
to the user. The actual design involves an entrance collimator and exit
beam stop that occlude 15° in the back direction and 3° in the forward
direction, so that in practice 3°-165° in 2θ are simultaneously observ-
able. One detector is mounted on a swinging arm; this feature then
allows the low-cost option of gathering ~60° of information at a time,

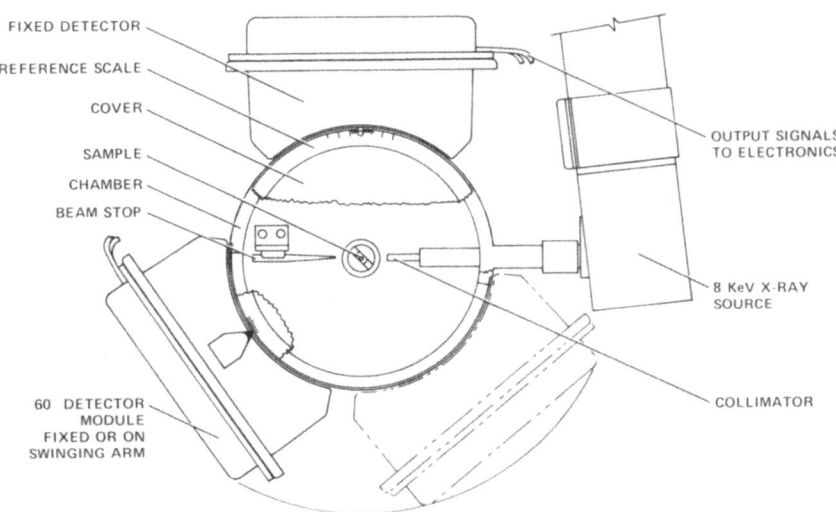

FIXED DETECTOR

REFERENCE SCALE

COVER

SAMPLE

CHAMBER

BEAM STOP

OUTPUT SIGNALS
TO ELECTRONICS

8 KeV X-RAY
SOURCE

60 DETECTOR
MODULE
FIXED OR ON
SWINGING ARM

COLLIMATOR

Figure 2. Diffraction geometry (Debye-Scherrer) and detector configuration. Three independent detectors, each sensitive to 60°, view virtually the entire spectrum simultaneously.

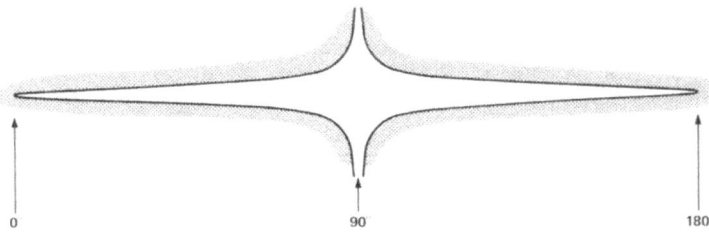

0   90   180

Figure 3. Aperture plate. This plate, inlayed into the vacuum chamber, limits the degradation of angular resolution due to the diffraction pattern curvature to 0.05° FWHM.

over an interval selected by the user. Sections of the full spectrum deemed uninteresting by the user are not viewed.

The chamber is filled with a static pressurized proportional gas mixture of Argon/CO_2. The gas pressure is empirically chosen to give good resolution at high efficiency. Nominal gas thickness is 11.5 mg/cm^2, so that 75% of Cu K_α radiation is detected. The detector vessel is designed to contain the gas mixture without appreciable loss or contamination due to either leakage or diffusion.

Although essentially a sealed detector, check and relief valves make gas renewal in the field possible when necessary.

Radiation is admitted to the detector through a 0.50 mm Beryllium window. Photoabsorption of the Cu K_α radiation in this thickness of Beryllium is 9.7%.

The detector, along with pre-amplifiers, is encapsulated in a sealed vessel that also serves as an electrostatic shield, and forms a completely modular device. The detector module can be conveniently installed and removed from the vacuum chamber, and in the event of failure can be readily replaced.

VACUUM CHAMBER

The powder sample, goniometer, and sample carousel reside in a cylindrical shell that is evacuated to forepump pressure to reduce absorption and scattering of the incident and reflected beams. The powder pattern emerges through a 25 μm thick mylar window (absorbing 5.9% of the radiation), impinges upon a 5.6 mm thick layer of air at ambient pressure (0.7% absorptive loss) before entering the detector through the Beryllium window. The vacuum chamber is designed to allow rapid and convenient entry to and closure of the sample area.

A special aperture plate is inlayed into the inner surface of the vacuum chamber for the purpose of restricting the observed curvature of the powder pattern rings. Since the detector measures the projection of an x-ray distribution on the axis of the anode wire, angular resolution is blurred at low angles in 2θ where the pattern curvature is large. Figure 3 shows the shape of an aperture that guarantees this geometrical blurring effect is no larger than 0.05° FWHM. This aperture, etched into a thin brass strip, is then indexed into the vacuum chamber at 0°. Software processing then corrects for this angle-dependent flux loss.

X-RAY SOURCE, INCIDENT BEAM COLLIMATION, SAMPLE, AND BEAM STOP

The diffractometer uses a standard 1.5 kwatt water cooled x-ray tube, with a nickel-filtered 1 mm x 10 mm copper target in line focus geometry. The characteristic spectrum emerges at a take-off angle of 6°, so that the apparent target size is ~100 μm

Samples were prepared to render the best possible resolution in 2θ, i.e., the powder was tamped into a capillary providing sample widths of the order of 200 μm. These narrow samples together with the line focus define the beam divergence to ~0.08° FWHM. The finite image of the sample at the detector contributes an additional 0.08° to resolution, so that a lower bound of 0.11° FWHM exists. This has been verified by exposing film strips located at the detector, and directly measuring line widths.

An entrance collimator is provided where larger samples are to be studied. Because data collection is digital, it is possible to perform background subtraction by collecting the empty sample spectrum.

A beam stop of high aspect ratio captures the beam after interaction in the sample.

READ-OUT

The ionization caused by an x-ray event is collected at the anode wire and an image charge is simultaneously formed on a conducting plane at virtual ground potential. This plane is sectioned into strips orthogonal to the anode wire, with 1° (2.35 mm) spacing. Each strip is tied to a tap of a multi-tapped delay line, so that the pulse observed at the end of the delay line is linearly correlated in time with the spatial coordinate of the event. The delay per tap must be chosen to reduce the read-out deadtime, yet be the order of, or larger than, the pulse risetime. We have found that 20 nsec/tap (1.4 μsec total delay) satisfies these requirements. The impedance of the delay line should be the order of the characteristic impedance of the chamber for maximum power transfer (nominally 200Ω). The delay lines used are custom fabricated to specifications.

The resolution of an event is the order of the size of the image charge formed divided by the S/N. It is easy to show that

$$\delta\theta = \frac{2G}{R\ S/N} \qquad (1)$$

where G is the anode-cathode gap, and R is the radius of curvature of the detector. For our detector, eqn. (1) predicts a FWHM of .08° given S/N = 50. The resolution can also be predicted using the observed pulse risetime, t_r. The time resolution is given by

$$\delta t = \frac{t_r}{S/N} \qquad (2)$$

the angular resolution is then

$$\delta\theta = \frac{\delta x}{R} = \frac{v}{R}\ \frac{t_r}{S/N} \qquad (3)$$

where v is the pulse velocity along the delay line. For our system, v = 0.1175 mm/nsec, R = 135 mm, t_r = 70 nsec, and given S/N = 50, the nominal expected resolution is

$$\delta\theta = 1.22\ \text{mrad} = .07°\ \text{FWHM} \qquad (4)$$

By combining this with other effects mentioned previously, an overall FWHM resolution of 0.15° is anticipated.

Delay line readout methods have been extensively developed by
several workers and reported in detail in the literature (9, 10).

ELECTRONICS

Figure 4 presents a block diagram of the electronics used to
digitize the position of x-ray events. Charge is pre-amplified by low
noise pre-amplifiers with 50 MHz bandpass and voltage gain of ~50 (34
db) referenced to 200Ω inputs. The resulting signals are filtered and
amplified, then presented to Constant Fraction Timing Discriminators
(CFTD) (11). The NIM-level logic pulses that are triggered serve as
START and STOP signals for a Time-to-Amplitude converter (TAC), set to
convert 0-1.4 μsec delays to 0-10 volt levels, of ~5 μsec width. The
START signal also triggers a set of monostable vibrators that strobe the
TAC output, and signal the ADC to begin conversion. The ADC then converts the analog level into a 12-bit word during a 4 μsec interval when
the TAC output level is stable. Twelve-bit conversion over a 60° interval results in a single channel resolution of 0.015°. Thus, a .1° FWHM
peak is fully represented in 7 channels. The cost of this level of
detail is of course the large memory required to store the spectrum.
Assuming a maximum count per channel of 65K, a single detector requires
4096 16-bit words.

The electronics are designed with a single set of amplifiers,
discriminators, TAC, and ADC--multiple detectors feed this processing
system in parallel. The 12-bit ADC output is then hardware flagged with
an additional 2 bits to identify in which detector the x-ray event
occurred. This method is cost-effective and does not multiplex data
(i.e., all detectors are viewed simultaneously), however, additional
memory is required. A complete three-fold detector spectrum is stored
as ~12K 16-bit words. it is clear that a more efficient encoding of
data such that spectral regions of little interest receive fewer bits
may promise to appreciably reduce memory requirements.

At the completion of the ADC conversion, the 14-bit data word is
then hardware added to a 16-bit starting register that allows the
spectrum to be stored in any block of the memory. The starting register
is selected by the software. The data word that results is placed on
the data-bus and an end-of-conversion pulse is sent to the computer,
which then hardware increments the count in the address in memory
specified by the data on the bus. This storage step requires .75 μsec.

DATA PROCESSING AND SOFTWARE ROUTINES

Data acquisition, reduction and display are controlled by software.
These routines are written in FORTRAN, and can be easily modified. All
routines are part of an integrated system which uses data files to
communicate between different parts. Figure 5 shows this software
architecture.

The user specifies by means of a simple keyboard dialog those
routines of interest. User input is in the form of two-character com-
mands that input parameters or initiate routines. Parameters are stored

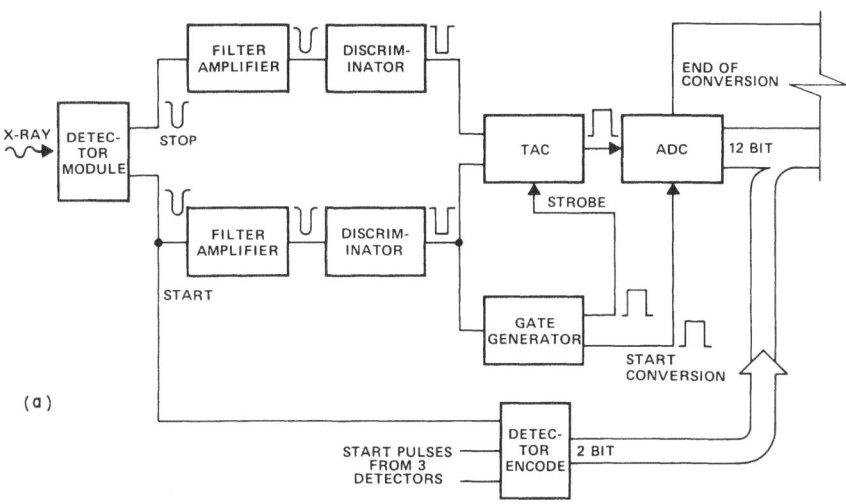

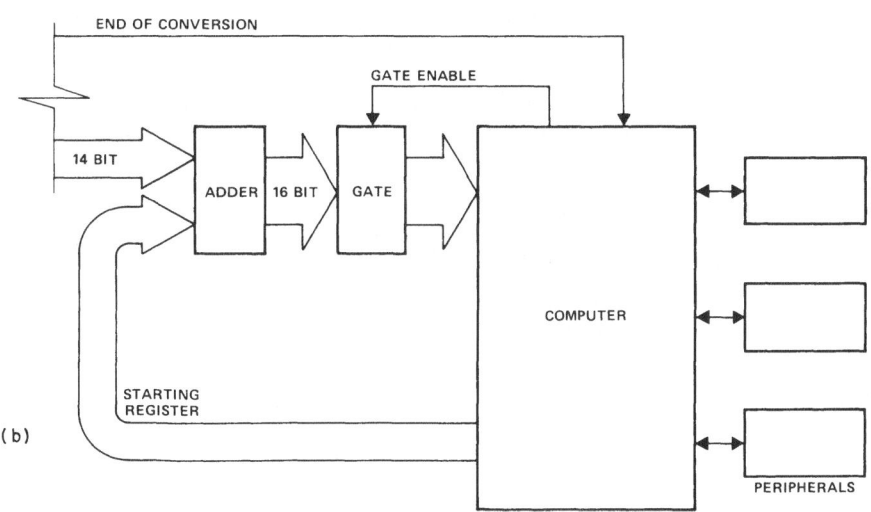

Figure 4. Signal processing electronics. (a) Signals from the detector modules are digitized to form a 14-bit word representing the angle of the detected x-ray; (b) a minicomputer increments and stores the acquired spectrum.

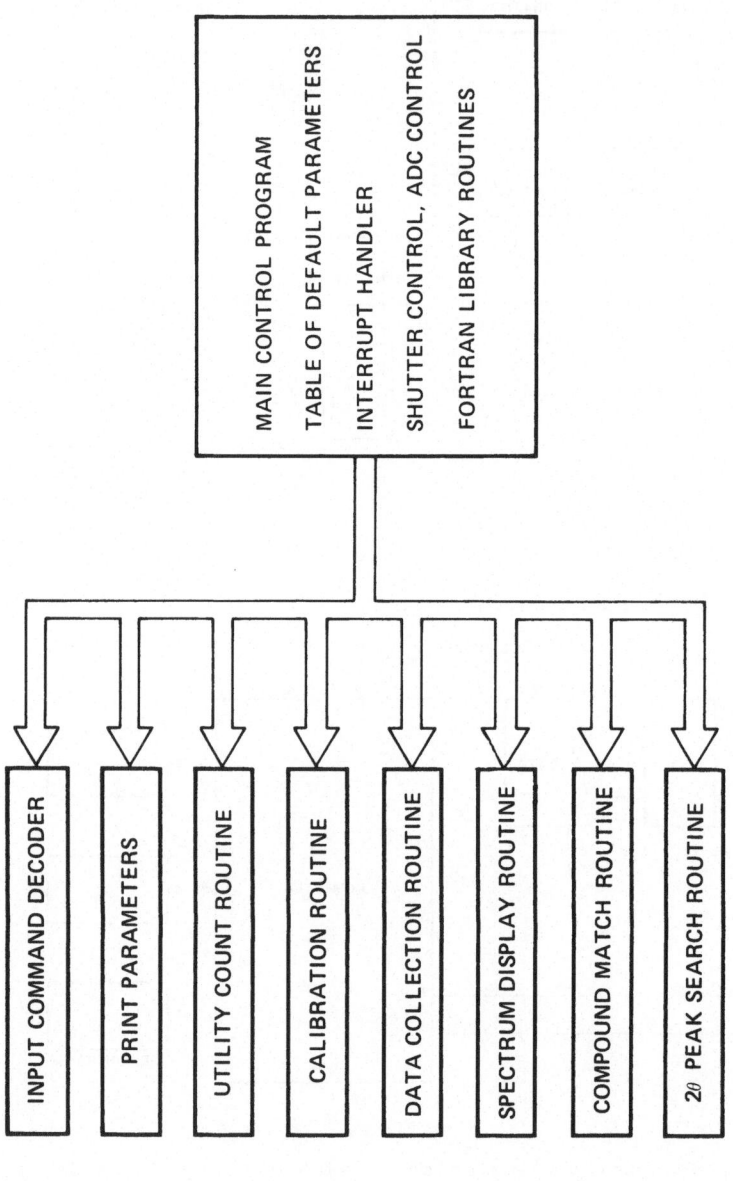

Figure 5. Software architecture.

in data files, so that the user interaction can in general be succinct. Output, however, is well labeled and understandable.

Existing routines are:
  i)   A utility count routine for test purposes that collects successive data sets and calculates deviations in the count rate.
  ii) Display routines that plot on either the input terminal or the line printer, with the ability to display only the region of interest. Scaling to maximum peak height is automatic.
  iii) The data collection routine can be performed to either preset time or count. Corrections for coincidence loss are made. Calibration by comparison with a standard sample is available.
  iv) The peak search routine scans the data for peaks significantly above background, subtracts the background, and calculates the integrated intensity under the peak. Output includes values of d-spacings as a function of I/I_{max} sorted in descending order, in JCPDS format. These values, along with 2θ, hkl, intensity, and peak width can be stored as a file at the user's discretion.
  v)   The compound match routine compares the sample with a library of known compounds, searching for matches on the strongest reflections. If a match is found the user may request the routine to attempt matching all remaining reflections.

SYSTEM PERFORMANCE

The performance of the detector was observed using both collimated ^{55}Fe radiation and the quartz reflections of Cu K_α radiation. Figure 6 presents the proportional behavior of the detector. The anode S/N is plotted vs. detector bias for both 5.9 KeV and 8.0 KeV x-rays, in 3.7 atm of Argon. From 2.5-2.9 KV charge collection and gain operate in the proportional mode, as evidenced by the characteristic exponential growth of gain vs. bias, and the constant factor between the two curves of differing energy. The factor of 1.9 is in good agreement with the ratio of deposited charge predicted by theory (1.8). At higher bias voltages, spurious pulses of long time constants are observed, along with a marked increase in the frequency of sparking, signaling the onset of the geiger mode. Our detector was biased to give the highest possible S/N prior to breakdown behavior, 2.8-2.85 KV.

The linear response of the detector is shown in Figure 7. By using the first 11 reflections of quartz, and matching their known 2θ values with their peak ADC channels, a plot is obtained that indicates the very good linearity available. The figure inset shows on a highly expanded vertical scale the deviations from linearity of a least-squares fit. The standard deviation from linearity is measured to be $0.14°$, which can be improved to $0.11°$ by using the weighted mean of the spectral peaks. It can be noticed that deviations from linearity may not be random, i.e., from the inset it appears that these deviations can be fit by, say, a sine function. The second order response function

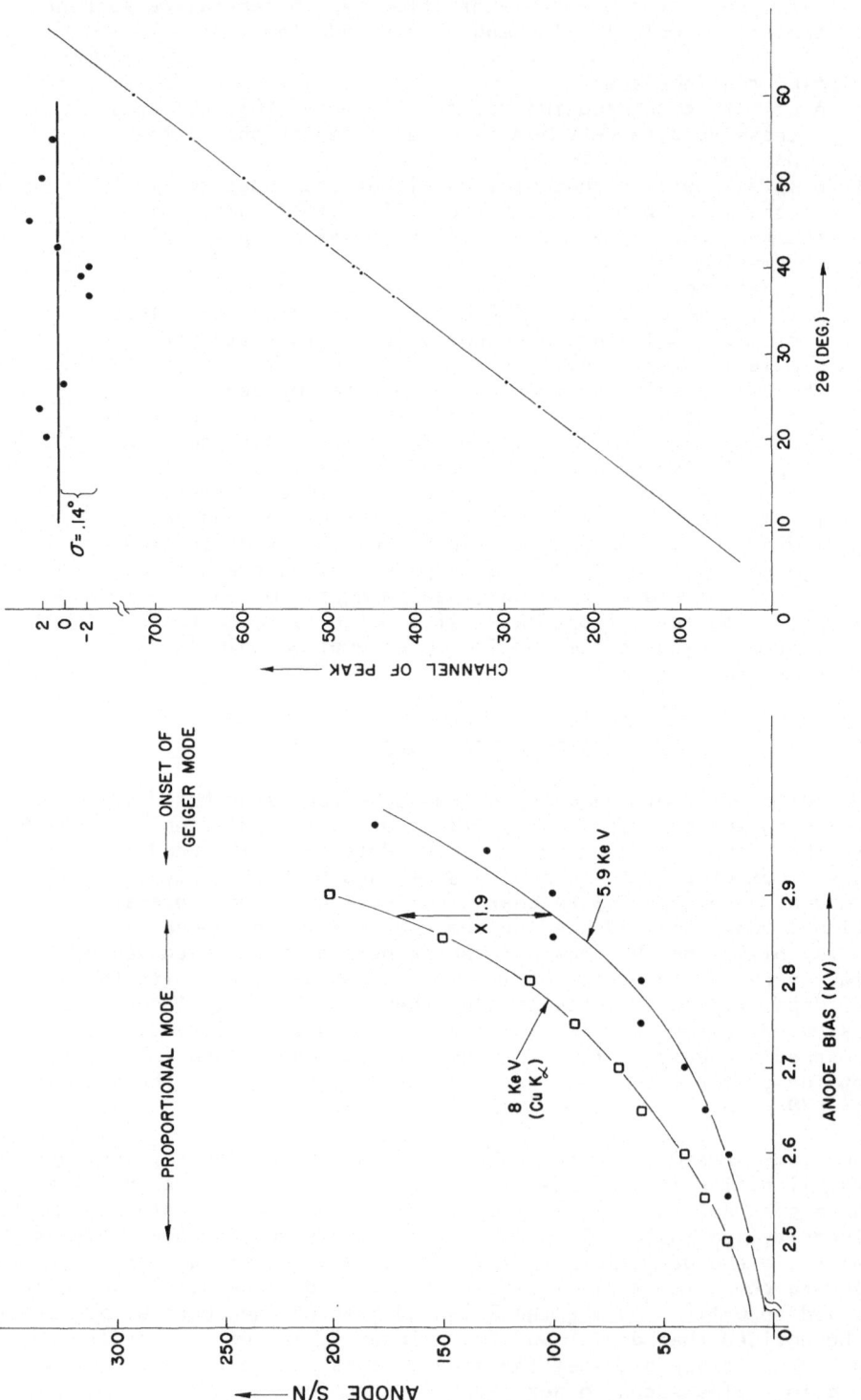

Figure 7. Detector linearity in 2θ.

Figure 6. Signal pulse height vs. high voltage.

$$2\theta(\text{degrees}) = .97826 \, N + 3.364$$

$$- 0.196 \sin(.01544 \, N - 4.46)$$

where N is the ADC peak channel then predicts the correct values of 2θ with a standard deviation of .086°. The corresponding d-spacing error at 4 Å is three parts per thousand. Using the weighted mean again may improve upon this.

Figure 8 presents the relative efficiency of the detector to 8 KeV x-rays over its range of angles. The 0-60° range is demonstrated. The uniform rate of loss in efficiency at large angles results from the attenuation of signal in the delayline. Corrections for the efficiency profile are made in the software. The region of depressed efficiency at 20° results from a non-uniformity in this particular anode wire and is not in general a feature of this detector. Fabrication methods are being devised to insure that wires are installed without these types of anomalies.

Figure 9 presents the spectrum of quartz taken in 120 seconds using a 200 μm sample rotated in an x-ray beam produced by a 1 KW Cu tube. About 30° are shown and include the 9 quartz reflections of smallest 2θ. Comparison of the intense 26.6° peak with its neighboring 24.0° peak demonstrate the exceedingly large dynamic range available. One channel represents 0.041°, so that the FWHM resolution = 0.20° in this spectrum. This is the resolution that is typically available in this time, and is chiefly limited by low S/N. We are presently developing state-of-the-art pre-amplifiers that are expected to improve S/N and hence resolution.

Figure 10 is the spectrum of Figure 9 with background subtracted. This was accomplished by replacing the sample with an empty glass capillary and subtracting counts for 120 seconds. Subtraction produces an essentially flat level of background free of artifacts.

Figure 11 presents the entire quartz spectrum (15°-165°) taken in 60 seconds. This was achieved by using a single detector in each of three positions and splicing the 3 spectra. However, the equivalence to a 3-detector system where data collection is simultaneous is obvious.

ACKNOWLEDGEMENTS

The authors wish to thank Richard Casper, Royce Essig and Karl Shill for their professional skills in the design and fabrication of the system, Arild Christensen for many helpful technical discussions, and J. Cohen of Northwestern University for clarifying a suitable avenue toward a product. The development of the system was enthusiastically sponsored by Tony Chan.

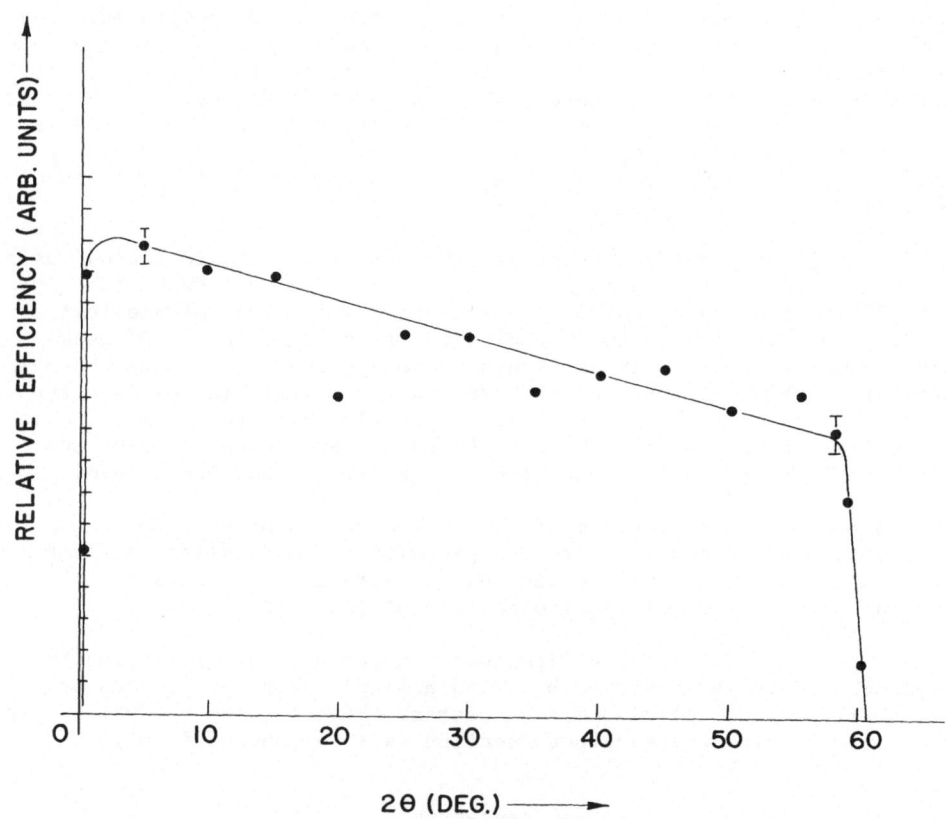

Figure 8.   Detector efficiency at 8 KeV vs. angle.   The loss of efficiency at large angles is due to delayline attenuation, and can be software corrected.

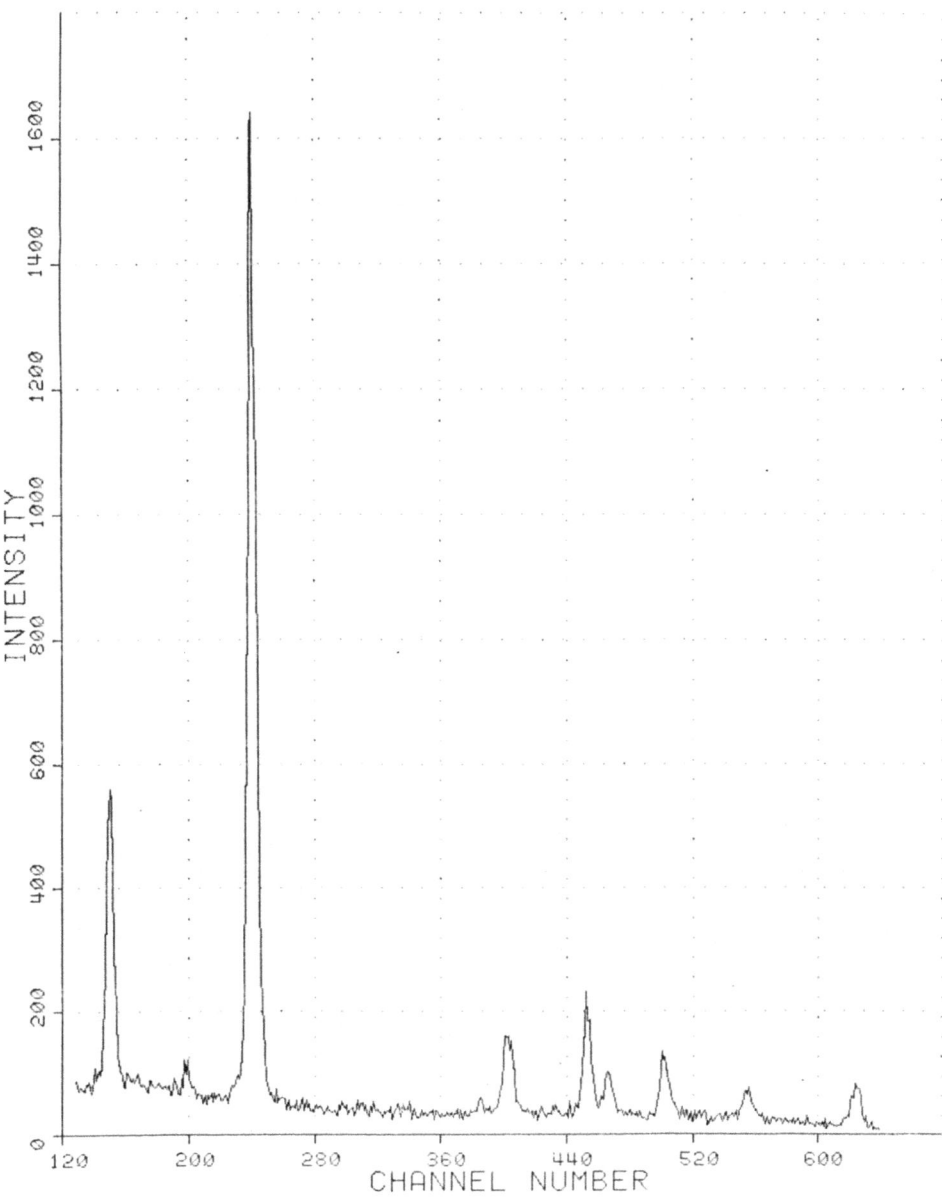

Figure 9.   Powder spectrum of quartz.   Collection time is 120 sec,
using a 200 μm rotating sample and a 1 KW Cu x-ray beam.   512 channels
are shown.   1 channel = 0.041° of 2θ.   Vertical axis is counts/channel.

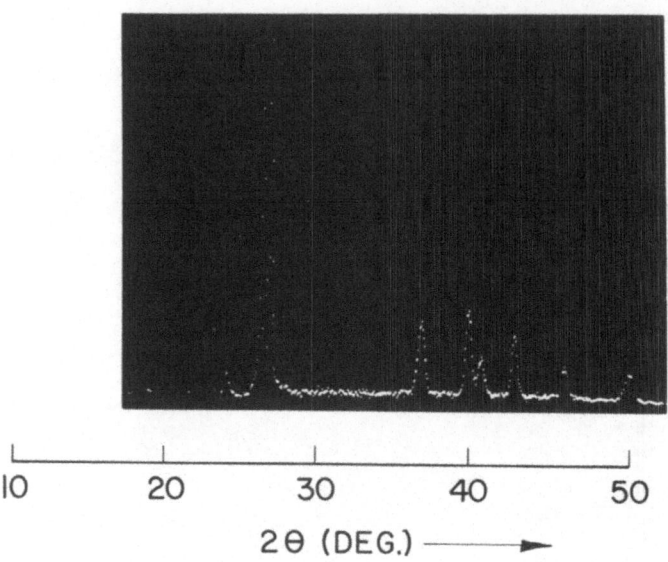

Figure 10. Quartz spectrum as in Figure 9, with background subtracted. An empty glass capillary was used to decrement the spectrum for 120 sec.

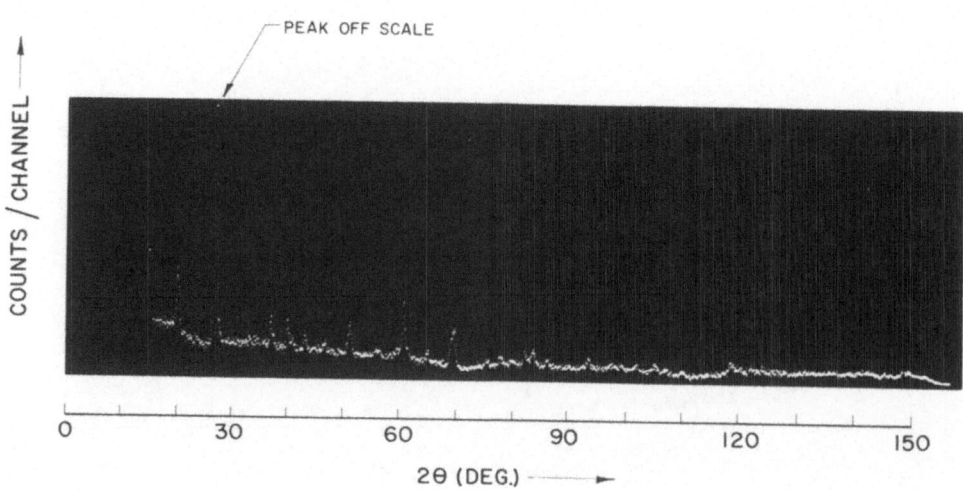

Figure 11. 60 second spectrum of quartz in the range of 15°-165° 2θ. The spectrum was acquired by using a single detector in 3 separate positions and displaying the 3 spectra simultaneously.

REFERENCES

1. G. W. Ewing (ed), "Topics in Chemical Instrumentation,"
    Chemical Education Publishing Co., 1971, p. 94.

2. G. C. Hanna, D. H. W. Kirkwood, B. Pontecorvo, "High Multi-
    plication Proportional Counters for Energy Measurements,"
    Phys. Rev., 75 (1949) 985.

3. S. Curran, J. Angus, A. L. Cockcroft, "Investigation of Soft
    Radiation by Proportional Counters," Phil. Mag., 40 (1949) 929.

4. T. Tabata, R. Ito, S. Okabe, "Generalized Semiempirical
    Equations for the Extrapolated Range of Electrons,"
    Nuc. Instr. & Meth., 103 (1972) 85.

5. M. R. James, J. B. Cohen, "The Application of A Position-
    Sensitive X-Ray Detector to the Measurement of Residual
    Stresses," Northwestern University, Dept. of Mat. Sci. and
    Mat. Res. Center, Technical Report No. 11, Aug. 1975.

6. G. Charpak, "Evolution of the Automatic Spark Chambers,"
    Ann. Rev. of Nucl. Sci., 20 (1970) 195.

7. R. E. Rand, E. B. Hughes, R. Kose, T. W. Martin, W. I.
    MacGregor, R. F. Schilling, "A Multiwire Proportional
    Chamber System for Use in a Storage Ring Experiment,"
    Nuc. Instr. & Meth., 118 (1974) 189.

8. M. R. James, J. B. Cohen, "Study of the Precision of X-Ray
    Stress Analysis," Advances in X-Ray Analysis, Vol. 20 (1976).

9. R. Grove, K. Lee, V. Perez-Mendez, J. Speriude, "Electro-
    magnetic Delay Line Readout for Proportional Wire Chambers,"
    Nucl. Instr. & Meth., 89 (1970) 257.

10. J. L. Lacy, R. S. Lindsey, "High-Resolution Readout of Multi-
    wire Proportional Counters Using the Cathode-Coupled Delay-
    Line Technique," Nucl. Instr. & Meth., 119 (1974) 483.

11. M. R. Maier, P. Sperr, "On the Construction of a Fast Constant
    Fraction Trigger with Integrated Circuits and Applications to
    Various Photomultiplier Tubes," Nucl. Instr. & Meth.,
    87 (1970) 13.

COUNTING RATE PERFORMANCE OF PULSED - TUBE SYSTEMS

A. O. Sandborg/J. C. Russ

EDAX INTERNATIONAL, INC.

Prairie View, Illinois 60069

Introduction

Energy - dispersive x-ray fluorescence systems are characterized by limited count rate capability, which is an important practical limitation on their application. This arises from two factors: 1) The amplifier time constant is many times greater than needed in wavelength-dispersive systems, in order to permit accurate measurement of pulse height, and reducing the time constant degrades system resolution; 2) x-ray pulses from all the major and minor elements, as well as scattered background and/or peaks from the incident radiation, must all be processed, reducing further the effective count rate for a single element of interest. Hence there has been a strong desire to find ways to increase the count rate capability of energy-dispersive systems.

A novel approach to this problem was presented by Jaklevic (1), who used an x-ray tube that could be pulsed under control of the amplifier. The tube was turned off while a pulse was being processed, to prevent pile up and losses in the amplifier, and to space the x-ray pulses in time to maximize thruput in the amplifier. We have used the same basic approach but with some differences in design and mode of operation.

Circuit Description

EDAX pulsed Xray generator uses the same high voltage supply (CPS Model 131P) as our continuous excitation systems. In fact, the system can also be operated in a continuous excitation mode by applying constant potential to the control grid and switching minor components, for example in the BLR circuit of the amplifier. Comparative data for this paper using continuous and pulsed excitation were obtained using the same system. The beam current is selected and controlled by a closed loop servo to ensure stable operation. The current that is set by the user or computer is the current that will flow when the tube is "ON", not the average current. All current values given in this paper are the "ON" current. Pulsing is accomplished by the use of an X-ray tube containing a control grid (Watkins Johnson 2703-2). The beam current is shut off each time an X-ray event is detected and allowed to return to the set current only after the x-ray event is processed and stored or at the end of the shaped linear pulse, whichever is longer.

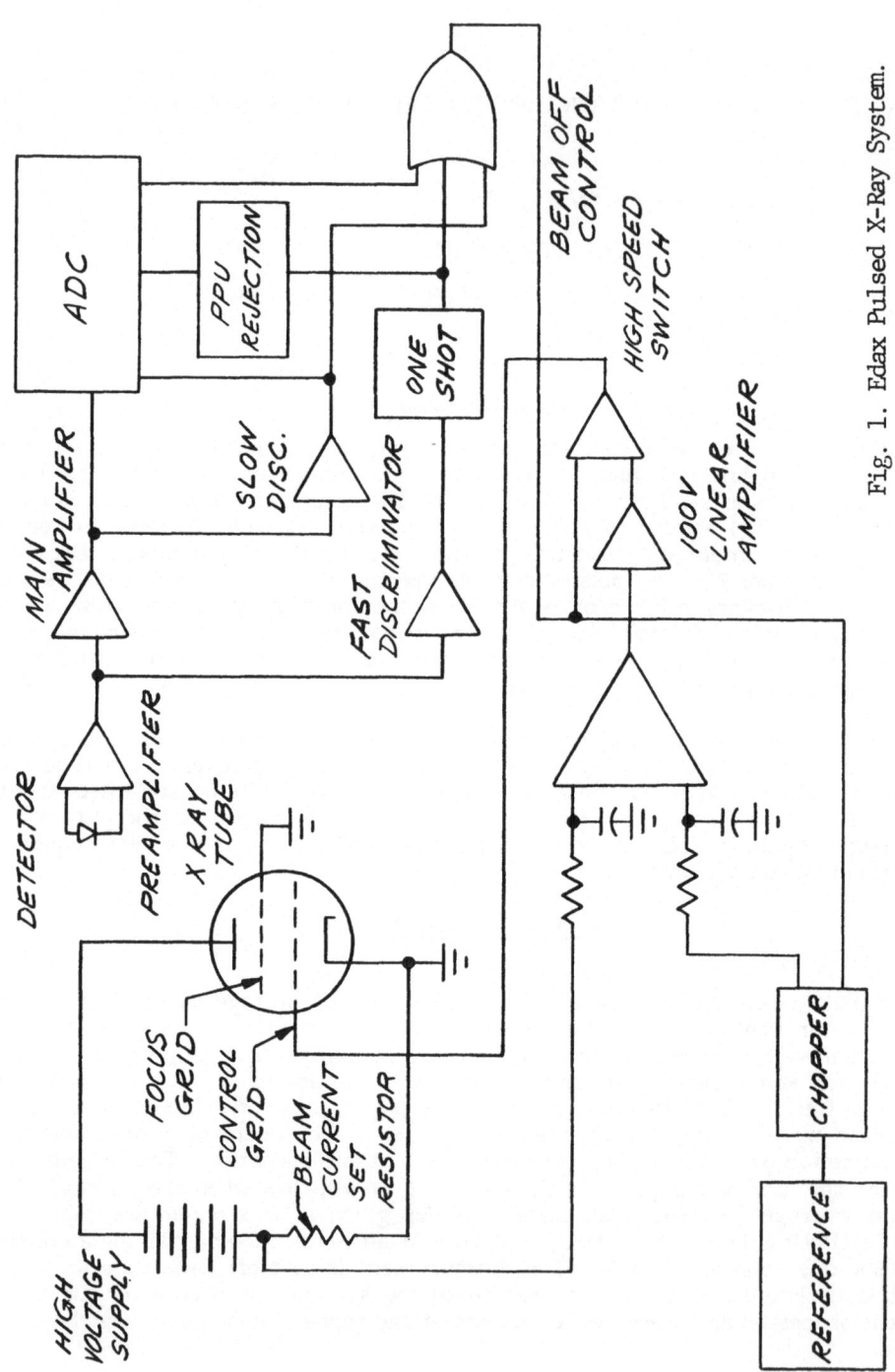

Fig. 1. Edax Pulsed X-Ray System.

The operation of the system can be explained by referring to the block diagram (Figure 1). An X-ray event detected by the silicon detector is processed in the preamplifier. The output of the preamplifier drives the main shaping amplifier (then to the ADC for processing) and the fast discriminator for detection of each X-ray event.

The fast discriminator triggers a retriggerable shot which is set for the output pulse width of the main amplifier (approximately 40 µsec for the 4 µsec time constant amplifier used for all data in this paper). The one shot output is "ORed" with the slow discriminator and the ADC busy to form the BEAM OFF signal. The one shot also drives the pulse pile up rejection logic. There is very little partial pulse pile up in a pulsed system since the beam is turned off at the detection of each X-ray event. The limiting factors here are the pulse pair resolution of the fast discriminator and the time required to turn the beam off.

The BEAM OFF signal drives a high speed analog switch which turns the control grid of the tube off upon detection of an X-ray event and allows the grid to return to the preset voltage at the conclusion of the X-ray processing time. The beam is turned off in less than 300 nanoseconds after detection of an event. Turn-on time is similar.

The BEAM OFF also drives a chopper on the reference in the servo system. The beam current servo is a closed loop system that utilizes an operational amplifier which drives the grid to the proper voltage to achieve the desired beam current. One input to the op-amp is a voltage derived from a resistor in the return path of the high voltage supply and the other input is a reference voltage.

Since the beam current is pulsed, the actual voltage on the resistor will seek same average value (because of integration in the supply) dependent on the sample and excitation conditions   Hence, the reference must also be switched and integrated identical to the resistor voltage. Therefore the output of the reference is applied to a chopper that is controlled by the BEAM OFF signal and the chopper output is applied to an integrator prior to the op. amp input. The actual beam current resistor voltage is also integrated in order to match time constants independent of high voltage supply variations.

The output of the control op. amp is applied to a 100V linear amplifier to develop the analog voltage required by the grid. The output of this linear amplifier is switched by the high speed analog gate circuit by the BEAM OFF as explained previously.

The result is a stable accurate switchable control of the beam current to achieve the desired current setting.

Performance Criteria - Stored Count Rate

The most important criterion is, of course, the practical increase in stored information in a given period of time. Many energy dispersive x-ray fluorescence application reports use the "live time" of analysis, which can be significantly greater than the actual elapsed time (clock time) at high count rates. To minimize this confusion, all data reported here will be specifically in terms of actual elapsed or clock time, which is of course the time required for analysis and hence of most interest to the user.

Table 1 shows the count rate (counts per clock second) entering the amplifier (input count rate) and being stored in the analyzer memory (stored count rate) as a function of tube current. The tube current includes a small leakage current that shifts the intercept from zero. The "efficiency" is defined as the "live" time set on the analyzer divided by actual elapsed clock time required for the analysis, expressed as percent. The sample was copper, excited by an unfiltered rhodium target x-ray take operated at 18 KV.

TABLE 1

Tube Current	Continuous			Pulsed		
	Input	Stored	Effic.	Input	Stored	Effic.
20	1810	1650	91%	1260	1240	99%
30	5200	3900	75	3280	3250	99
40	8300	5380	65	4900	4850	99
60	14700	6700	46	7500	7450	99
80	21250	6700	32	9550	9500	99
100	27500	6150	22	11050	11000	99
200	56600	2300	4	15600	15500	99
300	82800	630	1	17650	17500	99
400	106000	80	–	18900	18750	99
500	128000	30	–	19650	19500	99

From these data it is clear that operating the system in the pulsed mode increases the maximum (also shown in Figure 2) stored count rate from less than 7000 cps to nearly 20,000 cps. This will correspondingly reduce the total clock time needed to obtain results with a given level of statistical precision. Count rate increases, and hence analysis time reductions of 2½ - 3 times are typical. The tube current must be increased (during the time the tube is turned on) to achieve these results, so that a larger power supply is required.

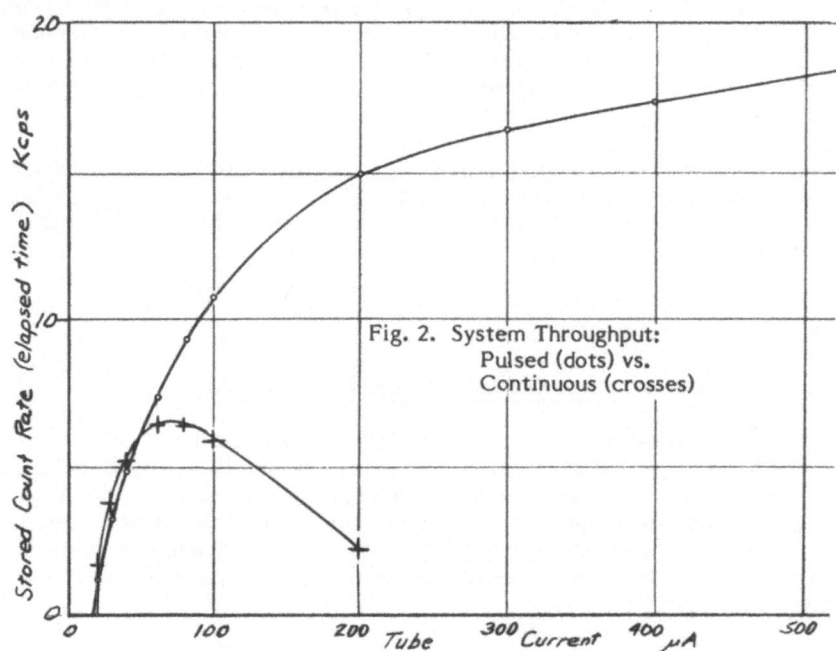

Fig. 2. System Throughput:
Pulsed (dots) vs.
Continuous (crosses)

For trace element analysis, the improvement in count rate achieved may be limited by generator capacity. For example, using a sample containing 112 ppm arsenic in an organic matrix, excited with a filtered rhodium tube operated at 35 KV and 990 μ A (the maximum current rating of our generator), we obtained an input count rate of 22800 cps and a stored count rate of 6140 cps in the continuous mode of operation, but only 10,500 cps in the pulsed mode. This still gives an improvement in detection limit, but less than could be achieved with a higher power generator. In a 300 second analysis using each method, the pulsed results gave an improvement greater than 20% in minimum detection limit, as shown in Table 2 and Figure 3.

TABLE 2

Mode	P+B	B	Min. Det. Line (3σ)
Cont.	4591	2388	7.5 ppm
Pulsed	8046	4300	5.9 ppm

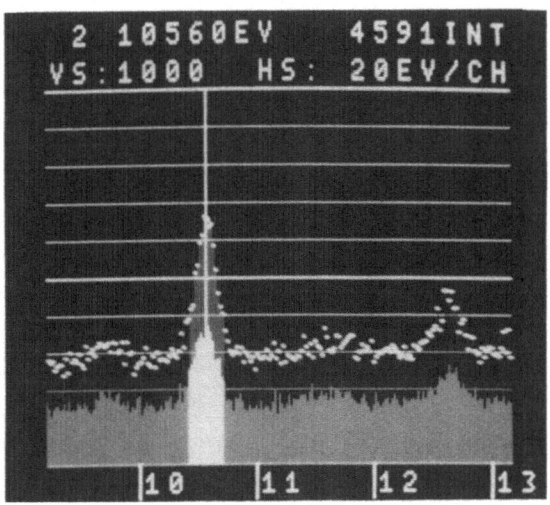

Fig. 3. Spectra from 112 ppm As in organic matrix
Pulsed (dots) vs Continuous (solid)

Other Criteria

The increased count rate capacity achieved with the pulsed tube cannot degrade other system parameters. Several of these which are particularly important are described below:

1. Count rate linearity. The specific count rate (counts per second per microampere of tube current) must be constant in order to perform quantitative analysis using different tube currents, in order to maximize stored count rate. To verify that the pulsing of the tube does not degrade this linearity data were measured as summarized in Figure 4 (iron sample, 18 KV, rhodium target tube). In this case the count rate is expressed in counts per live second since live time correction circuitry is still needed with the pulsed method. The data show a linear trend, so there are no counts being "lost" due to turning the tube on and off. The lines have an intercept of about 15 µA, which is the leakage current of the particular tube being used.

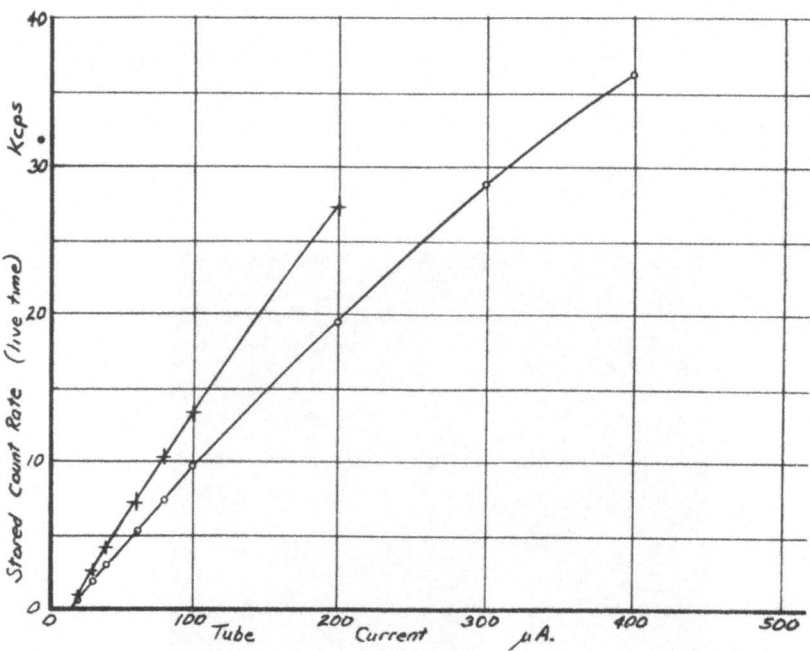

Fig. 4. System Linearity: Pulsed (dots)   Continuous (crosses)

2. Energy dependence. Logically, since the tubes "off time" and amplifier pulse width are independent of x-ray energy, the gain in stored count rate should be independent of energy. To check this we measured the Si, Ca, Fe, Sr, and Zr peaks from an AGV-1 mineral standard, using a Rh target tube operated at 35 KV, and at 80 µA continuous (22900 cps input, 6500 cps stored) and 500 µA pulsed (17,650 cps stored). The results (Table 3) show that while the average increase in stored count rate is about 2.7 times, the factor does vary with energy. We do not have any explanation for this behavior at present. It does not interfere in any way with quantitative analysis, but does mean that data from continuous and pulsed modes of operation cannot be intermixed. Similar results have been obtained with other samples and combinations of elements, and at lower count rates.

TABLE 3

Count Rate - CPS

Element / Energy		Continuous	Pulsed	Ratio
Si	1.74	504.73	1228.73	2.43
Ca	3.69	165.08	437.64	2.65
Fe	6.40	982.76	2639.33	2.69
Sr	14.14	106.27	293.81	2.76
Zr	15.74	89.36	251.20	2.81

3. Pileup rejection. Pileup rejection circuitry for a pulsed system is essentially identical to a continuous system. At high count rates, there is a finite probability of multiple pulses occurring before the tube can shut off. Table 4 shows the increase in pulse pileup with count rate for both modes of operation, using an iron sample excited with a Rh target tube operated at 18 KV. The rejection of partial pileup is shown by the ratio of background at 10-12 keV to the iron peak, and is very consistent to high count rates. The sum peak at 12.8 keV, again expressed as a ratio to the 6.4 keV peak, increases with count rate. In the pulsed mode of operation, as the upper limit of stored count rate is approached, the sum peak builds up rapidly. This emphasizes the importance of not operating at higher tube currents than needed to achieve adequate stored rates.

TABLE 4

Tube Current (μA)	Continuous				Pulsed		
	Input c/s	Stored c/s	bg/peak %	sum/peak %	Stored c/s	bg/peak %	Sum/peak %
20	2400	2100	.046	.203	1400	.048	.120
40	10300	5900	.055	.767	5600	.048	.403
60	18200	6900	.061	1.391	8500	.050	.739
80	26000	6400	.063	1.963	10600	.056	.973
100	33500	5400	.069	2.548	12300	.054	1.267
200	51500	2800	.084	3.909	16800	.068	2.644
300	68600	1300	.108	5.218	18800	.080	3.825
400					19900	.096	5.012
500					20600	.099	6.068
600					21000	.118	7.081
700					21400	.125	8.056
800					21600	.134	8.934
900					21800	.149	9.833
990					21900	.156	10.486

Conclusion

The operation of an energy dispersive xray fluorescence system with a pulsed x-ray tube to space x-ray pulses more uniformly in time gives an increase in stored count rate of about 2½ times, compared to conventional continuous excitation. This is an effective way to reduce analyzing time to about 40% of what would be needed with continuous excitation, to achieve the same level of statistical precision.

Practical implementation of such a system requires higher generator current capacity and some special circuitry compared to conventional systems, but imposes no limitations for routine use.

References

1. J. M. Jaklevic, D. A. Landis, F. S. Goulding, "Energy Dispersive X-Ray Fluorescence Spectrometry Using Pulsed X-Ray Excitation," Adv. in X-Ray Anal., vol. 19 (ed. R. W. Gould), 1976, p. 253-265.

A NEW METHOD FOR THE ELIMINATION OF THE WALL EFFECT IN PROPORTIONAL

COUNTER

H. Sipilä and E. Kiuru

Outokumpu Oy, Institute of Physics

Espoo, Finland

ABSTRACT

The wall effect present in the gas-filled proportional counters and the low-energy background caused by it are well-known disadvantages. In this paper a method is presented to define the radial position of a single X-ray absorption in ordinary proportional counters. The method utilizes the variation in detector pulse collection times due to the electron diffusion. This variation can be determined with a specially constructed risetime analyzer which effectively forms a dead zone of cylindrical shape inside the detector. The properties and performance of the method are demonstrated. In the experiment performed the signal to background ratio was improved by a factor of twenty.

INTRODUCTION

The energy dispersive properties of proportional counters are widely used in various X-ray analyzers. The detector is moderately good as to its resolution, and its efficiency can be tuned according to the particular problem. However, the wall effect and the low-energy background induced by it limit the performance of proportional counters. The background is mainly caused by the incomplete charge collection. The absorption trace terminates on the detector wall before the ionizing electron has lost all of its energy into the detector gas. An example of this phenomenon is illustrated in Fig. 1, which shows the energy spectrum of a detector constructed of pure materials. The usual impurity lines are absent except the Si Kα line from the quartz sheet used to filter the radiation from Cd-109 and a slight discontinuity in the region of the Ar absorption edge.

Kocharov and Petrov (1) have presented a procedure for the calculation of the background. Fink (2) has reviewed some methods to eliminate the background but these methods require elaborately constructed detectors, e.g. wall-less anticoincidence multiwire counters. In this paper a novel method is suggested which can be used for the elimination of the low-energy background from ordinary proportional counters.

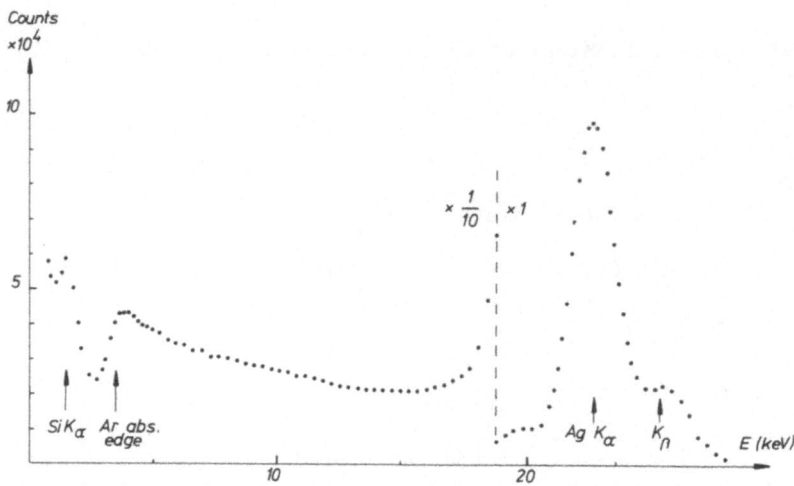

Figure 1.   The low-energy background in an Ar–CH$_4$ counter.
            Excitation source Cd–109.

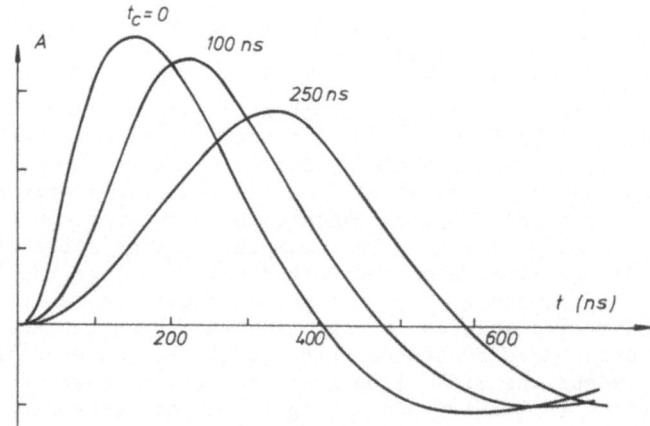

Figure 2.   Detector pulses doubly differentiated and once integrated
            in RC circuits with time constants of 100 ns. Collection
            times t$_c$ calculated from pulse shapes according to Mathieson
            and Harris. t$_c$ value 250 ns corresponds to the observed
            maximum.

METHOD

In principle the wall effect can be eliminated by forming a dead zone near the detector wall where the ionization events are not counted. The problem is to determine the radial position of the absorption track. The absorption of an X-ray quantum and the consequent photoelectron result in a rather short ionization track in the detector gas. The longer the distance between the absorption and the anode wire, the wider the electron cloud has diffused before reaching the anode. This naturally shows in the electron collection time and also in the pulse rise-time. Thus the electron diffusion is a measure of the distance of the absorption from the center wire. Fig. 2 shows the pulse shape variations in an experimental $Xe-CO_2$ counter. The calculated electron collection times are given for each pulse.

Mathieson and Harris (3,4) have discussed the effect of the diffusion to the risetime in proportional counters. In this paper a method is presented to derive position information from this effect.

THE EXPERIMENTAL COUNTER

The background elimination method was investigated using a proportional counter made of high purity aluminum with an inner diameter of 43 mm. The 25 μm diameter anode wire was made of tungsten and the detector was filled with a gas mixture 96:4 of $Xe-CO_2$ at 25 kPa. Some characteristics of the counter can be seen in Fig. 3. The efficiency of the detector is presented in Fig. 3a. The curve in Fig. 3b gives the range of electrons in the detector medium as a function of the electron energy (5). In Fig. 3c the electron drift time is shown as a function of the radial position of the absorption. This curve is computed using the electron drift velocities published by English and Hanna (6). Experimentally it was found that the ratio of the drift velocity and diffusion was more advantageous in xenon mixtures than in argon mixtures. Because the values of these velocities in xenon gas could not be found in the literature, the risetime differences caused by diffusion were measured and were already seen in Fig. 2. On the basis of this measurement the parameters of the risetime circuitry were fixed.

EXPERIMENTS AND RESULTS

Several authors have proposed methods for removing gamma induced background from the proportional counter (3,4,7-10). All of these are based on rejecting by risetime discrimination the long ionization tracks with low specific charge density. In this work the risetime analysis is applied to the collection time distribution due to electron diffusion. Because the risetime variation is relatively small, the circuitry suggested by Mathieson and Harris has been found to suit the experiment in question. They have proven that integrating the detector pulse once and differentiating it twice results in a bipolar pulse whose cross-over time is a rather linear function of the collection time. Fig. 2 shows this variation in the experimental counter.

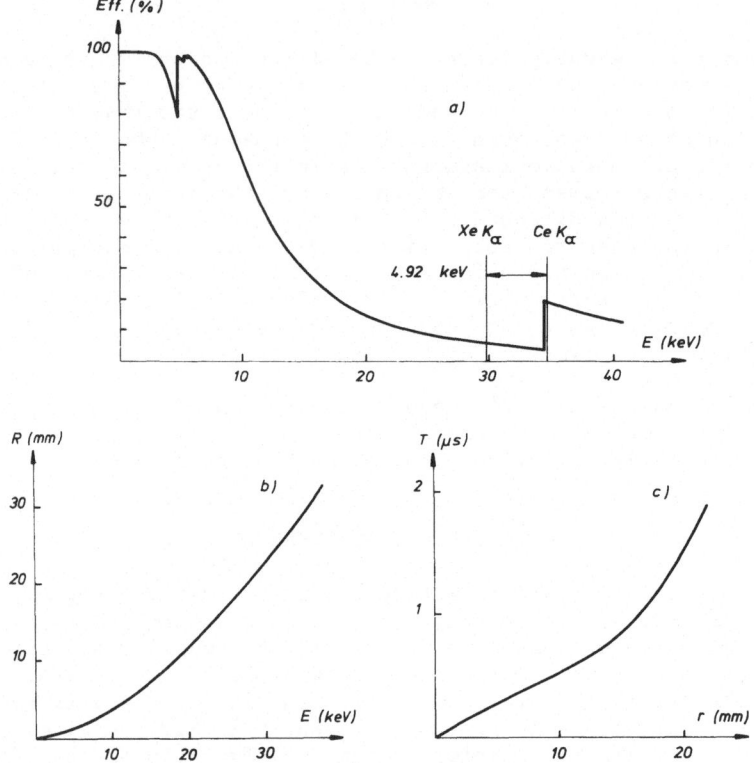

Figure 3.   Characteristics of the experimental detector:
a)   efficiency of the detector gas
b)   range R of electrons
c)   electron drift time T in the detector gas as a function
     of radial position r of absorption.

        The principle of the risetime discrimination circuitry can be
seen in Fig. 4. Parallel with the conventional energy measuring
channel there has been installed a risetime channel from which a
coincidence signal is derived for the multichannel analyzer. The
discriminator forms a bipolar signal from the detector pulse and
measures its zero-crossing time which is dependent on the collection
time of the original charge pulse. The zero-crossing time is compared
with the reference time T1. If the latter is longer than the measured
time the discriminator gives a coincidence mark.

        During experiments it was found that a simple system like this
cannot fully remove the background due to wall effect. The explanation
is that with certain probability two absorption tracks are produced
by one X-ray quantum due to detector gas fluorescence. The two pulses
may interfere with each other in the risetime discriminator with false
results. To prevent this a special temporary memory has been added to
the logic of the discriminator which rejects all pulses with a time
difference smaller than T2. This equals the maximum electron drift time
from the cathode to the anode. In the risetime channel the leading edge

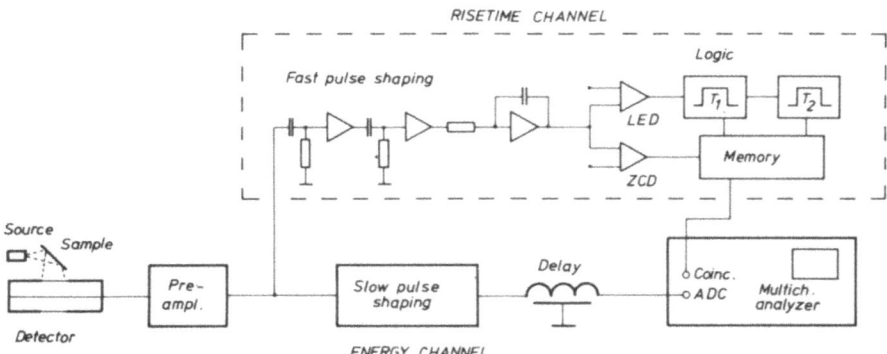

Figure 4.   Block diagram of the experimental risetime discriminating
            system.

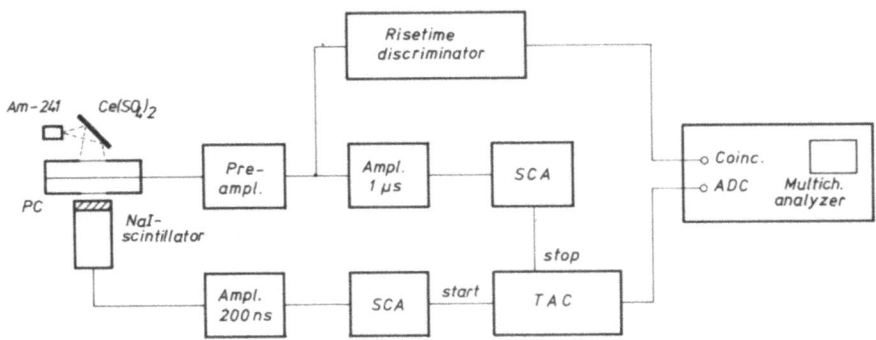

Figure 5.   Block diagram of the electron drift time spectrum measure-
            ment.

and zero-crossing detectors are constructed with conventional techniques.
A slight energy dependence is an inherent feature of this device. By
carefully adjusting the detection levels the energy dependence could
be kept within 10 percent in the energy range of 3-20 keV.

The drift time spectrum of electron clouds in the experimental
detector was investigated with a system similar to Fig. 5. From the
measured spectrum one can directly conclude the collection time
spectrum of detector pulses. However, the time scale is not consistent
because the linear dimension of the electron cloud grows as a square
root of time, that is, short times are not emphasized as in the drift
time spectrum.

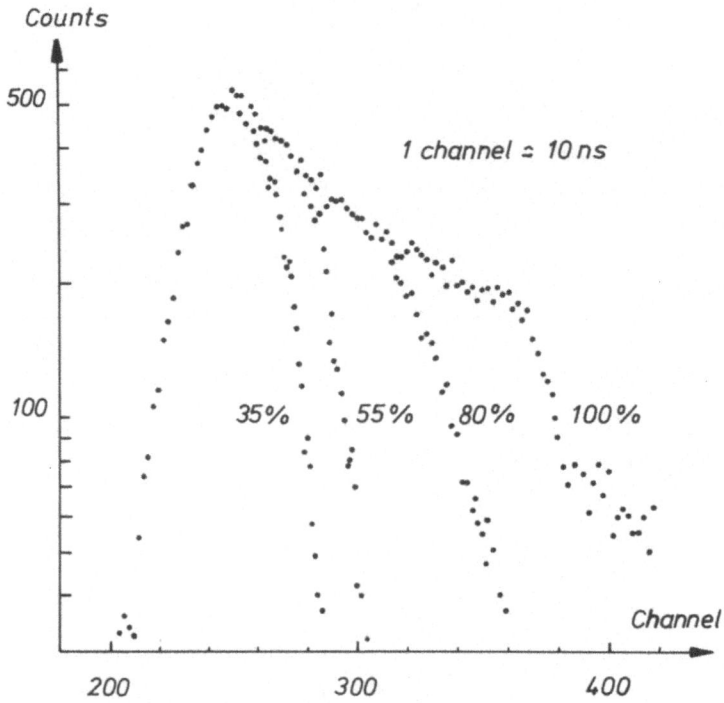

Figure 6.   Electron drift time spectrum with various grades of risetime
            discrimination applied on it. Percentages indicate trans-
            mitted pulses when using discrimination.

Am-241 was used to excitate Ce Kα (34.7 keV), which on its behalf
excitated Xe Kα (29.8 keV) inside the proportional counter. This is a
favourable situation, see Fig. 3a, because the counter gas is rather
inefficient at its own Kα energy, but has an absorption edge just below
Ce Kα energy. So Xe Kα escapes from the detector with a great probability
and an absorption of 4.9 keV remains in the detector. An accurate time
marker can be generated by detecting Xe Kα radiation with a NaI scintillation counter. From the proportional counter the corresponding
4.9 keV signal is received delayed by the electron drift time. Both
signals are fed via single channel analyzers to a time-to-amplitude
converter. Because of the poor energy resolution of the NaI counter
the Xe Kα and the scattered Am-241 lines overlap in the scintillation
counter. This results in a low rate random coincidence which is, however,
constant over whole time spectrum.

Measured spectra are seen in Fig. 6. Maximum drift time is about
1.8 μs which agrees with Fig. 3c. Changing risetime discrimination level
has a pronounced effect on the slow end of the spectrum without having
any effect on the short times. Thus the changes in the drift time
spectrum can be explained as the formation of a dead zone of cylindrical
shape inside the detector.

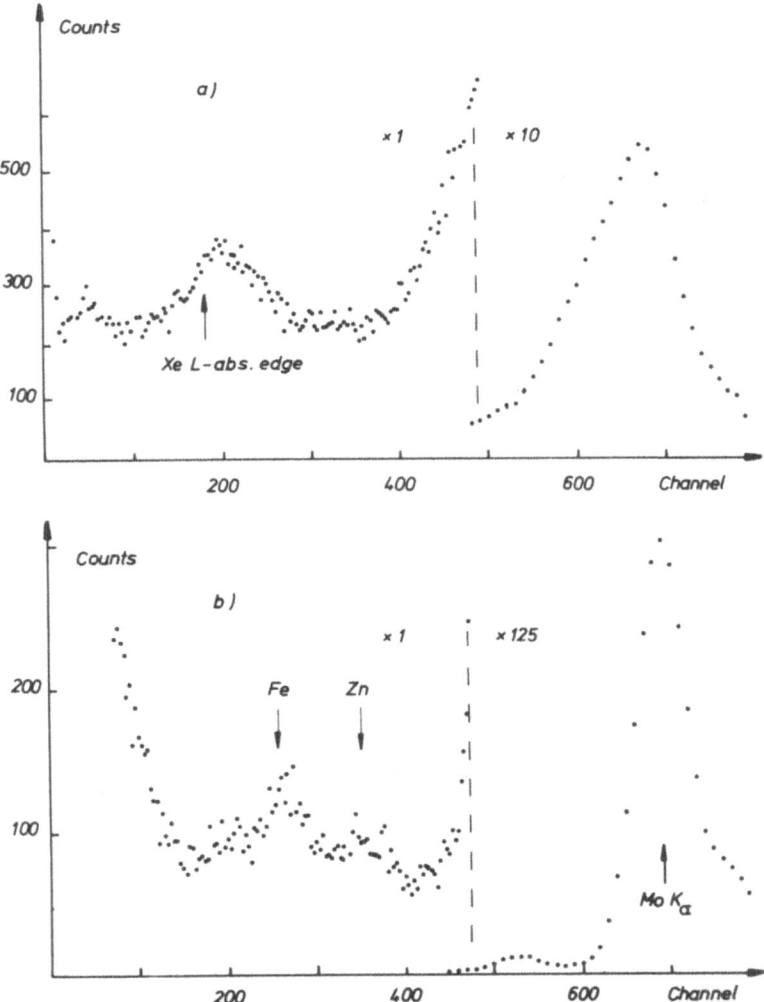

Figure 7.   Effect of the risetime discrimination on background:
            a)   original spectrum, measuring time 1 h
            b)   discrimination adjusted to 50 percent at Fe-55 5.9 keV
                 radiation, measuring time 18 h.

        The real background removing properties of the discriminator were
studied using the circuit of Fig. 4. The background inducing energy
was selected to be Mo Kα (17.7 keV) which was excited by Cd-109. This
arrangement was used because it removed a great part of the parasitic
88 keV line of Cd-109. The energy spectrum in Fig. 7a corresponds to
Fig. 1 with differences in detector gas and source. The L -absorption
edge of Xe is seen at 4.8 keV. In the discriminated spectrum of Fig. 7b
there are slight peaks of iron and zinc. These are supposed to be
originated from detector materials. On the low-energy side the Al Kα
line is visible.

In the experimental detector the charges generated near the window
are not collected completely. From the discriminated spectrum one can
see an impressive recovery in the resolution as incomplete charge collec-
tions have been eliminated due to their risetime.

It has to be noticed that among the rest of the background there
exists a short-traced background caused by other effects. So the improvement due to the risetime discriminator cannot be exactly defined.
Yet one can observe from Fig. 7 the improvement of the signal to background ratio to be about twentyfold.

CONCLUSIONS

As results indicate, the risetime discriminating system as described
can eliminate the low-energy background caused by the wall effect. In
order to reduce the remaining background,attention must first be paid
to the materials used in the detector. A considerable part of the remainder is short-trace Compton background originating from the excitation source.

Especially it was discovered that removing the dual tracks notably
helped to reduce the background. This method may be a useful feature
also in risetime circuits applied to the elimination of the gamma background.

The method presented is suitable to supplement the X-ray fluorescence analysis in measuring low concentrations. The effectiveness of
the detector has to be tuned so that the X-rays of interest do not absorb in the vicinity of the entrance window. A straightforward application of the system presented may be the low-background work with
gaseous sources.

ACKNOWLEDGEMENTS

The authors are grateful to Dr. P. Rautala for support and encouragement to investigate the properties of proportional counters.

REFERENCES

1. G. E. Kocharov and V. V. Petrov, "Wall Effect in Cylindrical Detec-
    tors," Kosm. Luchi, No. 10, 167-176 (1969), (in Russian).

2. R. W. Fink, Chap. 5 in B. Crasemann, Editor, Atomic Inner-Shell
    Processes, Vol. II, Academic Press (1975).

3. E. Mathieson and T. J. Harris, "Pulse Shape Discrimination in
    Proportional Counters - Theory of Electronic System," Nucl. Inst.
    Meth. 88, 181-192 (1970).

4. T. J. Harris and E. Mathieson, "Pulse Shape Discrimination in Pro-
    portional Counters - Experimental Results with an Optimised
    Electronic System," Nucl. Inst. Meth. 96, 397-403 (1971).

5. D. B. Brown, "Electron Range and Electron Stopping Power," in
    J. W. Robinson, Editor, Handbook of Spectroscopy, Vol. I, p. 249,
    CRC Press (1974).

6. W. N. English and G. C. Hanna, "Grid Ionization Chamber Measure-
    ments of Electron Drift velocities in Gas Mixtures," Canadian J.
    Phys. 31, 768-797 (1953).

7. L. L. Lewyn, "Pulse Shape Discrimination for X-Ray Proportional
    Counter Background Reduction," Nucl. Inst. Meth. 82, 138-140 (1970).

8. P. Gorenstein and S. Mickiewicz, "Reduction of Cosmic Background
    in X-Ray Proportional Counter through Risetime Discrimination,"
    Rev. Sci. Instr. 39, 816-820 (1968).

9. Y. Isozumi and S. Isozumi, "A Pulse Shape Discrimination for an
    X-Ray Proportional Counter and its Application to a Coincidence
    Experiment," Nucl. Inst. Meth. 96, 317-323 (1971).

10. K. Doi, "A Simple Risetime Discriminator for X-Ray Proportional
    Counter Background Reduction," Nucl. Inst. Meth. 125, 183-187
    (1975).

X-RAY INTENSITIES FROM COPPER-TARGET DIFFRACTION TUBES

M. A. Short

Ford Motor Company, Engineering & Research Staff

Dearborn, Michigan 48121

ABSTRACT

The relative intensities of the K_α characteristic radiation obtained from copper-target X-ray diffraction tubes have been calculated for a range of tube accelerating voltages and take-off angles. The calculations employ an over-voltage function, and absorption and atomic number corrections similar to those used in electron microprobe analysis. They apply only to constant potential X-ray generators. Measurements of actual intensities obtained on a Picker diffractometer using a sodium chloride monochromator gave relative intensities in close agreement with those calculated. The calculations and measurements show that there is an optimum tube voltage, with respect to intensity, for each take-off angle. This voltage increases with increasing take-off angle. The application of these results to the consideration of the relative intensities obtainable from broad, standard and fine focus copper-target X-ray diffraction tubes is discussed. It is concluded that, for many purposes, a fine focus tube set at a relatively high take-off angle, at least six degrees, gives both high intensity and good resolution.

INTRODUCTION

X-ray diffractometers are manufactured by several different companies, and for most of these it is possible to purchase fine focus, standard focus, and broad focus X-ray tubes. Two factors are important in comparing the merits of different diffractometers and X-ray tubes: the useful intensity of the characteristic radiation and the resolution produced under comparable conditions. It is usually impracticable for the crystallographer or analyst to make an experimental comparison of intensities and resolutions available from different combinations of diffractometers and X-ray tubes. To compensate for this, calculations are presented here of the useful intensity of characteristic radiation and the apparent size of the X-ray diffraction tube focus for fine, standard and broad focus copper-target X-ray tubes, when used in a variety of comparable experimental conditions. The principal tube parameters considered are: tube power, tube accelerating voltage, take-off angle and tube type. The validity of the calculations is

supported by experimental measurements. Two questions are answered by
these calculations. Firstly, for comparable resolution, is the higher
intensity obtained from a "high" power, broad focus tube at a low take-
off angle, or from a "low" power, fine focus tube at a high take-off
angle? Secondly, what are the optimum accelerating (tube) voltages
for copper-target X-ray tubes at take-off angles from 1° to 10°?

CALCULATION OF X-RAY INTENSITIES

When a beam of electrons is normally incident on a solid target,
only some of these will be absorbed within the target, the remainder
will be backscattered. R, the effective electron current factor,
denotes the fraction of the total possible ionizations (if all the
electrons remained within the target) that actually occurs. The
absorbed electrons will penetrate to a spectrum of depths; the mean
penetration depth of the electrons is dependent on E_o, their initial
energy, and S, the 'mass stopping power' of the target. The intensity
of the characteristic X-rays produced in the target depends upon the
initial energy of the electrons and the excitation potential of the
appropriate electron energy level. These characteristic X-rays, and
also the characteristic radiation produced by Bremsstrahlung, are
absorbed to some extent as they pass through the target to the surface.
The transmission factor for the characteristic X-rays through the target
is denoted by $f(x)$. This factor is dependent on the mean penetration
depth of the electron beam, the take-off angle, a, of the X-rays
utilized, and the mass absorption coefficient, μ/ρ, of the characteristic radiation in the target.

A considerable amount of work has been done in recent years,
summarized by Reed (1), on the calculation of characteristic X-ray
intensities resulting from electron bombardment for use in quantitative
electron microprobe analysis. The generation of characteristic X-rays
in an X-ray tube is similar in principle to the generation of X-rays
in an electron microprobe, and consequently it is possible to use the
same methods of calculating X-ray intensities from diffraction tubes
that are used for the electron microprobe.

Green and Cosslett (2) have derived an expression for the total
number of K quanta, N_K, produced per incident electron of initial
energy E_o for thick targets of pure elements. They show that

$$N_K = \omega_K(n_K(\text{direct}) + n_K(\text{indirect}))$$

where ω_K is the fluorescence yield of the K shell, $n_K(\text{direct})$ is the
number of direct ionizations of the atoms of the target material due
to the incident electron, and $n_K(\text{indirect})$ is the number of ionizations
caused by the continuum arising from the incident electron. The number
of direct ionizations is given by

$$n_K(\text{direct}) = 9.535 \times 10^4 (U_o \ln U_o - (U_o - 1))R/(Ac)$$

where U_0 is the overvoltage ($= E_0/E_K$), E_K being the excitation potential of the K shell, R is the effective electron current factor, A is the atomic weight (63.54 for copper), and c is the constant in the Thomson-Whiddington energy loss equation (3).

For the calculations described here, the values of R for copper were interpolated from Duncumb and Reed's Table (4) which was compiled from Bishop's values (5).

The Thomson-Whiddington constant c is related (2) to E, the instantaneous electron energy, and S, the mass stopping power of the target

$$c = 2ES.$$

Following common practice (1), E was taken as the mean of the initial electron energy, E_0, and the K shell ionization energy, E_K,

$$E = (E_0 + E_K)/2,$$

all electron energies being expressed in keV.

The expression obtained by Bethe (6) was used for S:

$$S = 7.85 \times 10^4 (Z \ln(kE/J))/(EA)$$

where Z is the atomic number, k is the constant 1.166 (7), and J is the "mean excitation energy" of the atom. Values of J to be used in calculating S have been discussed at length by Reed (1). The values used here are derived from Bloch's law (8) which states that $J/Z = $ a constant; Wilson's value (9) of 0.0115 keV was used for the constant. Before substituting in Bethe's expression, the values of J obtained from Bloch's law were modified for shell effects as suggested by Livingston and Bethe (10).

Combining these various equations, it can be shown that the number of direct ionizations per incident electron is given by:

$$n_K(\text{direct}) = 0.6073R(U_0 \ln U_0 - U_0 + 1)/(Z \ln(0.583(E_0 + E_K)/J))$$

where $R = 0.7949 + 1.742/E_0$ in the range 20 to 60 keV, $U_0 = E_0/E_K$ E_K being taken as 8.98 keV, Z = 29, and J was taken as 0.27 keV.

Green and Cosslett (2) show also that the number of indirect ionizations per incident electron is given by:

$$n_K(\text{indirect}) = 1.38 \times 10^{-6} Z E_K((r - 1)/r)(U_0 \ln U_0 - U_0 + 1)$$

where r is the K edge absorption jump ratio and, for copper, was taken to be 7.692.

The fraction of characteristic X-rays generated within the target that is transmitted to the surface, $f(x)$, has been given by Philibert (11) as:

$$f(x) = (1 + h)/((1+x/\sigma)(1 + h(1+x/\sigma))),$$

where $h = 1.2A/Z^2$, $x = (\mu/a)\csc a$ and

$$\sigma = 4.25 \times 10^5 /(E_o^{1.67} - E_K^{1.67}).$$

This expression for σ includes the modifications suggested by Duncumb and Shields (12, 13) and by Heinrich (14).

The product $\omega_K(n_K(\text{direct}) + n_K(\text{indirect})) f(x)$ will give relative intensities of characteristic radiation per incident electron (and per unit of electron beam current) of incident energy, that is tube voltage, E_o and target take-off angle a. For constant power input, it is necessary to divide by E_o giving a relative intensity of characteristic radiation per watt (or other unit of power) of

$$\omega_K(n_K(\text{direct}) + n_K(\text{indirect})) f(x) /E_o.$$

These relative intensities have been calculated for copper $K\alpha$ radiation from 20 keV to 60 keV for take-off angles, a, of 1°, 2°, 3°, 4°, 6°, 8°, and 10°. The results are shown as continuous lines in Figure 1. It can be seen that there is an optimum tube voltage with respect to intensity for each take-off angle. These optima (with interpolations) are given in Table 1. To obtain the maximum $CuK\alpha$ intensity, the tube must be run at the optimum kV for each take-off angle.

Table 1: Optimum kV for 1° to 10° take-off angles

Take-off angle:	1°	2°	3°	4°	5°	6°	7°	8°	9°	10°
Optimum kV:	26	32	36	39	42	45	48	50	52	54

EXPERIMENTAL VERIFICATION OF CALCULATIONS

Intensities were measured on a Picker X-ray diffractometer using a new fine focus copper-target diffraction tube, a sodium chloride single crystal specimen, and a scintillation detector with pulse height analysis to remove harmonics. The intensities were measured at 20 kV, 30 kV, 40 kV and 50 kV for take-off angles of 2°, 4°, 6°, 8°, and 10°. The accuracy of the take-off angle gauge was checked before the experiments were started; a correction of $- 1.0^o$ to the dial reading was found to be necessary. To avoid any possible effects due to heating of the target, all measurements were made at low power settings. Integrated intensities of the $CuK\alpha$ radiation were obtained; the mA setting of the tube was not altered during the experiment. Repeated intensity measurements were made at each setting of kV and take-off angle, care being taken to approach the take-off angle desired to avoid backlash and to approach the kV setting from the same direction to avoid any missetting of the tube voltage.

Three intensity measurements were made at each kV/take-off angle setting, normalized to unit power input, and averaged. Each average intensity was then ratioed to the corresponding calculated intensity.

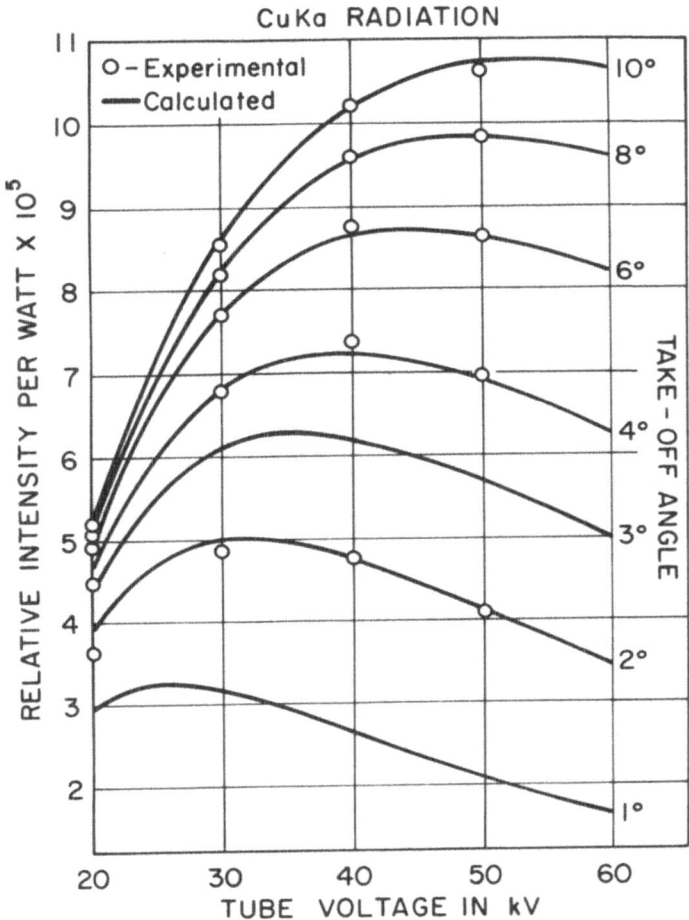

Figure 1: Experimental and calculated relative intensities of CuKα
           radiation as functions of tube voltage and take-off angle.

From these an average ratio was obtained. The experimental intensities were then divided by this average ratio so that they were placed on the same relative scale as the calculated intensities. These relative experimental intensities are shown as open circles in Figure 1. It can be seen that there is good agreement between the calculated and the experimental results. There is, however, substantial disagreement between these calculated and experimental results and the experimental results published by Parrish (15) in 1968. Parrish finds different voltage optima and different relative intensities between different take-off angles at the same tube voltage. His measurements are not supported theoretically.

APPLICATION TO X-RAY DIFFRACTOMETRY

With a knowledge of the relative intensities of the characteristic radiation as a function of tube voltage and take-off angle, a comparison can be made of the relative intensities obtainable from different X-ray diffraction tubes in X-ray diffractometers. The useful relative intensity of characteristic radiation that may be obtained is given by the product of the following four factors:

1. the relative intensity per watt as calculated and shown in Figure 1,
2. the watts available from the tube, which is a function of the operating voltage,
3. the transmission factor of the tube window,
4. a line height factor which depends on the length of the line focus and whether the diffractometer is designed to use some or all of this length.

Several assumptions are made in applying the results of the calculations to practical diffractometry:

a) that the tube is powered by a ripple-free constant potential generator,
b) that the focus is uniformly illuminated,
c) that the size of the focus is as specified by the manufacturer,
d) that the target surface is clean, smooth, and flat.

It is obvious that some of these assumptions will not be completely attained. In addition, it is assumed that the diffractometer will include a crystal monochromator because the optimization of the tube accelerating voltage is considered only from the viewpoint of maximizing the intensity of the characteristic radiation and does not take into account the intensity of the white radiation which is very much voltage dependent and might greatly influence the selection of the tube voltage in the absence of a monochromator.

Using the data shown in Figure 1, the useful relative intensities of $CuK\alpha$ radiation have been calculated from the four factors described above for Philips broad focus, standard focus, fine focus, and long fine focus X-ray tubes (note that the long fine focus tube requires a 14v filament power supply). The relative intensities, shown in Table 2, were calculated for three different take-off angles,

Table 2: Useful relative intensities from copper-target X-ray diffraction tubes available from Philips Electronic Instruments.

Tube specifications	14600340 broad 2.0 x 12 2000, 40/60 kV 0.94 (Be)			14500300 standard 1.0 x 10 1400, 42/60 kV 0.94 (Be)			14500320 fine 0.4 x 8 1200, 30/60 kV 0.94 (Be)			14600350 long fine 0.4 x 12 1800, 42/60 kV 0.94 (Be)		
For selected take-off angles												
Take-off angle	3°	6°	10°	3°	6°	10°	3°	6°	10°	3°	6°	10°
Apparent width of focus, mm	0.10	0.21	0.35	0.05	0.10	0.17	0.02	0.04	0.07	0.02	0.04	0.07
Optimum tube voltage, kV	40	45	54	42	45	54	36	45	54	42	45	54
CuKα useful relative intensity	0.12	0.16	0.20	0.08	0.11	0.14	0.07	0.10	0.12	0.10	0.15	0.18
For selected apparent focus width												
Apparent focus width, mm	0.03	0.05	0.07	0.03	0.05	0.07	0.03	0.05	0.07	0.03	0.05	0.07
Take-off angle	0.9°	1.4°	2.0°	1.7°	2.9°	4.0°	4.3°	7.2°	10.1°	4.3°	7.2°	10.1°
Optimum tube voltage, kV	40	40	40	42	42	42	40	48	54	42	48	54
CuKα useful relative intensity	0.047	0.068	0.089	0.055	0.079	0.095	0.087	0.10	0.12	0.13	0.16	0.18

3°, 6° and 10°, and for three different apparent focus widths, 0.03, 0.05 and 0.07 mm. The apparent focus width, w, is related to the actual focus width, f, and the target take-off angle, a, by w = f sin$a$. The optimum tube voltage is taken from Figure 1 for the appropriate take-off angle, interpolating when necessary. Where this voltage is lower than that at which the tube can be run at maximum power, then the optimum voltage listed is the lowest voltage for maximum power.

As might be expected, Table 2 indicates that at any given take-off angle the maximum intensities available from the different tubes are related directly to the maximum power available from the tubes. A more useful comparison, however, is made when the take-off angles are adjusted to give the same apparent focus width, which will correspond to the same degree of resolution. The results in this case indicate that, provided the tubes are run at their optimum voltages, the highest intensity is given by the long fine focus tube, the second highest by the regular fine focus tube, the third by the standard focus tube, and the lowest by the broad focus tube - the tube which has the highest nominal power output! More specific comparisons can be made using Table 2. For example, if a standard focus tube set at 42 kV and a 3° take-off angle is used as a baseline for both intensity and resolution, 25% more intensity and the same resolution can be obtained using a regular fine focus tube at 48 kV and a 7° take-off angle. Double the baseline intensity at the same resolution can be obtained if a long fine focus tube is used.

Similar calculations can be made on X-ray diffraction tubes available from other equipment suppliers.

CONCLUSIONS

It has been shown that, for maximum intensity of characteristic radiation, there is an optimum tube voltage which is dependent on take-off angle and the power curve of the tube. It has also been shown that, for some desired resolution on an X-ray powder diffracto-meter which will require the take-off angle to be adjusted appropriately for the tube, the maximum intensity of CuKa radiation will be obtained from a lower power fine focus tube rather than from a higher power broad focus tube.

If resolution is of no concern, an unlikely case in practice, then the maximum intensity is, of course, given by the highest power tube at a high take-off angle.

ACKNOWLEDGEMENTS

I would like to thank Mr. S. Bonfiglio for his help during the course of this work.

REFERENCES

1. S. J. B. Reed, Electron Microprobe Analysis, Cambridge
    University Press (1975).

2. M. Green and V. E. Cosslett, "The Efficiency of Production of
    Characteristic X-Radiation in Thick Targets of a Pure Element",
    Proc. Phys. Soc. 78, 1206-1214 (1961).

3. R. Whiddington, "Transmission of Kathode Rays through Matter",
    Proc. Roy. Soc. A 86, 360-370 (1912).

4. P. Duncumb and S. J. B. Reed, "The Calculation of Stopping Power
    and Backscatter Effects in Electron Probe Microanalysis", in
    K. F. J. Heinrich, Editor, Quantitative Electron Probe Micro-
    analysis, pp. 133-154, U. S. Department of Commerce, National
    Bureau of Standards, Special Publication 298 (1968).

5. H. E. Bishop, in R. Castaing, P. Deschamps, & J. Philibert,
    Editors, Optique des Rayons X et Microanalyse, p. 153, Paris:
    Hermann (1966); quoted in S. J. B. Reed, Electron Microprobe
    Analysis, Cambridge University Press (1975).

6. H. A. Bethe, "Zur Theorie des Durchgangs schneller Korpuskular-
    strahlen durch Materie", Ann. Phys. Leipz. 5, 325-400 (1930).

7. H. A. Bethe and J. Ashkin, Experimental Nuclear Physics, New
    York: Wiley (1953); quoted in S. J. B. Reed, Electron Micro-
    probe Analysis, Cambridge University Press (1975).

8. F. Bloch, Zeit. f Phys. 81, 363-373 (1933); quoted in
    S. J. B. Reed, Electron Microprobe Analysis, Cambridge University
    Press (1975).

9. R. R. Wilson, "Range and Ionisation Measurements on High Speed
    Protons", Phys. Rev. 60, 749-753 (1941).

10. M. S. Livingston and H. A. Bethe, "Nuclear Physics Part III"
    Rev. Mod. Phys. 9, 245-390 (1937).

11. J. Philibert, "A Method for Calculating the Absorption Correction
    in Electron-Probe Microanalysis", in H. H. Pattee, V. E. Cosslett,
    and A. Engstrom, Editors, X-Ray Optics & X-Ray Microanalysis,
    pp. 379-392, New York: Academic Press (1963).

12. P. Duncumb and P. K. Shields, "Effect of Critical Excitation
    Potential on the Absorption Correction", in T. D. McKinley,
    K. F. J. Heinrich and D. B. Wittry, Editors, The Electron
    Microprobe, pp. 284-295, New York: John Wiley (1966).

13. P. Duncumb and P. K. Shields, "Effect of Critical Excitation
    Potential on the Absorption Correction in X-Ray Microanalysis",
    Tube Investments Research Laboratories, Cambridge, Technical
    Report No. 181 (1964).

14. K. F. J. Heinrich, "Common Sources of Error in Electron Probe
     Microanalysis", in J. B. Newkirk, G. R. Mallett, and
     H. G. Pfeiffer, Editors, _Advances in X-Ray Analysis_, vol. 11,
     pp. 40-55, New York: Plenum Press (1968).

15. W. Parrish, "X-Ray Diffractometry Methods for Complex Powder
     Patterns", in H. van Olphen and W. Parrish, Editors, _X-Ray and
     Electron Methods of Analysis_, pp. 1-35, New York: Plenum Press
     (1968).

POLARIZED RADIATION PRODUCED BY SCATTER FOR

ENERGY DISPERSIVE X-RAY FLUORESCENCE TRACE ANALYSIS*

Richard W. Ryon

University of California

Lawrence Livermore Laboratory

Livermore, California 94550

ABSTRACT

Polarized x-radiation produced by scatter at 90° from boron carbide is shown to be a superior excitation source for the measurement of trace elements by energy dispersive x-ray fluorescence. In a close-coupled system which uses a high power x-ray tube as the primary source, the losses due to the geometric requirements and scattering efficiency of the polarizer can be more than compensated. With the system described here which uses a Mo anode x-ray tube, detection limits for the element between K and Sr in NBS orchard leaves are approximately 2 to 4.5 times lower using polarized excitation in comparison to direct excitation, and about <1 to 3 times lower in comparison to secondary excitation.

INTRODUCTION

Energy dispersive x-ray fluorescence is being used to meet the increasing demands upon the analytical chemist to measure smaller and smaller amounts of more and more elements for a host of applications. However, the source radiation used to excite fluorescence is scattered by the specimen into the detector. This scattered radiation is a limiting factor because of the following considerations:

a) <u>Background</u>. This is a signal-to-noise problem because of the random nature of the background. Even in the case of a monochromatic source, there is a significant background at all lower energies due to incomplete charge collection and Compton scatter in the detector.

b) <u>Count rate limitation</u>. In trace analysis, a very small fraction of the detected radiation is the fluorescent radiation of interest, the remainder being the scattered source radiation.

---

*This work performed under the auspices of the U. S. Energy Research & Development Administration, under contract No. W-7405-Eng-48.

The electronic system requires a finite time to process each
photon induced pulse, thus limiting the total count rate.
Once the source intensity is increased to give the maximum
count rate (or scatter), any further increase in the source
intensity only reduces the number of pulses accepted per unit
time. The maximum count rate is easily reached with conventional x-ray tubes.

Because of these limitations, the detection limits for most commonly
measured elements are on the order of a few parts per million in bulk
specimens.

One means to overcome the scatter problems is to use polarized
x-rays to excite the fluorescence. Polarized radiation is not scattered
isotropically. If the detector is placed in the plane of polarization
and at 90° to the beam incident on the specimen, very little scatter
will be detected (1). Fluorescent radiation is enhanced in proportion
to scattered source radiation and the signal-to-noise ratios are increased. Higher source intensities can also be used to increase the
fluorescent count rate because the total count rate is no longer over-
whelmingly determined by scatter.

There are several methods of obtaining polarized x-rays (2), including synchrotron radiation. Two reports were presented at this conference in 1974 (3,4) which clearly demonstrated the background reduction obtained when a polarized source was used to excite fluorescence.
Generally, polarization and intensity are inversely related. Therefore,
there is a trade-off between signal-to-noise improvement and the total
fluorescent counts obtained per unit time. Furthermore, multiple scatter within the polarizer and specimen sets an ultimate limit to the
background reduction obtainable when a highly polarized source is used.
Previous studies (3,4,5) have not clearly demonstrated whether or not
the intensity and multiple scatter limitations are too severe to allow
practical applications of polarized x-rays in energy dispersive x-ray
analysis. The goal of this work was to answer this question.

THEORETICAL AND GEOMETRICAL CONSIDERATIONS

Figure 1 is a sketch of the system geometry. A collimated beam
from the source defines the z-axis of a three-dimensional rectangular
coordinate system. The radiation scattered by the polarizer is described by classical theory as follows:

$$I_y \equiv \text{Intensity (y-axis)} \propto 1/2 \; I_0 \; \cos^2\theta$$

$$I_x \equiv \text{Intensity (x-axis)} \propto 1/2 \; I_0 \qquad\qquad (1)$$

where I_0 is the source intensity and θ is the angle between the incident
beam and the point of observation (along the y-axis in this case). Thus,
scattering at 90° produces a beam polarized normal to the yz scatter
plane. If the detector is placed perpendicular to the first (yz) scatter plane the $\cos^2$ function applies to the remaining component of the
source radiation and ideally no scattered radiation will be detected.
However, since each collimator has a finite solid angle, a range of

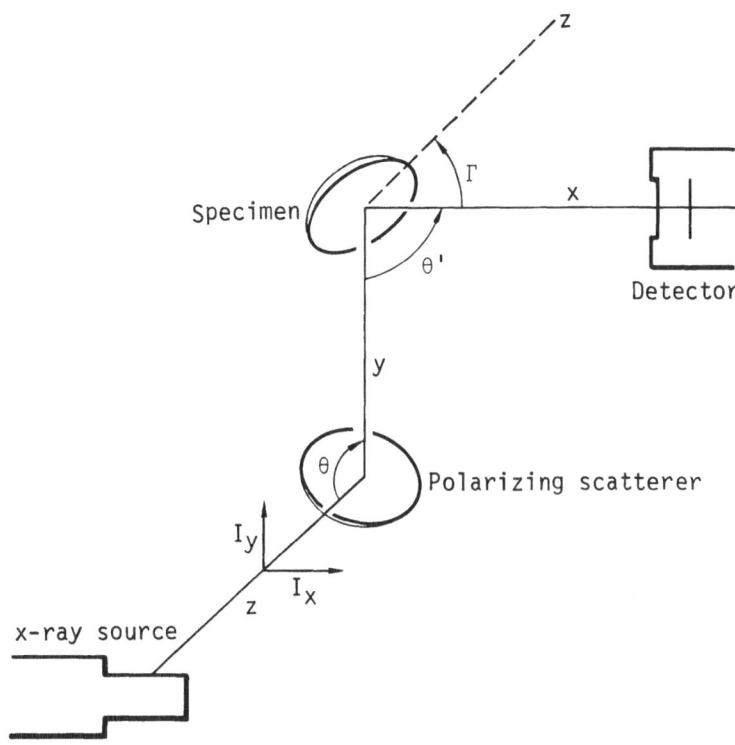

Figure 1. Sketch of system geometry. After scattering twice, the scatter intensity measured along the x-axis is much less than that measured along the z-axis.

angles about 90° is accepted by each, polarization is not complete, and some scatter will in fact be detected.

Polarization is defined by the geometry as

$$P_g = \frac{I_x - I_y}{I_x + I_y} = \frac{1 - \cos^2\theta}{1 + \cos^2\theta}.$$

(2)

If the angle θ can have the range of values $\pi/2 \pm \omega$, the polarization becomes

$$P_g \approx \frac{1 - 1/3\omega^2}{1 + 1/3\omega^2} \approx 1 - 2/3\omega^2$$

(3)

for small values of ω. There are three system scattering angles to be considered. They are:

$\theta \equiv$ first scattering angle

$\theta' \equiv$ second scattering angle

$\Gamma \equiv$ angle between the two scattering planes.

The angular divergences may be defined as

ω, ω', $\gamma \equiv 1/2$ the range of the scattering system planar angles about $\pi/2$.

When all three angular divergences are considered, they are summed in quadrature, and the geometrically defined polarization becomes

$$P_g = 1 - 2/3(\omega^2 + \omega'^2 + \gamma^2). \tag{4}$$

Howell and Pickles (3) have defined the second term on the right as the scatter fraction. The scatter fraction is related to measureable quantities as follows:

$R_g \equiv$ scatter fraction determined by geometry

$$R_g = 2/3(\omega^2 + \omega'^2 + \gamma^2)$$

$R \equiv$ experimentally measured scatter fraction

$$R = \frac{\left(I_{scattered}/I_{fluorescent}\right) \text{Polarized}}{\left(I_{scattered}/I_{fluorescent}\right) \text{Unpolarized}} \tag{5}$$

The angles ω, ω', and γ in the system used here are generally smaller than the angles defined by the collimators, particularly for geometries with large angular divergences. The effective angles are best found from detailed drawings of the system used.

In fluorescence experiments, the intensity of both the scattered and fluorescent radiation is proportional to the effective solid angle of each collimator. That is, the intensities are proportional to the square of the planar angular divergence of each collimator. In double scattering experiments as described, the scattered radiation has an additional factor of the interaction cross section, which to a first approximation is proportional to $\cos^2\theta$. Therefore, we may write the following expressions:

$$\left(I_{scatter}\right) \text{Polarized} \propto \omega^2\omega'^2\gamma^2\cos^2\theta \propto \bar{\omega}^8$$

$$\left(I_{fluorescence}\right) \text{Polarized} \propto \omega^2\omega'^2\gamma^2 \propto \bar{\omega}^6$$

$$\left(I_{scatter}\right) \text{Direct excitation} \propto \omega^2\omega'^2 \propto \bar{\omega}^4$$

$$\left(I_{fluorescence}\right) \text{Direct excitation} \propto \omega^2\omega'^2 \propto \bar{\omega}^4 \tag{6}$$

where $\bar{\omega}$ is some "average" angular divergence of the system angles. Because the background under the measured peaks is related to the scattered radiation, the detection limits are approximately:

$$MDL \propto \frac{(Scatter)^{1/2}}{Fluorescence}$$

$$MDL_{polarized} \propto \frac{(\frac{8}{\omega})^{1/2}}{\omega-6} \propto \omega^{-2}$$

$$MDL_{direct} \propto \frac{(\omega^4)^{1/2}}{\frac{4}{\omega}} \propto \omega^{-2} \tag{7}$$

We see that the detection limits for both direct and polarized excitation have the same functional dependence on collimation. Other geometric factors and the scattering efficiency of the polarizer make the expected detection limits for the polarized case almost 10 times worse than expected for direct excitation with the same collimation. In either case, wide open collimators are desired in order to obtain the maximum counting rate.

In practice, the count rate limitations of the energy dispersive system allow the use of less than 10^{-3} of the full power available from high intensity x-ray tubes. The ratio of detection limits for polarized excitation relative to direct excitation can be made close to unity with an increase in useable intensity of less than 100 (depending on the scatter efficiency of the polarizer) and the ratio becomes favorable to the secondary polarized source in actuality.

The intensity of the polarized beam is strongly dependent upon the energy of the scattered radiation and upon the scattering material. The scatterer must be of low average atomic number to insure that the scatter cross section is large compared to the photoelectric cross section. High density is also required; otherwise the scatter would be diffuse and fall outside the acceptance range of the collimators. Kaufman and Camp (4) experimentally determined that boron carbide is close to ideal.

A fundamental question is whether multiple scatter events restrict the achievable background reduction so much that the effort to obtain and use polarized x-rays is of little or no value. Measured polarization is always less than what is calculated from the system geometry because of multiple scattering which occurs in both the polarizer and specimen. If the first scatter occurs at some angle other than 90°, there will be a component which can scatter a second time and be detected at a right angle relative to the initial incident beam.

A relationship for the multiple scatter effect is difficult to obtain for the geometry described here. The functional relationships which would be useful in interpreting experimental results may be obtained from the expression for a spherical scatterer which was derived by Kirkpatrick (6) many years ago. The ratio of doubly scattered to singly scattered x-rays observed at 90° to the incident beam may be calculated from the expression

$$L\rho \left(\frac{1 - e^{-\mu\rho L}}{\mu\rho L}\right) \left(2\frac{I_{\shortparallel}}{I_{\perp}} + 9\right) \left(\frac{\pi}{5}\right) \left(\frac{e^2}{mc^2}\right)^2, \tag{8}$$

where

    L = radius of the scatterer

    $I_{\shortparallel}$ = intensity of the electric vector in the scatter
        plane

    $I_{\perp}$ = intensity of the electric vector perpendicular to
        the scatter plane

    ρ = density of the scatterer

    μ = mass absorption coefficient.

The other symbols have their usual physical meaning. Kirkpatrick further calculated that $I_{\perp}$ contributes 4.5 times more to double scatter
than does $I_{\shortparallel}$ so that the out-of-plane component persists even in double
scattering. In single scattering, the observed radiation is due exclusively to the out-of-plane component (Equation 1). Equation 8 shows
what is expected intuitively; that is, the thicker the polarizer and
specimen, the greater the multiple scatter. Consequently the polarization will be less. More precisely, the multiple scatter contribution
to the scatter fraction will increase nearly linearly with thickness (L)
for thin scatterers, and will approach a constant value for scatterers
of greater mass.

EXPERIMENTS AND DISCUSSION

    Two experimental arrangements were constructed for these studies
One had long intercomponent distances in order to obtain highly collimated beams while the other had closely coupled components in order to
obtain high counting rates. The large system had interchangeable
collimator tubes which could be varied in length from about 6 cm to as
long as desired. In the close coupled system, the intercomponent distances were ∿4 cm, with the collimators being about 2.5 cm in length.
In each apparatus, the collimator diameters could be varied between
zero and 1. cm. The angle between the scattering planes was variable.
With these two systems, the angular divergences about the system scattering angles (Figure 1) could be varied between approximately 0.10 and
0.27 radians so that the polarization as defined by geometry could be
varied between approximately 85% to 98%. The actual polarization in
the experiments was determined by measuring the scatter intensity in
and perpendicular to the first scatter plane. Unless otherwise stated,
the excitation source in the following experiments was an unfiltered
Molybdenum anode x-ray tube, operated at 45 KV and 40 mA.

    The magnitude of multiple scatter depolarization was experimentally
investigated in several ways. The results of changing the system collimation on the measured polarization from various specimens are shown in
Figure 2. When the geometry was fixed with a high degree of collimation
and the specimen thickness was varied, we obtained the results shown in
Figure 3. It was found from these studies that the scatter fraction is
limited to about 0.04 (96% polarization) for filter membrane specimens,
and to about 0.08 (92% polarization) for bulk specimens. If the maximum counting rate could be maintained with the collimation required to
obtain these scatter fractions, the detection limits obtained with the
polarized source relative to those obtained with direct excitation would

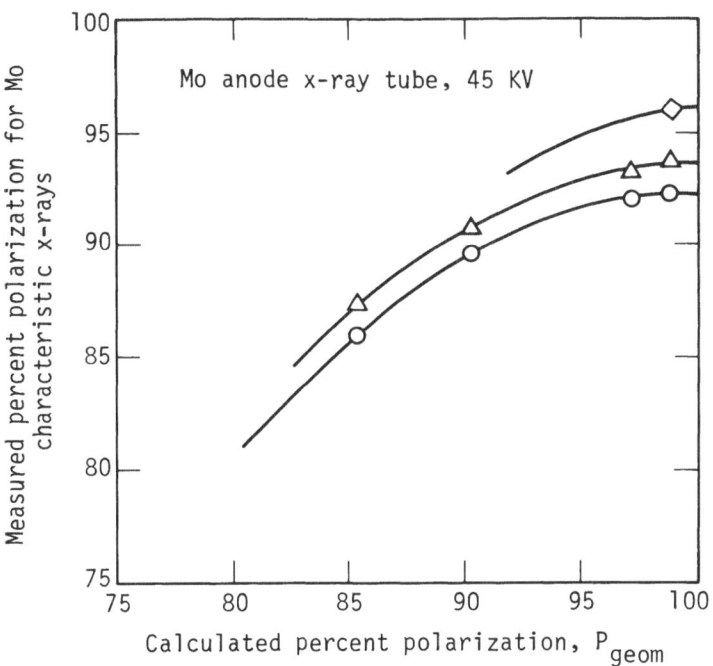

Figure 2. Measured polarization is a function of the system collimation and of the specimen thickness. Top curve: 0.0046 g/cm² membrane filter; middle curve: 0.13 g/cm² pressed orchard leaves; bottom curve: 0.63 g/cm² pressed orchard leaves.

be approximately equal to the scatter fraction. For instance, the detection limits for bulk specimens would be about 12 times lower than when using the polarized source. However, conventional x-ray tubes cannot be operated with enough power to maintain high counting rates when using the collimation required to obtain high polarization. Extrapolating from the present experiments, it is estimated that about 4500 watts would be required to maintain the counting rate at 20,000 counts/second at 90% polarization. Therefore, it is tube power rather than multiple scatter which is the limitation when a flat scatterer is used. We have already seen that, when power is the limitation, detection limits are proportional to $\bar{\omega}^{-2}$ and the best results will be obtained by opening up the collimators to bring back the maximum counting rate.

A similar study of multiple scatter depolarization was made by varying the thickness of the boron carbide polarizer (Figures 4a and 4b). The polarization again decreased as the scatterer was made thicker, in the same manner as was previously seen when the specimen thickness was increased. (Figure 4a shows some leveling off of the depolarization as predicted by Equation 8; however, this may also be due to the thickness of the polarizer approaching the diameter of the collimators.) Figure 4b shows that the emitted radiation intensities increase somewhat more slowly than the Mo scatter intensity. Lines arising from absorption edges just below the energy of the Mo scatter energy increase at nearly the same rate as the Mo characteristic scatter intensity as expected. Scatter of the lower energy continuum levels off more rapidly

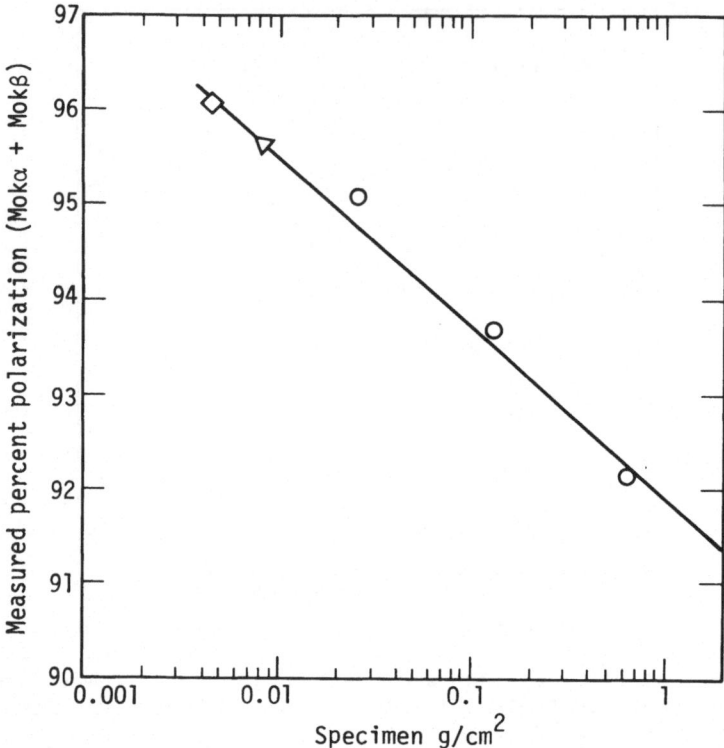

Figure 3. Measured polarization as a function of specimen mass, while holding the collimation constant ($P_g \approx 98\%$). Polarizer: 1.3 mm boron carbide. The diamond and triangle data points are for two different filter papers; the other data points are for pressed orchard leaves.

with thickness so that the intensities of the lines excited by the continuum also level off more rapidly. The net effect of the changes in background and fluorescence as measured by detection limits is shown in Table 1. The numbers indicate that the thickness of the boron carbide polarizer is of little importance, and of no significance when the polarizer is more than about 3 mm thick. As with the specimen thickness, the polarizer thickness does not place any practical limit on the achievable polarization due to multiple scatter.

Polarization was measured as a function of energy, using characteristic lines and continuum from various x-ray tubes. The results shown in Figure 5 indicate that the measured polarization from a thick specimen approaches the geometric limit at low energies. It might be anticipated from this observation that the best background reductions could be obtained for light elements. However, scattering efficiencies are small at low energies so that the intensity would be diminished in the polarized case and the scatter would be less in the unpolarized case. Further work needs to be done in order to determine the net effect on light element detectabilities.

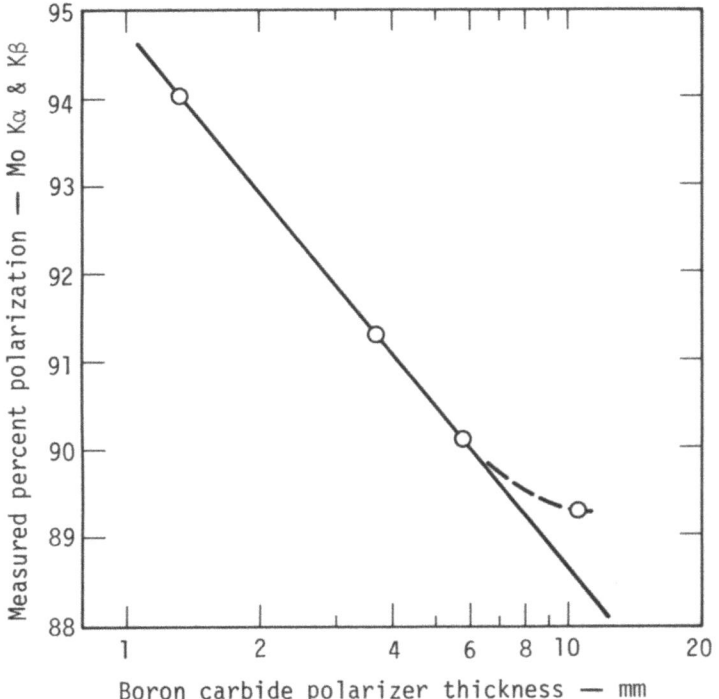

Figure 4a. The measured polarization is also influenced by the polarizer thickness. The collimator diameters were 9.5 mm, with P_g being about 97%. The specimen was boron carbide power, 0.23 g/cm^2.

The overall effect of system collimation on the relative detection limits is demonstrated in Figure 6. The relative improvement increases as the collimators are opened (less polarization) until the maximum count rate is approached. With the amplifier used in these experiments, the maximum count rate is about 20,000 counts/second. That rate is reached when the collimators are opened to give about 85% polarization.

Through the work of others and the considerations presented here, we have learned the essentials of how to optimize the secondary polarized source, given the physical and instrumental constraints over which we have little or no control. A test of the efficacy of the polarized source in comparison to other possible x-ray excitation methods is graphically demonstrated in Figure 7. The x-ray tube current was adjusted to give the same overall counting rate in each case, while holding the potential constant. The boron carbide polarizer gave lower detection limits than were obtained by other means, except for a limited region of cross-over with the nearly monochromatic yttrium secondary. The detection limits for the elements between K and Sr in a thick specimen of NBS orchard leaves were about 2 to 4.5 times lower in comparison to direct excitation, and about <1 to 3 times lower in comparison to yttrium secondary excitation. These results are further amplified in Table 2.

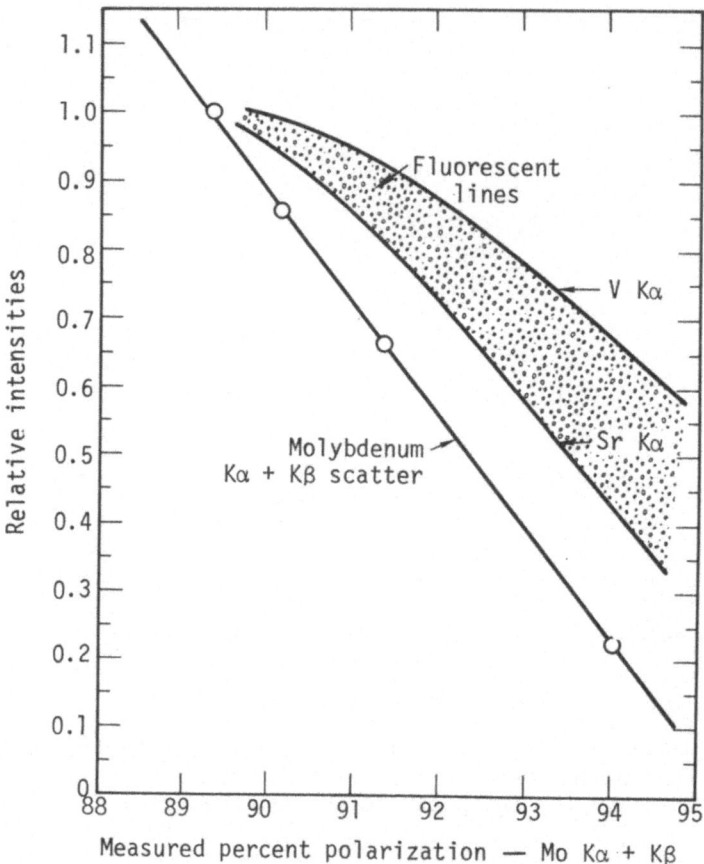

Figure 4b. These data were taken from the same experiments as those illustrated in Figure 4a, where the polarizer thickness was varied while holding the other parameters constant. The intensities are measured relative to those obtained with the 10 mm polarizer.

Table 1. Detection limits (1000 sec.) as a function of polarizer thickness.

Polarizer Thickness	Vanadium	Zinc	Strontium
1.3 mm	14 ppm	3.0 ppm	3.5 ppm
3.7 mm	15 ppm	2.5 ppm	3.0 ppm
5.7 mm	16 ppm	2.5 ppm	3.0 ppm
10.4 mm	16 ppm	2.5 ppm	3.1 ppm

(Note: These data were taken from the same experiments as illustrated in Figure 4a.)

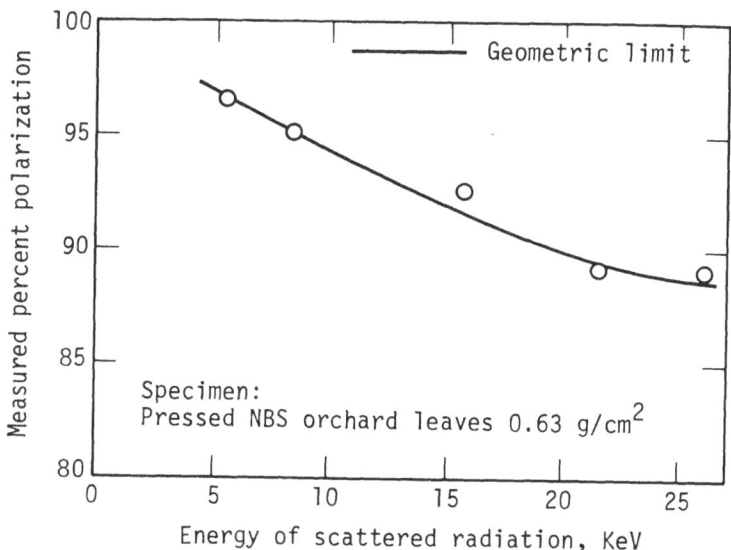

Figure 5. Measured polarization is energy dependent.

The spectra in Figure 8 show the reduced background and increased
sensitivity obtained with the partially polarized, non-monochromatic
source which make the improvement in detection limits possible.

CONCLUSION

A polarized secondary source of radiation produced by scatter at
90°C can be a superior exciter of fluorescence for energy dispersive
spectrometry. The polarizer material and system collimation must be
optimized. A high density, low atomic number scatterer such as
boron carbide is required to obtain good scatter efficiency (high intensity polarization), particularly at the intermediate energies used in
this investigation. Multiple scatter depolarization does place an ultimate limitation on expected improvements in detection limits if more
intense sources of polarized x-rays were available. However, even with
2500 watt x-ray tubes, power is the limiting factor when polarization is
produced by scattering from flat polarizers. Under this circumstance
detection limits are improved by sacrificing a high degree of polarization in favor of increased intensity by opening up the system collimators. With the system described here, significant improvement in detection limits and some extension of multielement capability were obtained
in comparison to direct, filtered beam, and secondary fluorescer
excitation.

A further advantage of a well-designed polarized system is that
exactly the same geometry is required for secondary fluorescence. If
a monochramatic (unpolarized) source is desired, all that need be done
is to exchange the polarizer-scatterer with the desired fluorescer.

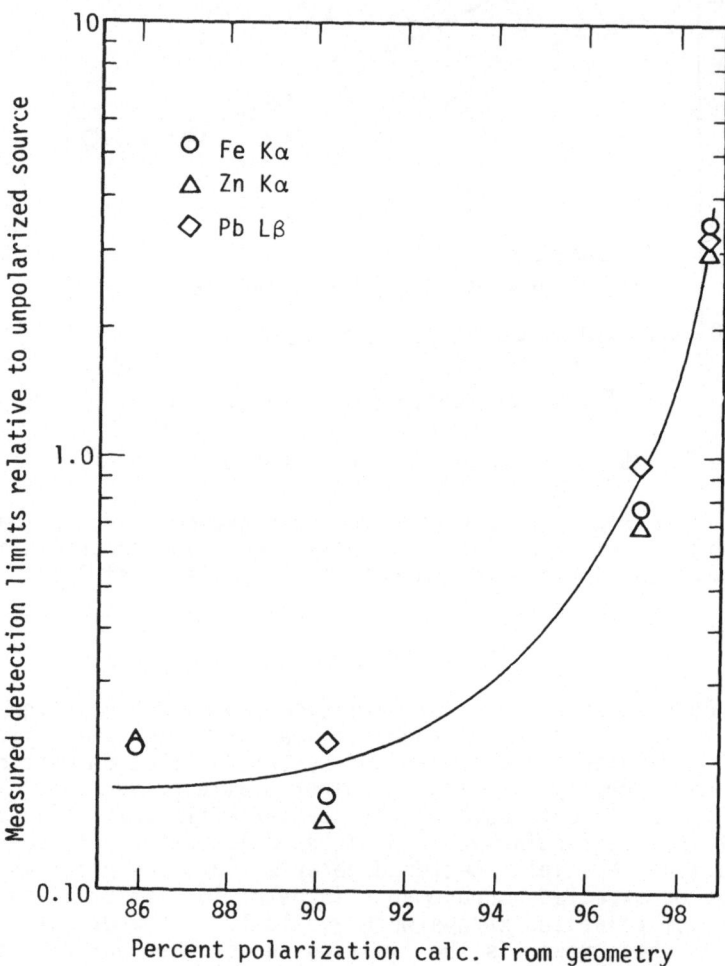

Figure 6. Measured detection limits, relative to those obtained with an
unpolarized source, as a function of the system collimation. As the
solid angles of the collimators are increased, the count rate increases
until the system maximum of about 20,000 c/s is obtained at about 85%
polarization.

Specimen: 0.63 g/cm^2 NBS orchard leaves. Polarizer: 1.3 mm boron
carbide.

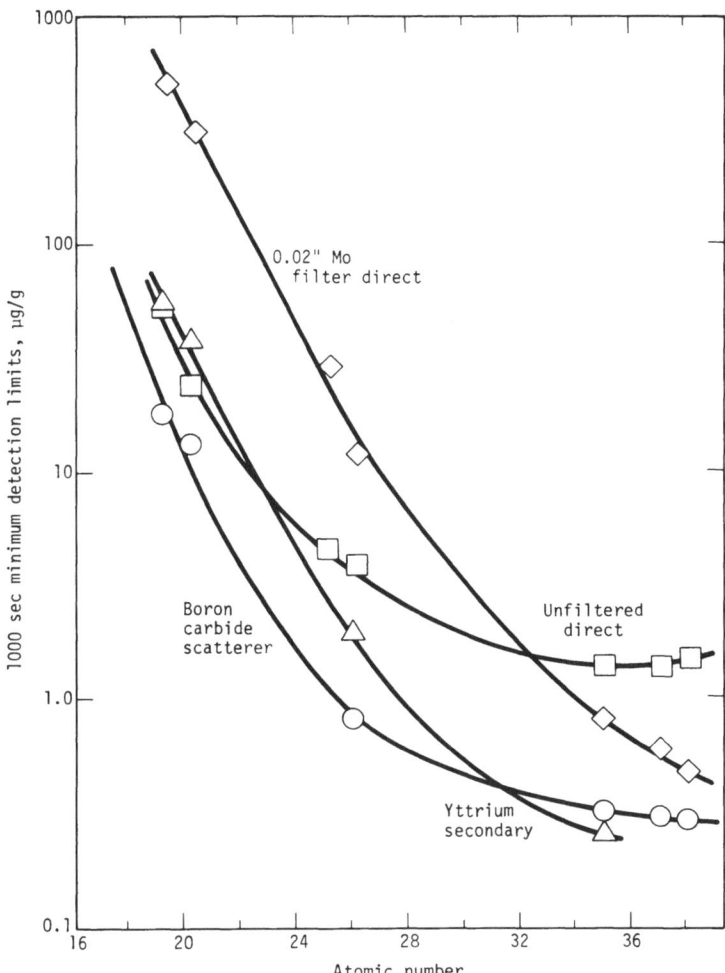

Figure 7. Comparison of Kα detection limits. Specimen: 0.63 g/cm^2 NBS orchard leaves. Source: Mo x-ray tube, 45 KV with current adjusted to give 40% instrument dead time. Boron carbide scatterer gave 85% polarization.

Table 2. Comparison of x-ray excitation methods.

Element & Line	Boron Carbide Polarizer		Yttrium Secondary		Mo Tube Direct		0.050 cm Mo Filter	
	Sensitivity cps/(µg/g)	Detection Limit* µg/g	Sensitivity cps/(µg/g)	Detection Limit* µg/g	Sensitivity cps/(µg/g)	Detection Limit* µg/g	Sensitivity cps/(µg/g)	Detection Limit* µg/g
K Kα	0.022	18	0.0040	54	0.0060	52	0.00065	500
Ca Kα	0.028	13	0.0058	37	0.012	24	0.0010	300
Fe Kα	0.37	0.84	0.087	1.9	0.11	3.9	0.017	13
Br Kα	1.5	0.33	1.1	0.26	0.61	1.4	0.21	.83
PB Lβ	0.55	0.98			0.20	4.5	0.097	1.9
Rb Kα	1.85	0.31			0.64	1.4	0.32	0.62
Sr Kα	2.1	0.30			0.62	1.5	0.37	0.58

*Detection Limit = 3(Background counts in 1000 sec.)$^{1/2}$ ÷ 1000/(cps/µg/g).

Specimen: Pressed NBS Orchard Leaves, Standard Reference Material 1571, 0.63 g/cm^2.

Source: Mo anode x-ray tube, 45 KV and with current adjusted to give 40% instrument dead time.

Polarizer: 1.3 mm boron carbide with collimation to give about 85% polarization.

Secondary: Yttrium replaced boron carbide in the same apparatus as used in the polarized mode.

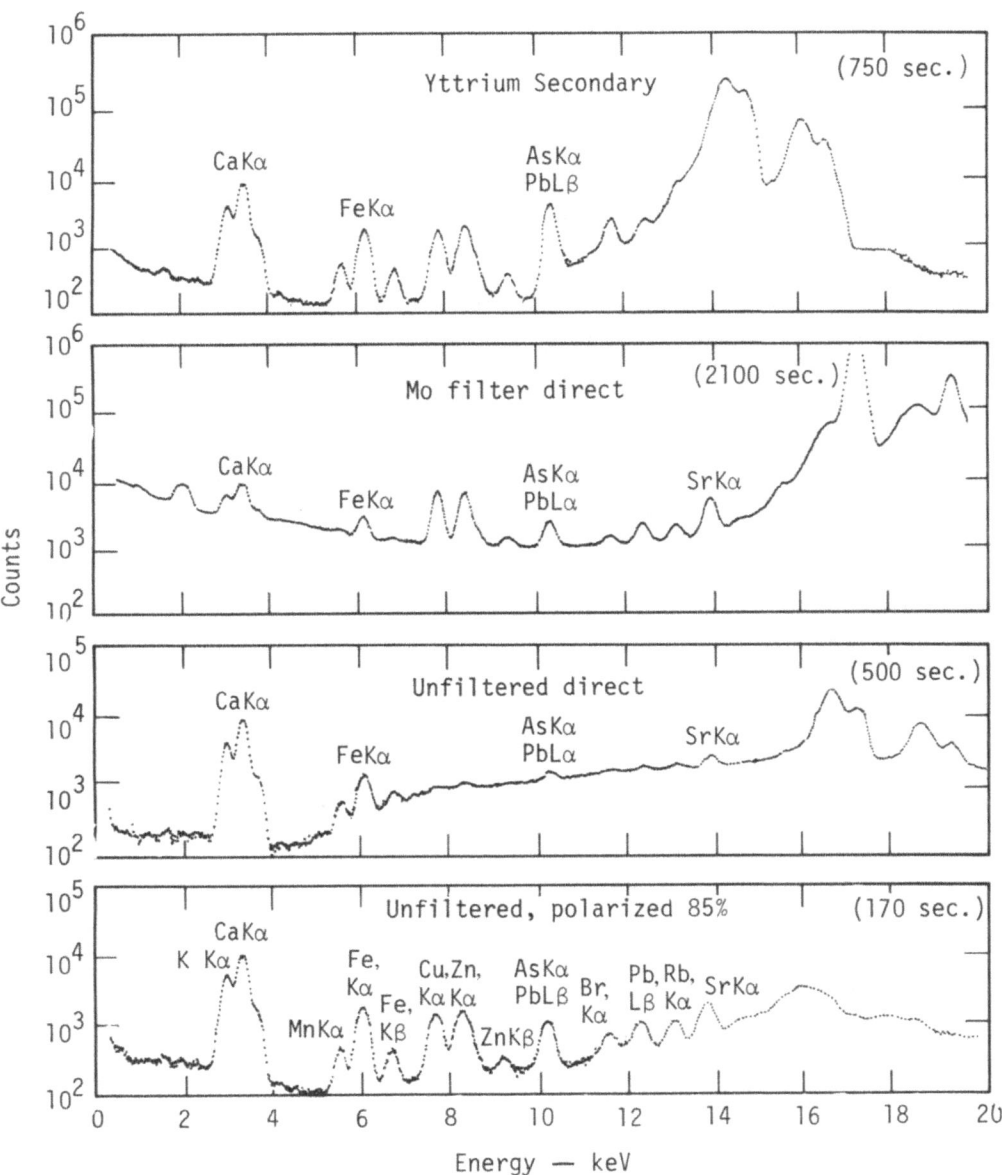

Figure 8.   Spectra from the excitation methods comparison (Figure 7 and Table 2).   Each spectrum was accumulated until Ca Kα had 10^4 counts in the peak channel.

ACKNOWLEDGEMENTS

I wish to thank David Camp, Richard Howell, and William Pickles for their help during the course of this work. They gave encouragement and insight for which I am very grateful. David Camp was also particularly helpful in the preparation of this manuscript.

REFERENCES

1. Arthur H. Compton and C. F. Hagenow, "A Measurement of the Polarization of Secondary X-Rays," Journal of the Optical Society of America and Review of Scientific Instruments, 8 (4), 487-91 (1924).

2. R. H. Howell and W. L. Pickles, "Possible sources of Polarized X-Rays for X-Ray Fluorescence Spectra with Reduced Backgrounds," Nuclear Instruments and Methods 120, 187-8 (1974) and Lawrence Livermore Laboratory Report UCRL 75651, May 8, 1974.

3. R. H. Howell, W. L. Pickles, and J. L. Cate, Jr., "X-Ray Fluorescence Experiments with Polarized X-Rays," Advances in X-Ray Analysis 18, 265-77 (1974).

4. L. Kaufman and D. C. Camp, "Polarized Radiation for X-Ray Fluorescence Analysis," Advances in X-Ray Analysis 18, 247-58 (1974) and Lawrence Livermore Laboratory Report UCRL 76024, 1974.

5. T. G. Dzubay, B. V. Jarret, and J. M. Jaklevic, "Background Reduction in X-Ray Fluorescence Using Polarization," Nuclear Instruments and Methods 115, 297 (1974).

6. Paul Kirkpatrick, "Double Scattering of Polarized X-Rays," The Physical Review 52 (12), 1201-1209 (1937).

AUTHOR INDEX

A

Adams, W. R., 403
Alexander, L. E., 1
Armstrong, R. W., 201

B

Baro, R., 187
Barrett, C. S., 329
Bartell, D. M., 423, 431
Bennett, J. M., 471
Boettinger, W. J., 207, 245
Burdette, H. E., 207, 245
Byram, S. K., 529

C

Caulfield, P. B., 283
Chrenko, R. M., 393
Claisse, F., 459
Cohen, J. B., 291, 355
Cooper, J. A., 423, 431
Cullity, B. D., 259

D

Dietrich, G., 321
Dismore, P. F., 113
Dowell, L. G., 471
Dzubay, T. G., 411

F

Farabaugh, E. N., 201, 207
Frevel, L. K., 15

Galindo, C. P., 337
Garcia, S. R., 497

G

Gazzara, C. P., 161
Gould, R. W., 153
Green, R. E., Jr., 221
Guerra, R. E., 139

H

Hakkila, E. A., 445
Han, B., 529
Hanawalt, J. D., 63
Hansel, J. M., Jr., 445
Hanss, R. E., 337
Hathaway, L. R., 453
Hauff, P. L., 103
Hearn, E. W., 273
Henslee, W. W., 139
Hubbard, C. R., 27

J

James, G. W., 453
James, M. R., 291
Jenkins, R., 125, 507

K

Keller, J., 481
Kirk, D., 283
Kiuru, E., 555
Kuriyama, M., 207, 245

L

Laguitton, D., 515
Lamothe, P. J., 411
Lear, R. D., 403
LeHouillier, R., 459
Leyden, D. E., 437
Lurio, A., 481

591